Engineering rock mechanics: part 2

Illustrative worked examples

CHILE: Continuous, Homogeneous, Isotropic and Linearly Elastic

DIANE: Discontinuous, Inhomogeneous, Anisotropic and Not-Elastic

Frontispiece

Part of the concrete foundation beneath a multi-storey car park on the Island of Jersey in the Channel Islands

Engineering rock mechanics: part 2

Illustrative worked examples

John P. Harrison
Senior Lecturer in Engineering Rock Mechanics
Imperial College of Science, Technology and Medicine
University of London, UK

and

John A. Hudson *FREng*
Professor of Engineering Rock Mechanics
Imperial College of Science, Technology and Medicine
University of London, UK

Pergamon

Elsevier
The Boulevard, Langford Lane, Kidlington, Oxford OX5 1GB, UK
Radarweg 29, PO Box 211, 1000 AE Amsterdam, The Netherlands

First edition 2000
Reprinted 2006, 2007

British Library Cataloguing in Publication Data
A catalogue record for this book is available from the British Library

Library of Congress Cataloging-in-Publication Data
A catalog record for this book is available from the Library of Congress

ISBN-13: 978-0-08-043010-2

Transferred to Digital Printing 2008

For all our past, present and future students and colleagues at Imperial College

About the authors

Dr J.P. Harrison

John Harrison graduated in civil engineering from Imperial College, University of London, and then worked for some years in the civil engineering industry for both contracting and consulting organisations. This was interspersed by studies leading to a Master's degree, also from Imperial College, in Engineering Rock Mechanics. He was appointed Lecturer in Engineering Rock Mechanics at Imperial College in 1986, then obtained his Ph.D. in 1993, and became Senior Lecturer in 1996.

He currently directs undergraduate and postgraduate teaching of engineering rock mechanics within the Huxley School of the Environment, Earth Sciences and Engineering. His personal research interests are in the characterisation and behaviour of discontinuous rock masses, an extension of his earlier Ph.D. work at Imperial College on novel mathematical methods applied to the analysis of discontinuity geometry.

Professor J.A. Hudson *FREng*

John Hudson graduated in 1965 from the Heriot-Watt University, U.K. and obtained his Ph.D. at the University of Minnesota, U.S.A. He has spent his professional career in engineering rock mechanics — as it applies to civil, mining and environmental engineering — in consulting, research, teaching and publishing and has been awarded the D.Sc. degree for his contributions to the subject. In addition to authoring many scientific papers, he edited the 1993 five-volume "Comprehensive Rock Engineering" compendium, and currently edits the International Journal of Rock Mechanics and Mining Sciences.

From 1983 to the present, Professor Hudson has been affiliated with Imperial College as Reader and Professor. He is also a Principal of Rock Engineering Consultants, actively engaged in applying engineering rock mechanics principles and techniques to relevant engineering practice worldwide. In 1998, he was elected as a Fellow of the Royal Academy of Engineering in the U.K.

Contents

Part B: Questions only

The Questions in Part A are reproduced here without the answers for those who wish to attempt the questions without the answers being visible.

Preface

This book can be used as a 'standalone' textbook or as a complement to our first book, *Engineering Rock Mechanics: An introduction to the Principles*. It contains illustrative worked examples of engineering rock mechanics in action as the subject applies to civil, mining, petroleum and environmental engineering. The book covers the necessary understanding and the key techniques supporting the rock engineering design of structural foundations, dams, rock slopes, wellbores, tunnels, caverns, hydroelectric schemes, mines.

In our first book, we presented the basic principles of engineering rock mechanics with strong emphasis on understanding the fundamental concepts. Because it is also important to consider the principles in action, to have practice in applying them, and to be able to link the principles with specific engineering problems, we prepared this second book containing the illustrative worked examples. We have adopted a question and worked answer presentation: the question and answer sets have been collated into twenty chapters which match the subject matter of our first book — Chapters 1–13 on rock mechanics principles and Chapters 14–20 on applications in rock engineering. Part A of this book can be read as a narrative consisting of sequences of text, questions and answers, or in Part B the same questions can be tackled without the answers being visible.

Chapters 1–20 have the same format:

Section 1. Introductory aide-memoire to the chapter subject.

Section 2. Questions with worked answers that illustrate the principles of the rock mechanics subject and the associated rock engineering design issues.

Section 3. Additional points, often reinforcing the most important aspects of the subject.

Not only will the question and answer sets enhance understanding of the rock mechanics principles, but they will also provide the reader with fluency in dealing with the concepts explained in our first book. Moreover, the question sets give examples of the procedures often encountered in practice. In this way, confidence in tackling practical problems will be developed, together with an improved creative ability for approaching all rock engineering problems. It is important to realize that engineering flair is only possible if the basic principles and techniques are understood and implementable.

There are three appendices. Appendix A contains a 3-D stress cube cut-out which can be copied and made into a model as an aide-memoire. Appendix B contains a hemispherical projection sheet which can be copied and used especially for the questions in Chapter 7. Appendix C contains *RMR* and *Q* rock mass classification tables.

Thus, the book serves as an illustrated guide and explanation of the key rock mechanics principles and techniques for students, teachers, researchers, clients, consulting engineers and contractors. We mentioned in the Preface to our first book that rock engineering occurs deep in the earth, high in the mountains and often in the world's wildest places. We engineer with rocks as we create structures, extract the primary raw materials for mankind and harness the forces of nature. It is the romance and the passion associated with rock engineering that has led us to try to communicate some of this excitement. 'Personal experience is everything'. So, we hope that you will be able to experience some of the science, art and romance of the subject by understanding and then implementing the principles and techniques described in this book.

The book contains the tutorial exercises for students who take the integrated engineering rock mechanics course at Imperial College, University of London, plus many extra examples to ensure that the book is comprehensive and is suitable for all reader purposes and backgrounds, whether academic or practical. Because the tutorial exercises have been incrementally refined, extended and corrected over the years by the rock mechanics staff and students at Imperial College, it is not possible to coherently acknowledge the origin of all individual questions. However, we express our profound appreciation to everyone who has contributed in different ways to the questions and answers contained herein.

The authors are especially grateful to their wives, Gwen Harrison and Carol Hudson, for all their support and for helping to improve the style and accuracy of the text. The final version is, of course, our responsibility. If there is anything that you do not understand in the following pages, it is our fault.

J.P. Harrison and *J.A. Hudson*

T.H. Huxley School of Environment, Earth Sciences and Engineering,
Imperial College of Science, Technology and Medicine,
University of London, SW7 2BP, UK

j.harrison@ic.ac.uk
jah@rockeng.co.uk

Our companion first book "Engineering Rock Mechanics — An Introduction to the Principles", also published by Pergamon, Elsevier Science, will be referred to throughout as "ERM 1".

Units and symbols

Units

There are two reasons why it is important to understand and use engineering rock mechanics units correctly:

- engineering rock mechanics calculations used for rock engineering design should be numerically correct; and
- to use engineering rock mechanics properly, an understanding of units is necessary.

We have used standard symbols and the SI (International System) of units. There are seven base units in the SI system: length, mass, time, electric current, thermodynamic temperature, amount of substance and luminous intensity. These base units are dimensionally independent.

Base units

For engineering rock mechanics, we consider just the length, mass and time base units.

Base quantity	Quantity symbol	Name of SI unit	SI unit	Dimensions of unit
Length	l	metre	m	L
Mass	m	kilogram	kg	M
Time	t	second	s	T

Derived units

From the three base units, all the other mechanical units are derived. Some of the main derived units are listed below.

Derived quantity	Quantity symbol	Name of SI unit	SI unit	Dimensions of unit
Area	A		m^2	L^2
Volume	V		m^3	L^3
Density	ρ		$kg\ m^{-3}$	$L^{-3}M$
Velocity	v		$m\ s^{-1}$	LT^{-1}
Acceleration	a		$m\ s^{-2}$	LT^{-2}
Weight	W	newton, N	$m\ kg\ s^{-2}$	LMT^{-2}
Force	F	newton, N	$m\ kg\ s^{-2}$	LMT^{-2}
Pressure	p	pascal, Pa	$N\ m^{-2}$, $m^{-1}\ kg\ s^{-2}$	$L^{-1}MT^{-2}$
Energy	E	joule, J	$N\ m$, $m^2\ kg\ s^{-2}$	L^2MT^{-2}

The name of the SI unit, e.g. newton, is written with an initial lower case letter, and its abbreviation, e.g. N, is written with an initial upper case letter.

Note that force is defined through the relation: force = mass × acceleration. A newton, N, is the force necessary to accelerate a one kilogram mass at a rate of one metre per second per second. This is clear for dynamic circumstances but the force definition also applies to the concept and the units used in the static case. When a static force exists, the force between two stationary objects, the units of force are still $m\ kg\ s^{-2}$ with dimensions LMT^{-2} because of the definition of force. Thus, other derived units, such as Young's modulus, have units of $m^{-1}\ kg\ s^{-2}$ and dimensions $L^{-1}MT^{-2}$, despite the fact that there may be no time dependency in their definition.

The most common prefixes used for decimal multiples of units in engineering rock mechanics are

10^{-6}	10^{-3}	10^3	10^6	10^9
micro	milli	kilo	mega	giga
μ	m	k	M	G

Symbols used in this book [1]

The main symbols used in this book are listed below, together with the name of the quantity they represent, the SI unit name (where appropriate), the SI unit and the dimensions [2] of the unit. Other symbols and abbreviations introduced for a specific question and answer have been defined 'locally' in those questions and answers.

[1] We follow the recommendations in *Quantities, Units and Symbols* prepared by the Symbols Committee of the Royal Society, 1975, 54pp.

[2] The term 'dimensions' is used here to mean the complete listing of the dimensions and exponents, as in $L^{-1}MT^{-2}$, rather than just the LMT components, or just their exponents, −1, 1, −2.

Symbol	Quantity	Name of SI unit	SI unit	Dimensions of unit
α	angle, specifically dip direction of a plane or trend of a line	radian, rad; degree, deg		
β	angle, specifically dip angle of a plane or plunge of a line	radian, rad; degree, deg		
β_w	orientation angle of plane of weakness	radian, rad; degree, deg		
γ	shear strain		1	L^0
γ	unit weight		$kg\ s^{-2}\ m^{-2}$	$L^{-2}MT^{-2}$
$\frac{\partial}{\partial x}, \frac{\partial}{\partial y}, \frac{\partial}{\partial z}$	partial differential operator			
$\Delta l, \delta x, \delta y, \delta z$	increment of distance, displacement		m	L
ε	linear strain		1	L^0
θ	angle	radian, rad; degree, deg		1
λ	fracture frequency		m^{-1}	L^{-1}
ν	Poisson's ratio		1	–
υ	kinematic viscosity		$m^2\ s^{-1}$	L^2T^{-1}
ρ	density		$kg\ m^{-3}$	$L^{-3}M$
σ	stress tensor	pascal, Pa	$N\ m^{-2}$, $m^{-1}\ kg\ s^{-2}$	$L^{-1}MT^{-2}$
σ	normal stress	pascal, Pa	$N\ m^{-2}$, $m^{-1}\ kg\ s^{-2}$	$L^{-1}MT^{-2}$
$\sigma_1, \sigma_2, \sigma_3$	principal stress	pascal, Pa	$N\ m^{-2}$, $m^{-1}\ kg\ s^{-2}$	$L^{-1}MT^{-2}$
σ_c	uniaxial compressive strength	pascal, Pa	$N\ m^{-2}$, $m^{-1}\ kg\ s^{-2}$	$L^{-1}MT^{-2}$
σ_h, σ_H	principal horizontal stress	pascal, Pa	$N\ m^{-2}$, $m^{-1}\ kg\ s^{-2}$	$L^{-1}MT^{-2}$
σ_t	uniaxial tensile strength	pascal, Pa	$N\ m^{-2}$, $m^{-1}\ kg\ s^{-2}$	$L^{-1}MT^{-2}$
σ^2	variance			
τ	shear stress	pascal, Pa	$N\ m^{-2}$, $m^{-1}\ kg\ s^{-2}$	$L^{-1}MT^{-2}$
ϕ	angle of friction	radian, rad; degree, deg		
ϕ_w	friction angle of plane of weakness	radian, rad; degree, deg		
A	area		m^2	L^2
c	cohesion	pascal, Pa	$N\ m^{-2}$, $m^{-1}\ kg\ s^{-2}$	$L^{-1}MT^{-2}$
c	hydraulic conductivity of a fracture		$m\ s^{-1}$	$L\ T^{-1}$
E	Young's modulus	pascal, Pa	$N\ m^{-2}$, $m^{-1}\ kg\ s^{-2}$	$L^{-1}MT^{-2}$
e	fracture aperture		m	L
E_m	elastic modulus of rock mass	pascal, Pa	$N\ m^{-2}$, $m^{-1}\ kg\ s^{-2}$	$L^{-1}MT^{-2}$
F	force	newton, N	$kg\ m\ s^{-2}$	LMT^{-2}
G	shear modulus	pascal, Pa	$N\ m^{-2}$, $m^{-1}\ kg\ s^{-2}$	$L^{-1}MT^{-2}$
G_m	shear modulus of rock mass	pascal, Pa	$N\ m^{-2}$, $m^{-1}\ kg\ s^{-2}$	$L^{-1}MT^{-2}$
GSI	geological strength index value			
i	hydraulic gradient			
i	asperity angle	radian, rad; degree, deg		
I_1, I_2, I_3	stress invariants			
k	constant of proportionality			
k	number of events			
k	coefficient of permeability		m^2	L^2

Symbol	Quantity	Name of SI unit	SI unit	Dimensions of unit
K	hydraulic conductivity		$m\ s^{-1}$	LT^{-1}
K	stiffness		$kg\ s^{-2}$	MT^{-2}
k_n, k_s	fracture normal stiffness, fracture shear stiffness,		$m^{-2}\ kg\ s^{-2}$	$L^{-2}MT^{-2}$
L	length		m	L
l, m, n	Cartesian axes			
m	coefficient in Hoek–Brown strength criterion			
N	number in sample			
p	pressure	pascal, Pa	$N\ m^{-2}$, $m^{-1}\ kg\ s^{-2}$	$L^{-1}MT^{-2}$
P_B, P_S	breakdown pressure, shut-in pressure	pascal, Pa	$N\ m^{-2}$, $m^{-1}\ kg\ s^{-2}$	$L^{-1}MT^{-2}$
PL	point load index value	pascal, Pa	$N\ m^{-2}$, $m^{-1}\ kg\ s^{-2}$	$L^{-1}MT^{-2}$
Q	flow rate		$m^3\ s^{-1}$	L^3T^{-1}
Q	rock mass quality rating			
r	radius		m	L
RMR	rock mass rating value			
RQD	rock quality designation, %			
RQD_t	rock quality designation for threshold value t			
S	elastic compliance	lacsap, Pa^{-1}	$N^{-1}\ m^2$, $m\ kg^{-1}\ s^2$	$LM^{-1}T^2$
S	elastic compliance matrix	lacsap, Pa^{-1}	$N^{-1}\ m^2$, $m\ kg^{-1}\ s^2$	$LM^{-1}T^2$
s	coefficient in Hoek–Brown strength criterion			
S	sample standard deviation			
t	threshold value for RQD		m	L
t	thickness		m	L
u	displacement		m	L
UCS	unconfined compressive strength	pascal, Pa	$N\ m^{-2}$, $m^{-1}\ kg\ s^{-2}$	$L^{-1}MT^{-2}$
v	displacement		m	L
W	weight		$kg\ m\ s^{-2}$	LMT^{-2}
x, y, z	Cartesian axes			
$xbar$	mean fracture spacing		m	L
$Xbar$	sample mean			
z	depth		m	L
z	standard normal variable			

The convention for writing symbols is as follows.

- Symbols for tensor quantities should be in sans serif bold italic form, e.g. ***S***.
- Symbols for vector quantities should be in bold italic form, e.g. ***F***.
- Symbols in Latin or Greek should be in italic form, e.g. *x*.

Part A:

Illustrative worked examples — Questions and Answers

1 Introduction

1.1 The subject of engineering rock mechanics

The term **engineering rock mechanics** is used to describe the engineering application of rock mechanics to civil, mining, petroleum and environmental engineering circumstances. The term **mechanics**, means the study of the equilibrium and motion of bodies, which includes statics and dynamics [1]. Thus, **rock mechanics** is the study of mechanics applied to rock and rock masses. 'Engineering rock mechanics' is this study within an engineering context, rather than in the context of natural processes that occur in the Earth's crust, such as folding and faulting. The term **rock engineering** refers to the process of engineering with rock, and especially to creating structures on or in rock masses, such as slopes alongside roads and railways, dam foundations, shafts, tunnels, caverns, mines, and petroleum wellbores.

There is an important distinction between 'rock mechanics' and 'rock engineering'. When 'rock mechanics' is studied in isolation, there is no specific engineering objective. The potential collapse of a rock mass is neither good nor bad: it is just a mechanical fact. However, if the collapsing rock mass is in the roof of a civil engineering cavern, there is an adverse engineering connotation. Conversely, if the collapsing rock mass is part of a block caving system in mining (where the rock mass is intended to fail), there is a beneficial engineering connotation. In the civil engineering case, the integrity of the cavern is maintained if the rock mass in the roof does not collapse. In the mining engineering case, the integrity of the mining operation is maintained if the rock mass does collapse.

Hence, rock engineering applies a subjective element to rock mechanics, because of the **engineering objective**. The significance of the rock mass behaviour lies in the eye and brain of the engineer, not in the mechanics.

---

[1] It is not always realized that the term 'mechanics' includes 'dynamics', but a book title such as 'River Mechanics' is correct. Similarly, 'rock dynamics', the topic of Chapter 13, is part of 'rock mechanics'.

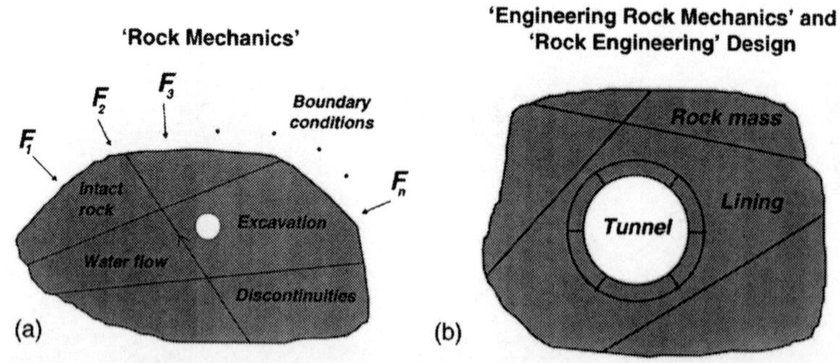

Figure 1.1 The distinction between 'rock mechanics' itself (a) and engineering applications of rock mechanics (b). In (a), $F_1 \ldots F_n$ are the boundary forces caused by rock weight and current tectonic activity. In (b) a tunnel is being constructed in a rock mass.

The distinction between 'rock mechanics' and 'rock engineering' illustrated in Fig. 1.1 is highlighted further in Fig. 1.2 which shows part of the concrete foundation illustrated in the Frontispiece. 'Rock mechanics' involves characterizing the intact rock strength and the geometry and mechanical properties of the natural fractures of the rock mass. These studies, together with other aspects of the rock mass properties such as rock stiffness and permeability, can be studied without reference to a specific engineering function. When the studies take on a generic engineering direction, such as the structural analysis of foundations, we are in the realm of 'engineering rock mechanics'. This is analogous to the term **engineering geology** in which geology is studied, not in its entirety but those aspects which are relevant to engineering.

'Rock engineering' is concerned with specific engineering circumstances: in this case (Fig. 1.2), the consequences of loading the rock mass via the concrete support. How much load will the rock foundation support under these conditions? Will the support load cause the rock to

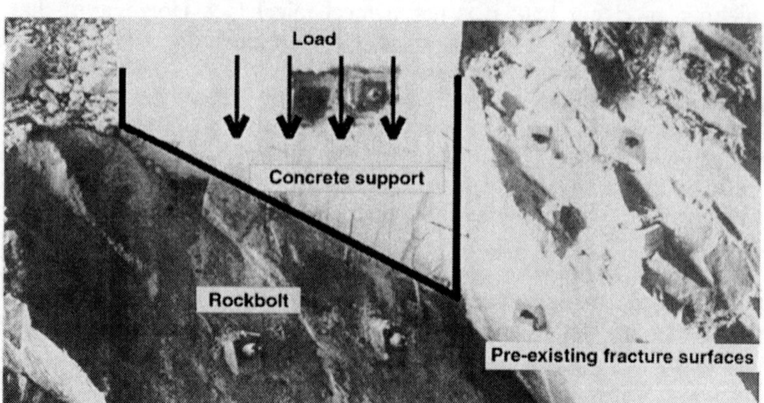

Figure 1.2 Portion of Frontispiece photograph illustrating loading of discontinuous rock mass by the concrete support of a multi-storey car park, Jersey, UK.

slip on the pre-existing fractures? Is the stiffness of the concrete support a significant parameter in these deliberations? If the rock mass is to be reinforced with rockbolts, where should these be installed? How many rockbolts should there be? At what orientation should they be installed? All these issues are highlighted by the photograph in Fig. 1.2.

Above the Frontispiece photograph, there are two acronyms:

CHILE — Continuous, Homogeneous, Isotropic and Linearly Elastic;
DIANE — Discontinuous, Inhomogeneous, Anisotropic and Not-Elastic.

These refer to two ways of thinking about and modelling the rock mass. In the CHILE case, we assume an ideal type of material which is not fractured, or if it is fractured the fracturing can be incorporated in the elastic continuum properties. In the DIANE case, the nature of the real rock mass is recognized and we model accordingly, still often making gross approximations. Rock mechanics started with the CHILE approach and has now developed techniques to enable the DIANE approach to be implemented. It is evident from Fig. 1.2 that a DIANE approach is essential for this problem, using information about the orientation and strength of the rock fractures. However, both approaches have their advantages and disadvantages, and the wise rock engineer will utilize each to maximal advantage according to the circumstances.

Modelling for rock mechanics and rock engineering should be based on ensuring that the relevant mechanisms and the governing parameters relating to the problem in hand have been identified. Then, the choice of modelling technique is based on the information required, e.g. ensuring an adequate foundation as illustrated in Fig. 1.2.

Accordingly, and to enhance an engineer's skills, the question sets in Chapters 1–13 are designed to improve familiarity with the main rock mechanics topics and the techniques associated with the topics, such as **stress analysis** and **hemispherical projection** methods. In Chapter 14, we emphasize the importance of considering the 'rock mass–engineering structure' as a complete system. Finally, in Chapters 15–20, the question sets are related to specific engineering activities and design requirements.

You can read the question and answer text directly in each of the chapters, as in Section 1.2 following, or you can attempt the questions first without seeing the answers, as in Question Set 1 in Part B. Whichever method you choose for reading the book, we recommend that you read the introductory text for each chapter topic before tackling the questions.

1.2 Questions and answers: introduction

In this introductory chapter, there are five questions concerned with the nature of engineering rock mechanics. In all subsequent chapters there are ten questions.

***Q1.1* Define the following terms:**
- **rock mechanics;**
- **engineering rock mechanics;**

- **rock engineering;**
- **structural geology;**
- **engineering geology;**
- **soil mechanics;**
- **geotechnical engineering.**

A1.1 *Rock mechanics* is the study of the statics and dynamics of rocks and rock masses.

Engineering rock mechanics is the study of the statics and dynamics of rocks and rock masses in anticipation of the results being applied to engineering.

Rock engineering involves engineering with rocks, especially the construction of structures on or in rock masses, and includes the design process.

Structural geology deals with the description and analysis of the structure of rock masses.

Engineering geology is the study of geology in anticipation of the results being applied to engineering.

Soil mechanics is the study of the statics and dynamics of soils.

Geotechnical engineering is the process of engineering with rocks and/or soils[2].

Q1.2 Explain the fundamental purposes of excavation in civil engineering, mining engineering, and petroleum engineering.

A1.2 *Civil engineering*. It is the rock opening, the space resulting from excavation, that is required in civil engineering — for railways, roads, water transport, storage and disposal of different materials — often designed for an engineering life of 120 years.

Mining engineering. It is the excavated rock itself that is required in mining engineering, plus the ability to transport the rock. Underground space is created when the rock is removed, e.g. the mine stopes in metal mines; separate underground space is required to transport the mined rock/ore to the surface. The design life of mine openings can vary from a few days (as in longwall coal mining), to some months or years, to many years, depending on the mine design, methods, and requirements.

Petroleum engineering. Wellbores (deep boreholes) are used to extract petroleum and so the excavated space is used for transport. The design life of the wellbores is similar to the mining circumstances: it will depend on the overall strategy and lifetime of the oil field. Note that, in contrast to civil and mining engineering, environmental problems such as surface subsidence and groundwater movement are not caused by the creation of underground space *per se*, but by the removal of oil from the reservoir rock where it is trapped.

[2] In the 1990s, the International Society for Soil Mechanics and Foundation Engineering changed its name to the International Society for Soil Mechanics and Geotechnical Engineering. The International Society for Rock Mechanics considered a complementary change to the International Society for Rock Mechanics and Geotechnical Engineering but did not go ahead with the change.

Q1.3 The photograph below illustrates construction of the 61 m span, 25 to 50 m deep, underground Gjøvik Olympiske Fjellhall (Olympic Mountain Hall) in Precambrian gneiss in Norway. This is the largest[3] roofspan public-access civil engineering cavern in the world. Describe the engineering rock mechanics factors that would have to be considered in the design and excavation of such a cavern.

A1.3 The main factors to be considered in excavation of such a cavern are the geological setting, the natural rock stress, the deformability and strength of the intact rock, the geometry and nature of the pre-existing fractures, the groundwater, variations in the rock properties, the use of a rock mass classification technique to indicate rock mass quality, time-dependent effects, and the excavation and support methods. The cavern is to be constructed in hard rock, but it has a large span (of 61 m compared to the usual 15–25 m) and is located close to the surface. Under these circumstances, we would need to consider in the first instance any instability that might arise from rock blocks falling by gravity from the cavern roof.

In fact, after considerable site investigation, the use of the *Q* rock mass classification scheme, associated numerical modelling and design work[4], the cavern was first excavated to a 36 m span and then, after installation of 6 m rockbolts and 12 m long cable bolts plus fibre-reinforced shotcrete, increased to the 61 m span. The long axis of the cavern

[3] This refers to the year 2000. It is likely that in the future this project will be superseded by even larger projects.

[4] Bhasin R. and Løset F. (1992) Norway's Olympic Cavern. *Civ. Eng.*, December, 60–61.

axis was orientated perpendicular to the maximum horizontal stress of 3.5–4.0 MPa which helped to stabilize the rock blocks in the roof. After excavation of 140,000 m^3 of rock and installation of the internal fittings, the Gjøvik Olympiske Fjellhall can seat 5300 people. The impression one has inside the cavern is the same as that in a building constructed above ground.

Q1.4 Why do you think that the techniques used in rock mechanics for site characterization, analysis and modelling are not the same as those used in soil mechanics?

A1.4 Although there is a significant overlap between the two subjects, for example both subjects make extensive use of stress analysis and elasticity theory, soil particles are several orders of magnitude smaller than the dimensions of the engineered structure, whereas rock blocks can be of a similar size to the engineered structure. This means that the discrete nature of the ground is more important in rock mechanics, and techniques such as hemispherical projection and dedicated computer modelling are required to assess the associated rock movements. Also, some support methods such as the rockbolts and cable bolts mentioned in A1.3 can only be used in rock masses.

In fact, the two main factors that cause the differences between rock mechanics and soil mechanics are (a) the importance in rock mechanics of the pre-existing *in situ* rock stress, and (b) the presence of the fractures which govern the rock mass stiffness, strength, failure behaviour and permeability. Understanding and modelling these two aspects alone require a dedicated rock mechanics approach.

Q1.5 How can the subject of 'engineering rock mechanics' be useful to organizations outside the civil and mining engineering professions, e.g. to the petroleum industry, to insurance companies, to environmental engineers?

A1.5 The subject is potentially useful to any person or organization concerned with the engineering behaviour of rock masses. In petroleum engineering, the engineer wishes to be able to predict the stability of wellbores and the conditions under which borehole breakout will occur (damage caused by high rock stress at the borehole walls), in addition to the overall rock mechanics behaviour of oil reservoirs. Similarly, insurance companies wish to evaluate the hazard to large structures built on or in rock masses, and this requires a knowledge of engineering rock mechanics. Environmental engineers need to understand processes such as coastal cliff degradation, water flow through rock masses, and the stability of disused mine workings.

1.3 Additional points

In 2000, the **largest underground excavation** for civil engineering purposes is in the Indian state of Himachal in the Himalayas. It is part

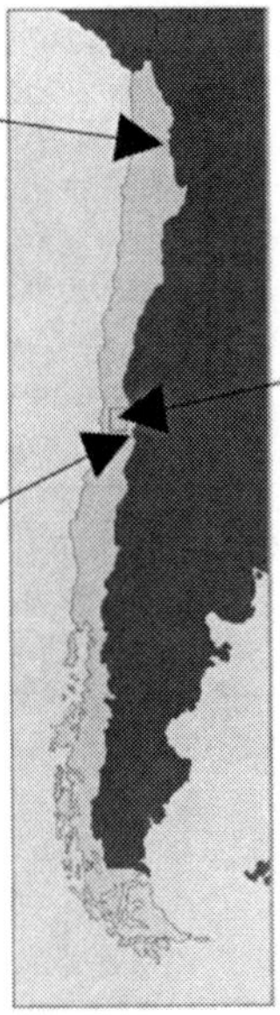

Figure 1.3 Location of the Chuquicamata and El Teniente copper mines.

of the Nathpa Jhakri hydro-electric power station and consists of four siltation chambers with dimensions 525 m long, 16 m wide and 27 m deep, built to exclude sediment particles above 0.2 mm from entering the headrace tunnel and hence the turbines. The Nathpa Jhakri construction project has many interesting features, including the Daj Khad shear zone through which the headrace tunnel was driven[5].

The **largest surface and underground mines** are in Chile (Fig. 1.3): the Chuquicamata open-pit copper mine and the underground El Teniente copper and molybdenum mine. The Chuquicamata surface mine is in the Atacama desert in northern Chile, and is several kilometres long and 750 m deep. The El Teniente mine in the foothills of the Andes is in a zone of complex geology and high rock stress, and produces 90,000 tonnes of ore per day.

The professional society for rock mechanics is the **International Society for Rock Mechanics** (ISRM) which was formed in 1963. The Secretariat is based at the Laboratório Nacional de Engenharia Civil in Lisbon, Portugal. There are about 5000 members. Each year, the Manuel Rocha ISRM prize is awarded for the best PhD thesis submitted to the ISRM Board. The ISRM distributes the ISRM News Journal which is a magazine containing news and technical articles.

The authors of the **first major textbook** in rock mechanics, 'Fundamentals of Rock Mechanics', were John C. Jaeger and Neville G.W.

[5] A discussion of some of the rock mechanics analyses for the Nathpa Jhakri project is contained in the paper of the Glossop Lecture given by Dr. E. Hoek to the Geological Society of London Engineering Group in 1998: Hoek E. (1999) Putting Numbers to Geology — An Engineer's Viewpoint, *Q. J. Eng. Geol.*, **32**, 1, 1–19.

Cook[6]. Professor Jaeger was a mathematician and engineer, working in Australia; Professor Cook was a seismologist and mining engineer, working in South Africa and later in the USA. The first edition of the textbook was published in 1969. This book has a strong emphasis on the theory of elasticity applied to rock masses, resulting from Professor Jaeger's expertise and the utility of the theory when applied to Professor Cook's working environment deep in the South African gold mines. In these gold mines, high rock stresses close the pre-existing fractures; thus, the rock mass can be modelled as a continuum and elastic calculations for stresses, displacements and energies are often good approximations. 'Fundamentals of Rock Mechanics' had a significant influence on the development of rock mechanics. For example, in China it was the only foreign book available on rock mechanics for many years. We will highlight in later chapters that elasticity theory is one of the major tools available to support rock engineering design. However, for near-surface rock engineering, where there are more fractures, often subjected to relatively low rock stress levels, we use additional techniques to study the rock mass behaviour.

For research work, there are **two main journals** in the engineering rock mechanics subject area.

(i) 'International Journal of Rock Mechanics and Mining Sciences' edited by J. A. Hudson and published by Pergamon Press, Elsevier Science. This Journal was started in 1964 and concentrates on original research, new developments and case studies in rock mechanics and rock engineering for mining and civil applications. Suggested Methods generated by the ISRM Commission on Testing Methods are published in the Journal; for example there are several new ones in Volume 36 for 1999. Also, the Journal publishes Special Issues on important topics (e.g. the one described in Footnote 6) and has published the proceedings of conferences in compact disk form. The web site of the Journal is http://www.elsevier.nl/locate/ijrmms.

(ii) 'Rock Mechanics and Rock Engineering' edited by K. Kovari and H. H. Einstein and published by Springer-Verlag. This Journal was started in 1968 and concentrates on experimental and theoretical aspects of rock mechanics, including laboratory and field testing, methods of computation and field observation of structural behaviour, with applications in tunnelling, rock slopes, large dam foundations, mining, engineering and engineering geology. The web site of the Journal is http://link.springer.de/link/service/journals/00603/about.htm.

We encourage you to consider **rock mechanics as a unique discipline**. Of course, there are many common factors with other subjects: Newton's

---

[6] Jaeger J. C. and Cook N. G. W. (1979, 3rd edn.) *Fundamentals of Rock Mechanics*. Chapman and Hall, London, 593pp. In 1998, a commemorative conference was held at the Ernest Orlando Lawrence Berkeley National Laboratory in California, USA, to honour Neville Cook's contributions to rock mechanics. The Neville Cook Special Issue of the International Journal of Rock Mechanics and Mining Sciences was published in 2000. This Special Issue contains reminiscing contributions and 30 papers presented at the conference on subjects pioneered by him.

laws will apply, the theory of elasticity remains the same, etc. Although much of the science will be common with other disciplines, rock is a natural material and so engineering rock mechanics is also an art. Thus, when working through the question and answer sets in this book, we recommend that you concentrate on developing a deeper understanding of the principles and hence to be capable of a more creative approach to this fascinating subject.

2 Geological setting

2.1 Rock masses

Rock masses are the natural structures that will host rock engineering projects. A road may pass through a rock cutting with rock slopes on each side; the foundations of a dam may rest on a rock mass; a tunnel or cavern can be completely contained within a rock mass; a borehole can be drilled several kilometres into the earth's crust; an underground mine can involve the excavation of large volumes of ore; a repository might be excavated underground for disposing of large volumes of radioactive waste.

In Figs. 1–6, we give examples of engineering projects where the geological features play a significant role in the overall stability and success of the project. In Fig. 2.1, there is an example of one of the cave

Figure 2.1 9th century monolithic Buddhist temples excavated in the Deccan basalts in India.

temples at the World Heritage site at Ellora in SW India. This temple has been created in the Deccan Traps by simple hand excavation of the volcanic basalt. The pillars that can be seen at the entrance are part of the *in situ* rock mass. Above the temple, natural rock **fractures**[1] are visible; such fractures are encountered in almost all rock masses and can lead to instability of engineered structures. Most of these temples have, however, remained stable well beyond a civil engineering design life of 120 years, the figure that we often use today for design purposes.

In Fig. 2.2, a road has been severely damaged by the sliding of a large block of rock on which the road had been built (to the right of the upper picture). The rock block was able to slide because there was a large-scale natural weakness, a shear zone, in the limestone formation as shown in the lower picture. The coefficient of friction on the limestone bedding planes was low because they were clay-filled, and this enabled the limestone block to slide and damage the road. For all rock engineering projects, it is crucial to be able to locate such significant geological features.

Fig. 2.3 shows two slopes. In Fig. 2.3a, a pre-split rock slope at the side of the A82 road in Scotland is shown. The **pre-splitting technique** is a rock-blasting process whereby the final rock slope is created as a fracture plane first by blasting in a row of parallel blastholes, with the rock subsequently being excavated up to this fracture plane. The fact that the blasting has been successful is evidenced by the visible traces of the half boreholes left on the rock surface as the whitish parallel lines.

However, the rock already contained fractures formed long ago when the rock was subjected to high stresses caused by tectonic activity. Because the fractures were formed as a result of the applied stresses, they tend to occur in sets of sub-parallel fractures with specific orientations. The sets of fractures can occur at several orientations because there were different phases of tectonic activity during the history of the rock mass. In Fig. 2.3a, two fractures from different sets have formed a **rock wedge** which has slid out of the excavation (and was removed during slope construction). The engineer standing on the top of the slope indicates the scale.

These natural fractures are an inherent feature of rock masses. Engineers cannot specify that the rock mass should be unfractured: the properties of the rock mass have to be established by site investigation and the design adjusted accordingly.

In the case of this road, the location of the road and hence the slope were determined by the overall topographic features, and there was little

[1] During the development of rock mechanics, the word 'discontinuity' was used to denote natural faults, joints, fissures, etc., because they are discontinuities in the rock continuum. The word 'fracture' was previously used mainly to denote man-made discontinuities. Nowadays, and especially in the USA, the word 'fracture' is used in place of 'discontinuity'. We have adopted this usage in this book.

Figure 2.2 Road instability in Spain. The displacement of the road, shown in the top photograph (a), was caused by movement of a large limestone block released by the shear zone, in the lower photograph (b), with sliding on clay-filled bedding planes. Note the engineer standing on the lip of the shear zone, in the black square.

Figure 2.3 Rock slopes: (a) at the A82 roadside near Loch Lomond in Scotland; (b) forming one side of the New Celebration open-pit gold mine in Western Australia.

opportunity to alter the location of the road to suit the rock engineering. Similarly, in mining engineering, the purpose of the mine is to extract the ore, which is in a specific location. The slopes in the gold mine in Fig. 2.3b are determined by the orebody geometry and economics. The large scale of this operation can be seen by the vehicles on the lowest level.

The type of failure on the roadside rock face shown in Fig. 2.3a, where instability was caused by pre-existing fractures forming a rock wedge, can also occur on a large scale, as illustrated in Fig. 2.4. In this case,

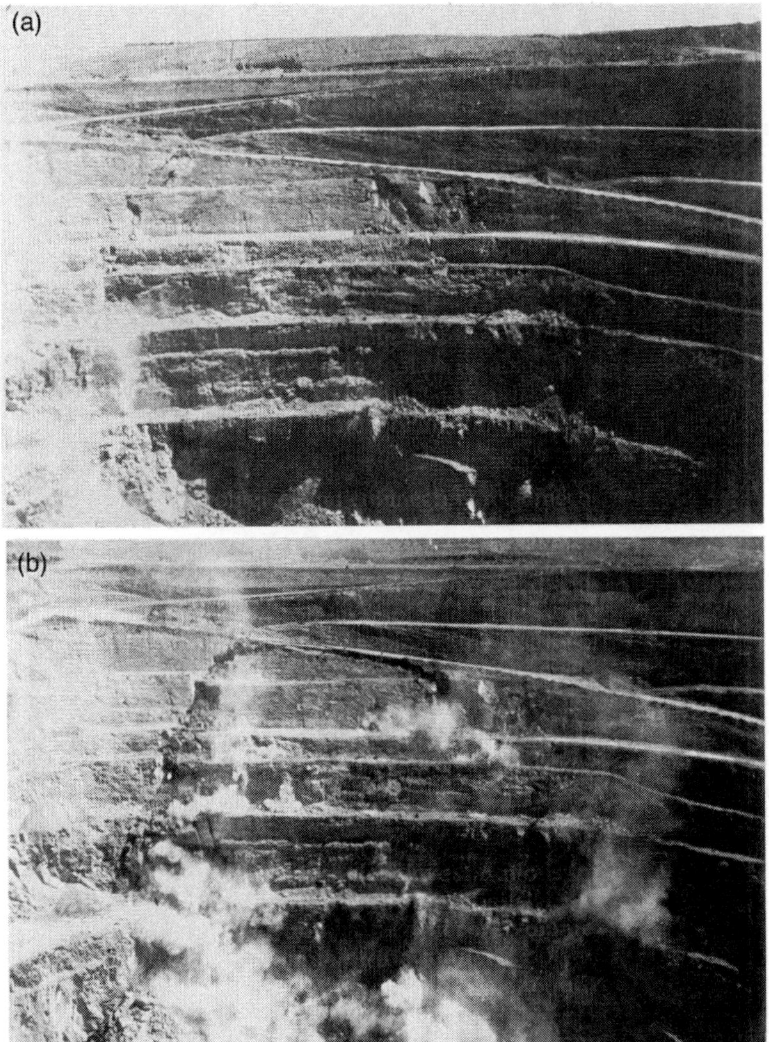

Figure 2.4 Initiation and propagation of a large wedge failure in an open-pit mine, Western Australia.

two major shear zones and the open-pit mine wall have formed a rock block which has slid downwards. The traces of dust in Fig. 2.4a and the clouds of dust in Fig. 2.4b were not caused by blasting: the unstable rock block slid down under its own weight, and the dust was generated by the rock surfaces sliding over one another and escaping through rock fractures.

Natural fractures in the rock mass, especially joints and faults, can also cause instabilities underground. Some large unsupported caverns may be stable, as in the cavern shown in Fig. 2.5, but often the rock

Figure 2.5 Construction of the underground facilities at the Alto Lindoso hydro-electric project in Portugal.

structure needs to be stabilized by engineered reinforcement or support.

In Fig. 2.6, the desert location of a potential repository for high-level radioactive waste is shown. Here, the purpose is waste isolation for long periods and there are many aspects of geology, hydrogeology, engineering and biospherics to consider. Again, it is the features of the rock mass resulting from its geological history that dominate the design, e.g. the presence of rock fractures, the permeability of the fracture array, and the age of the water currently in the rock mass.

The rock engineering projects that have been described are widely different in their locations and purposes but, whatever the purpose of the project, the rock mass is the host structure. Unlike other materials

Figure 2.6 View southwest towards Death Valley from the top of Yucca Mountain in Nevada, USA, the site of a potential radioactive waste repository in a dry region. Note Amargosa Desert to the left of the picture, and the volcanic craters in Crater Flats at the middle right.

used in engineering, such as steel, concrete and polymers, we cannot specify the material properties beforehand: the rock is already there and we have to find out what its properties are. We are interested in the stiffness and strength of the intact rock and mechanical weaknesses in the rock mass, such as bedding planes, faults, joints, and fissures, generically termed 'fractures' or 'discontinuities'. For rock mechanics analysis, we also need to know the natural stress state that is in the rock before engineering begins. This stress state is determined by gravity, tectonic forces operating and several other factors. So, for all these reasons and depending on the project, it is helpful, if not essential, to have a good understanding of the geological history of a site, especially the structural geology.

The subject of this chapter is explained further in Chapter 2 of ERM 1[2]. The introduction here in this section is intended as an *aide-memoire* to the subject before the questions and answers in the next section. This applies similarly to all subsequent chapters.

2.2 Questions and answers: geological setting

Q2.1 The picture below shows a limestone slope above a highway in Spain. Comment briefly on the geological factors that could influence rock slope stability at this location.

A2.1 The rock strata are folded and there is evidence of opening of the bedding planes. Generally, in limestones there will be two sets of joints perpendicular to each other and to the bedding planes. Thus, it is

[2] Throughout the text, we will refer to our earlier companion book 'Engineering Rock Mechanics: An Introduction to the Principles' as ERM 1.

possible that rock blocks could be formed and these might be unstable because of the steepness of the slope. Also, the folding is variable along the slope, meaning that some regions of the slope will be potentially more unstable than other regions. Such limestone masses are likely to contain shear zones, so the rock should be studied in order to anticipate problems of major instabilities such as that illustrated in Fig. 2.4.

Q2.2 The picture below shows the surface of a fault in a hard rock aggregate quarry on which a rock slide has occurred. Explain (a) why the existence of this fault could indicate that other similar features will be encountered as quarrying continues, and (b) why encountering an adverse geological feature such as this is likely to be less significant in a quarry than in a road cutting.

A2.2 (a) Faults and shear zones are caused by rock stresses: the presence of one fault is an indicator that others may be present in the same region where the mechanical conditions have been similar. (b) Unlike the rock slopes in a road cutting, the working rock slopes in a quarry are not permanent. So long as the fault does not affect excavation too much, the associated instability is acceptable.

Q2.3 The picture below shows tooth marks from the bucket of a mechanical excavator in the Carboniferous rocks of a near-surface slope in an opencast coal mine. What evidence is there here of geological disruption to the rock strata?

A2.3 The excavator tooth marks show that the rock is soft, but a much more important aspect is the evidence of glacial deformation. Note the irregular marker bands passing across the slope and below the hammer head. Such irregular near-surface strata are evidence of glacial perturbations and the possibility of slope instability problems.

Q2.4 A site investigation was conducted in a granitic rock mass (see picture on next page). One side of fracture #300 in the core is shown. What does this fracture indicate about the rock mass history and what significance does this have for rock mechanics design of slopes and tunnels in the rock mass?

A2.4 The alteration ring around the fracture (the thin and lighter zone at the base of the fracture in the photograph) indicates that some alteration has occurred because of circulating water or other fluids. The texture on the fracture, running from top left to bottom right, represents a 'slickensided' surface which occurs when the rock surfaces have moved over one another. Thus, fluid has travelled through this fracture and there has been shear movement on the fracture. These features indicate a connected rock fracture system in which the rock blocks have been

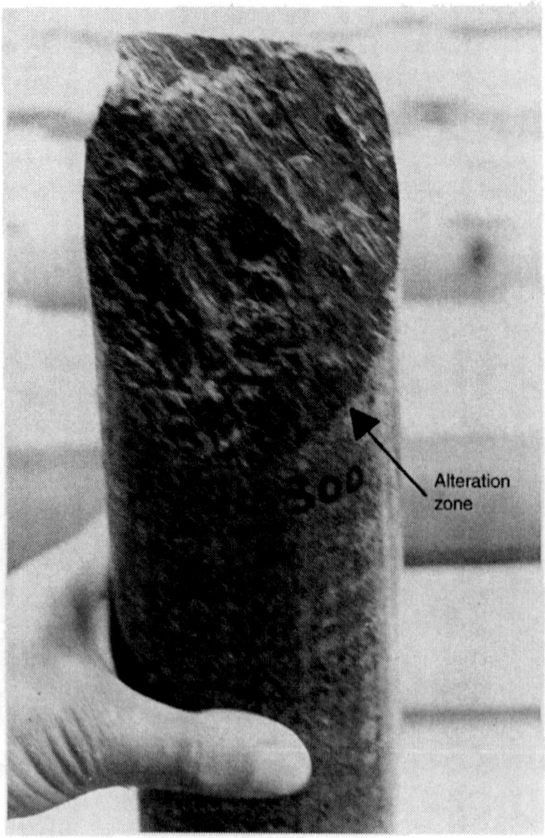

moved about. Thus, the rock blocks could be well developed and hence more likely to be unstable.

Q2.5 In the first Glossop Lecture[3] on geology for engineers, Fookes (1997) quotes from Glossop's 1968 Rankine Lecture: "What you look for should be suggested by the natural environment and by the nature of the construction problem to be solved." Explain why this quotation is critically important for rock mechanics and rock engineering.

A2.5 We have explained that engineering rock mechanics can be studied generically but, when there is a specific engineering project to consider, we need to establish the engineering objective. Then, the engineering structure located on or in the rock mass is tailored directly to the engineering objective. So the quotation refers to the fact that different

[3] Fookes P. G. (1997) Geology for Engineers: the geological model, prediction and performance. *Q. J. Eng. Geol.*, **30**, 293–424.

geological environments will have different characteristics and we can anticipate many of the features present. Also, certain factors will be critical for certain engineering projects. For example, for studies of storing pressurized gas in an unlined cavern, we are critically concerned with potential gas leakage via natural fractures in the rock mass.

There are two immediate corollaries:

(a) there can be no standardized geological information for different engineering projects, because there will be different emphases on different rock properties depending on the engineering objectives, a subject we cover in Chapters 11 and 12;

(b) it is important to consider the interaction between the rock mass and the engineering structure, a subject we cover in Chapter 14.

Emphasis is placed on these points in the definition of engineering geology (based on the Association of Engineering Geologists 1969 definition) given in the Glossary of Geology[4]: "The application of geologic data, techniques, and principles to the study of naturally occurring rock and soil materials or groundwater for the purpose of assuring that geologic factors affecting the location, planning, design, construction, operation, and maintenance of engineering structures, and the development of groundwater resources are properly recognized and adequately interpreted, utilized, and presented for use in engineering practice."

Q2.6 The quotation at the beginning of the Glossary of Geology[4] is that: "It is not really a mark of distinction for a geologist's writing to be so obscure that a glossary is required for its comprehension." Discuss this in the context of engineering rock mechanics.

A2.6 The authors agree with the general sentiment in this quotation, especially when the geological information is being supplied for engineering purposes. It is important that the reader with an engineering background can understand the content and will interpret the text with the same meaning as the writer having the geological background. At the same time, it is necessary to have a glossary so that all the technical words are defined.

Q2.7 When considering the geological setting for a rock engineering project, would an engineer expect information relating to all potential geological hazards to be available?

A2.7 Yes, nowadays the engineer should ensure that such information is made available. In the same paper[3] as referenced in Q2.5, Fookes explains that:

> "Sometimes you hear something said like, 'rocks and soils were not made to a BS [British Standard]'. I interpret this as meaning that every rock and soil is different and cannot be

[4] Bates R. L. and Jackson J. A. (eds) (1980) *Glossary of Geology*. American Geological Institute, USA, 749pp.

> relied upon. I also believed this to be the case and subscribed strongly to this view for many years. However, I have come to believe that by and large they are made, not to a man-made British Standard, but to Nature's equivalent to a BS which follows rules of physics, chemistry, biology, mathematics, engineering and so on. Further, I believe that in the context of site investigation and in the understanding of the site, application of geological education and experience should be able to make a moderately close approximation of the actual geological conditions from the desk study and that when this is supplemented by ground and laboratory investigation, there should be, ideally, nothing that has not been discovered ... it should be a realistic goal to be able to make a very close approximation to the actual site conditions, particularly if you know what you are looking for and what questions to ask".

Thus, from a knowledge of the geological and engineering rock mechanics principles, we ought to be able to predict what we will find underground and what will happen when the rock engineering takes place. The authors agree with this: there should be no 'unexpected' failures. Against such engineering rationality are examples like the saying in the tunnelling industry that 'the history of tunnelling is the history of the unexpected'. Indeed, all sorts of problems are experienced in tunnels, as Whittaker and Frith (1990)[5] have illustrated.

We can say for sure, however, that taking the advice of good engineering and structural geologists and implementing the principles of engineering rock mechanics will certainly reduce the likelihood of 'unexpected events' occurring during rock engineering.

Q2.8 In an article on geological and geotechnical investigations for tunnelling, Parker (1996)[6] estimates that "even comprehensive exploration programs recover a relatively miniscule drill core volume, less than 0.0005% of the excavated volume of the tunnel". Do you think that sampling only this proportion of the rock mass is enough?

A2.8 The 0.0005% sampling level is equivalent to one vertical 100 mm diameter borehole for every 100 m length along a 5 m diameter tunnel and, if we did not have other information to guide us, such a sampling proportion would certainly not be enough.

However, the reason for highlighting the sampling percentage is that (given we have supplementary information from regional geology, outcrops, previous engineering, etc., and hence some form of geological model already established) such spot sampling can often be enough, but only because the core provides the method for refining an existing

---

[5] Whittaker B. N. and Frith R. C. (1990) *Tunnelling: Design, Stability and Construction*. Institution of Mining and Metallurgy, London, 460pp.

[6] Parker H. W. (1996) Geotechnical Investigations, in *Tunnel Engineering Handbook* (J. O. Bickel, T. R. Kuesel and E. H. King, eds). Chapman and Hall, New York, 544pp.

model which is based on the geological and engineering rock mechanics principles.

Q2.9 A cavern (165 m long, 22 m wide and 15 m high) is to be excavated in chalk strata beneath the sea. The crown of the cavern will be 35 m below the seabed. What is the main geological information you would like to have before proceeding with the excavation?

A2.9 The main danger is that water from the sea will enter the cavern. Hence, much of the geological information should be directed to establishing whether water inflow will be a problem. In a chalk rock mass, the water will mainly travel through weathered chalk and the pre-existing fractures, so information on the degree of weathering and on the fractures and their characteristics is crucial.

This question was stimulated by the construction of the sub-sea crossover cavern on the UK side of the England–France Channel Tunnel. To paraphrase Warren and Varley (1995)[7]: "In this area there is a normal geological succession dipping gently northwards and affected by minor faulting. The cavern is located within carbonate clayey mudstones forming the lowermost part of the Cenomanian succession, namely lower and basal chalk, the more sandy glauconitic marl and a clay-dominant material 7 m thick of carbonate mudstone at the top of the Gault clay. Weathering at the seabed penetrates down through the overlying grey chalk and into the upper chalk marl strata to within 20 m of the roof. Formation mass permeabilities are generally low, i.e. 10^{-7} to 10^{-10} m/s, although higher permeabilities do exist in the glauconitic marl (owing to the presence of open joints) and in the upper chalk marl immediately above the cavern crown. The rock mass quality was fair to good with sub-horizontal and sub-vertical joints spaced at 1/m and greater than 2/m, respectively (average persistence 2 m). A number of minor faults were present, usually of an arcuate nature with downthrows less than 0.5 m and striking 40° to 220° parallel to one of the major joint directions."

Given this geological information, we might expect the engineering of the cavern to be difficult. In fact, few problems were encountered during construction. Water entering the cavern through fractures was limited to minor seepage with a maximum local inflow of 5 l/min at the crown, and a cavern inflow of 50 l/min.

Q2.10 The pre-existing stress state in a rock mass is caused by geological processes and is often a critical factor for rock engineering. Why do you think that quantifying the rock stress is important?

A2.10 Quantifying the rock stress is important because the pre-existing stress, concentrated around an excavation, can reach the rock strength and hence cause rock failure. We will be explaining in the next chapter

[7] Warren C. and Varley P. (1995) Geology, in *Engineering the Channel Tunnel* (C. J. Kirkland, ed). E and F N Spon, Chapman and Hall, London, p. 334.

that stress is a tensor quantity and six independent components are required to characterize the stress state. As an example, if the maximum stress component is horizontal, a simple and early design step in weak rock is to orientate tunnels so that they are parallel to this maximum stress component. This reduces the stress concentrations, a concept which has been used to great advantage in coal mining in the USA [8] and China [9].

2.3 Additional points

It is useful to read through engineering geology and structural geology textbooks and to become aware of the techniques used in these subjects. A good book in this context is Price and Cosgrove (1990) [10] which presents a "unified approach to the mechanistic analysis of geological structures". In the Preface to their book, the authors state that "... emphasis is placed on mechanical principles and the way in which they can be used to interpret and understand how and why a wide variety of geological structures develop."

Thus, rock mechanics and structural geology are inextricably linked. Whether the rock mass is disturbed by nature or by man, the rock mechanics mechanisms are similar in geology and engineering; they just differ in their emphases because of the different volume and time scales. Our emphasis in the engineering context is on the advantages of understanding the geometrical and mechanical features of rock masses which is further emphasized if one imagines slopes and tunnels constructed in the many rock masses illustrated in the Price and Cosgrove book.

[8] Su D. W. H. and Hasenfus G. J. (1995) Regional Horizontal Stress and Its Effect on Longwall Mining in the Northern Appalachian Coal Field. *Proceedings of the 14th International Conference on Ground Control in Mining* (S. S. Peng, ed.) West Virginia University, Morgantown, pp. 39–43.

[9] Wang Tongliang and Fan Qiuyan (1999) Optimization of Soft Rock Engineering with Particular Reference to Coal Mining. *Int. J. Rock Mech., Min. Sci.*, Rock Mechanics in China Special Issue.

[10] Price N. J. and Cosgrove J. W. (1990) *Analysis of Geological Structures.* Cambridge University Press, Cambridge, 502pp.

3 Stress

3.1 Understanding stress

To become at all proficient in engineering rock mechanics, it is essential to understand **the concept of stress**. Stress is not the same type of quantity as pressure or force because stress is not a scalar or vector: it is a **tensor quantity**. Since a tensor is a mathematical entity that obeys a certain set of rules, it is not difficult to develop a superficial understanding of stress. However, it is also important to have a deeper understanding of why these rules apply, and we have prepared this chapter to provide that understanding.

Pressure in a fluid is a **scalar** quantity: this means that it has a certain magnitude, which is independent of direction, e.g. the pressure in a car tyre[1] is 25 lbf/in^2 in all directions. **Force** is a **vector** quantity: it has magnitude and direction and must be specified by three components in the three-dimensional case, usually the three components in three mutually orthogonal (perpendicular) directions. However, the **stress** at a point inside a rock has three components acting perpendicular to the faces of a cube, and six stress components acting along the faces. The way in which these components vary as the cube is rotated means that stress is a **tensor** quantity and it must be specified in the three-dimensional case by six independent components. The normal and shear stresses acting on planes at different orientations inside the rock mass are required for rock engineering design studies and can be calculated using transformation equations, as we will illustrate.

The problem with perceiving stress is that humans deal with scalar and vector quantities during daily life, but not tensors. For example, temperature is a scalar quantity and wind velocity is a vector quantity, and so we have no difficulty relating to these concepts. However, stress, strain, permeability and moments of inertia, which are tensor quantities,

---

[1] We emphasize that it is the engineering rock mechanics principles that are important. Once you understand the principles, you can work with any units (such as the Imperial, or 'British', units used here) and with any symbols. Most of the time we will use SI (System International) units.

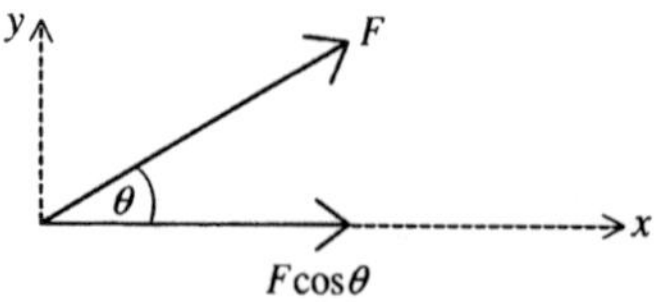

Figure 3.1 Resolution of a force.

are not consciously encountered in daily life. This means that we do not have intuitive feelings about quantities such as stress and strain, except in reduced one-dimensional forms (where the stress or strain is acting in only one direction) or when the values of the normal components are equal (so that stress becomes pressure and strain becomes a uniform contraction or expansion). An extra effort is required to develop an understanding of a tensor quantity and hence the nature of stress in a solid.

Readers who have already been exposed to some stress analysis and who may be somewhat sceptical about the lack of their intuitive understanding of stress might like to try Q3.10 in Part B of the book. In Part B, the questions can be tackled without the answers being visible. If you can solve problem Q3.10 without recourse to mathematics, you do have a feeling for stress!

Force is a vector quantity: it has magnitude and direction. For example, we might say that 'a force of 5 MN is acting horizontally in a northeasterly direction'. To specify a vector in two dimensions requires two pieces of information: either the magnitude and direction or the x and y components of the vector. In three dimensions, three pieces of information are required: either the magnitude and two directions or the x, y and z components of the vector. When a force, F, is resolved through an angle θ, the resultant is $F \cos \theta$, as the diagram in Fig. 3.1 illustrates.

The units of force are the newton (N) or the pound force (lbf) with dimensions LMT^{-2}.

Stress is a tensor quantity developed from the idea of normal forces and shear forces acting within the rock. Pushing your hand down on a table generates a normal force. Pushing your hand along the surface of the table generates a shear force. It is the fact that a solid can sustain a shear force which causes the stress field: otherwise, there would just be a scalar pressure, with the same value in all directions, and rock tunnels would float to the surface. The **first key to understanding stress** is understanding the existence of the shear force. The normal and shear forces are scaled by the areas on which they act, giving normal stresses and shear stresses[2]. The units of stress are newtons per metre squared, $N\ m^{-2}$, known as pascals, Pa, or pounds force per inch squared, lbf/in^2, with dimensions $L^{-1}MT^{-2}$.

When a plane is specified on which the stresses are acting, the normal and shear stress components have magnitude and direction. The normal stress acts normal to the plane; the shear stress acts along the plane.

[2] In fact the stress components are defined as the limiting values **at a point** when the area on which the forces act is reduced to zero.

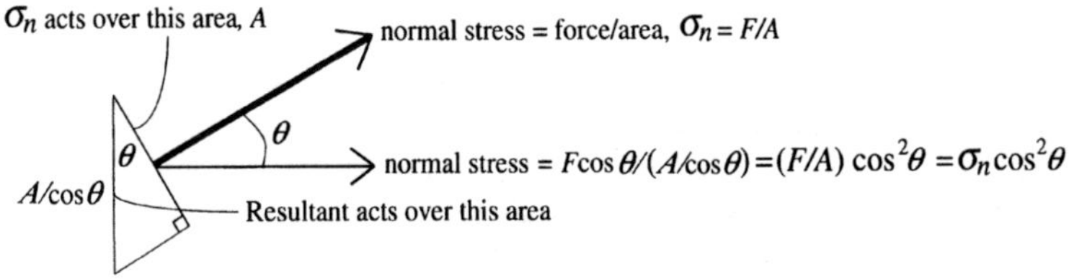

Figure 3.2 Resolution of a stress component, from the heavier arrow to the lighter arrow (for a prism of unit depth).

Because the shear stress on the plane can be resolved into two perpendicular components, there will be a total of three orthogonal stress components acting on the plane: the normal stress and two shear stresses. For example, on a vertical plane in the E–W direction, there could be a normal stress of 10 MPa acting due north, a shear stress of 5 MPa acting due west, and a shear stress of 7 MPa acting vertically downwards. Thus, to specify a tensor in **two dimensions** requires **three** pieces of information: either (a) two normal stresses acting in the specified x, and y directions, plus the shear stress; or (b) the two principal stresses (see Section 3.2, Q3.5) and their orientation. To specify a tensor in **three dimensions** requires **six** pieces of information: (a) either the three normal stresses and the three shear stresses acting on three specified orthogonal planes; or (b) the three principal stresses and their three directions.

We noted above that when a force, F, is resolved through an angle θ, the result is $F\cos\theta$. However, when a stress component, say a normal component σ_n, is resolved through an angle θ, the resultant is $\sigma_n\cos^2\theta$. The $\cos^2\theta$ term arises because a double resolution is required: i.e. a resolution of the force component and a resolution of the area on which it is acting. This is illustrated in Fig. 3.2, where the original normal stress component (represented by the heavy arrow) is transformed to the new stress component (represented by the lighter arrow) using the $\cos^2\theta$ term. **The second key to understanding stress** is understanding this double resolution.

When all the stress components are transformed in this way in two dimensions, the resulting equations for the stresses on a plane give the locus of a circle in normal stress–shear stress space. **The third key to understanding stress** is realizing that the $\cos^2\theta$ resolution of one normal stress and the $\sin^2\theta$ resolution of the other normal stress enables a graphical resolution of the stress components using Mohr's circle (see Q3.4). The circle occurs due to the $\cos^2\theta$ resolution of the first normal stress, the $\sin^2\theta$ resolution of the second normal stress ($\sin^2$ because the second normal stress is perpendicular to the first normal stress), and the fact that a circle with radius r is represented in (r, θ) space as

$$r^2\cos^2\theta + r^2\sin^2\theta = r^2.$$

These stress principles apply in all materials from chalk to cheese, and hence in all rock types. Moreover, in engineering rock mechanics, a knowledge of the values of the *in situ* natural stress components is required to understand the pre-engineering stress conditions. This is the subject of Chapter 4. The *in situ* stress field usually governs the boundary conditions for the rock mechanics problem. A knowledge of how the *in situ* stress field is then perturbed by changes in geometry (e.g. excavating a tunnel) or direct loading (e.g. the foundations of a dam) is crucial for engineering design. Understanding the fundamental nature of rock stress is an essential prerequisite, and so the questions in this chapter concentrate on this fundamental understanding.

3.2 Questions and answers: stress[3]

Q3.1 Show how the stress state in a solid can be described via the stress components (normal and shear) on an elemental cube of rock. Also, show how these components are listed in the stress matrix. What do the components in a row of the stress matrix have in common? What do the components in a column of the stress matrix have in common?

A3.1 The components on an elemental cube are shown below.

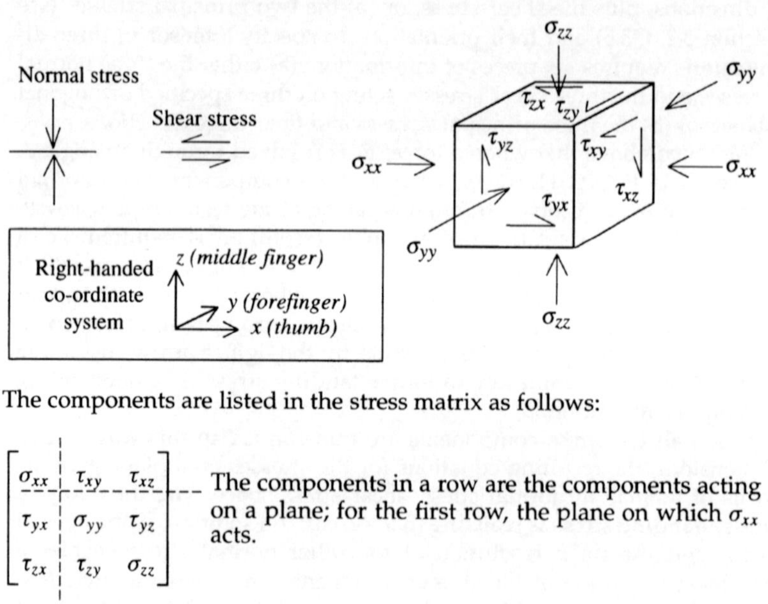

The components are listed in the stress matrix as follows:

$$\begin{bmatrix} \sigma_{xx} & \tau_{xy} & \tau_{xz} \\ \tau_{yx} & \sigma_{yy} & \tau_{yz} \\ \tau_{zx} & \tau_{zy} & \sigma_{zz} \end{bmatrix}$$

The components in a row are the components acting on a plane; for the first row, the plane on which σ_{xx} acts.

The components in a column are the components acting in one direction; for the first column, the x direction.

[3] *Note*: There are several conventions for expressing the stress components. We have used the most common notation in engineering rock mechanics, with σ for normal stress components and τ for shear stress components, and with compressive stress positive. The most important aspect is to understand the concept and manipulation of stress: these are independent of the symbol convention used. With this understanding, any convention can be used with equal facility. Similarly, if necessary one can learn to use the SI units (Pa) and Imperial or 'British' units (lbf/in^2) with equal facility: 1 MPa $\approx$ 145 psi.

Q3.2 When is a matrix symmetrical? Why is the stress matrix symmetrical?

A3.2 A term in the ith row and jth column of a matrix can be written as x_{ij}. A matrix is symmetrical when all the complementary terms x_{ij} and x_{ji} are equal.

The elemental cube shown in answer A3.1 is in equilibrium, and taking moments about the axes shows that the corresponding shear stresses must be equal.

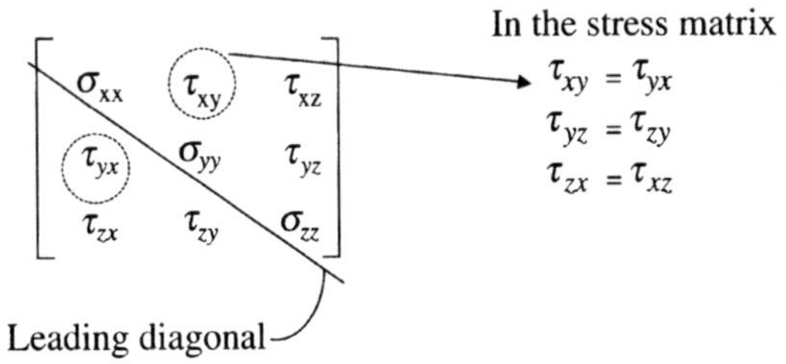

So, for a cube of side Δl, taking moments about the z-axis gives $\tau_{xy}\Delta l/2 - \tau_{yx}\Delta l/2 = 0$, or $\tau_{xy} = \tau_{yx}$. Because the complementary shear stresses are equal, the stress matrix has six independent components, e.g. the normal stresses on the leading diagonal and the three shear stresses above the leading diagonal: $\sigma_{xx}, \sigma_{yy}, \sigma_{zz}, \tau_{xy}, \tau_{xz}, \tau_{yz}$.

Q3.3 Explain the differences between scalar, vector and tensor quantities. Why is stress a tensor quantity?

A3.3 A scalar is a quantity with magnitude only, e.g. temperature or pressure. A vector is a quantity with magnitude and direction, e.g. velocity or force. A tensor is a multi-component quantity, each of the components having magnitude and direction, e.g. stress.

Stress is a tensor because the rules which govern the changes in the stress components as the reference axes are changed are those of a tensor. More mathematically, a tensor is a "multilinear differential form invariant with respect to a group of permissible co-ordinate transformations in n-space" (Borowski and Borwein, 1989 [4]).

Q3.4 How are normal and shear stress components plotted on Mohr's circle?

A3.4 The components are plotted as shown in the diagram below. Remember that compressive stresses are positive in rock mechanics, and notice the signs of the shear stresses. (A detailed explanation of Mohr's circle and its use may be found in ERM 1.)

[4] Borowski E. J. and Borwein J. M. (1989) *Dictionary of Mathematics*. Harper Collins, London, 659pp.

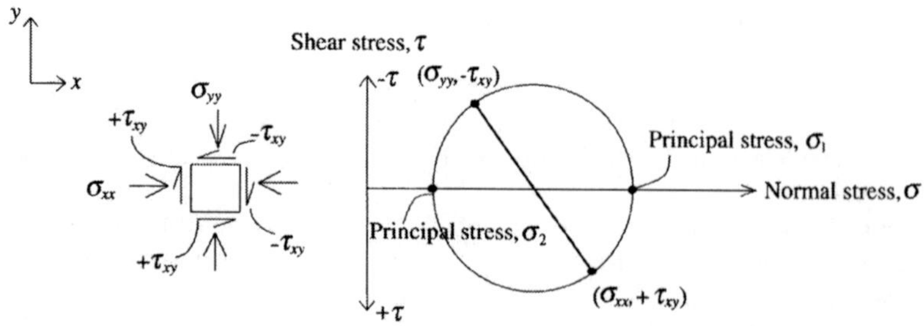

Q3.5 What is a principal stress plane? What is a principal stress?

A3.5 A principal stress[5] plane is a plane on which there are no shear stresses. A principal stress is the normal stress acting on a principal stress plane. In 3-D, the three principal stresses are written as σ_1, σ_2, and σ_3, in order of the largest value, intermediate value, smallest value, respectively. For any given stress state, the maximal and minimal normal stresses occur on the principal stress planes: see where σ_1 and σ_2 are plotted on the Mohr circle above for the 2-D case.

In Chapter 16 of ERM 1, we explained that **all unsupported rock excavation surfaces are principal stress planes**, because there are no shear stresses acting on them. Thus, one of the effects of excavation is to define locally the orientations of the principal stresses, i.e. they will be parallel and perpendicular to the excavation surfaces. Moreover, the principal stress component acting normal to unsupported excavation surfaces is also zero.

When the principal stresses at an excavation surface are shown in the stress matrix (with the z-axis perpendicular to the excavation surface), most of the terms have zero value.

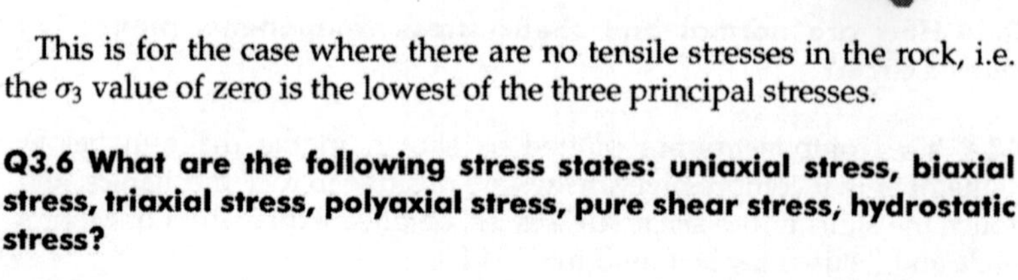

This is for the case where there are no tensile stresses in the rock, i.e. the σ_3 value of zero is the lowest of the three principal stresses.

Q3.6 What are the following stress states: uniaxial stress, biaxial stress, triaxial stress, polyaxial stress, pure shear stress, hydrostatic stress?

A3.6 These terms are commonly used to describe the stress states applied to laboratory test specimens.

[5] Do not use 'principle stress'! The correct term is 'principal stress'.

Uniaxial stress. One principal stress is applied, i.e. one principal stress has a non-zero value, $\sigma_1 \neq 0, \sigma_2 = \sigma_3 = 0$.

Biaxial stress. Two principal stresses are applied, i.e. two principal stresses have non-zero values, $\sigma_1 \neq 0, \sigma_2 \neq 0, \sigma_3 = 0$.

Triaxial stress. Three principal stresses are applied *but two have the same value.* This term came into use during the development of laboratory testing: a cylinder of rock is compressed by one principal stress along its axis and a fluid pressure is applied to the sides, equivalent to a stress state $\sigma_1 \neq 0$, $\sigma_2 = \sigma_3 \neq 0$. The term is correct in that three stress components are applied along three axes, but somewhat misleading because two of the components have the same value.

Polyaxial stress. Three principal stresses are applied, i.e. the three principal stresses have non-zero values, $\sigma_1 \neq 0$, $\sigma_2 \neq 0$, $\sigma_3 \neq 0$. These three stresses are usually unequal.

Pure shear stress. This term is used to denote a stress state where there is only shear stress on a plane (i.e. there is no normal stress on the plane). This occurs within a specimen when equal compressive and tensile stresses are externally applied, σ_1 = some value, say k, $\sigma_2 = -k$, $\sigma_3 = 0$. The plane with a state of pure shear is then at 45° to the applied stresses, as shown below.

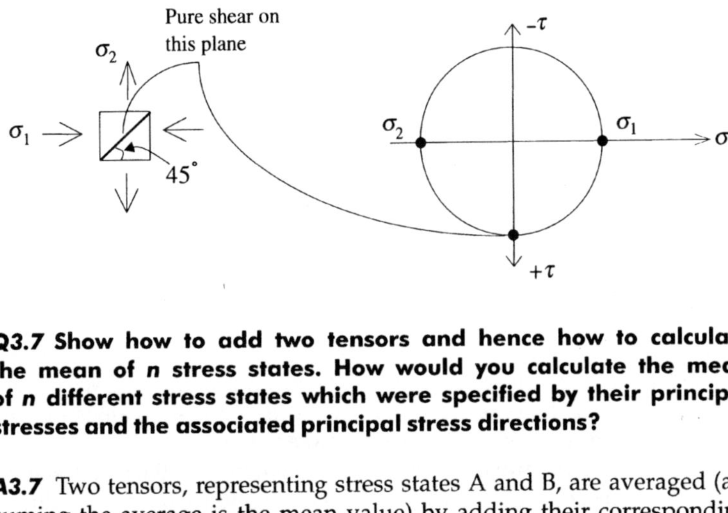

Q3.7 Show how to add two tensors and hence how to calculate the mean of *n* stress states. How would you calculate the mean of *n* different stress states which were specified by their principal stresses and the associated principal stress directions?

A3.7 Two tensors, representing stress states A and B, are averaged (assuming the average is the mean value) by adding their corresponding components and dividing by two.

$$\text{Mean} = \left\{ \begin{bmatrix} \sigma_{xx}^{A} & \tau_{xy}^{A} & \tau_{xz}^{A} \\ \tau_{yx}^{A} & \sigma_{yy}^{A} & \tau_{yz}^{A} \\ \tau_{zx}^{A} & \tau_{zy}^{A} & \sigma_{zz}^{A} \end{bmatrix} + \begin{bmatrix} \sigma_{xx}^{B} & \tau_{xy}^{B} & \tau_{xz}^{B} \\ \tau_{yx}^{B} & \sigma_{yy}^{B} & \tau_{yz}^{B} \\ \tau_{zx}^{B} & \tau_{zy}^{B} & \sigma_{zz}^{B} \end{bmatrix} \right\} /2$$

$$= \left\{ \begin{bmatrix} \sigma_{xx}^{(A+B)} & \tau_{xy}^{(A+B)} & \tau_{xz}^{(A+B)} \\ \tau_{yx}^{(A+B)} & \sigma_{yy}^{(A+B)} & \tau_{yz}^{(A+B)} \\ \tau_{zx}^{(A+B)} & \tau_{zy}^{(A+B)} & \sigma_{zz}^{(A+B)} \end{bmatrix} \right\} /2$$

where

$$\sigma_{xx}^{(A+B)} = \sigma_{xx}^{A} + \sigma_{xx}^{B},$$

and so on.

However, when n stress tensors have been specified by the magnitudes and directions of their three principal stresses, σ_{1i}, σ_{2i}, σ_{3i}, for $i = 1$ to n, it is not correct to average the principal stress values and average their directions. For example, if one stress state has the maximum principal stress acting due north with a value of 5 MPa and a second stress state has the maximum principal stress acting due west with a value of 10 MPa, the average stress state is not a stress state with the maximum principal stress acting northwest with a value of 7.5 MPa.

Understanding how to average stress fields is important for data reduction in stress measurement programmes. The components for each tensor must be first specified relative to three reference axes, x, y and z, and then the components averaged, as in the case above which represents averaging for $n = 2$. Once this averaging has been done, the principal stresses for the mean stress tensor can be calculated.

Q3.8 What are the first, second and third stress invariants?

A3.8 When the stress tensor is expressed with reference to sets of axes oriented in different directions, the components of the tensor change. However, certain functions of the components do not change. These are known as the stress invariants, expressed as I_1, I_2, I_3. (Recall that in A3.3, the mathematical definition of a tensor included '...invariant with respect to ... permissible co-ordinate transformations...'). The three invariants are:

$$I_1 = \sigma_{xx} + \sigma_{yy} + \sigma_{zz}$$

$$I_2 = \sigma_{xx}\sigma_{yy} + \sigma_{yy}\sigma_{zz} + \sigma_{zz}\sigma_{xx} - \tau_{xy}^2 - \tau_{yz}^2 - \tau_{zx}^2$$

$$I_3 = \sigma_{xx}\sigma_{yy}\sigma_{zz} + 2\tau_{xy}\tau_{yz}\tau_{zx} - \sigma_{xx}\tau_{yz}^2 - \sigma_{yy}\tau_{zx}^2 - \sigma_{zz}\tau_{xy}^2$$

The expression for the first invariant, I_1, indicates that for a given stress state, whatever the orientation of the x, y and z axes, i.e. whatever the orientation of the reference cube shown in A3.1, the values of the three normal stresses will add up to the same value I_1.

When the principal stresses have to be calculated from the components of the stress tensor, a cubic equation is used for finding the three values, σ_1, σ_2, and σ_3. This equation is

$$\sigma^3 - (\sigma_{xx} + \sigma_{yy} + \sigma_{zz})\sigma^2 + (\sigma_{xx}\sigma_{yy} + \sigma_{yy}\sigma_{zz} + \sigma_{xx}\sigma_{zz} - \tau_{xy}^2 - \tau_{yz}^2 - \tau_{zx}^2)\sigma - (\sigma_{xx}\sigma_{yy}\sigma_{zz} + 2\tau_{xy}\tau_{yz}\tau_{zx} - \sigma_{xx}\tau_{yz}^2 - \sigma_{yy}\tau_{zx}^2 - \sigma_{zz}\tau_{xy}^2) = 0$$

or

$$\sigma^3 - I_1\sigma^2 + I_2\sigma - I_3 = 0$$

Because the values of the principal stresses must be independent of the choice of axes, the coefficients in the equation above, i.e. I_1, I_2, and I_3, must be invariant with respect to the orientation of the axes.

Note that the components of the first stress invariant, $I_1 = \sigma_{xx} + \sigma_{yy} + \sigma_{zz} = \sigma_1 + \sigma_2 + \sigma_3$, are on the leading diagonal of the stress matrix (the terms from top left to bottom right). Since I_1 is the sum of the normal stresses, it is three times the mean normal stress. This first invariance indicates that there is a constant mean normal stress as the reference axes are changed.

Q3.9 By considering the rates of change of the stress components in the answer to Q3.1, establish force equilibrium in the *x*, *y* and *z* directions and hence write down in differential form the three equilibrium equations for an elemental cube.

A3.9 Consider the infinitesimal elemental cube shown in A3.1 and take the rate of change[6] of σ_{xx} in the x direction as $\partial\sigma_{xx}/\partial x$. Because of this change, the stress on one side of the cube is larger than on the other, as shown below.

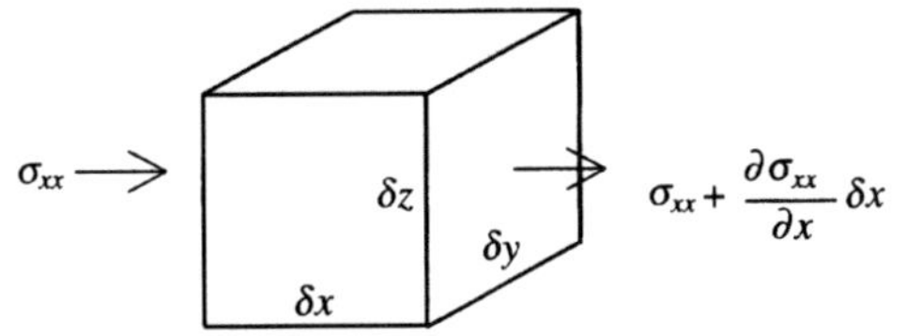

The net stress in the x direction due to this change across the elemental cube for a distance δx is the stress applied to one side of the cube minus the stress applied to the other side:

$$\left\{\left(\sigma_{xx} + \frac{\partial \sigma_{xx}}{\partial x}\delta x\right) - \sigma_{xx}\right\} \text{ or } \frac{\partial \sigma_{xx}}{\partial x}\delta x.$$

Assuming that the cube has side lengths of δx, δy and δz, and multiplying this stress increment by the area of the face on which σ_{xx} acts, i.e. $\delta y\delta z$, gives a force increment of $(\partial\sigma_{xx}/\partial x)\delta x\delta y\delta z$. Applying this principle to the three stress components acting in the x-direction (see the left-hand column of the stress matrix in A3.1), gives the three force increments:

$$\frac{\partial \sigma_{xx}}{\partial x}\delta x\delta y\delta z,\quad \frac{\partial \tau_{yx}}{\partial x}\delta x\delta y\delta z,\text{ and } \frac{\partial \tau_{zx}}{\partial x}\delta x\delta y\delta z$$

Because the infinitesimal cube is in equilibrium, these forces must also be in equilibrium (in the absence of any other forces) and hence have a sum of zero. Equating the sum of the force increments to zero, cancelling $\delta x\delta y\delta z$, and assuming that no body forces such as gravity are acting, provides one of the differential equations of equilibrium:

$$\frac{\partial \sigma_{xx}}{\partial x} + \frac{\partial \tau_{yx}}{\partial x} + \frac{\partial \tau_{zx}}{\partial x} = 0$$

[6] The mathematical operator, $\partial/\partial x$, represents differentiation with respect to x, all other variables being treated as constants.

By similar considerations in the y and z directions, we also have

$$\frac{\partial \tau_{xy}}{\partial y} + \frac{\partial \sigma_{yy}}{\partial y} + \frac{\partial \tau_{zy}}{\partial y} = 0$$

$$\frac{\partial \tau_{xz}}{\partial z} + \frac{\partial \tau_{yz}}{\partial z} + \frac{\partial \sigma_{zz}}{\partial z} = 0$$

The reason for including this question and answer is to demonstrate that equations such as these relating to stress analysis, and which are perhaps daunting at first sight, are not so difficult and are certainly easier to remember once they are understood.

Q3.10 Given an elemental cube with a normal stress component and two shear stress components acting on all its faces, it is always possible to find a cube orientation such that the shear stresses disappear on all faces and only normal stresses (the principal normal stresses) remain. Is it possible to find a complementary orientation such that the normal stresses disappear on all faces and only shear stresses (i.e. principal shear stresses) remain? Explain the reason for your answer.

A3.10 No, generally it is not possible. The answer is straightforward from the mathematical point of view. The first invariant I_1 explained in A3.8 provides the direct answer to the question. As the orientation of the cube changes, the normal and shear stresses on the cube faces change, but

$$I_1 = \sigma_{xx} + \sigma_{yy} + \sigma_{zz} = \sigma_1 + \sigma_2 + \sigma_3$$

and so, for any non-zero value of I_1, it is not possible to have $\sigma_1 = \sigma_2 = \sigma_3 = 0$.

There is the exceptional case when $I_1 = 0$. In 2-D, for the pure shear stress case, as defined in A3.6, where $I_1 = \sigma_1 + \sigma_2 = +k + (-k) = 0$, there is an orientation of the elemental square with the edges having no normal stresses, i.e. when the square is rotated by 45°. Note that this can only occur for a Mohr circle centred at the origin of the $\sigma - \tau$ axes, so that one diameter of Mohr's circle is along the τ axis where the normal stresses are zero.

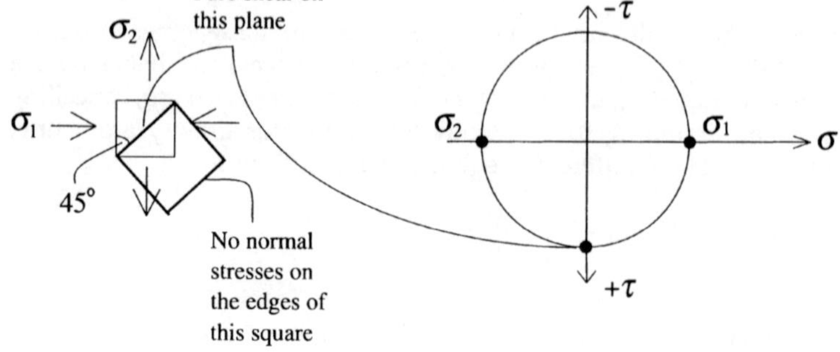

3.3 Additional points

Stress is important for many aspects of rock mechanics and is relevant to many of the chapter subjects in the book. In particular, further stress questions are included in Chapters 19 and 20.

We re-emphasize the point that principal stress planes are defined when new rock surfaces are created, e.g. by excavation of a slope, borehole, tunnel or cavern. By Newton's 3rd law — to every action there is an equal and opposite reaction — there can be no normal or shear stresses acting on such rock surfaces (neglecting atmospheric pressure). Not only is the rock surface a principal stress plane, but the principal stress acting on that plane is zero.

Thus, the process of creating a new rock surface causes the principal stresses in the rock mass to be locally oriented perpendicular and parallel to the surface. The principal stress perpendicular to the surface is zero and the maximum and minimum stress values occur in the rock in a direction parallel to the rock surface, i.e. the other two principal stresses. It is much easier to understand underground deformations and excavation-induced fracturing if this is borne in mind.

Remember that, although compressive stresses are usually reckoned as positive in rock mechanics, computer programs for numerical analysis are often developed from structural engineering codes in which tensile stresses are positive. Always check the sign convention.

4 *In situ* rock stress

4.1 The nature of *in situ* rock stress

Having explained the concept of stress in Chapter 3, we now consider why there is a **natural stress state** in the rock and some of the key aspects of the subject, especially measuring the *in situ* stress and predicting variation in the stress state from point to point in a rock mass. In Chapter 5, we will consider the concept of strain and how stress and strain can be linked.

Six independent components are required to specify the stress state at a point. The stress state can be presented as the matrix of stress components relative to specified x, y and z reference axes.

$$\begin{bmatrix} \sigma_{xx} & \tau_{xy} & \tau_{xz} \\ \tau_{yx} & \sigma_{yy} & \tau_{yz} \\ \tau_{zx} & \tau_{zy} & \sigma_{zz} \end{bmatrix} = \begin{bmatrix} \sigma_{xx} & \tau_{xy} & \tau_{xz} \\ & \sigma_{yy} & \tau_{yz} \\ \text{Symmetric} & & \sigma_{zz} \end{bmatrix}$$

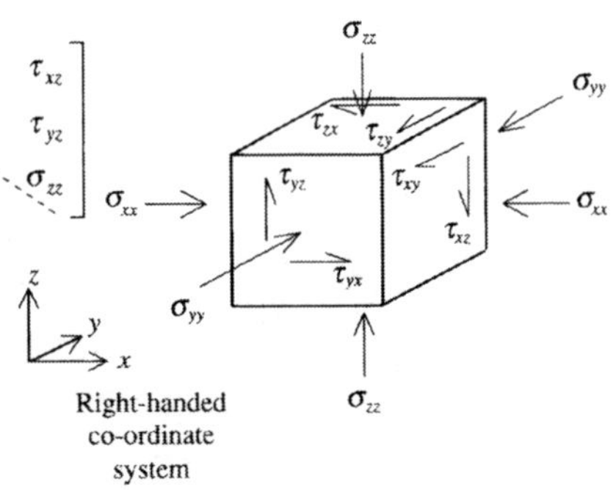

or in principal stress form

$$\begin{bmatrix} \sigma_1 & 0 & 0 \\ 0 & \sigma_2 & 0 \\ 0 & 0 & \sigma_3 \end{bmatrix} = \begin{bmatrix} \sigma_1 & 0 & 0 \\ & \sigma_2 & 0 \\ \text{Symmetric} & & \sigma_3 \end{bmatrix}$$

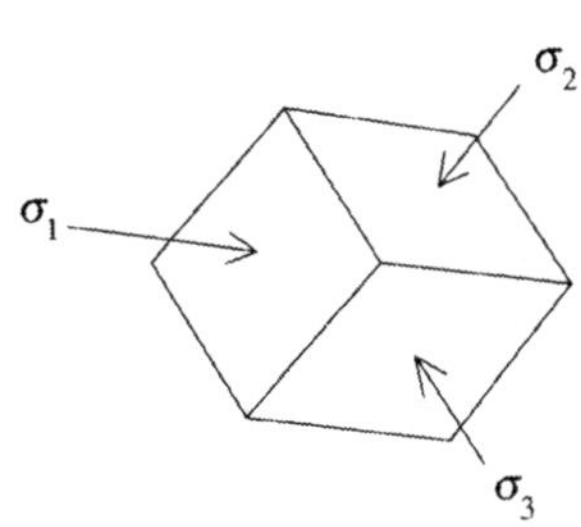

In the first case, the six components of the stress state are the six components of the stress tensor with respect to the reference axes. In the second case, the six components are given by the values of the three principal stresses, σ_1, σ_2, and σ_3, plus three values providing the orientation information needed to specify the three principal stress directions relative to the reference axes; for example, the trend and plunge[1] of the σ_1 direction and the trend of the σ_2 direction, remembering that σ_1, σ_2, and σ_3 are mutually orthogonal.

In situ stress states, as estimated from geological information or as determined from measurements during a site investigation, are usually reported using the principal stresses. An example stress state, plotted on a lower hemisphere projection, is shown below.

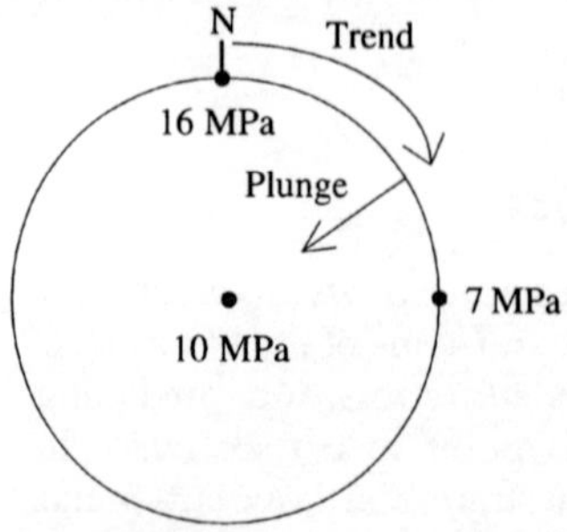

The diagram indicates a stress state
$\sigma_1 = 16$ MPa acting horizontally north–south,
$\sigma_2 = 10$ MPa acting vertically, and
$\sigma_3 = 7$ MPa acting horizontally east–west.

The use of hemispherical projection is explained in Appendix B of ERM 1 and in this book in Section 7.1.

To establish why there is a natural stress state in the rock, we must consider the geological circumstances. The stress state is caused by three main factors.

(a) *The weight of the rock.* In Imperial units, a cube of rock 1 ft × 1 ft × 1 ft has a base area of 12 in × 12 in = 144 in^2. Assume that the cube is on the floor and that it weighs 144 lbf, a representative value. The cube will then exert a stress of 144 lbf/144 in^2 = 1 lbf/in^2 = 1 psi on the floor. Thus, a useful rule of thumb is that in a rock mass the **vertical stress component** will increase by about 1 psi for every foot of depth. The equivalent rule in SI units is that the vertical stress will increase by 1 MPa for every 40 m of depth, because a representative value for rock unit weight is 25 kN/m^3. The exact values will depend on the local rock density.

The pressure in a car tyre, say 25 psi, is equivalent to a rock depth of 25 ft or about 8 m. At a depth of 2000 ft, or about 615 m, the vertical stress component is around 2000 psi or 15 MPa. This is equivalent to the weight of a car on every square inch of rock, or the weight of 1500 cars on every square metre.

[1] The terms 'dip direction' and 'dip angle' are used for the orientation of a **plane**: the 'dip angle' is the angle between the steepest line in the plane and the horizontal; the 'dip direction' is the compass bearing or azimuth of the dip line. The words 'trend' and 'plunge' are used for the orientation of a **line**: the 'trend' is the compass bearing of the line; the 'plunge' is the angle between the line and the horizontal. Thus, the orientation of a fracture will be specified by the dip direction and dip angle, whereas the orientation of the normal to a fracture, or a borehole axis or a principal stress, will be specified by the trend and plunge.

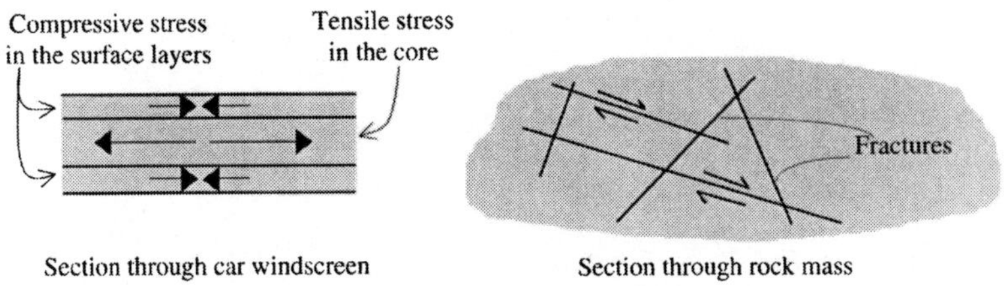

Figure 4.1 The concept of locked-in stresses, in a car windscreen and in a rock mass.

(b) *Current tectonic forces.* Due to the movement of tectonic plates, the earth's crust is subject to tectonic forces, as it has been during geological history. The rock mass may be strong enough to sustain these forces and the associated stresses. However, if the stresses are high enough, the rock will fracture and crumple causing the joints, faults (generically termed fractures) and folds, which are formed during high stress and orogenic events. In many regions of the world, the tectonic plates are in a state of limiting equilibrium and there are **high horizontal stresses** in the rock mass. In fact, it is the rule rather than the exception that the maximum principal stress in a rock mass will be acting sub-horizontally, not vertically.

(c) *Previous tectonic forces.* There may also be **residual stresses** from earlier tectonic events: when a fractured rock mass is compressed and then unloaded, stresses can be left locked in the rock mass. An analogue example of this effect is a pre-stressed car windscreen.

The section through the windscreen in Fig. 4.1 indicates how a layer on each surface in a compressive stress condition is balanced by the tensile stress in the central region. The purpose of this pre-stressing is to increase the windscreen's resistance to breakage by reducing the tensile bending stresses that can develop on either side during an impact. Similarly, and as a result of loading and unloading, the fractured rock mass may well contain significant residual stress. The subject of residual stress is a controversial one, but it is possible that many high horizontal stress components observed in rock masses are residual stresses.

All three of the main causes described above can contribute to the stress state at a point in a rock mass. In addition, other factors can alter the stress state, such as erosion which will reduce the vertical stress component more than the horizontal components. Fractures at all scales will perturb the stress field.

To understand how a fracture pattern perturbs the stress field, assume that in a horizontal section through a rock mass, as shown in Fig. 4.2, the stress state 'A' indicates the two horizontal components of the pervasive stress state in the rock. Nearer the fracture, states 'B' and 'C', the principal stress directions are rotated and the magnitudes of the principal stresses change. In the case of an open fracture, no normal or shear stress can be sustained perpendicular and parallel to the fracture surface, so the fracture surface becomes a principal stress plane with a principal stress value of zero (cf. A3.5). Imagining this

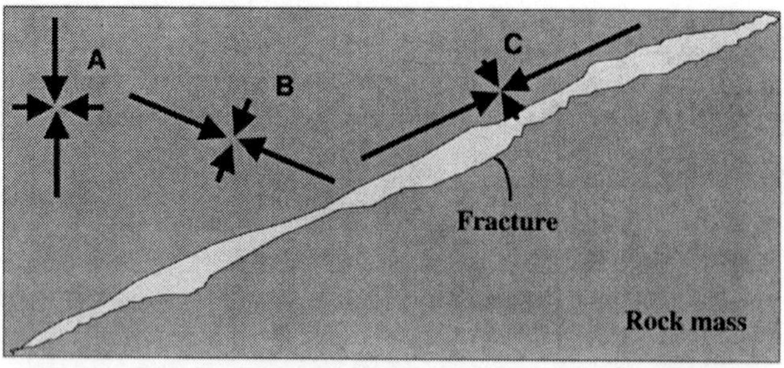

Figure 4.2 Perturbation of *in situ* rock stress state caused by a fracture.

effect adjacent to many fractures at all scales in a rock mass leads to the expectation that local values of *in situ* stress, and associated site investigation values, are likely to be variable. Also, the activities of the engineer when excavating the rock mass will perturb the stress state adjacent to excavations.

4.2 Questions and answers: *in situ* rock stress

Q4.1 There is no internationally agreed terminology for words describing the state of stress in a rock mass. However, describe in one sentence what you think is meant by each of the following terms.

Natural stress	**Thermal stress**
Induced stress	**Palaeostress**
Gravitational stress	**Near-field stress**
Tectonic stress	**Far-field stress**
Residual stress	**Local stress**

A4.1

Natural stress:	the *in situ* stress which exists prior to engineering.
Induced stress:	the natural stress state as perturbed by engineering[2].
Gravitational stress:	the stress state caused by the weight of the rock above.
Tectonic stress:	the stress state caused by tectonic plate movement.
Residual stress:	the stress state caused by previous tectonic activity.
Thermal stress:	the stress state caused by temperature change.
Palaeostress:	a previous natural stress that is no longer acting.
Near-field stress:	the stress state in the region of an engineering perturbation.
Far-field stress:	the stress state beyond the near-field.
Local stress:	the stress state in a region of interest.

[2] Some writers use 'induced stress' to mean the actual stress after engineering; others use the term to mean the stress changes caused by engineering. It is important to be sure which definition is being used in any particular context.

Q4.2 Add the following 2-D rock stress states, and find the principal stresses and directions of the resultant stress state.

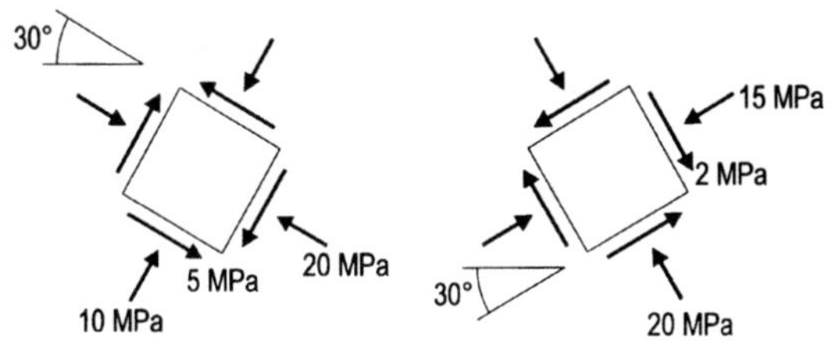

A4.2 Draw xy and lm axes for the first stress state, and then plot the corresponding Mohr circle.

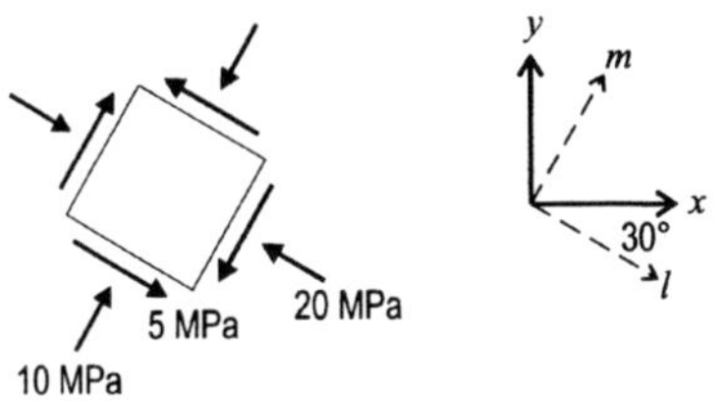

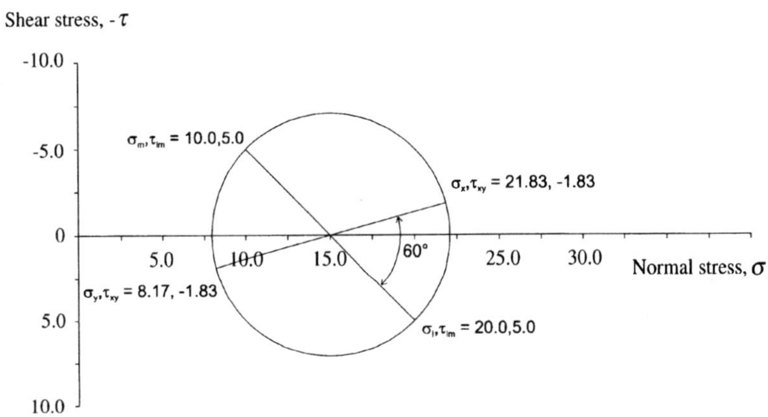

The stresses transformed to the xy axes are $\begin{bmatrix} 21.83 & -1.83 \\ -1.83 & 8.17 \end{bmatrix}$.

Draw xy and lm axes for the second stress state, and plot the corresponding Mohr circle.

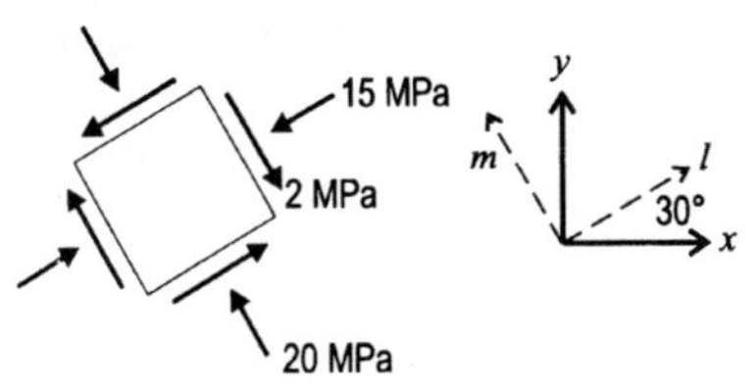

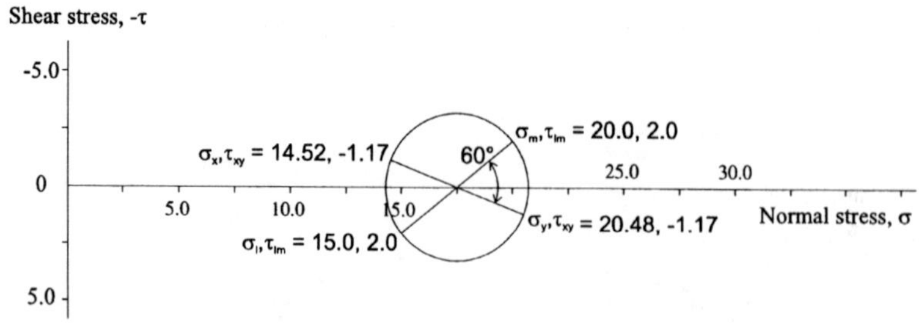

The stresses transformed to the xy axes are $\begin{bmatrix} 14.52 & -1.17 \\ -1.17 & 20.48 \end{bmatrix}$.

Adding the two xy stress states gives

$$\begin{bmatrix} 21.83 & -1.83 \\ -1.83 & 8.17 \end{bmatrix} + \begin{bmatrix} 14.52 & -1.17 \\ -1.17 & 20.48 \end{bmatrix} = \begin{bmatrix} 36.35 & -3.00 \\ -3.00 & 28.65 \end{bmatrix}.$$

Plotting the Mohr circle for the combined stress state and reading off the principal stresses and the principal directions gives the required values.

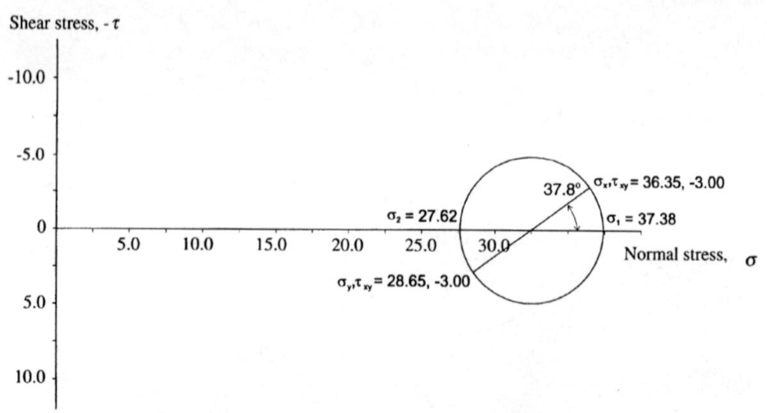

The principal stresses of the resultant stress state are $\sigma_1 = 37.38$ MPa and $\sigma_2 = 27.62$ MPa, with σ_1 being rotated 18.9° clockwise from the x-direction.

Q4.3 How many experimental set-ups are required to determine the 3-D state of stress in a rock mass, using each of the standard stress measurement methods[3] of flatjack, hydraulic fracturing, USBM gauge and CSIRO gauge? (Assume that the stress field is constant in the vicinity of the test site.)

[3] These stress measurement methods are described in Section 4.3 of ERM 1.

A4.3 The *flatjack* method measures one normal stress; so, given that the stress tensor has six independent components, six set-ups are required (at different orientations).

Conventional hydraulic fracturing in a vertical borehole measures two values, the principal stresses are assumed to be parallel and perpendicular to the borehole axis, and the vertical stress value is assumed on the basis of gravitational stress. As a result, with two measured values and the equivalent of four assumptions, only one set-up is required to establish the stress tensor components. For the *hydraulic testing of pre-existing fractures (HTPF)*, the borehole does not have to be parallel to a principal stress and one normal stress is measured each time, similar to the flatjack method; so six set-ups are required, but each must use a fracture at a different orientation.

The *USBM overcoring gauge* provides three values in a plane; so two set-ups are required in two boreholes at different orientations.

The *CSIRO overcoring gauge* provides at least six values; so one set-up is required.

Q4.4 Three flatjack tests have been made close to each other in the wall of a long, straight tunnel, the axis of which dips at 7°. The measurement position is approximately 250 m below the ground surface and it is assumed that the flatjacks are in the same stress field. The slots for the flatjacks were cut normal to the wall of the tunnel, and were oriented relative to the tunnel axis as shown. The cancellation pressure for each of the flatjacks A, B and C was 7.56 MPa, 6.72 MPa and 7.50 MPa, respectively. Compute the principal stresses and their directions, and ascertain whether they accord with worldwide trends.

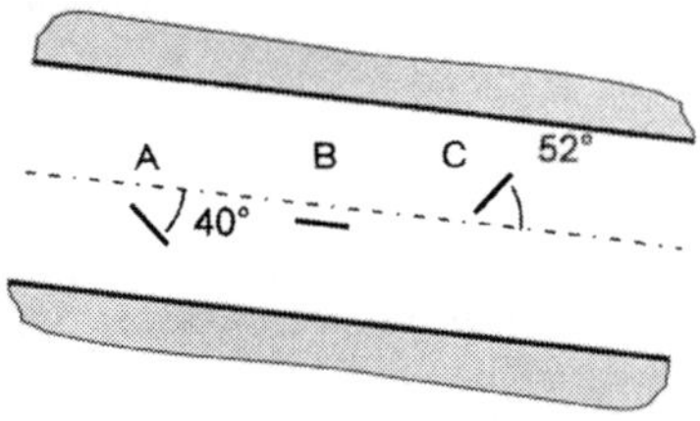

A4.4 One way of solving this problem is to use the stress transformation equations, i.e.

$$\sigma'_x = \sigma_x \cos^2\theta + \sigma_y \sin^2\theta + 2\tau_{xy}\sin\theta\cos\theta$$

where σ_x, σ_y and τ_{xy} are the global stress components, and θ is the angle between the global x axis and the direction of the stress in question.

Taking the x axis horizontal directed to the right, and the y axis vertical upwards, and all orientations measured anticlockwise positive from the positive x axis, we have the following dip angles:

$$\beta_{\text{tunnel}} = -7^\circ; \quad \beta_{\text{A}} = -40^\circ + \beta_{\text{tunnel}} = -47^\circ;$$

$$\beta_{\text{B}} = 0^\circ + \beta_{\text{tunnel}} = -7^\circ; \quad \beta_{\text{C}} = 52^\circ + \beta_{\text{tunnel}} = 45^\circ.$$

Finally, because each flatjack measures the normal stress component perpendicular to the flatjack, we add 90° to each of these directions to obtain the direction of the normal stress on each jack. Thus, the magnitude and direction of the normal stress on each jack is as follows:

Jack A $\quad \sigma_A = 7.56$ MPa; $\quad \theta_A = \beta_A + 90°$; $\quad \theta_A = 43°$

Jack B $\quad \sigma_B = 6.72$ MPa; $\quad \theta_B = \beta_B + 90°$; $\quad \theta_B = 83°$

Jack C $\quad \sigma_C = 7.50$ MPa; $\quad \theta_C = \beta_C + 90°$; $\quad \theta_C = 135°$

Assembling the stress transformation equation for all three jacks into matrix form gives

$$\begin{bmatrix} \sigma_A \\ \sigma_B \\ \sigma_C \end{bmatrix} = \begin{bmatrix} \cos^2\theta_A & \sin^2\theta_A & 2\sin\theta_A\cos\theta_A \\ \cos^2\theta_B & \sin^2\theta_B & 2\sin\theta_B\cos\theta_B \\ \cos^2\theta_C & \sin^2\theta_C & 2\sin\theta_C\cos\theta_C \end{bmatrix} \begin{bmatrix} \sigma_x \\ \sigma_y \\ \tau_{xy} \end{bmatrix} \quad \text{or } \boldsymbol{\sigma}_{\text{jack}} = \mathbf{R}\boldsymbol{\sigma}_{\text{global}}$$

which, upon evaluation, yields

$$\begin{bmatrix} 7.56 \\ 6.72 \\ 7.50 \end{bmatrix} = \begin{bmatrix} 0.535 & 0.465 & 0.998 \\ 0.015 & 0.985 & 0.242 \\ 0.500 & 0.500 & -1.000 \end{bmatrix} \begin{bmatrix} \sigma_x \\ \sigma_y \\ \tau_{xy} \end{bmatrix}$$

Inverting this matrix equation gives $\boldsymbol{\sigma}_{\text{global}} = \mathbf{R}^{-1}\boldsymbol{\sigma}_{\text{jack}}$ and this evaluates as

$$\begin{bmatrix} \sigma_x \\ \sigma_y \\ \tau_{xy} \end{bmatrix} = \begin{bmatrix} 1.093 & -0.952 & 0.860 \\ -0.134 & 1.021 & 0.113 \\ 0.479 & 0.034 & -0.514 \end{bmatrix} \begin{bmatrix} 7.56 \\ 6.72 \\ 7.50 \end{bmatrix}$$

from which we find $\boldsymbol{\sigma}_{\text{global}} = \begin{bmatrix} 8.31 \\ 6.70 \\ 0.00 \end{bmatrix}$ MPa.

We see that σ_x and σ_y are principal stresses, because $\tau_{xy} = 0$, and the principal stresses are vertical and horizontal, which is a reasonable result. In addition, we have the horizontal stress, σ_x, greater than the vertical stress, σ_y, which is usually the case.

Now, to compare the vertical stress with the weight of the overburden, compute a value for the vertical stress based on the depth of the tunnel and an assumed unit weight for the rock:

$$\gamma_{\text{rock}} = 27 \text{ kN/m}^3; \quad z = 250 \text{ m}; \quad \sigma_v = \gamma_{\text{rock}} \times z; \quad \sigma_v = 6.75 \text{ MPa}.$$

This compares well with the value for σ_y found above.

Q4.5 Two further flatjack measurements have been made in the wall of the tunnel considered in Q4.4. These dip at 20° and 90° relative to the tunnel axis, and produced cancellation pressures of 7.38 MPa and 7.86 MPa, respectively. Compute the best estimate of the principal stresses.

A4.5 Continuing from Q4.4, we determine the stress directions for all five jacks:

$$\beta_{\text{tunnel}} = -7^\circ; \quad \beta_{\text{A}} = -40^\circ + \beta_{\text{tunnel}} = -47^\circ; \quad \beta_{\text{B}} = 0^\circ + \beta_{\text{tunnel}} = -7^\circ;$$

$$\beta_{\text{C}} = +52^\circ + \beta_{\text{tunnel}} = 45^\circ; \quad \beta_{\text{D}} = -20^\circ + \beta_{\text{tunnel}} = -27^\circ;$$

$$\beta_{\text{E}} = -90^\circ + \beta_{\text{tunnel}} = -97^\circ$$

The magnitude and direction of the normal stress on each jack are then as follows:

Jack A	$\sigma_{\text{A}} = 7.56$ MPa;	$\theta_{\text{A}} = \beta_{\text{A}} + 90^\circ$;	$\theta_{\text{A}} = 43^\circ$
Jack B	$\sigma_{\text{B}} = 6.72$ MPa;	$\theta_{\text{B}} = \beta_{\text{B}} + 90^\circ$;	$\theta_{\text{B}} = 83^\circ$
Jack C	$\sigma_{\text{C}} = 7.50$ MPa;	$\theta_{\text{C}} = \beta_{\text{C}} + 90^\circ$;	$\theta_{\text{C}} = 135^\circ$
Jack D	$\sigma_{\text{D}} = 7.38$ MPa;	$\theta_{\text{D}} = \beta_{\text{D}} + 90^\circ$;	$\theta_{\text{D}} = 63^\circ$
Jack E	$\sigma_{\text{E}} = 7.86$ MPa;	$\theta_{\text{E}} = \beta_{\text{E}} + 90^\circ$;	$\theta_{\text{E}} = -7^\circ$

As in A4.4, we use $\boldsymbol{\sigma}_{\text{jack}} = \mathbf{R}\boldsymbol{\sigma}_{\text{global}}$ where

$$\boldsymbol{\sigma}_{\text{jack}} = \begin{bmatrix} \sigma_{\text{A}} \\ \sigma_{\text{B}} \\ \sigma_{\text{C}} \\ \sigma_{\text{D}} \\ \sigma_{\text{E}} \end{bmatrix}, \quad \mathbf{R} = \begin{bmatrix} \cos^2\theta_{\text{A}} & \sin^2\theta_{\text{A}} & 2\sin\theta_{\text{A}}\cos\theta_{\text{A}} \\ \cos^2\theta_{\text{B}} & \sin^2\theta_{\text{B}} & 2\sin\theta_{\text{B}}\cos\theta_{\text{B}} \\ \cos^2\theta_{\text{C}} & \sin^2\theta_{\text{C}} & 2\sin\theta_{\text{C}}\cos\theta_{\text{C}} \\ \cos^2\theta_{\text{D}} & \sin^2\theta_{\text{D}} & 2\sin\theta_{\text{D}}\cos\theta_{\text{D}} \\ \cos^2\theta_{\text{E}} & \sin^2\theta_{\text{E}} & 2\sin\theta_{\text{E}}\cos\theta_{\text{E}} \end{bmatrix}, \text{ and } \boldsymbol{\sigma}_{\text{global}} = \begin{bmatrix} \sigma_x \\ \sigma_y \\ \tau_{xy} \end{bmatrix}$$

This shows that we have five equations for only three unknowns and, in such cases where there are more equations than unknowns, we find the least-squares solution (Press et al., 1990[4]). Using the inverse matrix and finding the least-squares solution,

$$\boldsymbol{\sigma}_{\text{global}} = (\mathbf{R}^{\text{T}}\mathbf{R})^{-1}(\mathbf{R}^{\text{T}}\boldsymbol{\sigma}_{\text{jack}}), \text{ which gives } \boldsymbol{\sigma}_{\text{global}} = \begin{bmatrix} 7.99 \\ 6.90 \\ 0.07 \end{bmatrix} \text{ MPa.}$$

To compute the principal stresses, we use the equation for Mohr's circle. The centre of the circle on the normal stress axis is given by

$$c = \tfrac{1}{2}(\sigma_x + \sigma_y) = 7.45 \text{ MPa}$$

and the radius of the circle is

$$\frac{1}{2}\sqrt{(\sigma_x - \sigma_y)^2 + (2\tau_{xy})^2} = 0.55 \text{ MPa,}$$

from which

$$\sigma_1 = c + r = 9.00 \text{ MPa and } \sigma_2 = c - r = 6.90 \text{ MPa.}$$

[4] A description of least-squares methods is given in books from the series by Press W. H., Flannery B. P., Teukolosky S. A. and Vertterling W. T. (1990) *Numerical Recipes*. Cambridge University Press.

The expected stresses, σ_e, are given by

$$\sigma_e = \mathbf{R}\sigma_{global} \text{ which evaluates as } \begin{bmatrix} 7.56 \\ 6.94 \\ 7.37 \\ 7.19 \\ 7.96 \end{bmatrix} \text{ MPa:}$$

$\sigma_A = 7.56$ MPa, $\sigma_B = 6.94$ MPa, $\sigma_C = 7.37$ MPa, $\sigma_D = 7.19$ MPa, and $\sigma_E = 7.96$ MPa.

These are not the same as the measured values and so a least-squares solution should always involve computation of the errors involved, as described in statistical textbooks.

Q4.6 The stress in a granitic rock mass has been measured by the hydraulic fracturing technique. Two tests were conducted in a vertical borehole: one test at a depth of 500 m, and the other test at a depth of 1000 m. The results were as follows:

Depth (m)	Breakdown pressure, P_B (MPa)	Shut-in pressure, P_S (MPa)
500	14.0	8.0
1000	24.5	16.0

Given that the tensile strength, σ_t, of the rock is 10 MPa, estimate and list the values of σ_1, σ_2 and σ_3 at the two depths. State all of the assumptions you have to make in order to produce these estimates. Are any of them doubtful? State whether the two sets of results are consistent with each other, and justify your reasons for the statement. Are the results in agreement with trends exhibited by collated worldwide data?

A4.6 Using the hydraulic fracturing stress measurement technique, a portion of a borehole, say a 1 m length, is sealed off. Water in this borehole portion is pressurized until the rock fractures, giving the breakdown pressure. Then, the pressure required to just keep the fracture open is measured, giving the shut-in pressure (see sketch to the right of a cross-section through a borehole).

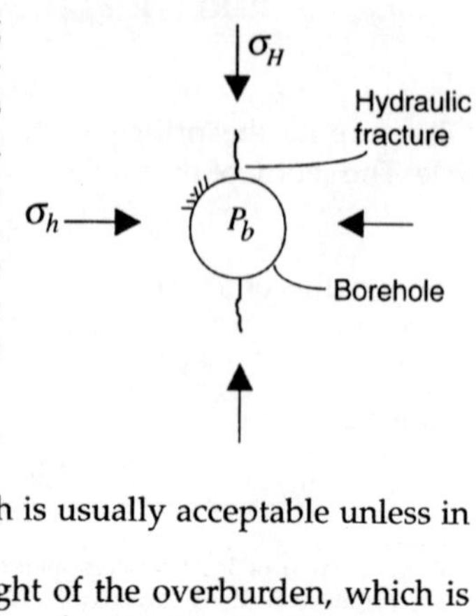

The assumptions required for hydraulic fracturing are:

(a) one principal stress is vertical, which is usually acceptable unless in hilly terrain;
(b) the vertical stress is due to the weight of the overburden, which is usually a reasonable approximation unless in local problem areas

such as regions of thrust faults where the stress is being transmitted through small areas of the fault;

(c) the fractures are formed in a vertical plane, which can be checked with an impression packer;

(d) the rock is impermeable, which is acceptable for rapid tests in granite, but we can incorporate permeability if required; and

(e) the stress concentration around the borehole can be estimated by the Kirsch equations (see ERM 1, Section 19.2); this depends on the rock being linearly elastic, and is an acceptable assumption in unweathered granite.

For a vertical fracture, a horizontal section through the borehole is as in the sketch above. From the Kirsch equations for the stress concentrations around a circular hole in an elastic material, the equation for the breakdown pressure is found to be: $P_B = 3\sigma_h - \sigma_H + \sigma_t$ where P_B is the breakdown pressure, σ_h is the minor horizontal stress, σ_H is the major horizontal stress, and σ_t is the tensile strength of the rock. This equation results because the water has to be pressurized to a value that will overcome the stress concentration of three times the minor horizontal principal stress and the tensile strength. However, because the maximum horizontal principal stress is aiding the breakdown pressure, it appears as a unit negative stress concentration in the equation.

Rearranging gives $\sigma_H = 3\sigma_h - P_B + \sigma_t$. We know that σ_t is 10 MPa, and we will assume that the density of the rock is 27 kN/m^3.

Test 1 $P_B = 14.0$ MPa, $P_S = 8.0$ MPa, depth, $z = 500$ m, vertical stress, $\sigma_z = \gamma z = 0.027 \times 500 = 13.5$ MPa, $\sigma_h =$ the shut-in pressure, $P_S = 8.0$ MPa

$$\sigma_H = 3\sigma_h - P_B + \sigma_t = 3 \times 8.0 - 14.0 + 10.0, \text{ and hence } \sigma_H = 20 \text{ MPa}.$$

Thus, $\sigma_H = 20 \text{ MPa} > \sigma_z = 13.5 \text{ MPa} > \sigma_h = 8$ MPa, which would produce a vertical fracture as assumed (because the fracture develops perpendicular to the least principal stress[5]).

The ratio, k, between the mean of the horizontal stresses and the vertical stresses is $\frac{1}{2}(\sigma_h + \sigma_H)/\sigma_z = \frac{1}{2}(8.0 + 20.0)/13.5$, giving $k = 1.04$.

Test 2 $P_B = 24.5$ MPa, $P_S = 16.0$ MPa, depth, $z = 1000$ m, vertical stress, $\sigma_z = \gamma z = 0.027 \times 1000 = 27$ MPa, $\sigma_h =$ the shut-in pressure, $P_S = 16.0$ MPa.

$$\sigma_H = 3\sigma_h - P_B + \sigma_t = 3 \times 16.0 - 24.5 + 10.0 = 33.5 \text{ MPa}.$$

Thus, $\sigma_H = 33.5 \text{ MPa} > \sigma_z = 27 \text{ MPa} > \sigma_h = 16$ MPa, which again would produce a vertical fracture, as assumed.

For this test, we have $k = \frac{1}{2}(\sigma_h + \sigma_H)/\sigma_z = \frac{1}{2}(16.0 + 33.5)/27.0$, giving $k = 0.92$.

---

[5] Note that for this question, the principal stresses have already been assumed to be vertical and horizontal. More advanced hydraulic fracturing methods using pre-existing fractures are available in which the orientation of the principal stresses can be determined from the measurements.

The values are consistent with each other and they are consistent with worldwide trends. The value of $k \approx 1$ is typical for these depths (noting that the averaging of the two horizontal stresses masks the fact that the maximum horizontal stress is a good deal higher than the vertical stress in both cases), and the value of k reducing with depth commonly occurs.

Q4.7 How are three-dimensional tensors transformed so that the stress components on any plane can be calculated?

A4.7 The matrix equation to conduct stress transformation is as follows:

$$\begin{bmatrix} \sigma_l & \tau_{lm} & \tau_{ln} \\ \tau_{ml} & \sigma_m & \tau_{mn} \\ \tau_{nl} & \tau_{nm} & \sigma_n \end{bmatrix} = \begin{bmatrix} l_x & l_y & l_z \\ m_x & m_y & m_z \\ n_x & n_y & n_z \end{bmatrix} \begin{bmatrix} \sigma_x & \tau_{xy} & \tau_{xz} \\ \tau_{yx} & \sigma_y & \tau_{yz} \\ \tau_{zx} & \tau_{zy} & \sigma_z \end{bmatrix} \begin{bmatrix} l_x & m_x & n_x \\ l_y & m_y & n_y \\ l_z & m_z & n_z \end{bmatrix}$$

where the stress components are assumed known in the xyz co-ordinate system and are required in another co-ordinate system lmn inclined with respect to the first. The term l_x is the direction cosine of the angle between the x axis and the l axis. Physically, it is the projection of a unit vector parallel to l on to the x axis, with the other terms similarly defined.

Expanding this matrix equation in order to obtain an expression for the normal component of stress in the l direction, for example, gives

$$\sigma_l = l_x^2\sigma_{xx} + l_y^2\sigma_{yy} + l_z^2\sigma_{zz} + 2(l_x l_y \sigma_{xy} + l_y l_z \sigma_{yz} + l_z l_x \sigma_{zx})$$

and a similar expansion for the shear component of stress on the l face in the m direction gives

$$\sigma_{lm} = l_x m_x \sigma_{xx} + l_y m_y \sigma_{yy} + l_z m_z \sigma_{zz} + \left(l_x m_y + l_y m_x\right)\sigma_{xy} + \\ \left(l_y m_z + l_z m_y\right)\sigma_{yz} + (l_z m_x + l_x m_z)\sigma_{zx}$$

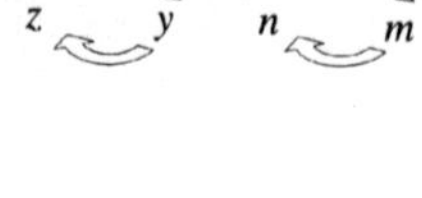

The other four necessary equations (i.e. for σ_y, σ_z, τ_{yz}, and τ_{zx}) are found using cyclic permutation of the subscripts in these equations. Cyclic permutation is shown graphically to the left.

So, if we know the orientations of each of the axes in the xyz and lmn co-ordinate systems, together with the stress tensor in the xyz co-ordinate system, we can calculate the transformed stress tensor.

On many occasions it will be convenient to refer to the orientation of a plane on which the components of stress are required to be known using the dip direction/dip angle (α, β) notation. The dip direction is given as a clockwise bearing from North and the dip angle as an angle between 0° and 90° measured downwards from the horizontal plane. If we use a right-handed co-ordinate system[6] with x = North, y = East,

[6] There are many systems of co-ordinate axes used in rock mechanics and it is important to become fluent in the use of all of them, as with the stress symbols and units. Three-dimensional co-ordinate axes may be either right-handed or left-handed: these are easily visualized using the thumb, index finger and middle finger of one's hand. If these three digits are held at right angles to each other, then the thumb represents the x-axis,

and z = down, and take direction n as normal to the desired plane, then

$$n_x = \cos\alpha_n \cos\beta_n;\ n_y = \sin\alpha_n \cos\beta_n,\ n_z = \sin\beta_n.$$

Note that when the z axis is directed downward, a clockwise bearing (i.e. measured from North) will always be a positive angle (i.e. anticlockwise measured from the x axis) when looking in the direction of the positive z axis. Hence, the rotation matrix is

$$\begin{bmatrix} l_x & l_y & l_z \\ m_x & m_y & m_z \\ n_x & n_y & n_z \end{bmatrix} = \begin{bmatrix} \cos\alpha_l \cos\beta_l & \sin\alpha_l \cos\beta_l & \sin\beta_l \\ \cos\alpha_m \cos\beta_m & \sin\alpha_m \cos\beta_m & \sin\beta_m \\ \cos\alpha_n \cos\beta_n & \sin\alpha_n \cos\beta_n & \sin\beta_n \end{bmatrix}.$$

Now, if we take direction l to be a horizontal line on the plane, then the following relations between the trends and plunges of the axes hold:

$$\alpha_l = 90 + \alpha_n \qquad \beta_l = 0$$

$$\alpha_m = 180 + \alpha_n \quad \beta_m = 90 - \beta_n$$

However, when we use these relations, we must ensure that α_l and α_m lie between 0° and 360°, by subtracting 360° if necessary. Using these relations, the matrix of direction cosines then becomes

$$\begin{bmatrix} l_x & l_y & l_z \\ m_x & m_y & m_z \\ n_x & n_y & n_z \end{bmatrix} = \begin{bmatrix} -\sin\alpha_n & \cos\alpha_n & 0 \\ -\cos\alpha_n \sin\beta_n & -\sin\alpha_n \sin\beta_n & \cos\beta_n \\ \cos\alpha_n \cos\beta_n & \sin\alpha_n \cos\beta_n & \sin\beta_n \end{bmatrix}.$$

Q4.8 Suppose that we have measured the pre-existing stress state in the ground by some means and that the results are as follows:

σ_1, magnitude 15 MPa, plunges 35° towards 085°;
σ_2, magnitude 10 MPa, plunges 43° towards 217°;
σ_3, magnitude 8 MPa, plunges 27° towards 335°.

Find the 3-D stress tensor in the right-handed xyz co-ordinate system with

x, horizontal to the east,
y, horizontal to the north,
z, vertically upwards.

Right-handed co-ordinate system

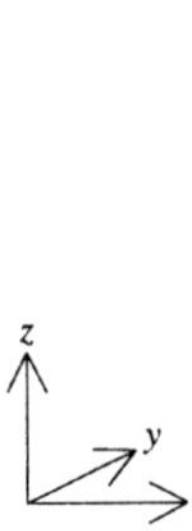

the index finger the y-axis and the middle finger the z-axis. A right-handed system is represented by the digits of the right hand, and a left-handed system is represented by the digits of the left hand. The use of a left-handed system of axes is convenient for rock mechanics: the x-axis represents East, the y-axis is North and the z-axis is down. However, right-handed systems are always used for mathematical work, and in order to maintain coherence with other mathematical work (as well as the mathematical functions provided in computers and calculators) we recommend the use of right-handed systems in rock mechanics. There are two obvious choices for a right-handed system of axes: x East, y North and z up; or x North, y East and z down. There are advantages and disadvantages to both of these systems, and so being adept with both is important. Accordingly, we will use both in this book.

A4.8 The stress transformation equations are given by

$$\begin{bmatrix} \sigma_l & \tau_{ml} & \tau_{nl} \\ \tau_{lm} & \sigma_m & \tau_{nm} \\ \tau_{ln} & \tau_{mn} & \sigma_n \end{bmatrix} = \begin{bmatrix} l_x & m_x & n_x \\ l_y & m_y & n_y \\ l_z & m_z & n_z \end{bmatrix} \begin{bmatrix} \sigma_x & \tau_{yx} & \tau_{zx} \\ \tau_{xy} & \sigma_y & \tau_{zy} \\ \tau_{xz} & \tau_{yz} & \sigma_z \end{bmatrix} \begin{bmatrix} l_x & l_y & l_z \\ m_x & m_y & m_z \\ n_x & n_y & n_z \end{bmatrix}$$

which can be written as $\boldsymbol{\sigma}_{lmn} = \mathbf{R}\boldsymbol{\sigma}_{xyz}\mathbf{R}^{\mathrm{T}}$.

This means that, if we know the stresses relative to the xyz axes (i.e. $\boldsymbol{\sigma}_{xyz}$) and the orientation of the lmn axes relative to the xyz axes (i.e. $\mathbf{R}$), we can compute the stresses relative to the lmn axes (i.e. $\boldsymbol{\sigma}_{lmn}$).

However, in this problem we have been given the principal stresses, which is a stress state relative to some lmn system (i.e. $\boldsymbol{\sigma}_{lmn}$), where the lmn axes correspond to the principal directions. As we know the principal directions relative to the xyz axes, we are able to compute $\mathbf{R}$. Thus, we need to evaluate $\boldsymbol{\sigma}_{xyz}$, and we do this using the inverse of the stress transformation equations: $\boldsymbol{\sigma}_{xyz} = \mathbf{R}^{\mathrm{T}}\boldsymbol{\sigma}_{lmn}\mathbf{R}$.

Notice that we have not had to use the inverse of $\mathbf{R}$, i.e. $\mathbf{R}^{-1}$. The rotation matrix is orthogonal, and this property means that $\mathbf{R}^{-1} = \mathbf{R}^{\mathrm{T}}$.

With the given data for the principal directions:

$$\alpha_l = 85^\circ \quad \alpha_m = 217^\circ \quad \alpha_n = 335^\circ$$

$$\beta_l = 35^\circ \quad \beta_m = 43^\circ \quad \beta_n = 27^\circ$$

the matrix $\mathbf{R}$ is computed as

$$\mathbf{R} = \begin{bmatrix} \cos\alpha_l\cos\beta_l & \sin\alpha_l\cos\beta_l & \sin\beta_l \\ \cos\alpha_m\cos\beta_m & \sin\alpha_m\cos\beta_m & \sin\beta_m \\ \cos\alpha_n\cos\beta_n & \sin\alpha_n\cos\beta_n & \sin\beta_n \end{bmatrix} = \begin{bmatrix} 0.071 & 0.816 & 0.574 \\ -0.584 & -0.440 & 0.682 \\ 0.808 & -0.377 & 0.454 \end{bmatrix}$$

and the matrix $\boldsymbol{\sigma}_{lmn}$ is given by

$$\boldsymbol{\sigma}_{lmn} = \begin{bmatrix} 15 & 0 & 0 \\ 0 & 10 & 0 \\ 0 & 0 & 8 \end{bmatrix} \text{ MPa}$$

thus, $\boldsymbol{\sigma}_{xyz} = \mathbf{R}^{\mathrm{T}}\boldsymbol{\sigma}_{lmn}\mathbf{R}$ is given by

$$\boldsymbol{\sigma}_{xyz} = \begin{bmatrix} 8.70 & 1.01 & -0.44 \\ 1.01 & 13.06 & 2.65 \\ -0.44 & 2.65 & 11.23 \end{bmatrix} \text{ MPa.}$$

A useful check of the calculations involved in stress transformation is obtained by computing the stress invariants. The first and third invariants are easily obtained using standard matrix functions. For example, the first invariant is computed as the sum of the leading diagonal elements (see A3.8); this sum is also known as the **trace** of the stress matrix. The third invariant is computed as the **determinant** of the stress matrix.

$$I_1 = \text{tr}(\sigma_{lmn}) = 33.0 \text{ MPa}, \text{ and } I_1 = \text{tr}(\sigma_{xyz}) = 33.0 \text{ MPa}.$$

$$I_3 = |\sigma_{lmn}| = 12000 \text{ MPa}^3 \text{ and } I_3 = |\sigma_{xyz}| = 1199.7 \text{ MPa}^3.$$

The orientations of the axes used above have been rounded to the nearest whole degree, and so the axes are not exactly orthogonal. The effect of this is to introduce errors in the computation, which can be seen in the values of the third invariant. Here we can use orientations that are correct to two decimal places of degrees, such that the orthogonality condition is improved and hence a more accurate answer is obtained, but in general these data will not be available from the field information.

The orientations correct to two decimal places are

$$\alpha_l = 85.50^\circ \quad \alpha_m = 216.65^\circ \quad \alpha_n = 334.81^\circ$$

$$\beta_l = 35.06^\circ \quad \beta_m = 43.15^\circ \quad \beta_n = 26.72^\circ$$

the matrix **R** is

$$\begin{bmatrix} 0.064 & 0.816 & 0.574 \\ -0.585 & -0.435 & 0.684 \\ 0.808 & 0.380 & 0.450 \end{bmatrix}$$

from which we find

$$\sigma_{xyz} = \begin{bmatrix} 8.71 & 0.88 & -0.54 \\ 0.88 & 13.04 & 2.69 \\ -0.54 & 2.69 & 11.24 \end{bmatrix} \text{ MPa}.$$

Calculating the invariants again gives

$$I_1 = \text{tr}(\sigma_{lmn}) = 33.00000 \text{ MPa}, \text{ and } I_1 = \text{tr}(\sigma_{xyz}) = 33.00000 \text{ MPa}.$$

$$I_3 = |\sigma_{lmn}| = 1200.000 \text{ MPa}^3 \text{ and } I_3 = |\sigma_{xyz}| = 1200.000 \text{ MPa}^3.$$

showing that the invariants now agree exactly to 7 significant figures.

Q4.9 A fault is present in the same rock (continuing from Q4.8) with an orientation of 295°/50°. Determine the stress components in a local co-ordinate system aligned with the fault. Assume for this question that the presence of the fault does not affect the stress field.

A4.9 Here we use the methodology given in A4.8 to find the 3-D stress tensor in an *lmn* co-ordinate system where the *n* axis coincides with the normal to the fault and the *l* axis coincides with the strike of the fault. We need to determine σ_{lmn}, where *lmn* are given by the orientation of the fault.

With the l axis parallel to the strike of the plane and the n axis normal to the plane, the m axis is the line of maximum dip of the plane. The trend and plunge of each of the axes are then as follows:

$$\alpha_l = 205^\circ \quad \alpha_m = 295^\circ \quad \alpha_n = 115^\circ$$

$$\beta_l = 0^\circ \quad \beta_m = 50^\circ \quad \beta_n = 40^\circ$$

The matrix **R**, computed as shown in A4.8, is

$$\begin{bmatrix} -0.906 & -0.423 & 0.000 \\ 0.272 & -0.583 & 0.766 \\ -0.324 & 0.694 & 0.643 \end{bmatrix}$$

and the matrix $\boldsymbol{\sigma}_{xyz}$ is

$$\begin{bmatrix} 8.70 & 1.01 & -0.44 \\ 1.01 & 13.06 & 2.65 \\ -0.44 & 2.65 & 11.23 \end{bmatrix}.$$

As a result, the matrix $\boldsymbol{\sigma}_{lmn} = \mathbf{R}\boldsymbol{\sigma}_{xyz}\mathbf{R}^{\mathrm{T}}$ or

$$\begin{bmatrix} 10.26 & 0.94 & -2.24 \\ 0.94 & 8.80 & 0.32 \\ -2.24 & 0.32 & 13.94 \end{bmatrix} \text{MPa.}$$

If we use the matrix $\boldsymbol{\sigma}_{xyz}$ resulting from the high-precision orientation data of A4.8 to compute the stresses on the fault, we obtain

$$\boldsymbol{\sigma}_{lmn} = \mathbf{R}\begin{bmatrix} 8.71 & 0.88 & -0.54 \\ 0.88 & 13.04 & 2.69 \\ -0.54 & 2.69 & 11.24 \end{bmatrix}\mathbf{R}^{\mathrm{T}} = \begin{bmatrix} 10.16 & 0.93 & -2.12 \\ 0.93 & 8.76 & 0.29 \\ -2.12 & 0.29 & 14.07 \end{bmatrix} \text{MPa.}$$

It is clear from these calculations that, although the stress invariants vary little between the two sets of input orientations, the same is not true for the values of the stress components calculated for the fault orientation. For example, the shear stress in the plane, τ_{mn}, is different by about 10% using the different precisions, i.e. 0.32 MPa to 0.29 MPa. Also, in practice it is likely that the stress field would be perturbed by the presence of the fault and so would differ from this theoretically computed stress state.

Q4.10 The plan below (Sugawara and Obara, 1999[7]) shows a horizontal section through a rock mass. Stress measurements were made from the gallery along the borehole line Ox using the Japanese CCBO technique. The measured principal stresses in the horizontal plane are plotted on the plan. What are the main conclusions that you can draw from the stress variations?

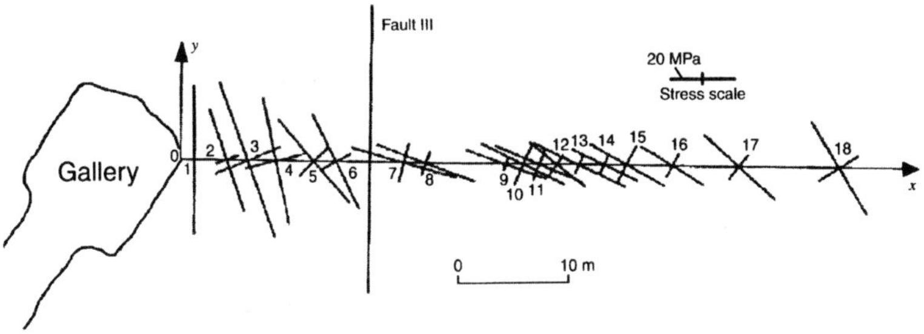

A4.10 An initial interpretation is as follows.

(a) The stress field is not homogeneous throughout the rock mass, as both the magnitudes and orientations of the principal stresses vary along Ox.

(b) The stress field near the excavation is as expected: the principal stresses should be parallel and perpendicular to the excavated rock surface, and the principal stress perpendicular to the excavation at measuring point 1 should be very low.

(c) The fault appears to be having an effect on the stress field because the maximum principal stress is rotating anticlockwise along the borehole from the gallery towards the fault, but then clockwise from the fault further into the rock mass.

(d) The fault appears to be transmitting stress; if it were an open fault, the maximum principal stress at measurement points 6 and 7 would be more sub-parallel to the fault.

(e) The changes in principal stress orientation in the regions to the left of the fault and to the right of the fault follow reasonably consistent trends (compared to random orientations at each measurement point). This implies that there is a consistent regional stress field which has been locally affected by the gallery excavation and the fault, and that there is a *prima facie* case for the measurement technique being credible.

(f) The magnitudes of the principal stresses in the region of points 9–16 are lower than those at points 7, 8 and 17, 18 indicating a relatively destressed region, probably caused by more than one fault or excavation in the location.

[7] This diagram is from the ISRM Draft Suggested Method. Sugawara K. and Obara Y. (1999) *In situ* stress measurement using the compact conical-ended borehole overcoring (CCBO) technique. *Int. J. Rock Mech. Min. Sci.*, **36**, 3, 307–322.

To interpret these results further, we would have to study the 3-D states of stress at the measurement points and obtain more information on other faults and excavations in the area.

4.3 Additional points

The ISRM has published Suggested Methods (Kim and Franklin, 1987[8]) for the four methods of rock stress measurement mentioned in Q4.3. In the overcoring method, an initial borehole is drilled, the gauge installed and a larger-diameter drill used to 'overcore' the gauge. A technique has now been developed[7] — the 'compact overcoring' technique (Q4.10) — in which drilling at only one borehole diameter is required.

Useful references which comprehensively describe the subject of *in situ* stress from the engineering viewpoint are given below.

- (Conference proceedings) — Stephansson O. (ed.) (1986) *Rock Stress and Rock Stress Measurements*, Centek, Stockholm, 694pp.
- (Book) — Amadei B. and Stephansson O. (1997) *Rock Stress and Its Measurement*. Chapman and Hall, London, 490pp.
- (Conference proceedings) — Sugawara K. and Obara Y. (eds) (1997) *Rock Stress*. Balkema, Rotterdam, 552pp.

An idea of the *in situ* stress state in the continental part of the Earth's crust can be obtained from The World Stress Map. In this project, which was initiated in 1986, stress measurement data have been systematically collected and mapped. The data are available from: http://www-wsm.physik.uni-karlsruhe.de/pub/Rel97/wsm97.html

[8] Kim K. and Franklin J. A. (1987) Suggested methods for rock stress determination. *Int. J Rock Mech. Min. Sci. Geomech. Abstr.*, **24**, 1, 53–73.

5 Strain and the theory of elasticity

5.1 Stress and strain are both tensor quantities

The concept of strain is required to understand and quantify how a rock mass has been deformed. There are many applications of strain analysis. For example, some *in situ* stress determination techniques measure strain and hence compute stress, so it is necessary to be able to manipulate the strain values, and understanding the anticipated deformation of structures built on and in rock masses is critical for rock engineering design and back analysis.

Strain is a measure of the deformation of a body and is the same type of mathematical quantity as stress: a second-order tensor. There are **normal strains** and there are **shear strains**, as shown below[1], which are directly analogous to normal stresses and shear stresses.

The nature of stress as a tensor is explained in Chapter 3, where we see that the stress state can be specified either as the six independent components of the stress matrix or as the magnitudes and orientations of the three principal stresses. The strain state is expressed using the symbol ε for normal strains and either the symbol ε or γ for shear strains. Note that the **engineering shear strains**, e.g. γ_{yx}, as illustrated in Fig. 5.1 are divided by two to give the components of the strain matrix, e.g. ε_{yx}. Intuitively, this is because engineering shear strain must be attributed equally to both τ_{yx} and its equal-value complement τ_{xy};

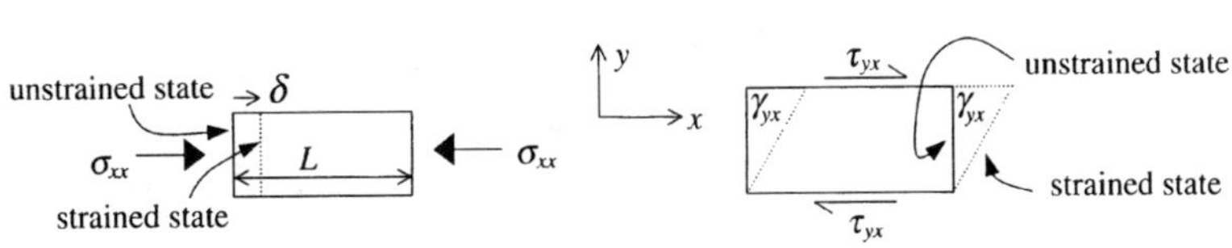

Figure 5.1 Normal strains and shear strains.

[1] In Fig. 5.1, the normal strain shown is positive, but the shear strain is negative. ERM 1 contains a detailed discussion of the sign of the shear strain.

Table 5.1 Components of the stress and strain tensors

	Components of the stress and strain matrices — relative to x, y and z reference axes	Components of the stress and strain matrices — relative to the three principal stress and strain directions.
Stress	$\begin{bmatrix} \sigma_{xx} & \tau_{xy} & \tau_{xz} \\ & \sigma_{yy} & \tau_{yz} \\ \text{symm} & & \sigma_{zz} \end{bmatrix}$	$\begin{bmatrix} \sigma_1 & 0 & 0 \\ & \sigma_2 & 0 \\ \text{symm} & & \sigma_3 \end{bmatrix}$
Strain	$\begin{bmatrix} \varepsilon_{xx} & \varepsilon_{xy} & \varepsilon_{xz} \\ & \varepsilon_{yy} & \varepsilon_{yz} \\ \text{symm} & & \varepsilon_{zz} \end{bmatrix}$ where $\varepsilon_{xy} = (\gamma_{xy})/2$	$\begin{bmatrix} \varepsilon_1 & 0 & 0 \\ & \varepsilon_2 & 0 \\ \text{symm} & & \varepsilon_3 \end{bmatrix}$

mathematically, the division by two is required to give similarity to both the stress and the strain transformation equations.

In Table 5.1, the correspondence between the stress and strain tensors is evident; the normal and shear strains correspond directly with the normal and shear stresses, and the principal strains correspond directly with the principal stresses. In fact, stress and strain are mathematically identical; only the physical interpretations are different. As a result, there are strain analogies to all the stress aspects described in Chapter 3.

The next step is to consider how stress and strain might be related. Historically, this has been considered using two approaches: one for finite strain and one for infinitesimal[2] strain. Both are useful in rock mechanics, but here we will highlight infinitesimal strain and the theory of elasticity because of their widespread use.

The **theory of elasticity** relates the stress and strain states for infinitesimal strains. Hooke's Law in its original form states that when, say, a wire is stretched, the strain is proportional to the stress or, as Hooke himself stated[3], "ut tensio sic vis". Given the tensors in Table 5.1, we can extend this concept to the **generalized Hooke's Law** relating all the components of the strain matrix to all the components of the stress matrix. The assumption is that each component of the strain matrix is linearly proportional to each component of the stress matrix by a factor S_{ij}. This means that the value of a specific strain component can be found from the six contributions that the stress components make to it. For example, the contribution to the strain component ε_{xx} (or ε_{11}) made by the stress component σ_{zz} (or σ_{33}) is $S_{13}\sigma_{zz}$. The relation between the stress and the strain components may not be a simple linear relation, but this is the first-order approximation made in the theory of elasticity.

[2] The word 'infinitesimal' means infinitely small, or approaching zero as a limit.

[3] This expression was first published by R. Hooke in The Times of London newspaper in anagram form in 1676.

The six strains are therefore expressed as

$$\varepsilon_{xx} = S_{11}\sigma_{xx} + S_{12}\sigma_{yy} + S_{13}\sigma_{zz} + S_{14}\tau_{xy} + S_{15}\tau_{yz} + S_{16}\tau_{zx}$$
$$\varepsilon_{yy} = S_{21}\sigma_{xx} + S_{22}\sigma_{yy} + S_{23}\sigma_{zz} + S_{24}\tau_{xy} + S_{25}\tau_{yz} + S_{26}\tau_{zx}$$
$$\varepsilon_{zz} = S_{31}\sigma_{xx} + S_{32}\sigma_{yy} + S_{33}\sigma_{zz} + S_{34}\tau_{xy} + S_{35}\tau_{yz} + S_{36}\tau_{zx}$$
$$\varepsilon_{xy} = S_{41}\sigma_{xx} + S_{42}\sigma_{yy} + S_{43}\sigma_{zz} + S_{44}\tau_{xy} + S_{45}\tau_{yz} + S_{46}\tau_{zx}$$
$$\varepsilon_{yz} = S_{51}\sigma_{xx} + S_{52}\sigma_{yy} + S_{53}\sigma_{zz} + S_{54}\tau_{xy} + S_{55}\tau_{yz} + S_{56}\tau_{zx}$$
$$\varepsilon_{zx} = S_{61}\sigma_{xx} + S_{62}\sigma_{yy} + S_{63}\sigma_{zz} + S_{64}\tau_{xy} + S_{65}\tau_{yz} + S_{66}\tau_{zx}$$

where the 36 S_{ij} coefficients in these equations are known as the elastic compliances, equivalent to the reciprocal of Young's modulus, E, in the basic Hooke's Law, $\sigma = E\varepsilon$ or $\varepsilon = (1/E)\sigma$.

This equation can be written in matrix form as

$$\boldsymbol{\varepsilon} = \mathbf{S}\boldsymbol{\sigma}$$

21 independent elastic constants

$$\text{where } \boldsymbol{\varepsilon} = \begin{pmatrix} \varepsilon_{xx} \\ \varepsilon_{yy} \\ \varepsilon_{zz} \\ \varepsilon_{xy} \\ \varepsilon_{yz} \\ \varepsilon_{zx} \end{pmatrix} \text{ and } \boldsymbol{\sigma} = \begin{pmatrix} \sigma_{xx} \\ \sigma_{yy} \\ \sigma_{zz} \\ \tau_{xy} \\ \tau_{yz} \\ \tau_{zx} \end{pmatrix} \text{ and } \mathbf{S} = \begin{pmatrix} S_{11} & S_{12} & S_{13} & S_{14} & S_{15} & S_{16} \\ S_{21} & S_{22} & S_{23} & S_{24} & S_{25} & S_{26} \\ S_{31} & S_{32} & S_{33} & S_{34} & S_{35} & S_{36} \\ S_{41} & S_{42} & S_{43} & S_{44} & S_{45} & S_{46} \\ S_{51} & S_{52} & S_{53} & S_{54} & S_{55} & S_{56} \\ S_{61} & S_{62} & S_{63} & S_{64} & S_{65} & S_{66} \end{pmatrix}$$

It can be shown from energy considerations that the **S** matrix is symmetrical, leading to 21 independent elastic constants.

This number of constants can be reduced if we assume that some of the strain–stress linkages are not significant. For example, omitting all shear linkages not on the leading diagonal — which means assuming that any contributions made by shearing stress components in a given direction to normal or shear strain components in other directions are negligible — causes all off-diagonal shear linkages to become zero. The compliance matrix then reduces to one with nine material properties, which is the case for an *orthotropic* material.

$$\begin{pmatrix} 1/E_x & -\nu_{xy}/E_y & -\nu_{xz}/E_z & 0 & 0 & 0 \\ & 1/E_y & -\nu_{yz}/E_z & 0 & 0 & 0 \\ & & 1/E_z & 0 & 0 & 0 \\ & & & 1/G_{xz} & 0 & 0 \\ & \textit{symmetric} & & & 1/G_{yz} & 0 \\ & & & & & 1/G_{zx} \end{pmatrix}$$

where E is a Young's modulus, ν is a Poisson's ratio, and G is a shear modulus.

Table 5.2 Number of elastic constants required to characterize different forms of rock mass symmetry

General anisotropic rock	**21 elastic constants:** all the independent S_{ij} in the **S** matrix. Because the matrix is symmetrical, there are 21 rather than 36 constants.
Orthotropic rock (three axes of symmetry, e.g. similar to a rock mass with three orthogonal fracture sets)	**9 elastic constants:** as in the matrix above — 3 Young's moduli, 3 Poisson's ratios and 3 shear moduli
Transversely isotropic rock (one axis of symmetry, e.g. similar to a rock mass with distinct laminations or with one main fracture set)	**5 elastic constants:** 2 Young's moduli, 2 Poisson's ratios, and 1 shear modulus (see Q5.4)
Perfectly isotropic rock	**2 elastic constants:** 1 Young's modulus, 1 Poisson's ratio

In this way, by considering the **architecture of the general elastic compliance matrix, S**, the number of different elastic constants required to characterize rock masses with different basic forms of symmetry can be established. This relates to the **anisotropy** of the rock mass, i.e. to what extent it has different properties in different directions. The results are in Table 5.2.

It is important to realize that most rock mechanics analyses have been conducted assuming that the rock mass is completely isotropic, i.e. assuming that the elastic moduli are the same in all directions, and that the rock mass can therefore be characterized by two parameters: a single value of Young's modulus, E, and a single value of Poisson's ratio, ν. A separate value for the shear modulus, G, is not required in the isotropic case because then G is a function of E and ν, see Q and A5.4.

In some cases, isotropy may be a useful simplifying engineering assumption; in other cases, especially where there is one dominant low-stiffness fracture set, the isotropic assumption is not appropriate.

5.2 Questions and answers: strain and the theory of elasticity

Q5.1 What is the meaning of the first stress invariant and the first strain invariant?

A5.1 The first stress invariant, I_1, is the sum of the leading diagonal terms of the stress matrix illustrated in Table 5.1: $I_1 = \sigma_{xx} + \sigma_{yy} + \sigma_{zz}$. It can be considered as proportional to the mean normal stress, and hence the mean pressure applied at a point.

The first strain invariant is the sum of the leading diagonal terms of the strain matrix illustrated in Table 5.1: i.e. $\varepsilon_{xx} + \varepsilon_{yy} + \varepsilon_{zz}$. Normal strain in one dimension is a measure of change in length, and so this invariant is a measure of the volumetric change, the contraction or dilatation.

Q5.2 The differential equations of force equilibrium were the subject of Q3.9. The equivalent equations for displacement and strain are the compatibility equations; these equations ensure that the normal and shear strains are compatible, so that no holes, tears or other discontinuities appear during straining. Show that the following compatibility equation is valid.

$$\frac{\partial^2 \varepsilon_{xx}}{\partial y^2} + \frac{\partial^2 \varepsilon_{yy}}{\partial x^2} = \frac{\partial^2 \gamma_{xy}}{\partial x \partial y}.$$

A5.2 The strain is the rate of change of the displacement, u, so

$$\varepsilon_{xx} = \frac{\partial u_x}{\partial x}, \quad \varepsilon_{yy} = \frac{\partial u_y}{\partial y}, \quad \text{and } \gamma_{xy} = \frac{\partial u_x}{\partial y} + \frac{\partial u_y}{\partial x}.$$

Double partial differentiation of $\varepsilon_{xx} = \partial u_x/\partial x$ and $\varepsilon_{yy} = \partial u_y/\partial y$ with respect to y and x gives

$$\frac{\partial^2 \varepsilon_{xx}}{\partial y^2} = \frac{\partial^3 u_x}{\partial x \partial y^2} \quad \text{and} \quad \frac{\partial^2 \varepsilon_{yy}}{\partial x^2} = \frac{\partial^3 u_y}{\partial y \partial x^2}.$$

Adding these,

$$\frac{\partial^2 \varepsilon_{xx}}{\partial y^2} + \frac{\partial^2 \varepsilon_{yy}}{\partial x^2} = \frac{\partial^3 u_x}{\partial x \partial y^2} + \frac{\partial^3 u_y}{\partial y \partial x^2} = \frac{\partial^2}{\partial x \partial y}\left(\frac{\partial u_x}{\partial y} + \frac{\partial u_y}{\partial x}\right)$$

The term in parentheses is the definition of γ_{xy}, and hence the right hand side can be written as $\partial^2 \gamma_{xy}/\partial x \partial y$. Thus, we arrive at the compatibility equation

$$\frac{\partial^2 \varepsilon_{xx}}{\partial y^2} + \frac{\partial^2 \varepsilon_{yy}}{\partial x^2} = \frac{\partial^2 \gamma_{xy}}{\partial x \partial y}.$$

Q5.3 Draw a Mohr circle for strain, indicating what quantities are on the two axes, how to plot a 2-D strain state, and the location of the principal strains, ε_1 and ε_2.

A5.3

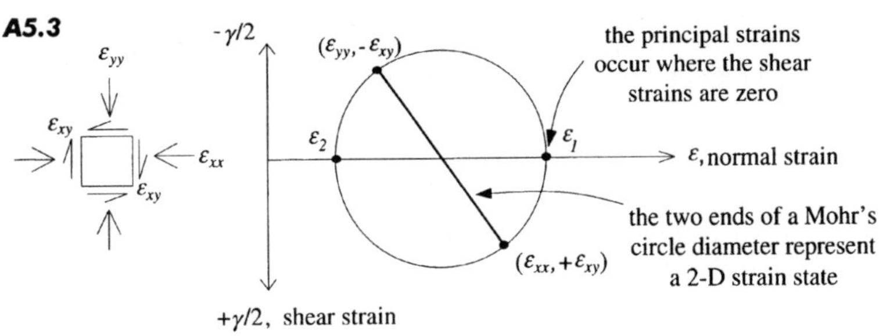

Q5.4 Show why the shear modulus, Young's modulus and Poisson's ratio are related as $G = E/2(1 + \nu)$ for an isotropic material. This equation holds for an isotropic material but not for an anisotropic

material — why? Hence explain why five elastic constants are required for a transversely isotropic material rather than six.

A5.4 We explain this using the diagram below for the case of pure shear, in which a square element of rock is subjected to compression in the x-direction and tension in the y-direction of equal amounts. However, the analysis is general in that it is applicable to all coupled stress and strain states.

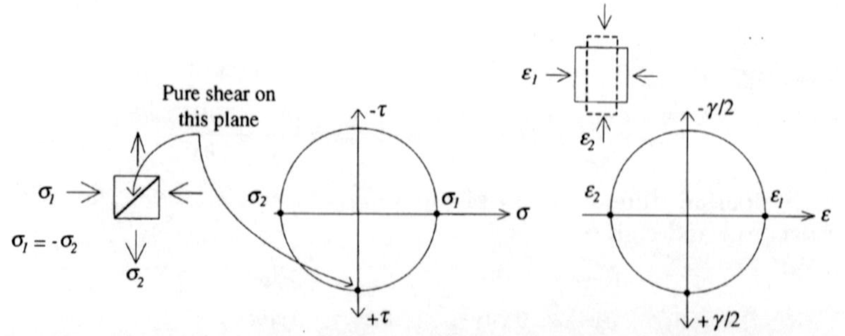

The two Mohr's circle diagrams show the stress and strain states for the square element. Because the circles are both centred at the origin, we can see that $\tau = \sigma_1$ and $\gamma/2 = \varepsilon_1$. Now, the shear modulus, G, is defined as $G = \tau/\gamma$ which, on substituting for the two terms, gives $G = \frac{1}{2}(\sigma_1/\varepsilon_1)$. For a 2-D stress state such as the one illustrated, we can write Hooke's Law as $\varepsilon_x = (1/E)(\sigma_x - \nu\sigma_y)$. In this case, we have $\varepsilon_x = \varepsilon_1$, $\sigma_x = \sigma_1$, and $\sigma_y = \sigma_2 = -\sigma_1$. Substituting these values gives $\varepsilon_1 = (1/E)(\sigma_1 + \nu\sigma_1)$, from which we find that $\sigma_1/\varepsilon_1 = E/(1+\nu)$.

Substituting these into the earlier expression for the shear modulus gives

$$G = \frac{1}{2}\frac{\sigma_1}{\varepsilon_1} = \frac{E}{2(1+\nu)}$$

as required.

This approach is only valid if the Young's moduli and Poisson's ratios in the σ_1 and σ_2 directions are the same, i.e. for an isotropic material having the same properties in all directions. If this were not the case, the strain ε_1 could not be expressed as simply as

$$\varepsilon_1 = \frac{1}{E}(\sigma_1 + \nu\sigma_1).$$

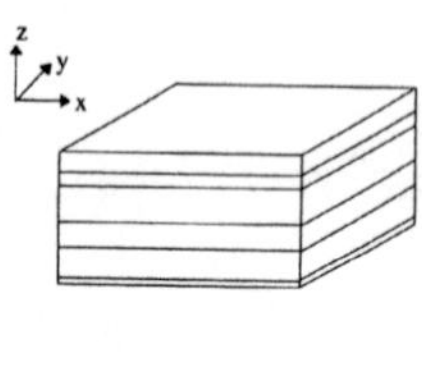

In the case of a transversely isotropic rock, say an unjointed rock with a set of bedding planes perpendicular to the z-axis, as shown in the adjacent sketch, the relation $G = E/2(1+\nu)$ will apply in the x–y plane but not in the x–z and y–z planes, because the rock has different properties in different directions in these planes. Thus, the six elastic constants required to characterize a transversely isotropic rock, $E_x = E_y$, E_z, ν_{xy}, $\nu_{xz} = \nu_{yz}$, G_{xy}, and $G_{yz} = G_{zx}$, are reduced to five, E_x, E_z, ν_{xy}, ν_{xz}, and G_{yz} because in the x–y plane of isotropy $G_{xy} = E_x/2(1+\nu_{xy})$ (see diagram below).

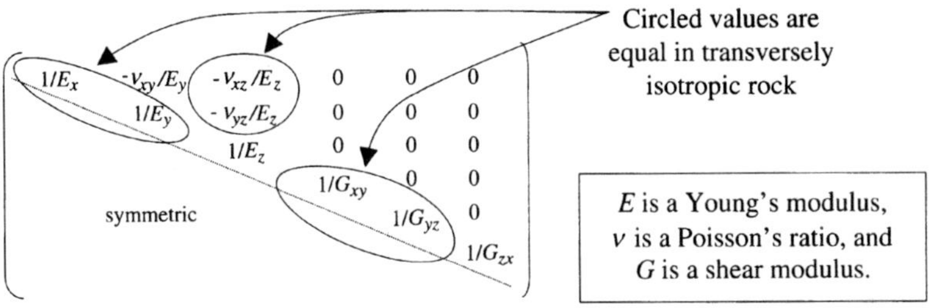

Q5.5 (a) How can the strain in a particular direction be found from the strain matrix components and hence how can a strain gauge rosette be used to estimate the state of strain at a point, and hence the state of stress at a point?

(b) Assume that strains measured by a strain gauge rosette are $\varepsilon_P = 43.0 \times 10^{-6}$, $\varepsilon_Q = 7.8 \times 10^{-6}$ and $\varepsilon_R = 17.0 \times 10^{-6}$, and that the gauges make the following angles to the x-direction: $\theta_P = 20°$, $\theta_Q = 80°$ and $\theta_R = 140°$. Determine the principal strains and their orientations and then, using values for the elastic constants of $E = 150$ GPa and $\nu = 0.30$, determine the principal stresses and their orientations.

A5.5 (a) In 2-D, the strain in a particular direction can be found from the strain matrix components, ε_{xx}, ε_{yy}, ε_{xy}, using the Mohr's circle approach or using the strain transformation equations. A strain gauge is a device for measuring the normal strains in three directions in a plane, as indicated by the sketch to the right. Foil electrical resistance strain gauges can be glued to the rock surface, or wire or rod extensometers can be used. The rosette device provides three normal strains at known orientations, from which the Mohr's circle can be constructed, the principal values found, and the normal strains in any other required direction evaluated.

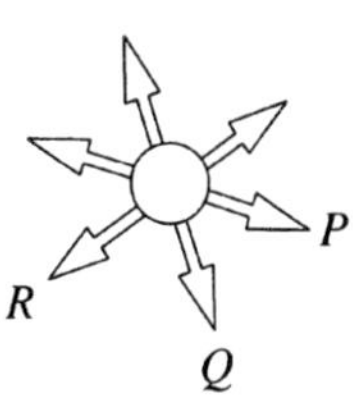

Knowing the elastic properties of the rock and making assumptions about the type of rock anisotropy, the stress state can be established through the generalized Hooke's Law equations.

(b) In order to use the strain transformation equations to determine the 2-D state of strain from measurements made with strain gauges, we firstly determine the angle each gauge makes to the x-axis: say, for gauges P, Q and R, these are θ_P, θ_Q and θ_R. The strains measured by the gauges are ε_P, ε_Q and ε_R. Thus, the strain transformation equation linking each of the measured strains ε_P, ε_Q and ε_R to the strains ε_x, ε_y and γ_{xy} are

$$\varepsilon_P = \varepsilon_x \cos^2\theta_P + \varepsilon_y \sin^2\theta_P + \gamma_{xy} \sin\theta_P \cos\theta_P$$
$$\varepsilon_Q = \varepsilon_x \cos^2\theta_Q + \varepsilon_y \sin^2\theta_Q + \gamma_{xy} \sin\theta_Q \cos\theta_Q$$
$$\varepsilon_R = \varepsilon_x \cos^2\theta_R + \varepsilon_y \sin^2\theta_R + \gamma_{xy} \sin\theta_R \cos\theta_R$$

or, in matrix form,

$$\begin{bmatrix} \varepsilon_P \\ \varepsilon_Q \\ \varepsilon_R \end{bmatrix} = \begin{bmatrix} \cos^2\theta_P & \sin^2\theta_P & \sin\theta_P\cos\theta_P \\ \cos^2\theta_Q & \sin^2\theta_Q & \sin\theta_Q\cos\theta_Q \\ \cos^2\theta_R & \sin^2\theta_R & \sin\theta_R\cos\theta_R \end{bmatrix} \begin{bmatrix} \varepsilon_x \\ \varepsilon_y \\ \gamma_{xy} \end{bmatrix}.$$

We invert these equations to find the strains ε_x, ε_y and γ_{xy} as

$$\begin{bmatrix} \varepsilon_x \\ \varepsilon_y \\ \gamma_{xy} \end{bmatrix} = \begin{bmatrix} \cos^2\theta_P & \sin^2\theta_P & \sin\theta_P\cos\theta_P \\ \cos^2\theta_Q & \sin^2\theta_Q & \sin\theta_Q\cos\theta_Q \\ \cos^2\theta_R & \sin^2\theta_R & \sin\theta_R\cos\theta_R \end{bmatrix}^{-1} \begin{bmatrix} \varepsilon_P \\ \varepsilon_Q \\ \varepsilon_R \end{bmatrix}.$$

Inverting the matrix of trigonometrical functions used to be difficult, and is the reason why strain gauge rosettes of specific geometry (e.g. rectangular and delta rosettes) were developed: the inverse of the matrix for these special geometries has a particularly simple form. However, it is now easier to perform the inversion of a general matrix on modern spreadsheets and scientific calculators, and so this constraint on rosette geometry is diminishing.

The solution to the problem is found using the matrix equation given above. In this case we have $\theta_P = 20^\circ$, $\theta_Q = 80^\circ$ and $\theta_R = 140^\circ$, and so the matrix equation is as follows:

$$\begin{bmatrix} \varepsilon_x \\ \varepsilon_y \\ \gamma_{xy} \end{bmatrix} = \begin{bmatrix} 0.883 & 0.117 & 0.321 \\ 0.030 & 0.970 & 0.171 \\ 0.587 & 0.413 & -0.492 \end{bmatrix}^{-1} \cdot \begin{bmatrix} 43.0\times10^{-6} \\ 7.8\times10^{-6} \\ 17.0\times10^{-6} \end{bmatrix}$$

$$= \begin{bmatrix} 0.884 & -0.293 & 0.449 \\ -0.177 & 0.960 & 0.218 \\ 0.857 & 0.456 & -1.313 \end{bmatrix} \cdot \begin{bmatrix} 43.6\times10^{-6} \\ 7.9\times10^{-6} \\ 17.0\times10^{-6} \end{bmatrix} = \begin{bmatrix} 41.6\times10^{-6} \\ 3.6\times10^{-6} \\ 18.1\times10^{-6} \end{bmatrix}.$$

The principal strains and their orientations are then computed from ε_x, ε_y, and γ_{xy}. In this case we find $\varepsilon_1 = 43.7\times10^{-6}$ and $\varepsilon_2 = 1.52\times10^{-6}$, with the angle between the x-direction and the major principal strain being 12.7°.

To compute the stress state from the strain state we use the stress–strain relations for an isotropic material, i.e.

$$\begin{bmatrix} \varepsilon_x \\ \varepsilon_y \\ \gamma_{xy} \end{bmatrix} = \frac{1}{E} \begin{bmatrix} 1 & -\nu & 0 \\ -\nu & 1 & 0 \\ 0 & 0 & 2(1+\nu) \end{bmatrix} \cdot \begin{bmatrix} \sigma_x \\ \sigma_y \\ \tau_{xy} \end{bmatrix}$$

which when inverted gives

$$\begin{bmatrix} \sigma_x \\ \sigma_y \\ \tau_{xy} \end{bmatrix} = \frac{E}{\nu^2-1} \begin{bmatrix} -1 & -\nu & 0 \\ -\nu & 1 & 0 \\ 0 & 0 & \frac{1}{2}(1-\nu) \end{bmatrix} \cdot \begin{bmatrix} \varepsilon_x \\ \varepsilon_y \\ \gamma_{xy} \end{bmatrix}.$$

From these we find that $\sigma_x = 7.04$ MPa, $\sigma_y = 2.65$ MPa and $\tau_{xy} = 1.04$ MPa. Computing the principal stresses and their orientations from these values gives $\sigma_1 = 7.28$ MPa and $\sigma_2 = 2.41$ MPa, with the angle between the x-direction and the major principal stress being 12.7°.

Notice that because this is an isotropic material, the orientations of the principal stresses and the principal strains are identical. Note also that we can perform the complement of this calculation, i.e. determine the principal strains from the measured strains, and then determine the principal stresses from the principal strains: the results will be identical to those shown here, and can be used as a useful check of the calculations.

Q5.6 Explain clearly why an isotropic rock has two independent elastic constants, a transversely isotropic rock has five independent constants and an orthotropic rock has nine independent constants (compared to the general anisotropic case where there are twenty-one independent elastic constants).

A5.6 Adding further detail to Table 5.2 provides the answers.

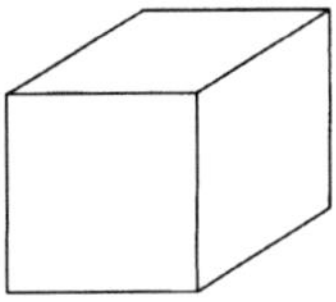

Perfectly isotropic rock. If all properties are equal in all directions, all the Young's moduli and Poisson's ratios in the different directions are equal. Similarly, all the shear moduli in different directions are equal and can be calculated from the single E and ν values (A5.4). Thus, there are *two elastic constants*: one Young's modulus, one Poisson's ratio.

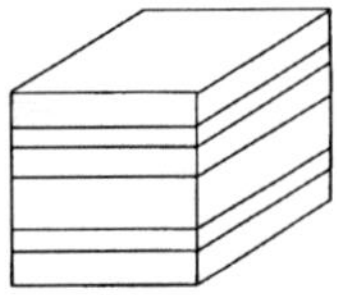

Transversely isotropic rock. Say, a rock mass with one main fracture set so that the properties are the same in a plane but different perpendicular to the plane. *Five elastic constants*: two Young's moduli, two Poisson's ratios, and one shear modulus (not two, see A5.4).

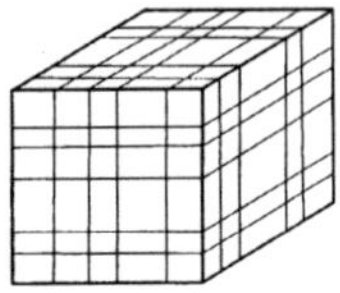

Orthotropic rock. Say, a rock mass with three mutually perpendicular fracture sets. Assuming there is no linkage between (a) normal and shear components, and (b) different shear components (see the **S** matrix in Section 5.1), this leaves *nine elastic constants*: three Young's moduli, three Poisson's ratios and three shear moduli.

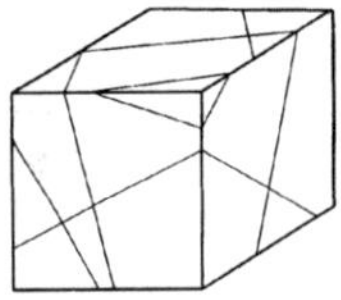

General anisotropic rock. The 6 × 6 elastic compliance matrix is symmetrical, reducing the number of independent constants to 21. No further reduction can be achieved. Thus, there are *21 elastic constants*: all the S_{ij} in the matrix.

Q5.7 Each of the following four rock masses is to be modelled using elasticity theory. State whether you think that an isotropic rock assumption is justified, or whether one of the anisotropic assumptions would be more appropriate. Assume that the fractures have a significant effect on the rock deformability.

(a) A limestone with effectively three fracture sets, i.e. the bedding with mean strata thicknesses of 1 m, plus fracture set 1 (perpendicular to the bedding) with two fractures/m, and fracture set 2 (perpendicular to both the bedding and fracture set 1) with five fractures/m.

(b) A welded volcanic tuff with five fracture sets.
Fracture set 1: dip direction 089°; dip 50°; frequency 2.9/m.
Fracture set 2: dip direction 278°; dip 88°; frequency 1.3/m.
Fracture set 3: dip direction 224°; dip 08°; frequency 0.9/m.
Fracture set 4: dip direction 169°; dip 23°; frequency 2.1/m.
Fracture set 5: dip direction 113°; dip 70°; frequency 0.7/m.

(c) A strong sandstone with almost no fractures present.
(d) A granodiorite with three fracture sets.
Fracture set 1: dip direction 314°; dip 35°; frequency 1.2/m.
Fracture set 2: dip direction 048°; dip 43°; frequency 1.3/m.
Fracture set 3: dip direction 089°; dip 79°; frequency 0.9/m.

A5.7 (a) For the limestone, the presence of three orthogonal fracture sets with significantly different fracture spacings (and probably different normal and shear fracture stiffnesses) indicates that an orthotropic model assumption with nine constants would be most appropriate. We would try to estimate the nine constants from the intact rock and fracture characteristics. This is the assumption from the elasticity point of view but it will have to be tempered in a practical case by the availability of analytical solutions and numerical codes that will support such an assumption.

(b) For the volcanic tuff, with five fracture sets having widely different orientations and frequencies, the first assumption will be an isotropic model, with the two effective elastic constants approximating the contributions from all fracture sets. There is no method for dealing with this more complex type of anisotropy in the theory of elasticity, apart from aggregating all the deformation contributions.

(c) For the intact sandstone, the first assumption would be the transversely isotropic model. We know that sandstone is a sedimentary rock and is likely to have different properties perpendicular and parallel to the bedding. However, if these properties turned out to be similar, we could then revert to the isotropic model.

(d) The granodiorite has three fracture sets which are not even close to being mutually orthogonal. There is no directly suitable simplified elasticity assumption for this type of rock mass because there is no orthogonal symmetry. One could use an isotropic model, although there is no coherent way to estimate the two effective elastic constants, and so the results would be unreliable.

Q5.8 (a) At the time of writing this book, most elastic analyses that have been conducted for rock engineering design purposes have assumed that the rock is perfectly isotropic with two elastic constants. Why do you suppose that is, given that most rock masses are clearly not isotropic?

(b) Conversely, no elastic analysis for rock mechanics has been conducted assuming that the rock mass is fully anisotropic with 21 elastic constants? Why is that?

(c) In this context, what do you think will happen in future analyses?

A5.8 (a) In the laboratory, the intact rock is studied, not the rock mass with all the fractures of different types it contains. Also, it is difficult to estimate, let alone measure, the *in situ* rock anisotropy. Moreover, measuring the elastic properties is time-consuming and expensive. Finally, few elastic solutions exist for non-isotropic elastic rocks. Thus, it is convenient to assume that the rock is isotropic and that the laboratory values, perhaps adjusted by some factor, are representative of the rock mass. However, many rock masses are significantly anisotropic because of their contained fractures, and so the isotropy assumption will often be invalid, as illustrated in Q and A 5.4 and 5.7.

(b) Because we do not have solutions to elastic problems for the fully anisotropic case, and because measuring the elastic properties is time-consuming and expensive, no one has yet felt justified in attempting to measure and use all twenty-one elastic constants for engineering application.

(c) In the future, it is likely that there will be a move towards transversely isotropic and orthotropic characterization where the elastic constants of the rock mass will be estimated either from the properties of the intact rock and the fractures or by back analysis from *in situ* displacement measurements. This is because there is increasing experience that the isotropic assumption does not adequately represent reality. For some years now, the strain values obtained using the CSIRO gauge for stress measurement have been interpreted using a transversely isotropic model for the intact rock, as appropriate. There is little doubt that this trend towards improved rock property characterization will continue.

Q5.9 (a) How do you know if a material is elastic?
(b) How is time taken into account in the theory of elasticity?

A5.9 (a) A material is elastic if, after deforming, it returns to its original shape and size and all the mechanical energy used in deforming the material can be recovered.

(b) Time is not taken into account in the theory of elasticity. Time-dependent effects and their interactions are complex but can be modelled to a first approximation by **the theory of visco-elasticity**. The complexities of time-dependent rock behaviour are described in the book by Cristescu and Hunsche (1998)[4].

Q5.10 When a rock specimen fails in a compressive test in the laboratory or when an *in situ* rock mass fails due to natural compression, is 'stress' or 'strain' the cause of failure?

A5.10 When a rock specimen is loaded in a laboratory compression test, a testing machine is used and nowadays these are servo-controlled testing machines. This type of machine is programmed to correct a feedback signal via a hydraulic jack (or actuator) so that either the force, and hence the stress on the specimen, is increased at a constant rate, or the displacement, and hence the strain of the specimen, is increased at a constant rate. Other feedbacks are also possible, in particular circumferential strain.

[4] Cristescu N. D. and Hunsche U. (1998) *Time Effects in Rock Mechanics*. Wiley, New York, 342pp.

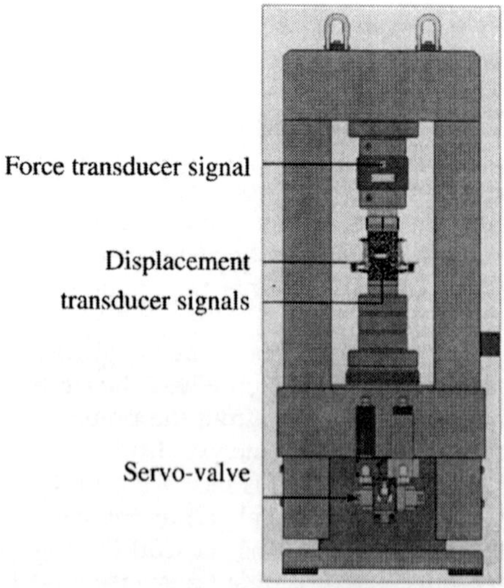

Laboratory rock testing controlled using load or displacement feedback (courtesy of MTS Systems Corp.).

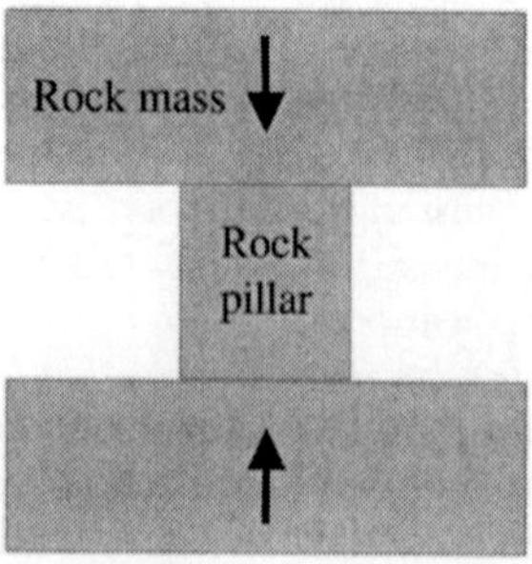

An underground rock pillar being stressed or strained?

If force is being controlled, then stress is the independent variable and strain is the dependent variable. Conversely, if displacement is being controlled, then strain is the independent variable and stress is the dependent variable. In each case, the independent variable in the experiment is controlled by the operator.

When *in situ* rock masses are compressed by tectonic forces, the situation is not so clear. Also, in rock masses hosting engineering projects, both stress and strain can be the causative variable. For example, the stability of a mine pillar can depend on the unloading stiffness of the rock strata both above and below it, as illustrated in the diagram. Stress will control failure if, for example, there is a constant loading, as in the weight of a rock block sliding on a surface. Strain will generally control failure if the displacement of a stiff rock is causing strain in a soft rock.

The message in this answer is that the controlled or independent variable is determined by the operator in the laboratory, but the causative variable in the *in situ* failure of rock masses may not be obvious, or indeed a single variable. However, a certain amount of energy is required for the rock to fail and so it could be argued that the energy (a product of stress and strain) is the cause.

5.3 Additional points

We have emphasized elasticity theory and the different types of elastic symmetry because of the widespread use of the theory in analytical and numerical solutions for the stresses and strains around rock excavations. When elasticity theory is used in engineering rock mechanics, it is important to realize the following points.

(1) Elasticity theory has been developed assuming that the strains are infinitesimal. There is a **theory of finite strain** which can be used when large strains are involved. This theory is used widely in structural geology analyses, e.g. Ramsay and Huber (1983)[5]. Methods of incorporating strains and block movements into numerical calculations are discussed by Pan (1988)[6].

(2) Most calculations are conducted assuming that the rock is elastically isotropic. One should always consider whether this will introduce a significant error in the calculations, and whether this error is important in the particular engineering context under investigation.

(3) Most calculations are also conducted assuming linear relations between stress and strain, i.e. the S_{ij} elastic compliances explained in Section 5.1 are indeed constants. Often, the stress–strain relation can be appreciably nonlinear, and certainly will be when rock failure occurs.

(4) Remember that all energy put into an elastic material can be recovered. Thus, if a rock mass exhibits significant hysteresis (e.g. a rock mass with low-modulus fractures which is being loaded and unloaded), elasticity theory may be inappropriate.

(5) There is no time component in the theory of elasticity. If *in situ* displacements are time-dependent, elasticity theory is theoretically invalid.

(6) Other factors can affect the stresses and strains besides their relation to each other through the theory of elasticity. For example, there can be thermal strains and complications introduced by fluids present in the rock mass.

Despite all these caveats, elasticity theory is useful and is used ubiquitously to support rock engineering design. Our emphasis is on ensuring that the application of elasticity theory to a particular problem is appropriate, bearing in mind the assumptions that have been made, either explicitly or implicitly. This type of check is part of the technical auditing procedure explained in Chapter 14.

---

[5] Ramsay J. G. and Huber M. I. (1983) *The Techniques of Modern Structural Geology. Vol. 1: Strain Analysis*. Academic Press, London, 307pp.

[6] Pan X. D. (1988) *Numerical Modelling of Rock Movements Around Mine Openings*. PhD thesis. University of London, 375pp.

6 Intact rock: deformability, strength and failure

6.1 Intact rock

The two main mechanical components of a rock mass are (a) the **intact rock** material, and (b) the **fractures**, e.g. the faults, shear zones, joints and bedding planes. This chapter is concerned with the mechanical properties of the unfractured rock.

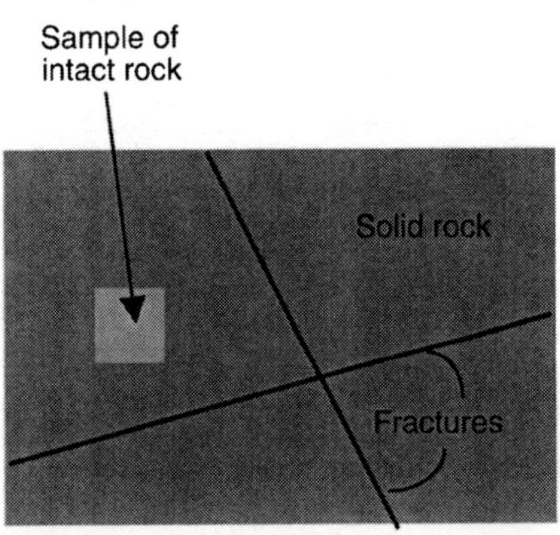

Samples of the intact rock can be obtained from surface outcrops or from borehole cores obtained during a site investigation. Using a coring drill bit, a cylindrical sample of the intact rock can be obtained from boreholes as shown in Fig. 6.1.

The testing of intact rock, which has been developed from the 1960s to the present day, has concentrated on the testing of rock cylinders because cylindrical rock cores are produced during site investigation drilling. The intact rock can be tested in compression, tension or shear, or a combination of these. For rock mechanics and rock engineering, we are interested in the deformability, strength and failure properties of rock under different loading conditions.

A force–displacement curve obtained during the uniaxial compressive testing of an intact rock specimen of marble is shown in Fig. 6.2[1].

Figure 6.1 Rock core obtained from a site investigation borehole provides samples of intact rock for testing. Note also the fractures intersecting the borehole core.

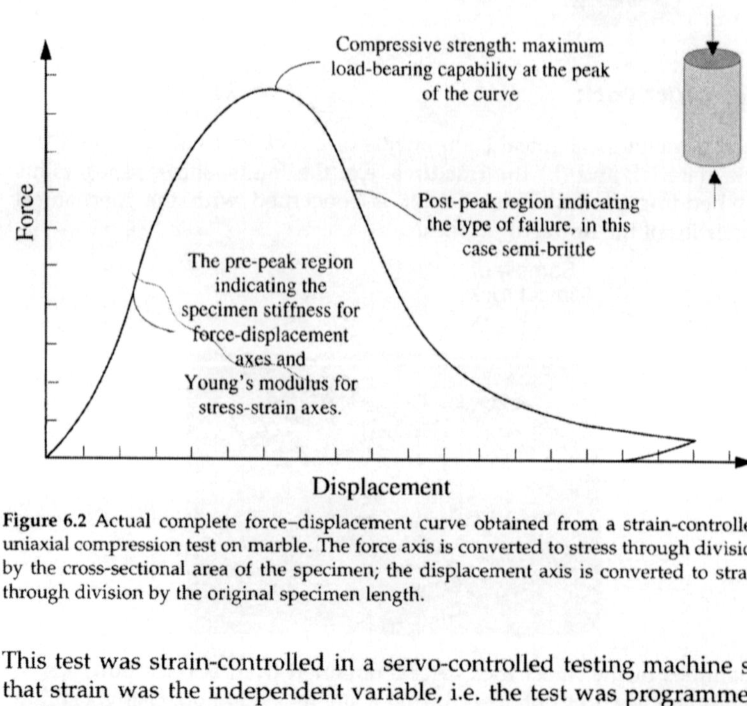

Figure 6.2 Actual complete force–displacement curve obtained from a strain-controlled uniaxial compression test on marble. The force axis is converted to stress through division by the cross-sectional area of the specimen; the displacement axis is converted to strain through division by the original specimen length.

This test was strain-controlled in a servo-controlled testing machine so that strain was the independent variable, i.e. the test was programmed so that the strain increased at a constant rate. Stress was therefore the dependent variable (see Q and A5.10 in Chapter 5). Under these conditions, the **complete force-displacement curve** shown in Fig. 6.2 was obtained. This curve represents the response of the intact rock to continued compressive displacement of the specimen ends, from initial

[1] When the axes in Fig. 6.2 are scaled to stress and strain using the original specimen area and length, the result is an 'engineering' stress–strain curve representing the structural breakdown of the specimen. Some researchers consider that the 'true' stress–strain curve for the rock material reduces less sharply after the peak than shown, because the effective force-transmitting area of the specimen reduces after the peak, but this is not accounted for in the calculation of the stress.

loading, through a relatively linear elastic zone, to the compressive strength (the peak of the curve), and then downwards as the specimen structure disintegrates and the load can no longer be sustained.

It is helpful to remember that this type of complete force–displacement curve occurs for intact rock under any of the conditions of compression, tension and shear (or any combination of these). Also, polyaxial and dynamic stress states can be applied to a rock specimen. For all these cases, different engineering stress–strain curves will be obtained, depending on the stress or strain states applied, and different curves will be obtained if the tests are programmed for stress control, strain control, or some other type of control.

The linear response in the ascending pre-peak region is used to provide the elastic constants discussed in Chapter 5. The strength of the rock in any of the loading conditions is given by the peaks of the curves, e.g. the compressive, tensile and shear strengths[2]. The form of the curve in the descending post-peak region indicates the nature of failure but the shape of the post-peak curve will depend on how the test is programmed: for example, if stress is programmed to increase as the independent variable, the test will become uncontrollable at the peak because the specimen cannot sustain a higher stress than its strength. Three rock failure criteria are explained in Chapter 6 of ERM 1 (namely, the **Mohr–Coulomb**, plane **Griffith**, and **Hoek–Brown** criteria) and all these predict failure as a function of the applied stress components.

In addition to the changes caused by direct mechanical loading, the intact rock can be degraded by or be resistant to **exposure** to water, freeze–thaw cycles, and chemical effects, as illustrated in Fig. 6.3 for the case of the rocks at Stonehenge in the UK. The standing stones are sandstone with a siliceous matrix which explains why they have been able to resist 5000 years of British weather.

At the surface of the Earth, there is a dynamic weathering environment within which rocks are subjected to a variety of thermal, hydrological, mechanical and chemical (THMC) gradients. The term '**weathering**' is used to describe the effect of these processes on the intact rock as it responds, changing towards a state which is in equilibrium with its environment. All exposed intact rock is subjected to these processes and will only retain its mechanical integrity if it has a high resistance to the processes, such as the granite used for building veneers, paving stones and gravestones.

The THMC processes also operate on unexposed underground rocks and on the fractures within rocks, although at a slower rate because the gradient changes and associated energy changes cannot occur as rapidly as on the surface. However, the natural processes have had millions of years in which to operate and so underground rocks can often be significantly weathered, especially in the vicinity of fractures

---

[2] The rock properties measured in such tests should be regarded as engineering, rather than material, properties. A **material property** does not depend on the specimen geometry and loading conditions of the test, but the strengths and failure properties of rocks do depend on these factors, and therefore they are not material properties.

Figure 6.3 The trilithons at Stonehenge in the UK.

where fluids have circulated. When the engineer excavates rock and creates new surfaces, the THMC processes are accelerated, because the temperature changes, water flow is enhanced (possibly with pH changes), and the rock stress is reduced to zero (see A3.5) on the excavation surfaces resulting in a high stress gradient in the proximate rock mass. Thus, the intact rock is in a state of change, the rate of which depends on the rock type and the applied THMC gradients. The rock has already undergone a variety of changes, and will do so again on excavation or when adjacent to an excavation.

6.2 Questions and answers: intact rock

Q6.1 Predict how the volume of a specimen will change during a uniaxial compression test, and sketch the variation of volumetric strain versus axial strain. What physical processes are occurring in the sample as the curve manifests significant gradient changes?

A6.1 We consider the volume change during the complete stress–strain curve as illustrated on the next page (i.e. as it occurs during generation of the curve in Fig. 6.2).

Initially, the compressive load closes any open, pre-existing cracks, and so there is contraction of the specimen. Following this is a region of elastic contraction during the linear portion of the pre-peak region of the curve. From about 50% of the peak load onwards, new cracks form at an increasing rate, causing a dilatation component to the volume change and slowing down the contraction rate. Eventually, the dilatation associated with new cracks is greater than the elastic contraction, and the volume change versus displacement curve alters from a negative

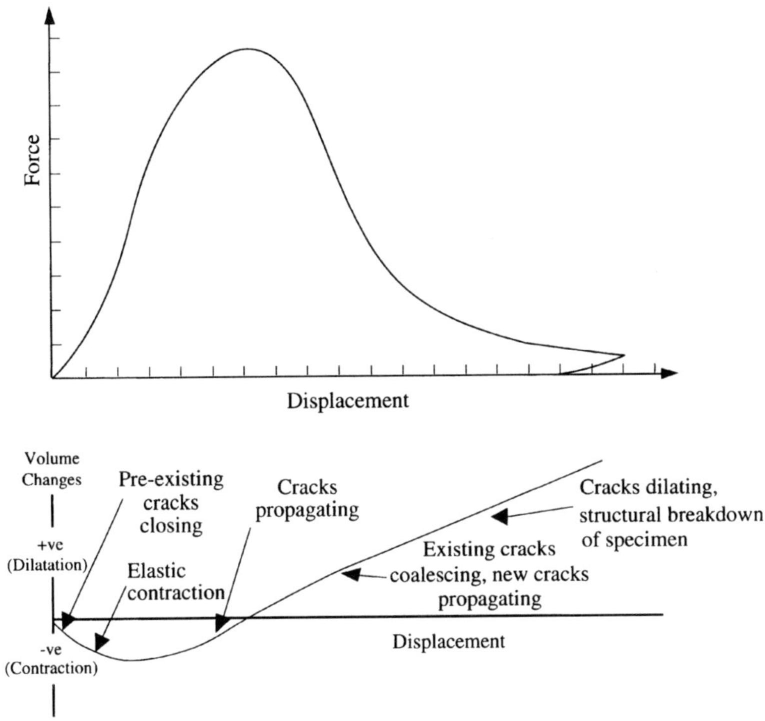

to a positive slope. The specimen continues to dilate at an increasing rate until shear planes form, which allow the specimen to break into separate mobile pieces, with the result that the dilatation rate is reduced.

Q6.2 What causes a rock to break when it is compressed uniaxially to failure? Is it when the stress reaches a certain value or when the strain reaches a certain value? Or is it when some other parameter reaches a critical value, such as the energy input per unit volume, or the microcrack density per unit volume?

A6.2 Over the years, many failure criteria have been proposed, and all the parameters mentioned in the question have been used to develop failure criteria at one time or another. Because the collapse of the rock microstructure is a complex process (see Q and A6.1), it is not clear which is the correct parameter, or combination of parameters, to use in a failure criterion. The short answer to the question is that no one knows the answer. As a result, the most widely used failure criteria are empirical criteria based on the stress components (especially the Mohr–Coulomb and Hoek–Brown failure criteria).

Q6.3 The geometry of the linear Mohr–Coulomb envelope is such that a number of useful relations between strength parameters can

be drawn from it. Derive an expression for the uniaxial compressive strength of rock in terms of the cohesion and angle of internal friction.

A6.3 The part of the Mohr envelope in which we are interested is shown below.

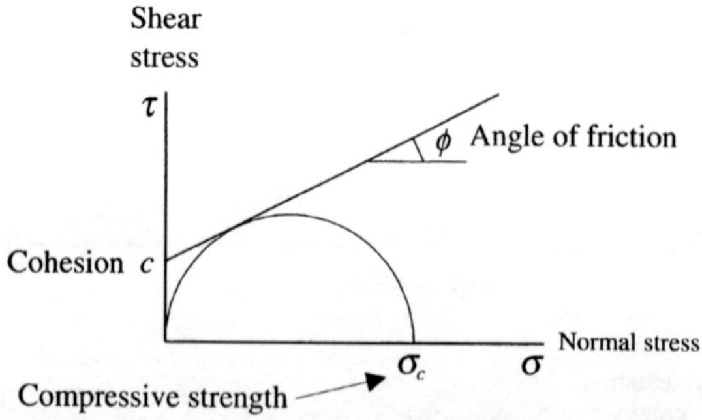

If we study the geometry of the left-hand half of the Mohr circle and the failure envelope, we find the following:

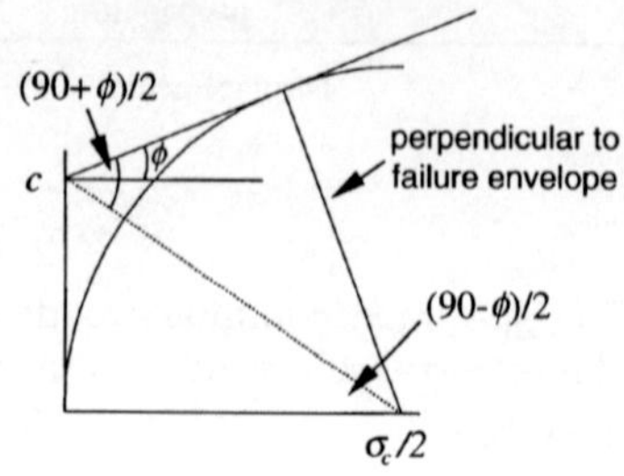

The geometry of the lower right-angled triangle then gives us $\tan(45 - (\phi/2)) = c/(\sigma_c/2)$ which, upon expanding and rearranging, results in $\sigma_c = 2c/\tan(45 - (\phi/2))$. This relation can be expanded and rearranged in a number of ways. For example, the friction angle in terms of the cohesion and uniaxial strength is also given by $\tan(\phi/2) = (\sigma_c - 2c)/(\sigma_c + 2c)$.

This shows how we can use typical laboratory test results to confirm the applicability of the Mohr–Coulomb criterion, or compute unknown values from known values.

Q6.4 The linear Mohr–Coulomb envelope with a tensile cut-off sets a definite limit on the maximal uniaxial tensile strength of a material. By considering the largest uniaxial tensile Mohr circle that can be drawn, determine this tensile strength limit in terms of σ_c and ϕ.

A6.4 The geometry of the Mohr–Coulomb criterion, together with the Mohr circles representing the uniaxial compressive and the uniaxial tensile strengths, are shown below.

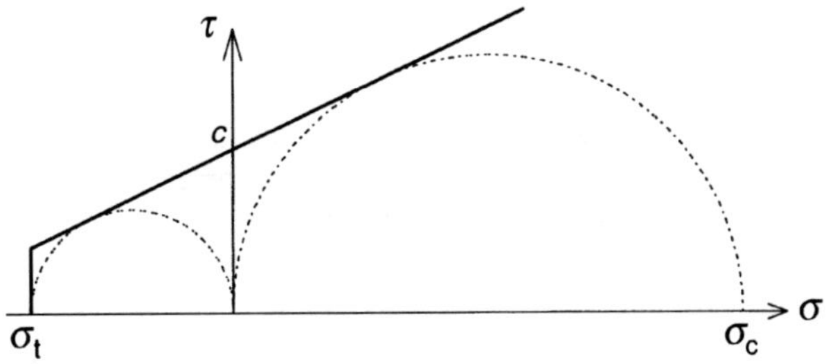

Next, for each Mohr circle, draw the radius that is perpendicular to the tangent Mohr–Coulomb envelope line representing shear failure, where r_t and r_c represent the radii of the uniaxial tension circle and the uniaxial compression circle, respectively.

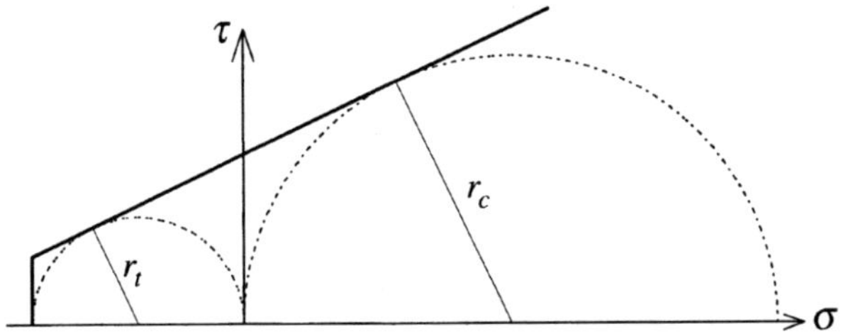

If we now draw a line from the centre of the circle representing uniaxial tension, parallel to the shear failure criterion line, to the previously drawn radial line in the circle representing uniaxial compression, we obtain the diagram below.

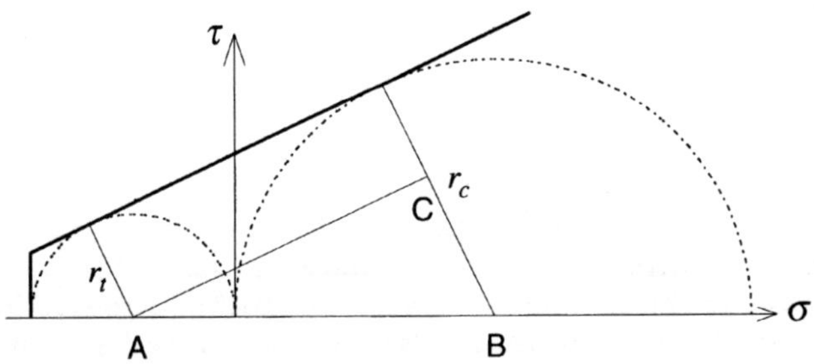

Inspection of this diagram reveals that, for the triangle ABC, the following lengths and angles apply:

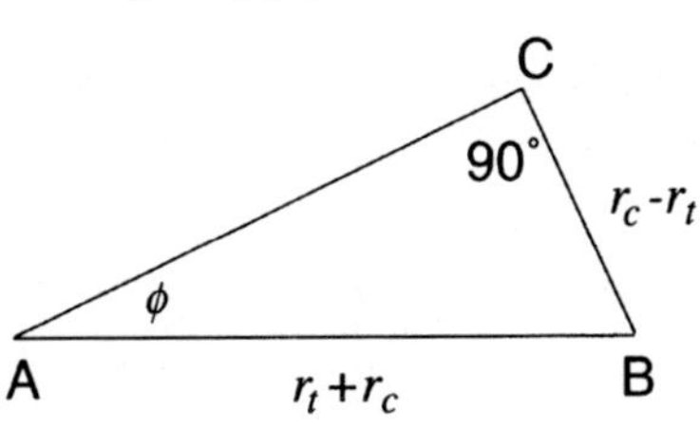

From this, we see that $(r_c - r_t)/(r_c + r_t) = \sin\phi$, and upon rearrangement we obtain $r_c/r_t = (1 + \sin\phi)/(1 - \sin\phi)$.

Finally, as each of r_c and r_t is one half of the corresponding compressive and tensile strengths, we find that $\sigma_c/\sigma_t = (1 + \sin\phi)/(1 - \sin\phi)$.

Clearly, the commonly held belief that the compressive strength of rock is about ten times the tensile strength will only be true for a particular value of the friction angle. Indeed, we can graph the equation above, as shown below.

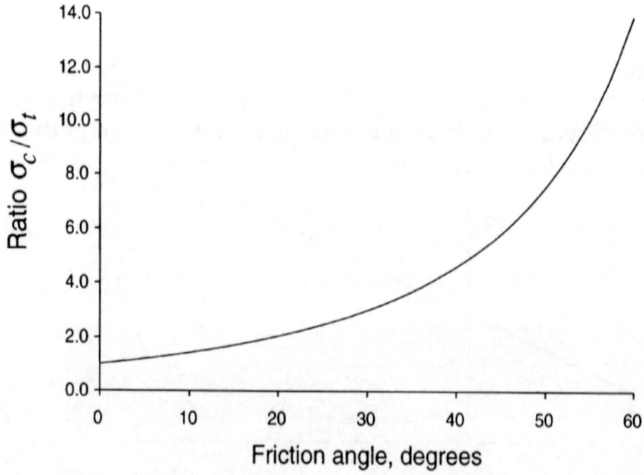

This shows that $\sigma_c/\sigma_t = 1$ for $\phi = 0^\circ$, and $\sigma_c/\sigma_t \Rightarrow \infty$ for $\phi = 90^\circ$. Furthermore, we can see that $\sigma_c/\sigma_t = 10$ for the specific case of $\phi = 55^\circ$, and that for the more frequent case of $\phi = 35^\circ$ we have $\sigma_c/\sigma_t = 3.7$. The difficulty with the relation arises because rock failure modes in compression, tension and shear are different[3].

Q6.5 A firm whose judgement has been questioned on a previous occasion has been entrusted with the strength testing of rock in a site investigation project. During their first uniaxial compression test, the equipment failed to measure the peak axial load, but the technician did record that the specimen failed by the formation of a single fracture inclined at 20° to the loading axis. In a subsequent triaxial test, as the confining pressure was being increased before application of the axial stress, the specimen failed prematurely when the confining pressure in the Hoek cell was 85 MPa. On the basis of these results, propose a failure criterion of the form $\sigma_1 = a\sigma_3 + b$ for the rock.

A6.5 The two tests, although incorrectly performed, do give us information about the strength of the rock in terms of the linear Mohr–Coulomb

[3] Cohesion is the resistance to failure in shear; adhesion is the resistance to failure in tension. Under uniaxial loading, rocks break perpendicular to the least principal stress — perpendicular to the loading in tension and parallel to the loading in compression. Thus, we see that both cohesion and adhesion play a part in determining the strength of the rock.

envelope. Firstly, if a fracture forms at 20° to the loading axis, then the angle between the major principal stress (i.e. the axial stress) and the normal to this plane is 70°. The point on a Mohr circle associated with the stress state on the fracture surface will then make an angle of 140° to the major principal stress. The geometry of the failure envelope shows that the friction angle is then 50°.

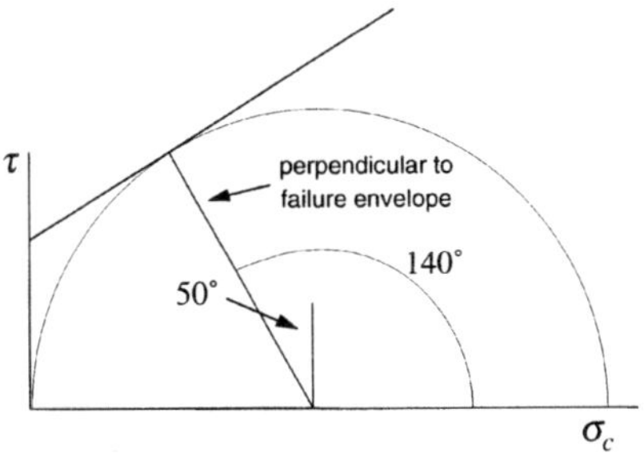

In the second test, the axial stress is zero and the confining pressure is the stress at failure, i.e. the compressive strength. The compressive strength is simply 85 MPa. Knowing that the friction angle is 50° and the compressive strength is 85 MPa, we can find the failure criterion in the required form. The Mohr–Coulomb failure criterion when written in its principal stress form is $\sigma_1 = a\sigma_3 + b$, and so we find that

$$\sigma_1 = \frac{1+\sin\phi}{1-\sin\phi}\sigma_3 + \sigma_c \text{ with } a = \frac{1+\sin\phi}{1-\sin\phi} = 7.55 \text{ and } b = \sigma_c = 85 \text{ MPa}.$$

Q6.6 A servo-controlled compression test has been conducted on a weak soapstone such that the specimen length remained unchanged throughout: as the axial stress, σ_a, was increased, so the confining pressure, p, was increased so that no net axial strain resulted. A plot of axial stress (vertical axis) against confining pressure (horizontal axis) gave an initial straight line passing through the origin. At a critical confining pressure of p = 85 MPa (when σ_a = 39.1 MPa), the slope of the $\sigma_a - p$ plot suddenly changed to 29° and remained constant for the remainder of the test. This change in slope may be taken to represent the onset of yield.

(i) Determine an elastic constant from the slope of the initial portion of the $\sigma_a - p$ curve.
(ii) Assuming that the Mohr–Coulomb criterion is applicable, determine σ_c, c and ϕ for the rock.

A6.6 In this question we are given information regarding two different types of behaviour: elastic behaviour up to yield, and strength behaviour after yield (it is a strain-controlled test, and so post-peak behaviour can be obtained). We are told that before yield — when the specimen is behaving elastically — the $\sigma_a - p$ plot passes through the origin, and the point ($p = 85$, $\sigma_a = 39.1$). After yield (i.e. when the specimen has, in some sense, failed) the slope of the $\sigma_a - p$ plot is 29°.

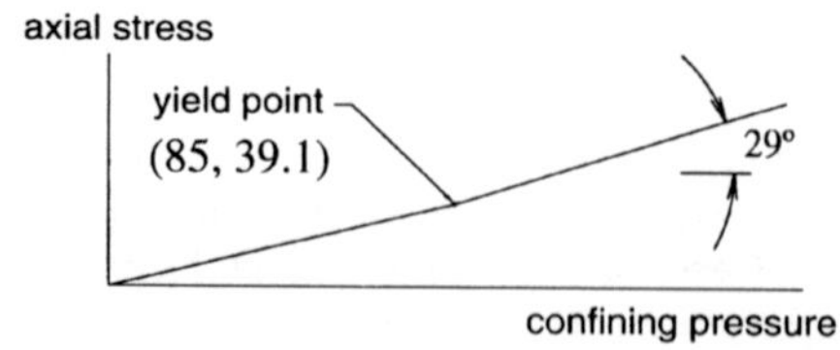

(i) We are given information regarding the behaviour of the specimen in a three-dimensional stress state: a relation between the axial stress, the axial strain and the confining pressure is indicated. To use this information, we start with Hooke's Law in three dimensions for an isotropic material,

$$\varepsilon_1 = \frac{1}{E}\left[\sigma_1 - \nu\left(\sigma_2 + \sigma_3\right)\right]$$

and substitute $\varepsilon_1 = \varepsilon_a$, $\sigma_1 = \sigma_a$ and $\sigma_2 = \sigma_3 = p$ to obtain

$$\varepsilon_a = \frac{1}{E}\left[\sigma_a - \nu\left(p + p\right)\right] = \frac{1}{E}\left[\sigma_a - 2\nu p\right].$$

We are told that in this test the axial strain is controlled to be zero. As a result, the equation above reduces to $0 = (1/E)\left[\sigma_a - 2\nu p\right]$ or $\sigma_a = 2\nu p$.

We now have an equation linking σ_a and p which, on comparison with the standard form of a straight line, $y = mx + c$, is found to pass through the origin and has a gradient of 2ν (when σ_a is plotted on the vertical axis and p is plotted on the horizontal axis). The gradient of this part of the $\sigma_a - p$ plot can either be measured, or computed. Using co-ordinate geometry to compute the gradient, we obtain

$$m = \frac{y_2 - y_1}{x_2 - x_1} = \frac{39.1 - 0}{85 - 0} = 0.46 = 2\nu$$

and hence $\nu = 0.23$.

Thus, from the first part of the curve we find that we can compute Poisson's ratio.

(ii) If we assume that yield represents peak strength being reached, then after yield the $\sigma_a - p$ curve represents a plot of the principal stress form of the Mohr–Coulomb criterion, with $\sigma_a \equiv \sigma_3$ and $p \equiv \sigma_1$. Thus we have

$$p = \frac{2c\cos\phi}{1 - \sin\phi} + \sigma_a\frac{1 + \sin\phi}{1 - \sin\phi} = \sigma_c + \sigma_a\frac{1 + \sin\phi}{1 - \sin\phi}.$$

It is worth redrawing the curve with σ_a on the horizontal axis. Now, if we extrapolate the failure criterion back to $\sigma_a = 0$, then the point where it intersects the p axis is the compressive strength, σ_c. Using straight line geometry, the compressive strength is found from

$$m = \frac{y_2 - y_1}{x_2 - x_1} \Rightarrow \tan 61^\circ = \frac{85 - \sigma_c}{39.1 - 0} \Rightarrow \sigma_c = 85 - 39.1\tan 61,$$

giving $\sigma_c = 14.5$ MPa.

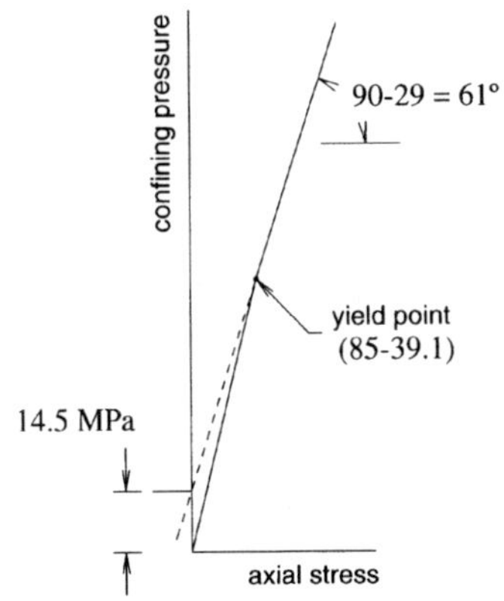

The principal stress form of the Mohr–Coulomb criterion tells us that

$$\tan 61^\circ = \frac{1+\sin\phi}{1-\sin\phi} \quad \text{or} \quad \sin\phi = \frac{\tan 61 - 1}{\tan 61 + 1},$$

giving $\phi = 16.7^\circ$.

And finally we have $\sigma_c = (2c\cos\phi)/(1-\sin\phi)$, from which we obtain $c = \sigma_c(1-\sin\phi)/(2\cos\phi)$, with the result that $c = 5.4$ MPa.

Q6.7 Laboratory tests on specimens of a limestone have produced unconfined compressive and tensile strengths of 80 MPa and 10 MPa, respectively. Using the Hoek–Brown and plane Griffith criteria, estimate the maximum principal stress at failure for two biaxial tests in which $\sigma_2 = 20$ MPa and $\sigma_2 = 40$ MPa. Which of these two criteria would best predict peak strength under these conditions?

A6.7 We are told that $\sigma_c = 80$ MPa and $\sigma_t = -10$ MPa (noting that our sign convention is compression-positive).

Hoek–Brown criterion

If we substitute $\sigma_3 = \sigma_t$ and $\sigma_1 = 0$ into the Hoek–Brown criterion, $\sigma_1 = \sigma_3 + (m\sigma_c\sigma_3 + s\sigma_c^2)^{0.5}$, and rearrange the resulting equation, we find

$$m = \frac{\sigma_t^2 - \sigma_c^2}{\sigma_t\sigma_c}.$$

Here, we are dealing with intact rock, and so $s = 1$. Consequently we find that $m = 7.88$.

For the biaxial tests, we have been given values for σ_2 and asked to find values for σ_1. As the tests are biaxial, we can presume $\sigma_3 = 0$. However, the Hoek–Brown criterion ignores the effect of the intermediate principal stress.

We could estimate the magnitude of the major principal stress at failure either assuming $\sigma_2 = \sigma_3 = 0$ or $\sigma_2 = \sigma_3 = p$. The first of these represents uniaxial compression, the second represents triaxial compression. The case of uniaxial compression will simply give the

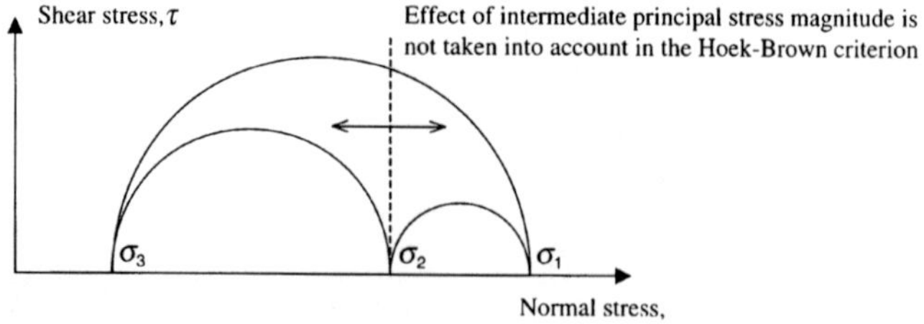

unconfined compressive strength, and so it is worthwhile investigating the triaxial compression case.

Thus, for $\sigma_2 = \sigma_3 = 20$ MPa we find that

$$\sigma_1 = 20 + \sqrt{7.88 \times 20 \times 80 + 1 \times 80^2} = 157.8 \text{ MPa}$$

and for $\sigma_2 = \sigma_3 = 40$ MPa we find that

$$\sigma_1 = 40 + \sqrt{7.88 \times 40 \times 80 + 1 \times 80^2} = 217.8 \text{ MPa}$$

However, it is likely that these will both be over-estimates of the major principal stress at failure.

Griffith criterion

There are two expressions for the Griffith criterion in compression:

$$(\sigma_1 - \sigma_3)^2 = 8T_o(\sigma_1 + \sigma_3) \quad \text{when } \sigma_1 + 3\sigma_3 > 0$$
$$\sigma_3 = -T_o \quad \text{when } \sigma_1 + 3\sigma_3 < 0$$

and in both of these compression is positive and $T_o = -\sigma_t$ is also positive.

As with the Hoek–Brown criterion, the Griffith criterion ignores the effect of the intermediate principal stress, and so neither of these criteria is valid for biaxial compression. Again, we will need to make the assumption that the biaxial conditions in the tests may be represented by triaxial conditions in the Griffith criterion. Rearranging the first of the Griffith criterion equations gives us

$$\sigma_1 = \sigma_3 + 4T_o \pm 4\sqrt{T_o\sigma_3 + T_o^2}$$

from which, on substitution of $T_o = 10$ MPa and the appropriate value of σ_3, we can determine the major principal stress at failure.

Thus, for $\sigma_2 = \sigma_3 = 20$ MPa we find that

$$\sigma_1 = 20 + 4 \times 10 \pm 4\sqrt{10 \times 20 + 10^2} = 129.3 \text{ MPa}$$

and for $\sigma_2 = \sigma_3 = 40$ MPa we find that

$$\sigma_1 = 40 + 4 \times 10 \pm 4\sqrt{10 \times 40 + 10^2} = 169.4 \text{ MPa.}$$

Assessment

Determining which of these criteria is the best predictor of peak strength is difficult. Strictly, neither of the criteria are valid for biaxial conditions, but the Hoek–Brown criterion is for strength (i.e. a structure breakdown criterion), whereas the Griffith criterion relates to the onset of fracturing. As the onset of fracturing occurs before peak strength and usually well before complete failure, perhaps the Hoek–Brown criterion is the better of the two. However, it is more likely that the peak strength will lie somewhere between the two sets of results.

Q6.8 Comment on the applicability of each of the Griffith, Mohr–Coulomb, and Hoek–Brown criteria for the following triaxial test results on quartzite:

$(\sigma_1+\sigma_3)/2$	**−6.65**	**100**	**135**	**160**	**200**	**298**	**435 MPa**
$(\sigma_1-\sigma_3)/2$	**6.65**	**100**	**130**	**150**	**180**	**248**	**335 MPa**

A6.8 The first step is to convert the given data into principal stress values:

$(\sigma_1+\sigma_3)/2$	−6.65	100	135	160	200	298	435
$(\sigma_1-\sigma_3)/2$	6.65	100	130	150	180	248	335
σ_1	0	200	265	310	380	546	770
σ_3	−13.3	0	5	10	20	50	100

From the final two rows of the table we can see that $\sigma_t = -13.3$ MPa and $\sigma_c = 200$ MPa. The values of σ_1 and σ_3 are plotted at the end of this answer.

Griffith criterion

The relevant expression for the Griffith criterion in compression is

$$(\sigma_1-\sigma_3)^2 = 8T_0(\sigma_1+\sigma_3)$$

where compression is positive and T_o is also positive. For given values of σ_3 we need to compute values of σ_1, using a value of $T_o = 13.3$ MPa. The result is

σ_3	−13.3	0	5	10	20	50	100
σ_1	39.9	106.4	120.6	133.6	157.4	219.3	308.5

Mohr–Coulomb criterion

The data are clearly curvilinear, whereas the Mohr–Coulomb criterion is linear. As a result, the Mohr–Coulomb criterion will not be a good fit to the data, at least not over their entire range. However, if we wish to fit a straight line to the data, the results are

σ_3	−13.3	0	5	10	20	50	100
σ_1	118.6	201.0	232.0	263.0	324.9	510.9	820.7

Hoek–Brown criterion

We can either simply take appropriate values for the parameters in the criterion from a table or attempt to statistically fit the criterion to the data. Using the former method, we have $s = 1$ and $m = 15$, and with $\sigma_c = 200$ MPa we obtain:

σ_3	−13.3	0	5	10	20	50	100
σ_1	−3.3	200.0	239.5	274.6	336.2	485.9	683.1

Using the latter method, we rearrange the criterion to

$$\left(\frac{\sigma_1}{\sigma_c} - \frac{\sigma_3}{\sigma_c}\right)^2 = m\frac{\sigma_3}{\sigma_c} + s$$

which corresponds to the linear form $y = mx + c$, and obtain values for m and s by linear regression. In this case, we know that $s = 1$, and so we perform a linear regression on

$$\left(\frac{\sigma_1}{\sigma_c} - \frac{\sigma_3}{\sigma_c}\right)^2 - 1 = m\frac{\sigma_3}{\sigma_c}$$

while constraining the solution to pass through the origin. The result is a parameter value of $m = 20.5$, with corresponding stress values of

σ_3	−13.3	0	5	10	20	50	100
σ_1	−10.4	200.0	251.0	294.6	369.3	545.1	770.9

The data generated by considering the three failure criteria are plotted below.

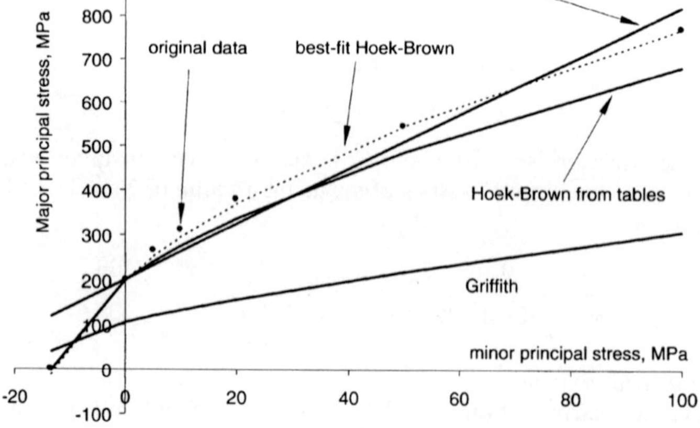

Assessment

As the plot above shows, the statistical fit of the Hoek–Brown criterion is the best of the three criteria. The curvilinear nature of the results means that the Mohr–Coulomb criterion will never be a good fit over the entire stress range, and the fundamental behaviour of rock over large stress ranges means that this is a general conclusion.

Q6.9 In 1926, in a paper on tensile tests for cotton yarns (Pierce, 1926[4]), Pierce stated, "it is a truism, of which the mathematical implications are of no little interest, that the strength of a chain is that of its weakest link". What is the relevance of this statement to the tensile strength of intact rock?

A6.9 The statement is relevant because intact rock failure is initiated by flaws in the rock microstructure. There is a distribution of such flaws (microcracks and grain boundaries) and so a larger specimen contains a larger sample of flaws than a smaller specimen, meaning that there is a greater probability that the large sample will contain the largest flaw.

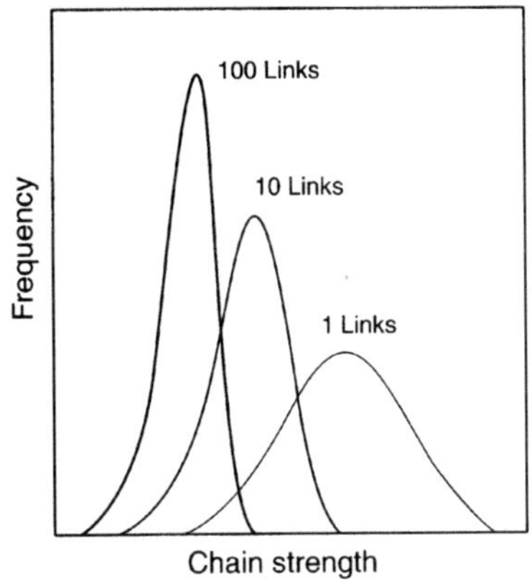

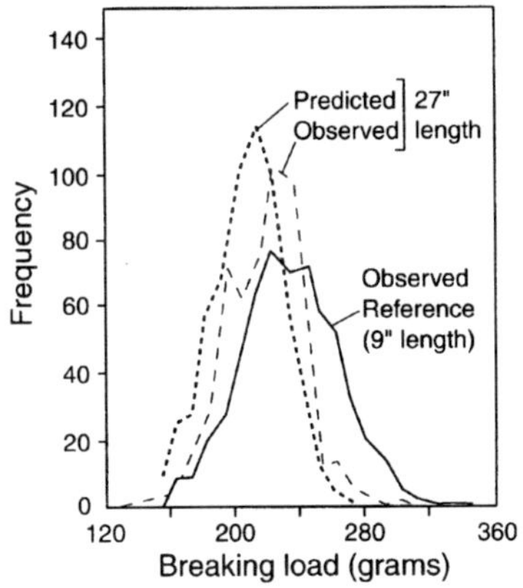

In terms of the chain strength, two of Pierce's diagrams are shown above. The left-hand one indicates how the chain strengths will vary as a function of the number of links, and the right-hand one presents tensile strength results for two different lengths of cotton yarn. The same phenomenon applies to the tensile strength testing of different sized specimens of intact rock.

We note that

(a) when testing many rock specimens under the same conditions, there will be a distribution of tensile strengths, i.e. a within-test statistical variation,
(b) on average, larger specimens will exhibit lower tensile strengths, i.e. a within-test scale effect, and
(c) different methods of testing the tensile strength (see diagram on the next page) will give different results, because there will be different critically stressed volumes in each test specimen, i.e. a between-test variation. In the direct tension test, the whole specimen is subjected to the same tensile stress, and so the whole volume is critically stressed. In the ring test, only the zones directly above and below the hole are subjected to the tensile stress.

[4] Pierce F. T. (1926) Tensile tests for cotton yarns, V. The weakest link, theorems on the strength of long and composite specimens. *J. Tex. Inst.*, **17**, 355–368.

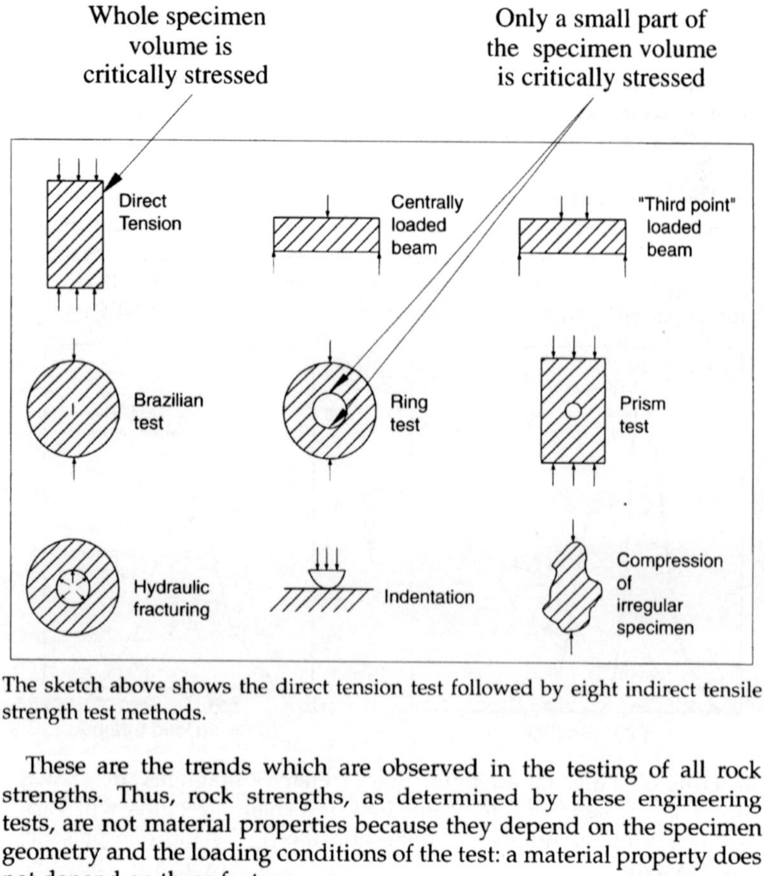

The sketch above shows the direct tension test followed by eight indirect tensile strength test methods.

These are the trends which are observed in the testing of all rock strengths. Thus, rock strengths, as determined by these engineering tests, are not material properties because they depend on the specimen geometry and the loading conditions of the test: a material property does not depend on these factors.

Q6.10 The two marble panels illustrated below are from the Greek Parthenon frieze which depicts the four-yearly procession of the 'Great Panathenaia'. They were sculpted in light relief[5] under the direction of Pheidias and placed in position around 450 BC. One panel is from the north side of the Parthenon and one panel is from the south side. Which is which?

A6.10 The Parthenon is situated in Athens in Greece, at a latitude of 38°N. At such a latitude, the southern side of a building is subjected to more intense light, temperature gradients, wind and rain than the northern side. After about 2450 years of such exposure, we would expect rock facing south to be more severely weathered than rock facing north. Thus, Panel A is from the south frieze and Panel B is from the north frieze.

[5] Howard Staunton used these sculptures as the source for his 19th century design of the knight in the classic Staunton chess pieces.

Panel A

Panel B

6.3 Additional points

Although a failure criterion based on the physical mechanisms of failure might be expected to exist, this has not yet been developed because of the structural complexities of both intact rock and rock masses, and the progressive nature of failure. Because the development of such a criterion is not yet practical, the empirically based failure criteria expressed in terms of stress are now used ubiquitously.

The empirical Hoek–Brown criterion has become the most widely used failure criterion for practical use. The criterion relates to the rock mass, i.e. the intact rock plus the fractures, and we will discuss it further in Chapter 8. Three journal publications that review the use of the criterion for various projects and comment on the reliability of the estimates are as follows:

Hoek E. and Brown E. T. (1997) Practical estimates of rock mass strength. *Int. J. Rock Mech. Min. Sci.*, **34**, 8, 1165–1186.

Hoek E. (1998) Reliability of Hoek–Brown estimates of rock mass properties and their impact on design. *Int. J. Rock Mech. Min. Sci.*, **35**, 1, 63–68.

Hoek E. (1999) Putting Numbers to Geology — An Engineer's Viewpoint. *Q. J. Eng. Geol.*, **32**, 1–19.

7 Fractures and hemispherical projection

7.1 Natural, pre-existing fractures

During its geological history a rock mass has been lithified and subjected to a variety of tectonic forces and fluid pressures. When the naturally applied stresses, strains and fluid pressures cause failure, the rock fractures in a brittle or semi-brittle manner. Initially, one or two **sets of fractures** develop. The fractures in each of the initial sets are usually planar, persistent and parallel because they were caused by tensile stresses and hydraulic pressures. As subsequent geological disturbances occurred, especially the major orogenic events, a suite of fracturing arrays was introduced into the rock mass by tensile, compressive and shear stresses (Fig. 7.1). The exact fracturing mode or combination of modes in each case would have been a function of the applied stresses and the fractures already present. Moreover, ductile rock displacements resulting in folding will also have occurred and these will have complicated the picture.

Thus, when today an engineer surveys a rock mass, it contains fractures. The **fracture pattern** can be either clear or rather daunting from the viewpoints of structural geology analysis and engineering charac-

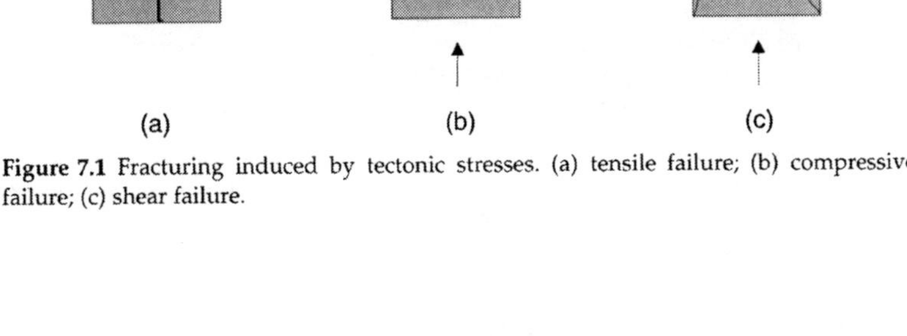

Figure 7.1 Fracturing induced by tectonic stresses. (a) tensile failure; (b) compressive failure; (c) shear failure.

Figure 7.2 (a) The Niagara dolomite at Lannon, Wisconsin, USA, with clear horizontal bedding and perpendicular vertical fractures. Note that the foreground vegetation is growing in the filling of the major joints.

terization, as illustrated by the examples in Fig. 7.2, showing quarry faces and associated line drawings. In the Fig. 7.2a case, the dolomite fractures are clear, and even the shape of the quarry faces occurs because of the dominant perpendicular vertical joints. In the Fig. 7.2b case, the fracturing is more intractable.

It is important, however, in all cases to be able to characterize the **geometry and mechanical properties** of the fractures because they are

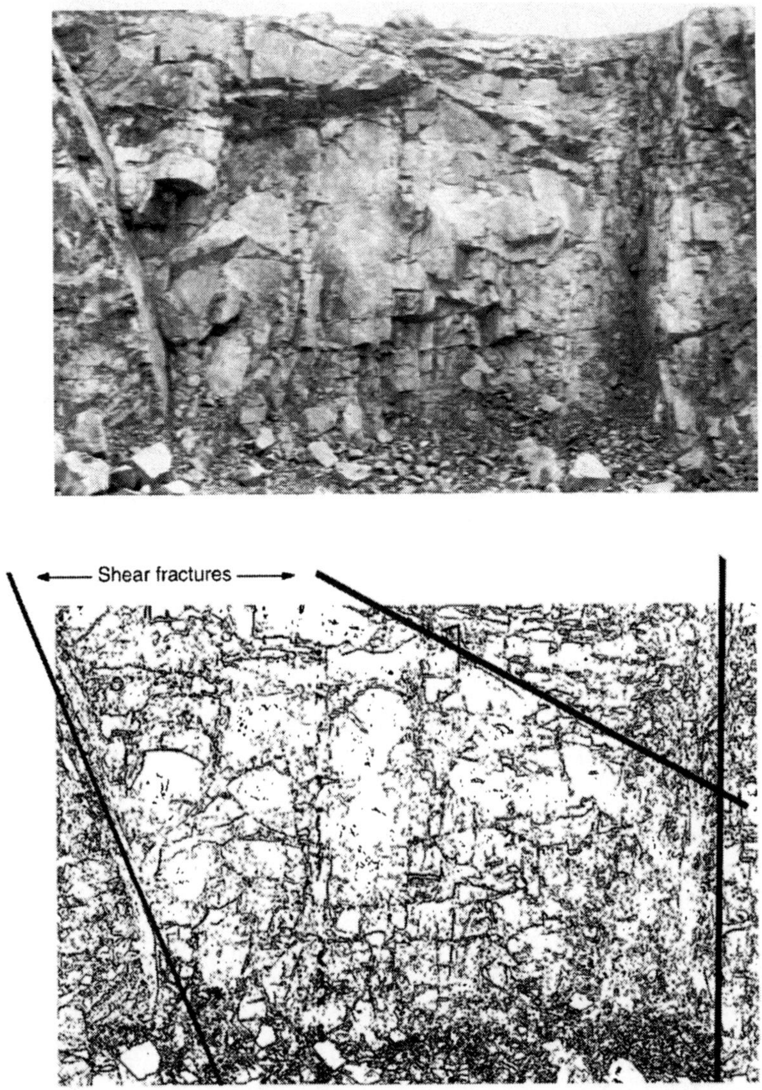

(b)

Figure 7.2 (b) The Mountsorrel granodiorite, Leicestershire, UK, with complex fracturing. Note the vertical fracturing and the presence of sub-vertical and transverse shear fractures.

the mechanical weaknesses in the rock mass and they have a significant effect on the rock mass deformability, strength, failure and permeability. Moreover, failure often occurs along the pre-existing fractures.

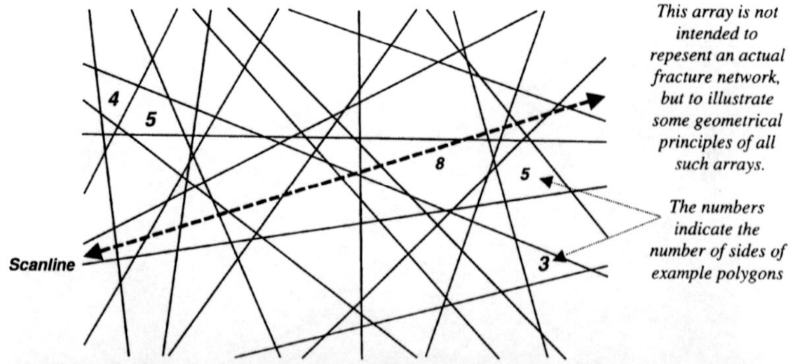

Figure 7.3 Array of lines illustrating two geometrical principles: there is a relatively large number of triangular areas compared to the number of other polygons, and there are more shorter lengths of intact rock intersected by the scanline than longer lengths.

Although the fracture[1] patterns can appear daunting (as in Fig. 7.2b), they exist because of definite mechanical and hydraulic principles: it is this fact that enables structural geologists to interpret their origin. Also, we have the necessary tools for engineering characterization, i.e. measurement equipment on site, understanding of the geometrical and mechanical features, and associated graphical and computer techniques. Thus, given a fracture array, such as one of those in Fig. 7.2, the geotechnical engineer wishing to know the frequency and orientation of the fractures, the mechanical properties of typical fractures and the mechanical and hydrogeological properties of the whole-rock mass does have the necessary tools available.

It is helpful to understand aspects of the **basic geometry** associated with the rock fracture arrays. Assume that on a sheet of paper 20 straight lines have been drawn, each at a random orientation, right across the page (see Fig. 7.3), representing rock fractures. Then assume that a straight line is drawn at any orientation through the array (the dashed line in Fig. 7.3), representing a sampling scanline on a fractured rock exposure or a borehole through a fractured rock mass. We can consider first all the areas formed by the fracture lines: these are equivalent to cross-sections through the rock blocks formed by the fractures. Some areas will be triangles, some quadrilaterals, some five-sided figures, and so on. It is possible to find the mean number of sides that all the areas have, both for a specific array and theoretically for all such arrays.

The number of sides of all the closed polygons in the fracture diagram in Fig. 7.3 was counted and a histogram (Fig. 7.4) prepared.

---

[1] *Note.* The terms 'fracture' and 'discontinuity' are both used to denote the pre-existing bedding planes, faults, shears, joints and fissures in a rock mass. We use the term 'fracture' without any structural geology or engineering genetic connotations, to indicate a discontinuity in the mechanical rock continuum. A fracture has little or no tensile strength. If we are discussing an engineering-induced fracture, rather than a natural fracture, this will be made clear.

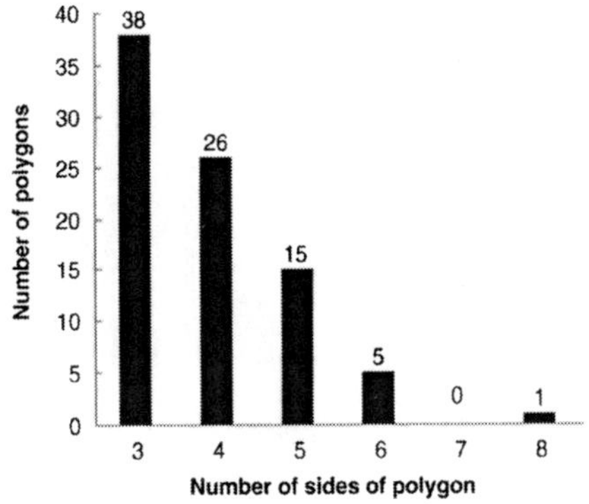

Figure 7.4 Histogram of polygons in Fig. 7.3.

The mean number of sides of the polygons is given by

$$\frac{(3 \times 38 + 4 \times 26 + 5 \times 15 + 6 \times 5 + 8 \times 1)}{(38 + 26 + 15 + 5 + 1)} = \frac{331}{85} = 3.9 \text{ sides.}$$

The theoretical answer is exactly 4, found from the theory (Miles, 1964[2]) of 'Poisson flats', which is beyond the scope of this book. However, for the average value to be equal to 4, there have to be many triangles present, see Fig. 7.4. This provides an indication of why there are many triangular areas bounded by fractures on a rock exposure.

In a similar way, a histogram of the lengths along the scanline in Fig. 7.3 is shown below in Fig. 7.5, together with the shape of the negative exponential distribution. The intact rock lengths are relative to a total scanline length in Fig. 7.3 of 23.4 units. Despite the small sample size, the negative exponential trend is evident in this example.

Thus, the existence of **many triangular areas** bounded by fractures on a rock exposure, or **many small lengths** along a scanline is a function of the basic geometry and does not necessarily indicate anything about the fracture genesis. Similarly, the negative exponential distribution of lengths of scanline between fractures, or lengths of intact rock in borehole core, occurs because the superimposition of a series of spacing distributions (in this case, the successive fracturing events through geological time) tends to a negative exponential distribution, whatever the types of original spacing distributions[3].

Knowing that the probability density distribution of intact lengths, x, can be well approximated by the negative exponential distribution,

[2] Miles R.E. (1964) Random polygons determined by random lines in a plane. *Proc. Natl. Acad. Sci. USA*, **52**, 901–907.

[3] This was conjectured, and proved for certain cases, by Karlin and Taylor in their 1973 book, *A First Course in Stochastic Processes*, p. 221, providing a theoretical basis for the existence of negative exponential distributions of spacings in rock masses. For a more advanced treatment of the geometry and engineering assessment of fracture occurrence, the reader is referred to the paper by Zhang Lianyang and Einstein H. H. (2000), Estimating the Intensity of Discontinuities, *Int. J. Rock Mech. Min. Sci.*, 37, 819–837.

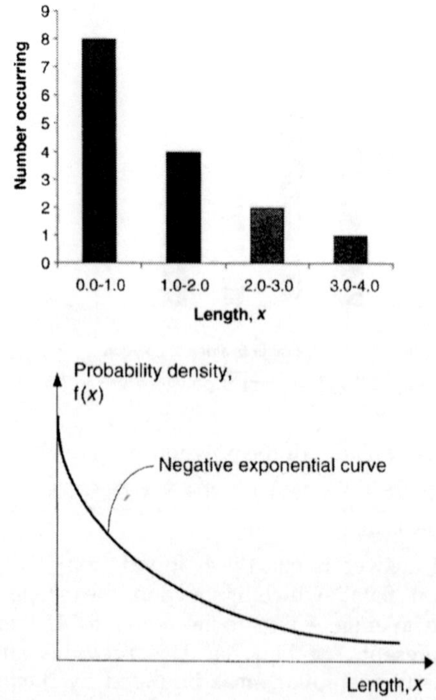

Figure 7.5 Distribution of lengths between fracture intersections along the scanline in Fig. 7.3.

$f(x) = \lambda e^{-\lambda x}$ where λ is the mean frequency, the mean and standard deviation values for the intact rock lengths can be established. The definition of the mean, $\bar{x}$, using probability theory for continuous distributions is $\bar{x} = [\int x f(x) dx]/[\int f(x) dx]$. In the case of the negative exponential distribution with $f(x) = \lambda e^{-\lambda x}$, with the limits of integration being taken as zero and infinity, and knowing that the integral $\int_0^\infty f(x) dx = 1$ (because the total probability is equal to unity), we find that $\bar{x} = \int_0^\infty x\lambda e^{-\lambda x} dx = 1/\lambda$.

The variance, σ^2, of a probability distribution is defined as $\int (x - \bar{x})^2 f(x) dx$, so for the case of a negative exponential distribution we have

$$\sigma^2 = \int_0^\infty (x - (1/\lambda))^2 \lambda e^{-\lambda x} dx$$

which reduces to $1/\lambda^2$.

Thus, the standard deviation[4], σ, is also $1/\lambda$. This means that **for the negative exponential distribution the mean and standard deviation are**

[4] Note that the symbols used here for the mean and standard deviation of a population are $\bar{x}$ and σ. Thus, the symbol σ is used in this book for both a stress component and the standard deviation of a statistical population. Because σ is used here in a specific statistical context, there should be no confusion — we will not be discussing the standard deviation of stress components.

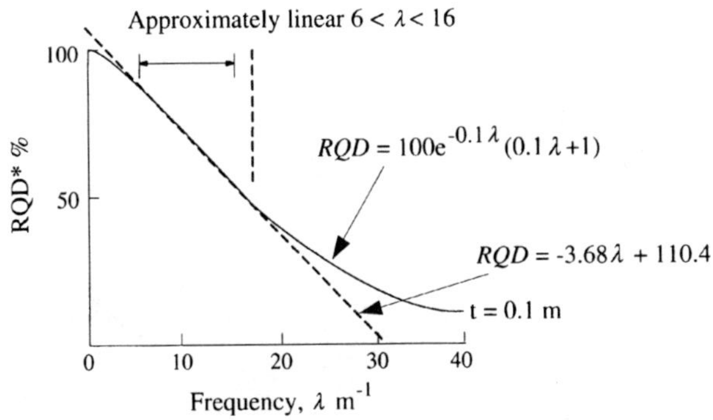

Figure 7.6 Theoretical relation between *RQD* and fracture frequency.

equal, with a value equal to the reciprocal of the fracture spacing. Along a scanline in the field, the sample mean and sample standard deviation are often found to have similar values, especially when 50 or more fractures have been intersected.

We emphasize that the negative exponential distribution is not being invoked because it characterizes a random or Poisson occurrence of fractures, but because it is the distribution to which a series of superimposed spatial distributions converges. Because most rock masses contain sets of fractures from a number of different fracturing episodes, and these sets can all have different distributions, the negative exponential distribution is a good description of the overall distribution.

The **Rock Quality Designation (*RQD*)** for a fractured borehole core or along a scanline such as that in Fig. 7.3 is defined as $100\sum_{i=1}^{N} X_i/L\%$, where X_i = intact core lengths > 0.1 m, N is the number of these in the core, and L is the total core length. This is the percentage of core length containing pieces greater than 0.1 m long. For the case of the usual *RQD* threshold value of 0.1 m, and with the assumption of a negative exponential distribution of intact rock lengths, it can be shown that the Rock Quality Designation (*RQD*) is related to the mean fracture frequency, λ, as $\boldsymbol{RQD = 100e^{-0.1\lambda}(0.1\lambda + 1)}$.

Fig. 7.6 illustrates the relation between *RQD* and fracture frequency and indicates that there is an approximately linear relation for $\lambda = 6\ \text{m}^{-1}$ to $16\ \text{m}^{-1}$. The equation of this line is $RQD = -3.68\lambda + 110.4$.

The *RQD* concept with a threshold value of 0.1 m can be generalized for an arbitrary threshold, t. Then the relation between *RQD* and t is $\boldsymbol{RQD = 100e^{-t\lambda}(t\lambda + 1)}$.

There are many characteristics of rock mass fractures that are important for engineering rock mechanics. The International Society for Rock Mechanics (ISRM) has recommended the measurement of the **ten fracture characteristics** illustrated in Fig. 7.7. These are: orientation, spacing, persistence, roughness, wall strength, aperture, filling, seepage, number of sets, and block size (Barton, 1978; Hudson, 1989) [5].

[5] Barton N. 1978. Suggested Methods for the Quantitative Description of Discontinuities

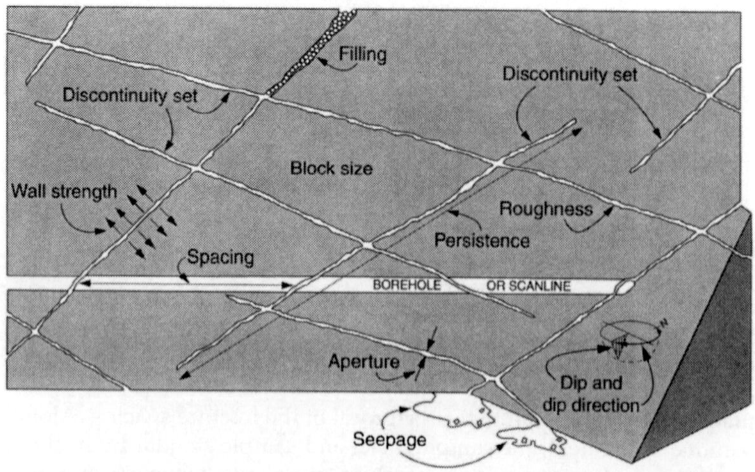

Figure 7.7 Important fracture characteristics for engineering rock mechanics.

One of the most important characteristics is the extent or **persistence** of the fractures because this governs whether the rock mass can act as a series of discrete blocks and whether water can travel through the fracture array. However, the estimation of fracture persistence can be subject to severe measurement bias; in particular, the distribution of tracelengths of fractures intersected by a scanline is not the same as the actual population of fracture tracelengths. As a scanline traverses a rock mass, it is more likely to sample fractures with longer tracelengths than with shorter tracelengths. If the fracture traces in Fig. 7.3 have the same location and orientation but are limited in their extents, a fracture array such as that shown in Fig. 7.8 might result. It is evident from this array that a borehole or scanline is more likely to intersect the longer traces than the shorter ones, with the result that the actual and sampled distributions of tracelengths are different, because the longer tracelengths are preferentially sampled.

If the actual distribution of tracelengths is negative exponential, the sampled distribution of tracelengths (of the fractures intersected along a scanline) will be of a different form, as indicated in Fig. 7.9. Note that the greatest discrepancy occurs for a tracelength close to zero, because the scanline will miss almost all of them. Needless to say, sampling surveys should take the sampling errors into account[6].

In this chapter, we will concentrate on the fracture frequency and geometrical aspects of fractures. The mechanical properties of fractures are included in Chapter 8.

---

in Rock Masses. *Int. J. Rock Mech. Min. Sci. Geomech. Abstr.*, 15, 6, 319–368. The diagram of the ten fracture characteristics in Fig. 7.7 was first published in Hudson J. A. (1989) Rock Mechanics Principles in Engineering Practice'. CIRIA Report, Butterworths, London, 72 pp.

[6] References to appropriate software packages for this are included in Section 7.3.

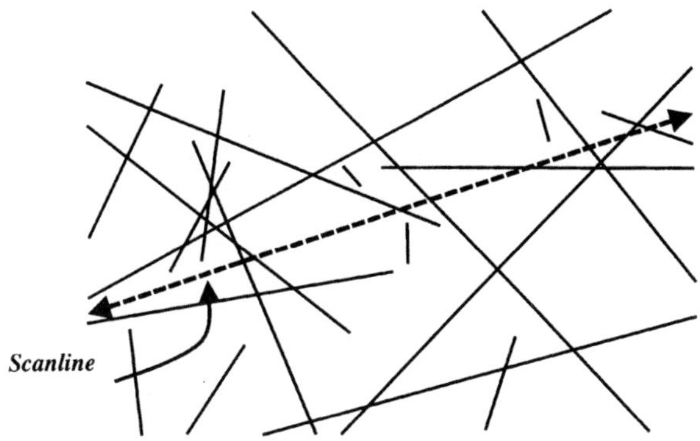

Figure 7.8 Longer fractures are more likely to be intersected by a scanline, borehole or underground excavation than shorter fractures.

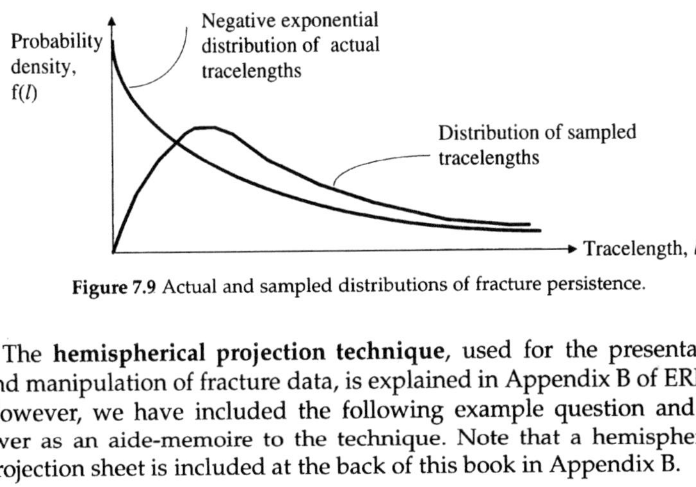

Figure 7.9 Actual and sampled distributions of fracture persistence.

The **hemispherical projection technique**, used for the presentation and manipulation of fracture data, is explained in Appendix B of ERM 1. However, we have included the following example question and answer as an aide-memoire to the technique. Note that a hemispherical projection sheet is included at the back of this book in Appendix B.

Example question and answer

Question **Two boreholes are drilled from the face of a quarry, one at an orientation[7] of 298/38, and the other at an orientation of 055/72. On a sheet of tracing paper over a hemispherical projection, plot the points corresponding to these boreholes, and then determine:**

(a) the orientation of the plane containing the two boreholes;
(b) the acute and obtuse angles between the two boreholes;
(c) the orientation of a borehole which bisects the acute angle; and
(d) the orientation of a borehole which is perpendicular to the two holes already drilled.

Answer In order to show the steps involved in solving typical hemispherical projection problems, we present diagrams here showing each

[7] The orientation is given as trend/plunge.

stage of the solution of the example question. When using the hemispherical projection method, a sheet of tracing paper is placed over the projection and rotated around a drawing pin protruding from underneath through both the projection and the tracing paper. The diagrams are of the working on the tracing paper used over the hemispherical projection, with the perimeter of the net included for reference.

(a) *The orientation of the plane containing the two boreholes*

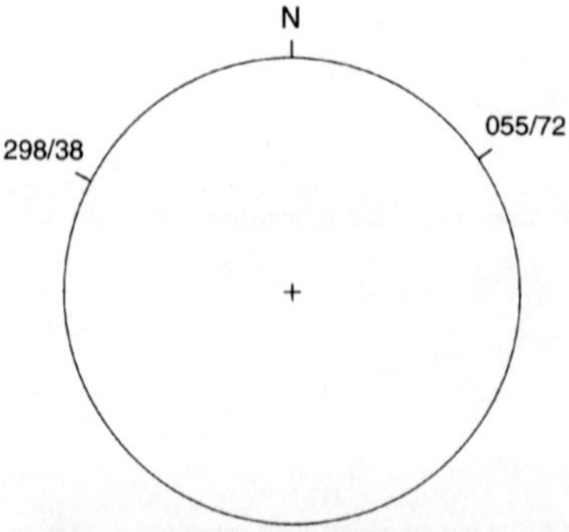

Mark the north point, draw tick marks on the periphery at the borehole azimuths of 298° and 055°, and write down the borehole orientations.

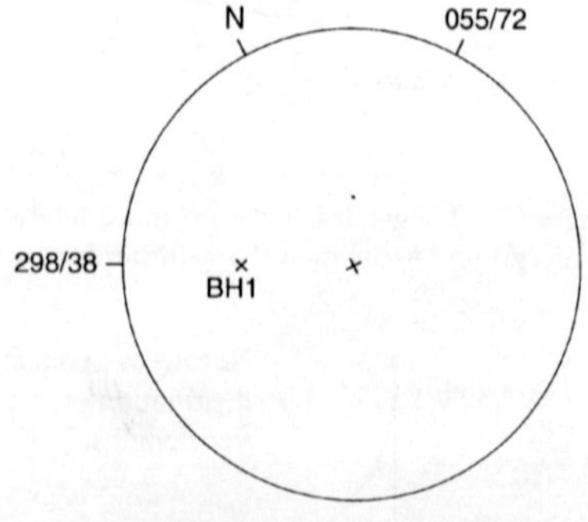

Rotate the tracing paper so that the tick for Borehole 1, BH1, is on the E–W line. Count in 38°, mark and label a cross for this borehole as BH1.

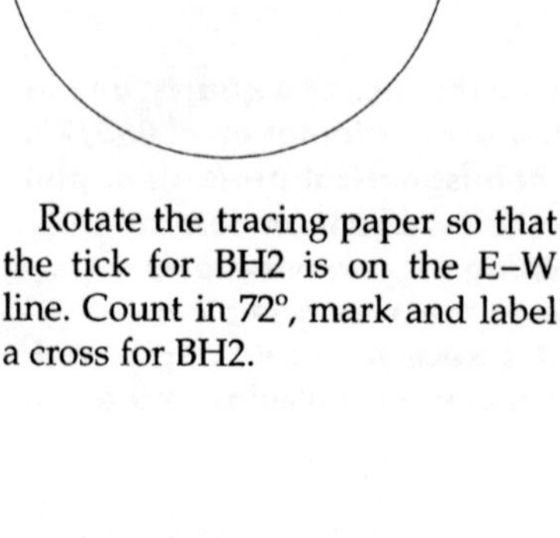

Rotate the tracing paper so that the tick for BH2 is on the E–W line. Count in 72°, mark and label a cross for BH2.

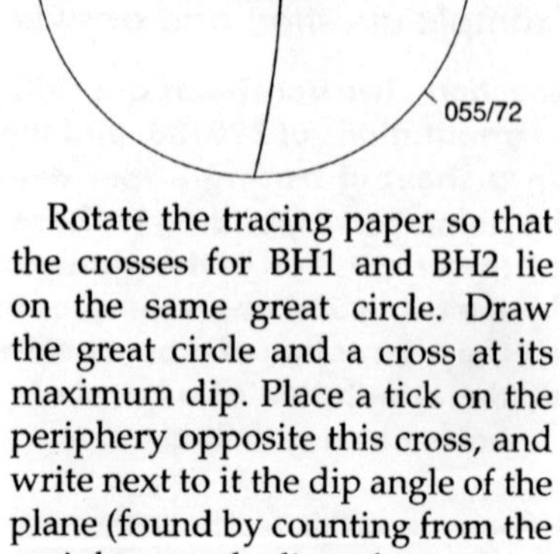

Rotate the tracing paper so that the crosses for BH1 and BH2 lie on the same great circle. Draw the great circle and a cross at its maximum dip. Place a tick on the periphery opposite this cross, and write next to it the dip angle of the plane (found by counting from the periphery to the line of maximum dip).

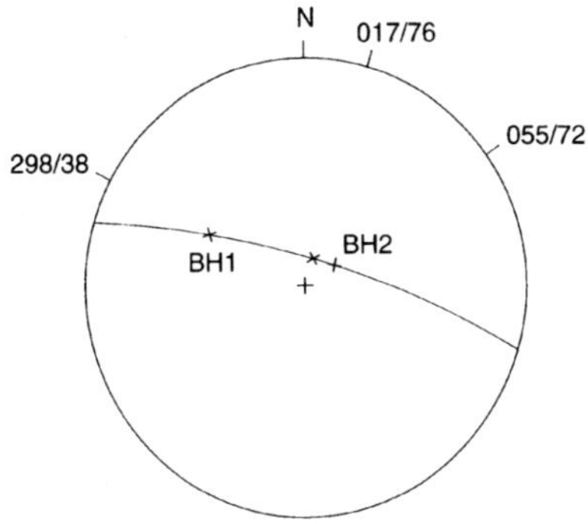

Finally, rotate the tracing paper back to north, measure the dip direction of the plane and write it in front of the dip angle. The orientation of the plane is therefore found to be 017/76.

(b) *Acute and obtuse angles between the boreholes*

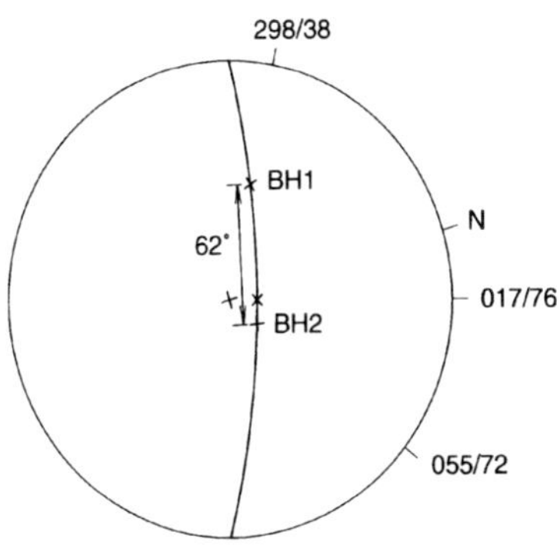

Rotate the tracing paper so that the plane containing BH1 and BH2 lies on the great circle. Measure along the great circle, using the small circles to count, to determine the angle between BH1 and BH2. It is 62°, and as this is less than 90° it is the acute angle. The obtuse angle is the complement of this, i.e. 118°. This figure can also be found by adding together the angles between BH1 and the upper periphery, and BH2 and the lower periphery.

(c) *Orientation of a borehole which bisects the acute angle*

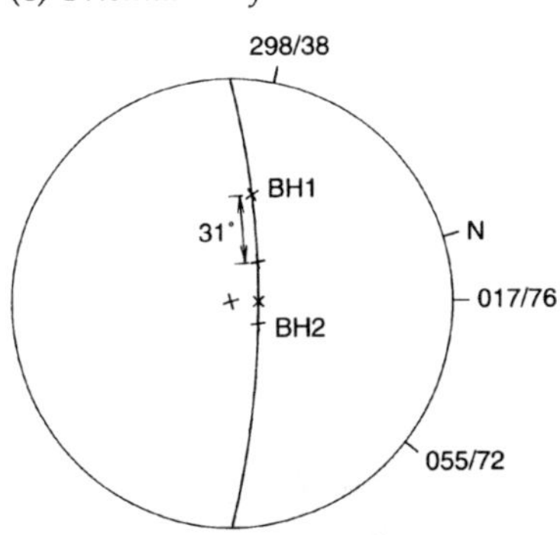

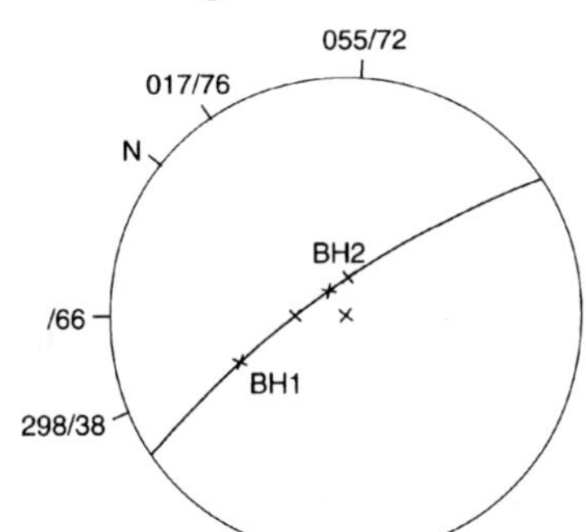

The acute bisector is midway between BH1 and BH2 along the great circle representing the plane 017/76. With the tracing paper rotated as shown, mark a cross 31° from either BH1 or BH2.

Rotate the tracing paper so that the bisector lies on the E–W line, mark a tick on the periphery and write the plunge of the line next to it.

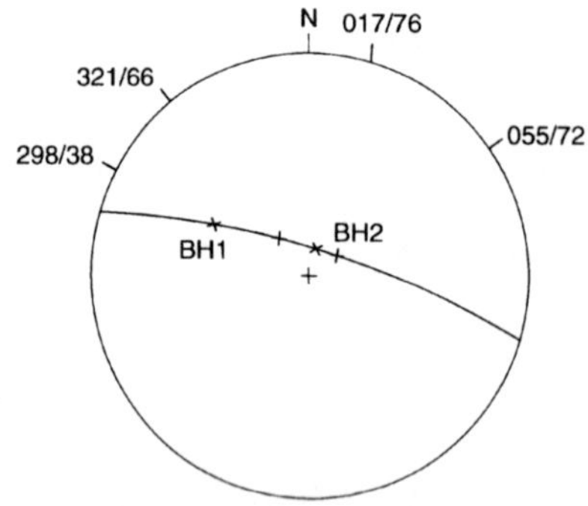

Finally, rotate the tracing paper back to N, and read off the trend of the bisector. The orientation of the bisector is therefore found to be 321/66.

(d) *Orientation of a borehole which is perpendicular to the two holes already drilled*

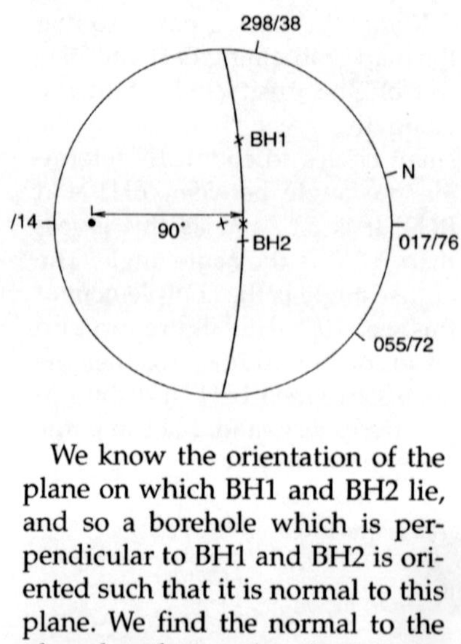

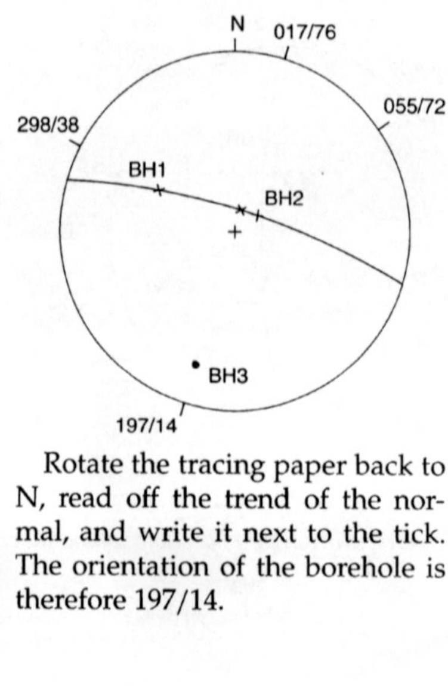

We know the orientation of the plane on which BH1 and BH2 lie, and so a borehole which is perpendicular to BH1 and BH2 is oriented such that it is normal to this plane. We find the normal to the plane by placing the plane on its great circle, counting across the E–W line 90° from the line of maximum dip, and marking the normal. Also mark a tick and write down the plunge of the normal.

Rotate the tracing paper back to N, read off the trend of the normal, and write it next to the tick. The orientation of the borehole is therefore 197/14.

7.2 Questions and answers: fractures and hemispherical projection

Q7.1 The overall *RQD* for 134 m of borehole core was found to be 58%.

(a) Compute estimates for the total number of pieces of core, and the total length of those pieces of core that could be expected to be greater than 0.1 m long.

(b) How many pieces of core could be expected to have a length greater than 0.2 m, and what is their mean length?

A7.1 (a) We know the length of core, and so in order to calculate the number of pieces we have to determine the fracture frequency (because the number of pieces, N, is found from $N = \lambda L$). As we know the *RQD* of the core and the threshold value at which this was measured, we can compute the frequency from the fundamental formula for *RQD*, i.e.

$$RQD = 100(\lambda t + 1)\exp(-\lambda t).$$

This formula for *RQD* is implicit in λ (which means we cannot rearrange it to leave λ by itself on one side of the equation), and so can only be solved iteratively (or by trial and error). Thus, with $RQD = 58\%$ and $t = 0.1$ m, we find that $\lambda = 14.344\ \text{m}^{-1}$. Having found a value for the frequency, we calculate the total number of fractures present in the core as

$$N_t = \lambda L = 14.344 \cdot 134 = 1922.2$$

and, as a part of a piece cannot exist, this result should be rounded down to 1922.

The definition of *RQD* is "the ratio of the total length of those pieces of core longer than 0.1 m to the total length of the core recovered, expressed as a percentage". In this case, the *RQD* is 58%, which directly represents the proportion of core composed of pieces longer than 0.1 m. As the overall length of the core is 134 m, the total length of these pieces is then $0.58 \times 134 = 77.72$ m.

(b) The number of pieces whose individual lengths are greater than 0.2 m is computed from the proportion of the total number that comprises pieces longer than 0.2 m. In turn, this proportion is found by computing the probability of finding an interval of core with no fractures over a length of 0.2 m.

If we sample the entire fractured core, the probability of finding a fracture must be 1. This is stated mathematically as the area under the probability density distribution, which is

$$\int_0^{\infty} \lambda \exp(-\lambda x)\,\mathrm{d}x = 1.$$

If we are only interested in a particular range of spacing values, then this integral is evaluated using limits that represent the range. So, for spacing values between 0 and some value b we have

$$\Pr(x \le b) = \int_0^{b} \lambda \exp(-\lambda x)\,\mathrm{d}x = 1 - \exp(-\lambda b)$$

and, if we are only interested in spacing values larger than b, then

$$\Pr(x > b) = 1 - \Pr(x \le b) = 1 - [1 - \exp(-\lambda b)] = \exp(-\lambda b).$$

In the case under consideration we have $b = 0.2$ m, giving

$$\Pr(x > 0.2) = \exp(-\lambda \cdot 0.02) = 0.057,$$

and from this we find that the number of pieces longer than 0.2 m is

$$N_{0.2} = 0.057 \cdot N_t = 0.057 \cdot 1922 = 109.5,$$

which rounds down to 109.

To compute the mean length of these pieces, we need to know their overall length. This is found by evaluating the *RQD* for a threshold value of 0.2 m,

$$RQD_{0.2} = 100(14.344 \cdot 0.2 + 1)\exp(-14.344 \cdot 0.2) = 22.0\%,$$

and then computing the overall length as

$$L_{0.2} = RQD_{0.2} \cdot L_t = (22.0/100) \cdot 134 = 29.4 \text{ m}.$$

The mean length of the pieces is then

$$\bar{x}_{0.2} = \frac{L_{0.2}}{N_{0.2}} = \frac{29.4}{109} = 0.27 \text{ m}.$$

This mean is much greater than the mean of all pieces (which is the reciprocal of the frequency, or 0.07 m), and so it is interesting to see how this mean length varies with the threshold we choose. The plot below shows this.

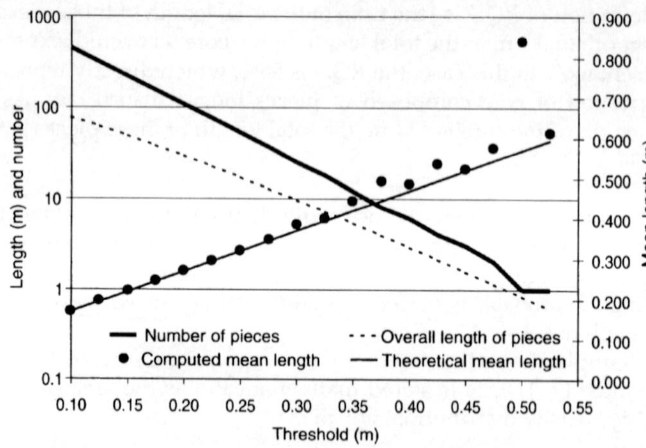

In this plot the mean length has been computed as shown above, but because the number of pieces has been rounded down, the points do not lie on the theoretically derived curve of $\bar{x} = t + (1/\lambda)$. Notice that the error becomes larger as the number of pieces reduces. This is because changing from 2 pieces to 1 is a much larger percentage change than from 500 pieces to 499. Also, the threshold cannot be increased beyond that shown, because the number of pieces would reduce to zero.

Q7.2 Based on a sample of 128 fracture spacing values which gave a mean spacing of 0.215 m, estimate the range of the population mean fracture spacing and frequency at the 80% confidence level. How many fractures should be in the sample for an error of ±10% at the 90% and 95% confidence levels?

A7.2 Using the negative exponential distribution, for which the mean and standard deviation are equal, the standard form of the error band around the mean, $\bar{x} \pm \sigma\varepsilon$, can be simplified to $\bar{x}(1 \pm \varepsilon)$. Here, σ is the standard deviation, not stress, and ε is the proportional error, not the strain!

At the 80% confidence level, the standard normal variable, z, is found from tables: $\phi(z) = 0.80 \Rightarrow z = 1.282$. This is the value of the normally distributed random variable, z, which has a mean of 0 and a standard deviation of 1, with 80% of the population within $\pm z$.

Hence

$$\varepsilon = \frac{z}{\sqrt{N}} = \frac{1.282}{\sqrt{128}} = 0.113.$$

The range for the mean spacing value is then given by $\bar{x}(1 \pm \varepsilon) = 0.215(1 \pm 0.113)$. This gives $0.191 \text{ m} \leq x \leq 0.239$ m, with the range of frequency being the reciprocal of these figures: $4.184 \text{ m}^{-1} \leq \lambda \leq 5.236 \text{ m}^{-1}$. Note that the mean of these two frequency values is 4.710 m^{-1}, whereas the frequency given by the mean spacing is $\lambda = 1/\bar{x} = 1/0.215 = 4.651 \text{ m}^{-1}$. The correct value for the mean frequency is the latter one (because the reciprocal of the average of two reciprocal values is not the same as the average of the two values).

To find the sample size required at the 90% and 95% confidence levels, we use the same technique to find N given ε.

Thus, at the 90% confidence level, we have

$$\phi(z) = 0.90 \Rightarrow z = 1.645$$

and hence

$$N = \left(\frac{z}{\varepsilon}\right)^2 = \left(\frac{1.645}{0.10}\right)^2 = 270.6.$$

This value has to be rounded up to the next integer (rounding down will give a sample size that is too small to reach the 90% confidence level), giving 271 fractures.

At the 95% confidence level, we have

$$\phi(z) = 0.95 \Rightarrow z = 1.960$$

and hence

$$N = \left(\frac{z}{\varepsilon}\right)^2 = \left(\frac{1.960}{0.10}\right)^2 = 384.2 \text{ or } 385 \text{ fractures.}$$

This shows how the size of the sample increases significantly as the required confidence level rises.

Q7.3 The mean fracture frequency in a vertical direction in a sandstone rock mass is 1.22 m^{-1}, and a total of 500 vertical 3 m long rockbolts are to be installed to stabilize the roof of an underground excavation in this rock mass. How many rockbolts would you expect to:

- **intersect no fractures;**
- **intersect less than 3 fractures; and,**
- **intersect more than 4 fractures?**

What length should the rockbolts be if 95% of them are required to intersect at least 3 fractures, i.e. extend to the fourth rock block back into the rock mass?

A7.3 If we assume that fracture occurrence is a stochastic process, then we can use the Poisson process which states that the probability of k events occurring in an interval, x, is given by $P(k, x) = e^{-\lambda x}(\lambda x)^k/k!$. In this question, $x = 3$ m (the length of a rockbolt), $\lambda = 1.22\ \mathrm{m}^{-1}$, and we have to determine the probability for different values of k.

Intersect no fractures
Here we have $k = 0$ (i.e. no fractures over the length of the rockbolt), and so $P(0, 3) = e^{-1.22\times3}(1.22 \times 3)^0/0! = 0.026$ (0! is conventionally taken as 1).
The number of bolts is then $500 \times 0.026 = 12.9$, which rounds down to *12 bolts*. We round down because we cannot have a fractional part of a bolt.

Intersect less than 3 fractures
Less than 3 fractures includes 0, 1 and 2 fractures, and so we have

$$P(<3, 3) = P\ (0, 3) + P\ (1, 3) + P\ (2, 3) = 0.026 + 0.094 + 0.172 = 0.292.$$

The number of bolts is $500 \times 0.292 = 146.1$, which rounds down to *146 bolts*.

Intersect more than 4 fractures
Theoretically, a bolt can intersect any number of fractures between zero and infinity. The probability of it doing so is 1 (i.e. certainty). This tells us that the probability of intersecting *more than* 4 fractures can be computed from the probability of intersecting 4 or fewer fractures as

$$\begin{aligned} P(>4, 3) &= 1 - P(\leq 4, 3) \text{ which expands as} \\ &= 1 - [P\ (0, 3) + P\ (1, 3) + P\ (2, 3) + P\ (3, 3) + P\ (4, 3)] \\ &= 1 - [0.026 + 0.094 + 0.172 + 0.210 + 0.192] \\ &= 1 - 0.695 = 0.305 \end{aligned}$$

The number of bolts $= 500 \times 0.305 = 152.5$, which rounds down to *152 bolts*.

What length of rockbolt is required?
The answer is given by the solution to the equation $0.95 = P(\geq 3, x)$. Expanding and rearranging

$$\begin{aligned} 0.95 &= P\ (\geq 3, x) = 1 - P\ (\leq 2, x) \\ &= 1 - P\ (0, x) - P\ (1, x) - P\ (2, x)\,. \end{aligned}$$

Substituting $w = \lambda x$ and expanding further gives

$$= 1 - \frac{e^{-w}\,(w)^0}{0!} - \frac{e^{-w}\,(w)^1}{1!} - \frac{e^{-w}\,(w)^2}{2!} = 1 - e^{-w} - e^{-w}w - e^{-w}\frac{w^2}{2}$$

$$= 1 - e^{-w}\left(1 + w + \frac{w^2}{2}\right).$$

This equation cannot be solved explicitly, and so we use trial-and-error. The result is $w = 6.296$, from which, using $\lambda = 1.22\ \text{m}^{-1}$, we obtain $x = 5.16$ m. This is a substantial increase on the previous length of 3 m.

Q7.4 A fault plane with orientation 234/56 has been discovered during a site investigation. Closer inspection shows that it has surface lineations which have a pitch of 78° measured from the northwest strike line. What is the trend and plunge of these lineations?

A7.4 The illustration below shows the final tracing paper diagram obtained when answering this question. The steps followed in its construction are given as follows.

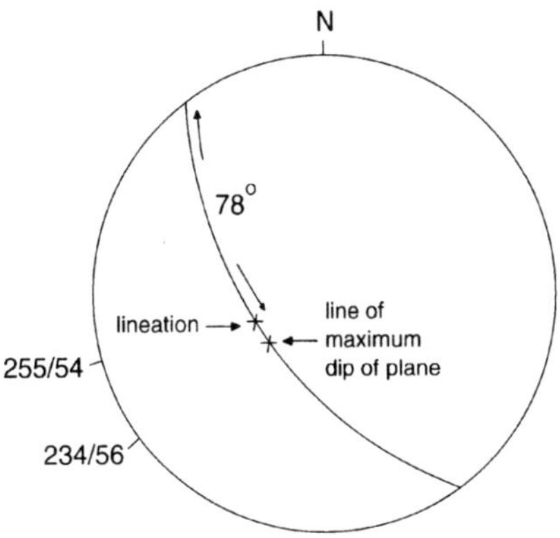

(1) Plot and label a tick mark for the fault plane. Rotate the tick mark to the east–west line, count in 56° and mark a cross for the line of maximum dip. Draw the great circle corresponding to the fault plane. Rotate the tracing paper back to north.

(2) The pitch of the lineation is measured from the northwest strike line. Rotate the tracing paper so that the northwest end of the great circle is north. Measure the pitch of the lineation (i.e. 78°) along the great circle, using the small circles as a scale and mark a cross on the great circle; this represents the orientation of the lineation.

(3) Rotate the tracing paper so that the lineation is on the east–west line, mark a tick on the periphery opposite this, read off and mark next to the tick the plunge of the lineation; it is 54°.

(4) Rotate the tracing paper to north, and read off the trend of the lineation; it is 255°. The orientation of the lineation (given as trend/plunge) is therefore 255/54.

Q7.5 The line of intersection between two planes trends approximately northwest and plunges at 38°. The orientation of one of the planes is 256/50, and the strike of the other is 132°. What is the trend of the line of intersection and what is the dip of the second plane?

A7.5 The illustration below shows the final tracing paper diagram obtained when answering this question. The steps followed in its construction are given as follows.

(1) Plot and label tick marks for the maximum dip of plane 1 and the strike of plane 2. Rotate the tick mark for plane 1 to the east–west line, count in 50° and mark a cross for the line of maximum dip. Draw the great circle corresponding to plane 1, and rotate the tracing paper back to north.

(2) Sketch a circular arc around the northwest quadrant of the tracing paper, centred on the centre of the tracing paper at a distance in from the periphery equal to the plunge of the line of intersection, 38°. This arc represents the locus of lines with a pitch of 38° and trend in the northwest quadrant.

(3) Rotate the tracing paper so that the intersection of the circular arc and the great circle to plane 1 is on the east–west line, and mark a tick on the periphery. This represents the trend of the intersection.

(4) Rotate the tracing paper so that the strike of plane 2 is on the north–south line, and draw the great circle that passes through the intersection. This is the great circle to plane 2. Mark a tick on the periphery opposite the line of maximum dip, measure the dip of plane 2 and mark it next to the tick. It is 80°.

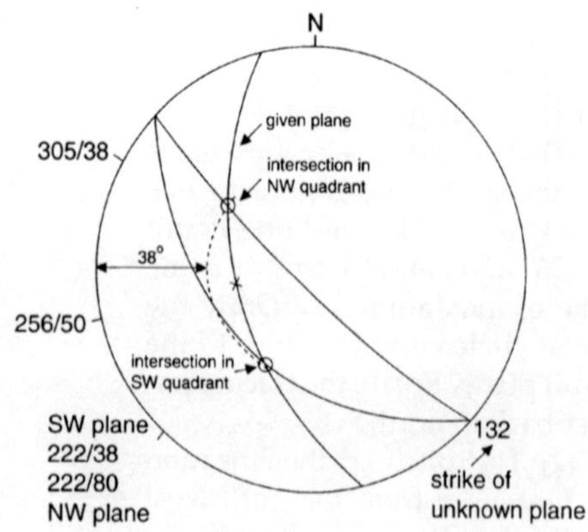

(5) Rotate the tracing paper back to north and read off the trend of the intersection and the dip direction of plane 2; this gives the answers 305° and 222°, respectively.

The orientation of plane 2 is then 222/80, and the orientation of the line of intersection of plane 1 and plane 2 is 305/38.

The diagram shows that there is a second line of intersection between the circular locus and the given plane, and the plane that passes through this has an orientation of 222/38. However, the intersection is formed in the SW quadrant, not the NW quadrant as stated, and so can be discounted.

Q7.6 Surveys have revealed that a rock mass contains 3 fracture sets, the dip directions/dip angles and fracture frequencies of which are 161/23 and 7.72 m^{-1}, 218/58 and 3.07 m^{-1}, and 100/75 and 5.34 m^{-1}.

(a) What will be the fracture frequency and mean length of the recovered pieces of core in:

- **a vertical borehole;**
- **a horizontal tunnel heading due north; and**
- **an inclined borehole with a trend/plunge of 280/35?**

Consider the engineering implications of this variation in terms of site investigation procedures and subsequent engineering design.

(b) Using the frequency values determined above, compute the theoretical *RQD* values that would be encountered in this rock mass using the customary threshold value of 0.1 m. Adopting a threshold value given by

$$t^* = \frac{2}{\lambda_{max} - \lambda_{min}} \ln\left(\frac{\lambda_{max}}{\lambda_{min}}\right)$$

will maximize the range of *RQD* values[8]. How do the *RQD* values computed with this threshold compare to the earlier values?

A7.6 (a) This question requires computation of fracture frequency in different directions. The fundamental equation to use (see ERM 1) is $\lambda_s = \sum_{i=1}^{n} \lambda_i |\cos\theta_i|$, where λ_s = the fracture frequency in a given scanline direction, the λ_i are the individual set frequencies of the n sets, and the θ_i are the angles between the set normals and the scanline.

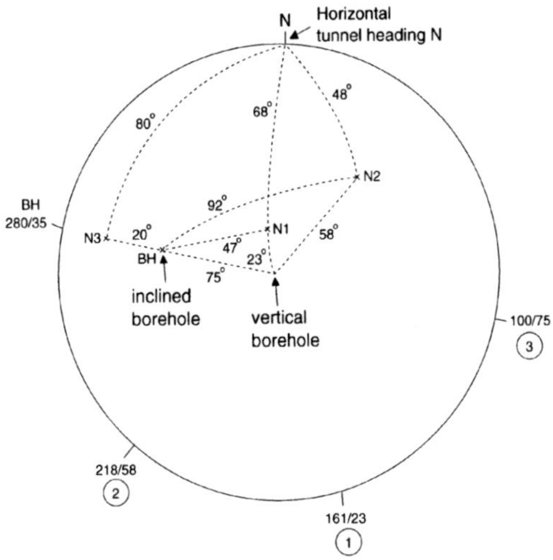

Firstly, we determine the angles between the normals to the three fracture sets and the three directions required in the question. For this, we can either use the hemispherical projection, or compute the angles vectorially. To find the angle between two lines on the hemispherical projection, plot the points representing the two lines, rotate the tracing paper so that the two points lie on a great circle, and then read off the angle where the small circles intersect the north–south line on the tracing paper. The figure shows how this is done for all of the angles required.

For a vertical borehole

The table below shows the required angle, the calculation of $\lambda|\cos\theta|$ for each set, and the sum of these contributions.

Set	λ	Angle between set normal and required direction	$\lambda\lvert\cos\theta\rvert$
1	7.72	23	7.106
2	3.07	58	1.627
3	5.34	75	1.382
Sum			10.115

[8] The method of choosing the optimal value of the *RQD* threshold value for optimizing the *RQD* sensitivity is given in Harrison J. P. (1999), Selection of the *RQD* threshold value in *RQD* assessments. *Int. J. Rock Mech. Min. Sci.*, **36**, 5, 673–685.

The frequency is then 10.115 m^{-1}, and the reciprocal of this is the mean length of the recovered pieces (because $\bar{x} = 1/\lambda$), i.e. 0.10 m.

For a horizontal tunnel heading north

Set	λ	Angle between set normal and required direction	$\lambda\lvert\cos\theta\rvert$
1	7.72	68	2.892
2	3.07	48	2.054
3	5.34	80	0.927
Sum			5.873

The frequency is then 5.873 m^{-1}, and the mean length of the recovered pieces is 0.17 m.

For a borehole trending 280/35

Set	λ	Angle between set normal and required direction	$\lambda\lvert\cos\theta\rvert$
1	7.72	47	5.265
2	3.07	92	0.107
3	5.34	20	5.018
Sum			10.390

The frequency is then 10.390 m^{-1}, and the mean length of the recovered pieces is 0.10 m (coincidentally similar to the vertical borehole value).

The engineering ramifications of the results are that, firstly, the fracture frequency in different directions through a rock mass can be significantly different. Secondly, this variation can be calculated if the set orientations and frequencies are known. Thirdly, measurements of fracture frequency in a vertical borehole will probably not correctly predict the fracture frequency in a horizontal tunnel (although the difference will depend on the properties of the fracture sets present).

(b) In order to use the equation for the threshold value that maximizes the *RQD* range, we need to know the maximum and minimum values of fracture frequency in the rock mass for the directions considered. On the basis of the results obtained earlier, we can say $\lambda_{max} = 10.39\ m^{-1}$ and $\lambda_{min} = 5.87\ m^{-1}$, from which we find

$$t^* = \frac{2}{\lambda_{max} - \lambda_{min}} \ln\left(\frac{\lambda_{max}}{\lambda_{min}}\right) = \frac{2}{10.39 - 5.87} \ln\left(\frac{10.39}{5.87}\right) = 0.25\ \text{m}.$$

We can now construct a table of frequency values and corresponding *RQD* values, using the relation $RQD = 100(\lambda t + 1)e^{-\lambda t}$. The results are as follows:

Frequency, m^{-1}	*RQD*, % ($t = 0.1$)	*RQD*, % ($t = 0.25$)
5.87	88.2	56.9
10.39	72.1	26.8
Range	16.1	30.1

This shows how, by adopting the larger RQD threshold, we have almost doubled the range of RQD in the rock mass. As a result of using this new threshold value, RQD would be a much more discriminatory measure of anisotropy than when the customary threshold is used.

With the discontinuity sets given, the complete variation in frequency in all directions in three dimensions is as shown on the hemispherical projection below. This shows that the global minimum frequency is 1.42 m^{-1} and the global maximum frequency is 12.48 m^{-1}, and these results indicate a threshold value of $t^* = 0.39$ m.

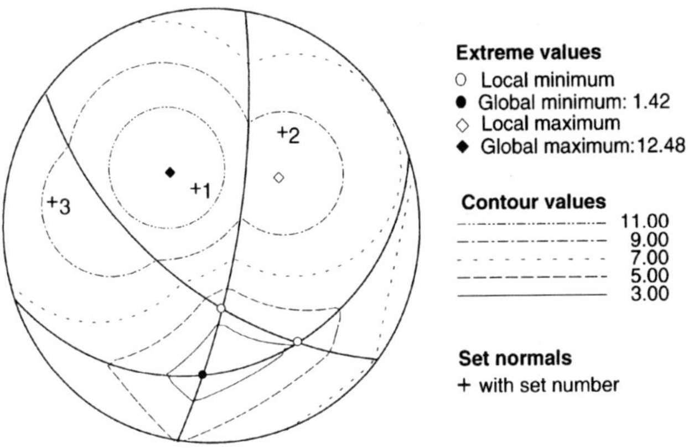

3-D variation of fracture frequency.

The two hemispherical projection plots below show how RQD varies with a threshold value of 0.1 m and a threshold value of 0.39 m. For the threshold value of 0.1 m, the global range is about 35%, whereas with a threshold value of 0.39 m the range is about 85%; an increase of nearly two and a half times. The contours in these plots show how the use of the higher threshold value has increased the discrimination of RQD in the region of the global frequency minimum.

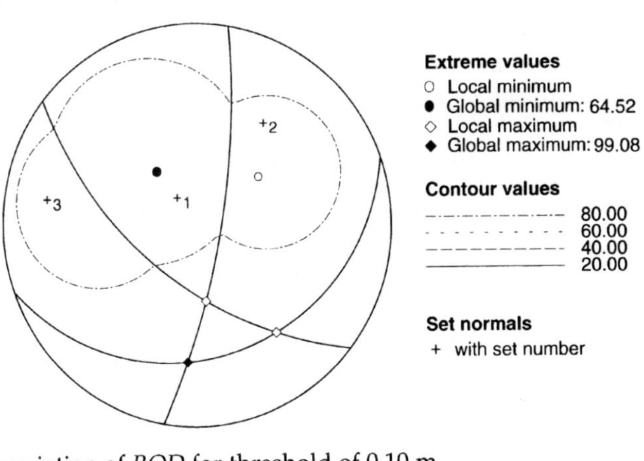

3-D variation of RQD for threshold of 0.10 m.

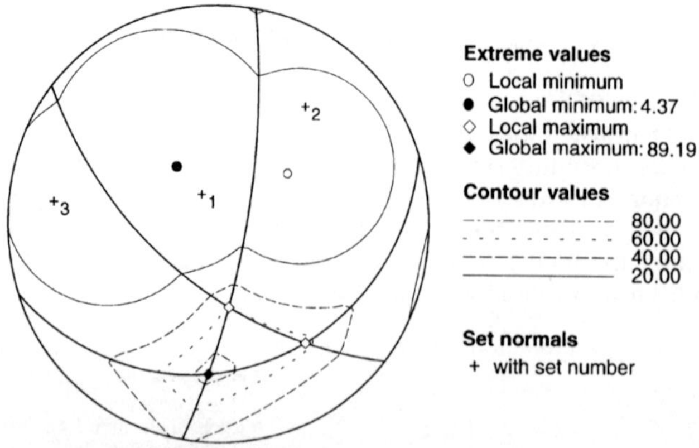

3-D variation of *RQD* for threshold of 0.39 m.

Q7.7 A rock mass is known to contain two sets of fractures, the orientations of which are 265/42 and 003/69. What borehole orientations will allow one to drill through the rock mass and intersect all of the fractures at an angle of 45° (measured between the borehole axis and the normals to the fractures)? What is the minimum angle that all fractures can be intersected at?

A7.7 The projection below shows the final solution to this question. The steps to follow are described below.

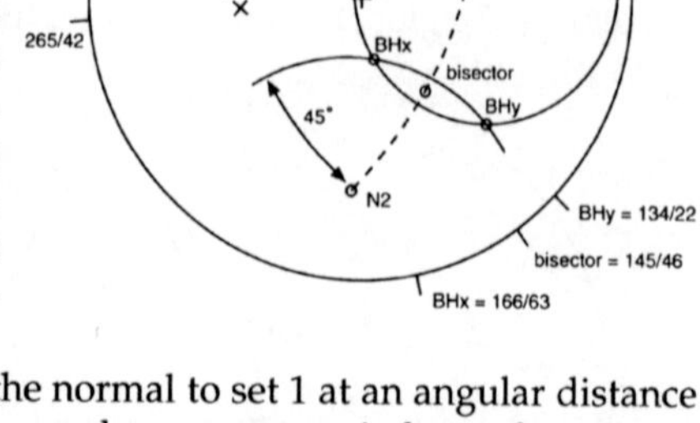

(1) Plot and label tick marks for the two sets. Rotate the tick mark for set 1 to the east–west line, count in 42° and mark a cross for the maximum dip of set 1. Count across the east–west line 90° and mark and label the position of the normal to set 1. Rotate the tick mark for set 2 to the east–west line, count in 69° and mark a cross for the maximum dip of set 2. Count across the east–west line 90° and mark and label the position of the normal to set 2.

(2) Construct a circle around the normal to set 1 at an angular distance of 45°. To do this, place the normal on a great circle, and count in either direction along the great circle an angle of 45°. Rotate the tracing paper a few degrees, place the normal on another great circle, and measure and mark two more points. Continue in this way until sufficient points have been marked to allow the circular locus around the normal to be drawn (note that the small circles on the projection net can be used as circle templates if the projection sheet is lifted off the drawing pin!).

(3) Using the normal to set 2, sketch those parts of the circle at an angular distance of 45° from this normal which intersect the circle around normal 1. Label the two intersection points of these circles BHx and BHy, rotate the tracing paper so that each is brought to the east–west line in turn, and read off the orientation of the lines. These points are the orientations of the two boreholes which will subtend an angle of 45° to both fracture sets, and are 166/63 and 134/22.

(4) Rotate the tracing paper so that normal 1 and normal 2 are on the same great circle. Measure the angle between the normals, 80°, divide by 2, and plot the bisector. Rotate the tracing paper so that the bisector is on the east–west line, and then read off its orientation: 145/46. This is the orientation of the borehole that makes the minimum angle to each fracture set, and this angle is 40°.

Q7.8 A petroleum reservoir is known to contain numerous fractures which are highly conductive. These fractures dip almost vertically in a northeasterly direction. To maximize production from the reservoir, the production wells are to be deviated to run sub-horizontally, so that they intersect the fractures as close to perpendicular as possible.

In order to determine the optimal orientation of the production holes, two test holes have been drilled to orientate the fractures. The orientations of these test holes within the reservoir are 011/09 and 087/03, and impression packers which were run down these holes show that in both holes the angle between the normal to the fractures and the hole axis is 40°. Determine the orientation of the fractures, and the required trend of the production holes.

A7.8 To find the orientation of the normal to the fractures, we draw circles at an angular distance of 40° away from each of the two borehole orientations, and locate the intersections of these circles. Only one of these two intersections will be at the appropriate orientation, but this will be apparent once we have found them.

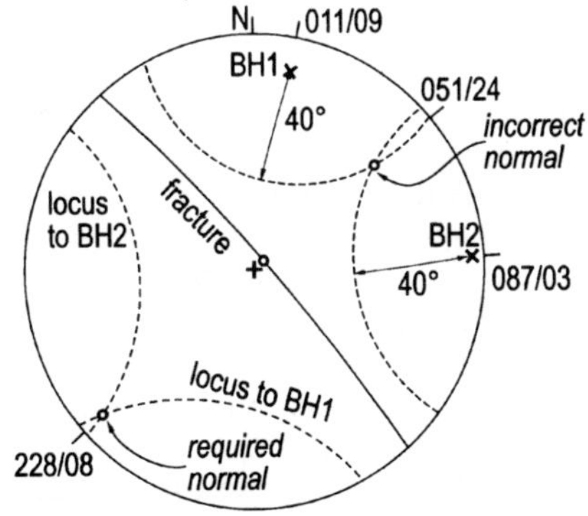

(1) Draw tick marks for the two boreholes, rotate each tick mark to the east–west line in turn, and mark the position of the boreholes on the tracing paper.

(2) For each borehole in turn, sketch as much as possible of a circle at an angular distance of 40°; the intersection of these two circles is in the northeast quadrant of the projection. Rotate the tracing paper so that the intersection is on the east–west line, draw a tick mark on the periphery and write next to it the plunge of the intersection, 24°.

(3) Rotate the tracing paper back to north and read off the trend of the intersection, 051°.

(4) A normal with an orientation of 051/24 measured represents a plane which dips towards the southwest at low angle (the orientation of the plane is 231/66). However, we know that the fractures dip towards the northeast and are almost vertical, which is clearly not the orientation just found. To resolve this problem, we must draw the portions of the circles around the boreholes which are at the west and south sides of the projection.

(5) Carefully sketch the remaining portions of the circles by counting 'across' the projection. This entails counting towards the periphery along a great circle, and then counting along the complementary great circle from the opposite side of the projection by the remaining angle, such that the total distance is 40°. When this is complete for both circles, the intersection will be found towards the southwest of the projection.

(6) Rotate the tracing paper so that the intersection is on the east–west line, mark a tick on the periphery and write down the plunge of the intersection, 08°.

(7) Rotate the tracing paper back to north and read off the trend of the intersection, 228°. Thus, the orientation of the normals to the fractures is 228/08 and hence the dip direction and dip angle of the fractures are 048/82. This accords well with the known orientation of the fractures: dipping towards the northeast at a steep angle.

(8) The trend of the production holes is that of the trend of the normals to the fractures, and is either 048° or 228°. Two orientations are possible, because the holes are sub-horizontal and can therefore run in two directions. In either case, the angle between the normals to the fractures and the axes of the boreholes will be 8°.

Q7.9 A length of core, from a borehole whose orientation is 143/68, contains a fracture plane of 204/47. The core has rotated through a clockwise angle (looking down the borehole) of 140° during retrieval. What will be the apparent orientation of the fracture as it emerges from the borehole?

A7.9 To solve this question, we incline the projection so that its centre coincides with the axis of the borehole. This allows us to apply the rotation, before considering the inclination and determining the orientation of the fracture.

(1) Mark ticks on the periphery for the plane and the borehole axis.

(2) Rotate each tick mark in turn to the east–west line, and mark the position of the borehole (BH) and the line of maximum dip of the plane (D). From this latter position, count across the projection by 90° to plot the position of the normal to the plane; call this position N.

(3) Rotate the tracing paper so that the borehole axis is on the east–west line, and incline the borehole to the centre of the projection; this represents an angular movement of 22°.

With the tracing paper in the same position, move the normal to the plane (i.e. point N) along its small circle in the same direction and by the same amount (22°) as the borehole inclination; call this new position N′. We have effectively inclined the projection so that the centre now represents the axis of the core, rather than the vertical.

(4) Rotate the tracing paper so that N′ is on the east–west line and mark a tick on the periphery. Measure the plunge of N′ as 30°. To apply the rotation undergone by the core, count round the periphery from this tick by the amount of rotation, i.e. 140° clockwise. Mark another tick. Rotate the tracing paper so that this new tick is on the east–west line and count in by the previously measured plunge of 30°; mark this point N'_R to indicate that it is an inclined normal which has been rotated. Marking the circular arc actually followed in rotating from N′ to N'_R helps to make the procedure clearer.

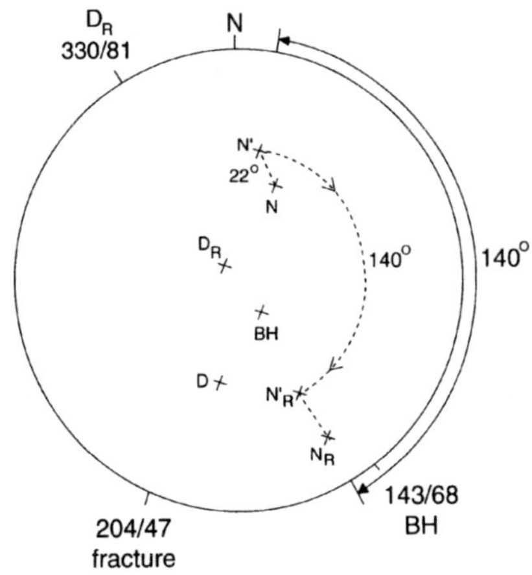

(5) Rotate the tracing paper so that the borehole is on the east–west line, and remove the inclination. This means that N′ moves 22° along its small circle to N, the point at the centre of the tracing paper moves 22° back to BH, and N'_R moves along its small circle 22° to N_R. With this step, we have effectively inclined the projection back to vertical.

(6) Rotate the tracing paper so that N_R is on the east–west line, count across the projection 90° to D_R and mark a cross. Mark a tick on the periphery, and write next to it the measured dip of D_R, 81°.

(7) Rotate the tracing paper back to north, and write next to the tick for D_R its measured dip direction, 330°. The orientation of the fracture plane is then 330/81.

Although relatively time consuming, this process is useful for determining the orientations of fractures from measurements made on core lying in a core box. A succession of inclinations and rotations can be used to effectively take a horizontally lying core box and position it in the attitude of the borehole itself.

Q7.10 A borehole of orientation 136/55 intersected six fractures belonging to the same sub-parallel set, the orientations of which were 201/39, 213/50, 215/63, 230/52, 247/42 and 253/28. Compute the mean orientation of the set,
(a) without accounting for sampling bias, and
(b) accounting for sampling bias

A7.10 This is an application of the process whereby we compute the mean orientation as the mean of the components of the normals to the fractures. Problems such as these are conveniently solved using a computer spreadsheet.

For the first part of the calculation, the various formulae used are: $\alpha_n = \text{mod}(180 + \alpha, 360)$; $\beta_n = 90 - \beta$; $\alpha_n^{\text{rad}} = \alpha_n{}^{\circ} \times (\pi/180)$;

$\beta_n^{rad} = \beta_n{}^{o} \times (\pi/180)$; $l = \sin(\alpha_n)\cos(\beta_n)$; $m = \cos(\alpha_n)\cos(\beta_n)$; and $n = \sin(\beta_n)$. The mod function is a 'remainder' function, and ensures that the result lies between 0° and 360°. Notice that all angles are converted to radians, as these are the units for trigonometrical functions in most computer packages. The calculations are shown in the table below.

Borehole:

Degrees		Radians		Components		
α_s	β_s	α_s	β_s	l_s	m_s	n_s
136	55	2.374	0.960	0.398	−0.413	0.819

Fracture data:

Degrees				Radians		Components					Weighted comp.		
α	β	α_n	β_n	α_n	β_n	l	m	n	$\cos\theta$	w	l'	m'	n'
201	39	21	51	0.367	0.890	0.226	0.588	0.777	0.484	2.066	0.466	1.214	1.606
213	50	33	40	0.576	0.698	0.417	0.642	0.643	0.428	2.338	0.975	1.502	1.503
215	63	35	27	0.611	0.471	0.511	0.730	0.454	0.274	3.645	1.863	2.660	1.655
230	52	50	38	0.873	0.663	0.604	0.507	0.616	0.536	1.866	1.127	0.945	1.149
247	42	67	48	1.169	0.838	0.616	0.261	0.743	0.746	1.340	0.825	0.350	0.996
253	28	73	62	1.274	1.082	0.449	0.137	0.883	0.846	1.183	0.531	0.162	1.044
Sum						2.822	2.865	4.116			5.787	6.834	7.952
Mean						0.470	0.478	0.686			0.964	1.139	1.325
Normalized mean						0.490	0.498	0.715			0.483	0.571	0.664

The mean of the components as calculated is not an orientation vector, as its magnitude is not equal to unity: $\sqrt{0.470^2 + 0.478^2 + 0.686^2} = 0.959$. The final row normalizes the mean by dividing the components by 0.959 so that the magnitude is equal to unity: $\sqrt{0.490^2 + 0.498^2 + 0.715^2} = 1.000$.

Because sampling bias correction is not required in the first part of the calculation, having found the components of the mean we calculate the orientation of the mean using the formulae $\alpha_n = \text{atan2}(m, l)$ and $\beta_n = \text{asin}(n)$. The result is $\alpha_n = \text{atan2}(0.498, 0.490) = 44.5°$ and $\beta_n = \text{asin}(0.715) = 45.6°$. Converting these to a mean dip direction and dip angle gives an orientation of 224.5/44.4, or more appropriately 225/44.

To compute the corrected mean, we weight each fracture orientation according to the cosine of the angle it makes to the borehole axis. This cosine is given as the scalar product of the normal vector and the borehole vector, i.e. for fracture i it is $\cos\theta_i = l_i \cdot l_s + m_i \cdot m_s + n_i \cdot n_s$, and the reciprocal of this is the weighting factor, i.e. $w = 1/\cos\theta$. The weighted components are given by the product of this factor and the unweighted components. Again, the mean of these components has to be normalized before it can be used to calculate the orientation, but once this has been done we find that the corrected mean normal has an orientation of $\alpha_n = \text{atan2}(0.571, 0.483) = 40.2°$ and $\beta_n = \text{asin}(0.664) = 41.6°$. In terms of dip direction and dip angle this is an orientation of 220.3/48.4, or more appropriately 220/48.

To find the angular error that would exist if the correction were not made, we compute the angle subtended by the uncorrected vector and the corrected vector. This is given by the scalar product of the two vectors, and is computed as

$$\mathrm{acos}(0.490 \cdot 0.483 + 0.498 \cdot 0.571 + 0.715 \cdot 0.664) = 5.1^{\circ}.$$

Depending on the relative orientations of the fractures and the sampling line, errors appreciably greater than this can be found, and so we recommend that this correction is always applied to fracture orientation data.

7.3 Additional points

There is now a considerable body of theory available that can be used for studying fracture geometry in the engineering rock mechanics context, and there are computer programs available for analyzing and presenting the fracture data. The use of the hemispherical projection, demonstrated through the answers in Section 7.2, enables many geometrical problems to be solved by hand to an accuracy of 1°. Further advantages are that it provides an enhanced understanding of the geometrical principles and is a useful spot check of solutions obtained from 'black box' computer programs.

For further reading about the origin of fractures, we recommend the book by Price and Cosgrove (1990)[9]. For further reading about the engineering characterization of discontinuities, we recommend the book by Priest (1993)[10]. An example of research into the link between the geometrical fracture properties and the hydrogeological properties of rock masses is provided in the paper by Panda and Kulatilake (1999)[11].

The most well-known software package for reducing and presenting fracture data is DIPS, available from Rocscience, see www.rocscience.com. A more extensive package, FRACNTWK, has been developed by Kulatilake at the University of Arizona (kulatila@u.arizona.edu).

---

[9] Price N. J. and Cosgrove J. W. (1990) *Analysis of Geological Structures*. Cambridge University Press, Cambridge, 502pp.

[10] Priest S. D. (1993) *Discontinuity Analysis for Rock Engineering*. Chapman and Hall, London, 473pp.

[11] Panda B. B. and Kulatilake P. H. S. W. (1999) Effect of joint geometry and transmissivity on jointed rock hydraulics. *J. Eng. Mech.*, 41–50.

8 Rock masses: deformability, strength and failure

8.1 The nature of rock masses

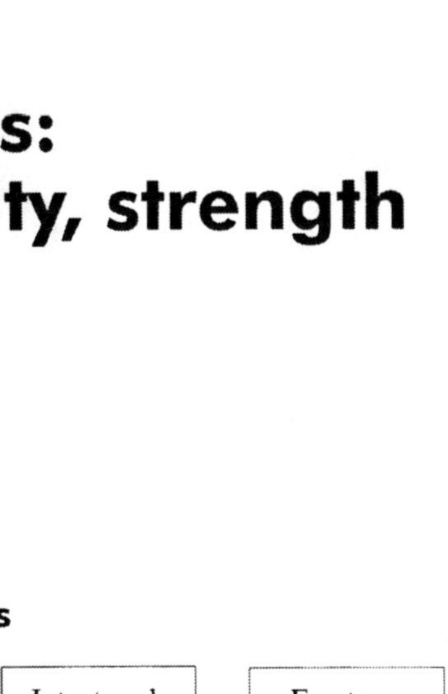

In Chapters 6 and 7, we have posed and answered questions concerning the properties of the intact rock and the fractures. In this chapter, we consider the deformation, strength and failure of a fractured rock mass, i.e. a discontinuum comprising intact rock dissected by fractures.

A **rock mass** is a complex geometrical and mechanical assemblage resulting from a long history of tectonic forces and other natural environmental effects. Additionally, the rock mass properties can be significantly affected by engineering activities, especially blasting. The following series of photographs in Fig. 8.1 illustrates different rock masses in different conditions.

The **deformation, strength and failure properties of a rock mass** are determined by the mechanical properties of the intact rock (Chapter 6), the geometrical properties of the fractures (Chapter 7), and the mechanical properties of the fractures (this chapter).

We can consider a fracture being stressed by a normal stress and two shear stresses, as shown in Fig. 8.2.

The compression of the fracture due to the normal stress component σ_{zz} produces a **normal displacement** δ_z. It is awkward to convert this directly into strain because there is no clear gauge length over which the displacement has taken place. Thus, assuming initially a linear relation between σ_{zz} and δ_z, we can define the fracture normal deformation modulus as $\sigma_{zz}/\delta_z = E_d$, where the units of E_d are stress/metre, or $L^{-2}MT^{-2}$ (*cf.* the units of Young's modulus which are stress units, or $L^{-1}MT^{-2}$). From the **shear displacements**, we can define shear deformation moduli in a similar way.

Within the local co-ordinate system illustrated in Fig. 8.2, the two **shear fracture stiffnesses** and the **normal stiffness** can be denoted as

(a)
(b)
(c)

Figure 8.1 (left and above) Examples of fractured rock masses. (a) Folded rock mass in Crete illustrating that the mechanical behaviour will be a function of both the intact rock and the fracture geometry. (b) Fractured limestone rock mass, relatively unaffected during excavation by tunnel boring machine. (c) Jointed granitic rock mass with joints dilated by bulk blasting. (d) Heavily jointed chalk rock mass, with joint dilation resulting from weathering.

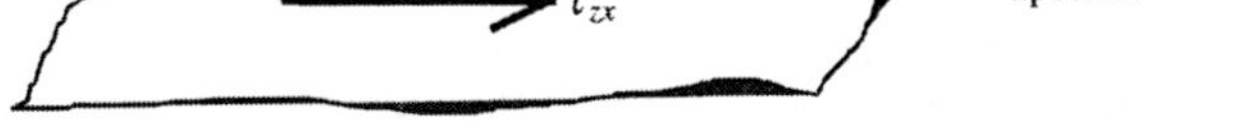

Figure 8.2 Normal and shear stresses applied to a fracture.

k_{xx}, k_{yy}, and k_{zz}. The reciprocals of these are the **fracture compliances,** s_{xx}, s_{yy}, and s_{zz}. There will also be interactions between the axes; for example a shear stress in the x-direction can cause a displacement in the z-direction, requiring a k_{xz} or s_{xz} term. This leads to nine stiffnesses or compliances and the relations

$$\begin{aligned} \delta_x &= s_{xx}\tau_{zx} + s_{yx}\tau_{zy} + s_{zx}\sigma_{zz} \\ \delta_y &= s_{xy}\tau_{zx} + s_{yy}\tau_{zy} + s_{zy}\sigma_{zz} \\ \delta_z &= s_{xz}\tau_{zx} + s_{yz}\tau_{zy} + s_{zz}\sigma_{zz} \end{aligned} \quad \text{or} \quad \begin{bmatrix} \delta_x \\ \delta_y \\ \delta_z \end{bmatrix} = \begin{bmatrix} s_{xx} & s_{yx} & s_{zx} \\ s_{xy} & s_{yy} & s_{zy} \\ s_{xz} & s_{yz} & s_{zz} \end{bmatrix} \times \begin{bmatrix} \tau_{zx} \\ \tau_{zy} \\ \sigma_{zz} \end{bmatrix}.$$

Usually, these nine stiffnesses (or compliances) are reduced to two: a

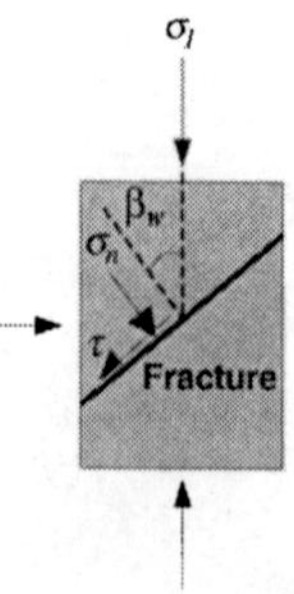

Figure 8.3 The presence of a fracture in a stressed rock specimen.

normal stiffness k_n and a shear stiffness k_s (or a normal compliance, s_n, and a shear compliance, s_s).

By studying the way in which the intact rock properties and the fracture properties combine to determine the properties of the whole-rock mass, an estimate of the importance of the various contributions can be established. However, because of the geometrical and mechanical complexity of the fractures, it is not possible to generate the exact rock mass properties from such information. If the deformation modulus [1], E_m, of a rock mass has not been measured directly and is required for calculation and design purposes, the value can either be estimated by combining the intact rock and fracture components to provide a composite modulus, or the modulus can be estimated empirically.

Although the **empirical estimations of rock mass deformation modulus** seem over-simplified, they can give reasonable predictions and are often the only practical method of estimating the modulus. The relation, $E_m = (2RMR - 100)$ where E_m has units of GPa and $RMR > 50$ (where RMR is the Rock Mass Rating, see Chapter 12), can give surprisingly accurate results (Bieniawski, 1989 [2]). Another empirical expression is $E_m = 10^{(RMR-10)/40}$ GPa which covers the complete RMR range, i.e. from low values to 100.

The effect of a single fracture on rock strength can be studied using the **single plane of weakness** theory. Assume that a fracture is present in a rock specimen as shown in Fig. 8.3. The strength of the specimen will then depend on the orientation of the principal stresses relative to the fracture orientation. Assuming that failure is induced when the normal and shear stress components on the fracture satisfy the Mohr–Coulomb failure criterion, we can develop an expression for the specimen strength as a function of β_w, the angle between the major principal stress and the normal to the fracture.

The normal stress on the fracture is given by

$$\sigma_n = \tfrac{1}{2}(\sigma_1 + \sigma_3) + \tfrac{1}{2}(\sigma_1 - \sigma_3)\cos 2\beta_w$$

[1] The term 'deformation modulus', is used in rock mechanics to indicate the apparent elastic modulus of an *in situ* rock mass.

[2] Bieniawski Z. T. (1989) *Engineering Rock Mass Classifications.* Wiley, Chichester, 251pp.

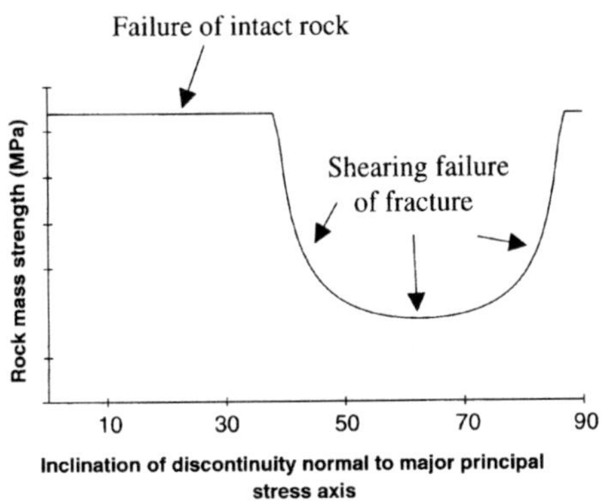

Figure 8.4 Effect of a fracture on the strength of a rock specimen.

and the shear stress on the fracture by

$$|\tau| = \tfrac{1}{2}(\sigma_1 - \sigma_3)\sin 2\beta_w.$$

Substituting these values in the Mohr–Coulomb criterion, $|\tau| = c_w + \sigma_n \tan\phi_w$, where c_w and ϕ_w are the cohesion and angle of friction of the fracture, gives for the 2-D case

$$(\sigma_1 - \sigma_3) = 2\,(c_w + \sigma_3 \tan\phi_w)\,/\,(1 - \cos\beta_w \tan\phi_w)\sin 2\beta_w.$$

This results in the type of locus illustrated in Fig. 8.4.

Thus, when the rock mass initially reaches its peak strength due to the application of stresses, either the intact rock or a fracture will begin to fail[3]. Hence, the properties that govern the **failure of rock masses** are the failure properties of the intact rock and the fractures. The orientation of the applied stresses relative to the fractures governs how failure occurs, bearing in mind that the failure of a rock mass is a complex structural breakdown process following a similar curve to that in Fig. 6.2 for the intact rock. The Hoek–Brown empirical failure criterion applies to rock masses through the appropriate choice of the governing parameters (Hoek and Brown, 1997[4]).

Finally, the behaviour of the rock mass after initial failure will be a function of the loading conditions. Where the load is constant, as in a rock block sliding down a slope, once failure initiates, it will usually propagate. However, in underground circumstances, the loading may reduce as failure occurs, e.g. when failure occurs in a weak, soft mine pillar loaded by stronger, stiff roof and floor strata.

There are other factors influencing rock mass failure, notably rock deterioration due to exposure and other time-dependent effects, which will not be included here. We will, however, discuss the general principles of water flow through rock masses in the next chapter and note the effects of excavation in Chapter 15.

---

[3] We have used the word 'failure' here to mean that some kind of limit state has been reached.

[4] Hoek E. and Brown E. T. (1997) Practical estimates of rock mass strength. *Int. J. Rock Mech. Min. Sci*, **34**, 8, 1165–1186.

8.2 Questions and answers: rock masses

Q8.1 For a simple sedimentary rock mass in which the only effective fractures are the bedding planes, the elastic modulus of the rock mass can be found from the addition of the displacements due to both the intact rock and the fractures, noting that the rock mass can comprise more than one stratum, each containing bedding plane fractures with different frequencies.

Unfractured strata **Consider firstly the case of n strata of intact rock, each with a thickness t_i and modulus of elasticity E_i. Derive an expression for the composite elastic modulus, E_m, of the rock mass in a direction normal to the strata by considering the total displacement (and hence strain) of the total thickness of the rock mass due to the applied stress. Write the expression in terms of E_i and t_i, and assume that the interfaces between adjacent units have no mechanical effect.**

Strata with bedding plane fractures. **Now consider the case where each stratum of rock contains a set of bedding plane fractures parallel to the stratum boundaries. The fracture frequency of the set within each stratum is unique — stratum i possesses a frequency λ_i; similarly, the modulus of deformation (i.e. applied stress/displacement) within each stratum is unique and for unit i is E_{di}. Extend the expression for E_m to include t_i, E_i, λ_i and E_{di}.**

A8.1 *Unfractured strata*

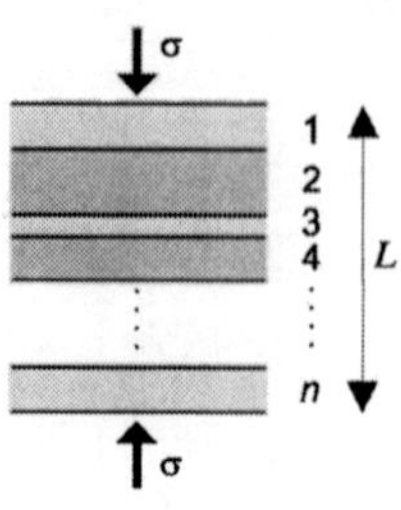

The model comprises n strata of rock, each of thickness t_i and elasticity modulus E_i. The total thickness of rock is $L = \sum_i t_i$, where the summation in this and succeeding equations is from 1 to n.

The strain in each unit is given by $\varepsilon_i = \delta_i/t_i = \sigma/E_i$ and hence the displacement of each unit is $\delta_i = t_i\varepsilon_i = t_i(\sigma/E_i)$.

This expression for δ_i then gives the total displacement as

$$\delta_T = \sum_i \delta_i = \sum_i t_i \frac{\sigma}{E_i}.$$

The overall strain is $\varepsilon = \delta_T/L$, and the results we have derived here for δ_i then give the overall strain as

$$\varepsilon = \frac{\delta_T}{L} = \frac{\sum_i t_i \dfrac{\sigma}{E_i}}{\sum_i t_i}.$$

Thus, the expression for the rock mass modulus is

$$E_m = \frac{\sigma}{\varepsilon} = \frac{\sigma}{\left(\sum_i t_i \dfrac{\sigma}{E_i} \Big/ \sum_i t_i\right)} = \frac{\sigma \sum_i t_i}{\sigma \sum_i \dfrac{t_i}{E_i}} = \frac{\sum_i t_i}{\sum_i \dfrac{t_i}{E_i}}.$$

Note that the t_i cannot be cancelled in the numerator and denominator of the right-hand side of this equation, as is evident when the expression is written out,

$$\frac{t_1 + t_2 + t_3 + \dots}{\dfrac{t_1}{E_1} + \dfrac{t_2}{E_2} + \dfrac{t_3}{E_3}\dots}.$$

Fractured strata

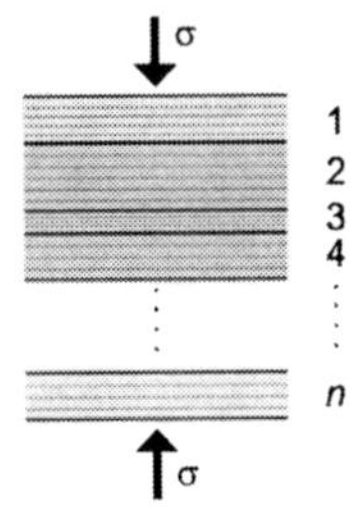

Although the total thickness of the rock mass is the sum of the thickness of each stratum, $L = \sum_i t_i$, because the fractures are now present the equation should also include a term to account for the thickness (i.e. aperture) of each fracture. However, this is assumed to be negligible compared to the thickness of the intact rock and has been ignored here.

As before, the strain in the intact rock of each unit is $\varepsilon_i = \delta_i / t_i = \sigma / E_i$, and so the displacement of each unit due to straining of the intact rock is $\delta_i = t_i(\sigma/E_i)$. For any fracture, its normal displacement is $\delta_d = \sigma/E_d$, and so, for n fractures in unit i, the displacement is $\delta_{di} = n_i\delta_d = \lambda_i t_i(\sigma/E_{di})$. The total normal displacement is the sum of the displacements due to the intact rock and the fractures and is thus $\delta_T = \sum_i \delta_i + \sum_i \delta_{di}$ which, on substitution of the expressions for δ_i and δ_{di}, becomes

$$\delta_T = \sum_i t_i \frac{\sigma}{E_i} + \sum_i \lambda_i t_i \frac{\sigma}{E_{di}} = \sigma \left[\sum_i \frac{t_i}{E_i} + \sum_i \frac{\lambda_i t_i}{E_{di}} \right].$$

The total strain is thus

$$\varepsilon = \frac{\delta_T}{L} = \frac{\sigma \left[\sum_i \dfrac{t_i}{E_i} + \sum_i \dfrac{\lambda_i t_i}{E_{di}} \right]}{\sum_i t_i} = \frac{\sigma \sum_i t_i \left(\dfrac{1}{E_i} + \dfrac{\lambda_i}{E_{di}} \right)}{\sum_i t_i}$$

and the rock mass modulus is

$$E_m = \frac{\sigma}{\varepsilon} = \frac{\sum_i t_i}{\sum_i t_i \left(\dfrac{1}{E_i} + \dfrac{\lambda_i}{E_{di}} \right)}.$$

The influences of the fracture frequency and fracture deformation moduli values on this composite rock mass modulus, E_m, are illustrated below.

Note that in the diagram below for the conditions assumed, the fracture deformation modulus is more important than the fracture frequency. The value of E_m does not reduce much when many fractures are present providing the fracture deformation modulus is high (the $E_d = 1000$ GPa curve in the diagram below). On the other hand, just one or two fractures every few metres will have a significant effect on E_m if E_d is, say 10 GPa/m or less. This diagram also shows that it is critical to make an assessment of fracture normal stiffness in order to be able to understand the rock mass modulus.

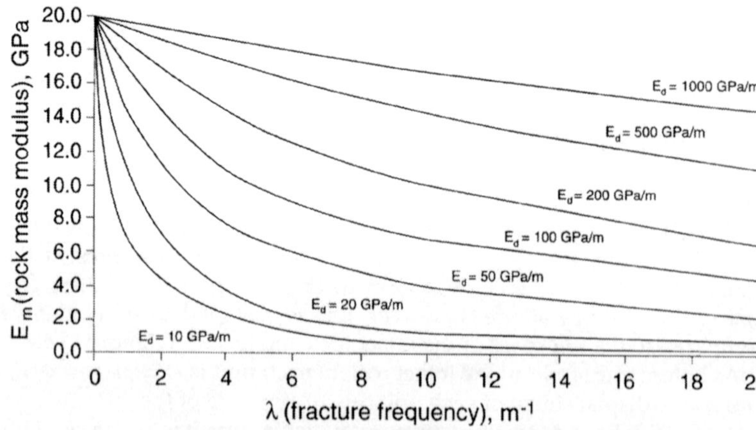

Q8.2 For the unfractured and fractured stratified rock mass geometries described in Q8.1, develop expressions for the composite shear modulus of a rock mass, G_m, using a shear stress τ and the parameters t_i, G_i, λ_i, and G_{di}.

A8.2 *Unfractured strata*

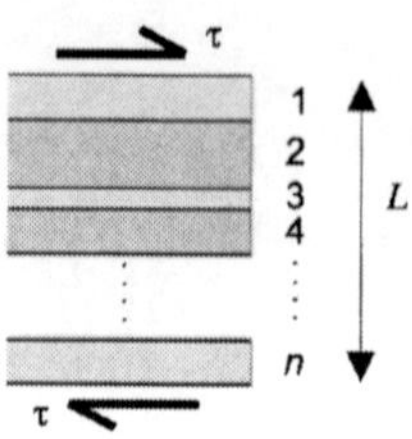

We are investigating the case of simple shear. If we consider the displacement of the rock under the action of a shear stress, we can consider γ and δ as in the sketch below.

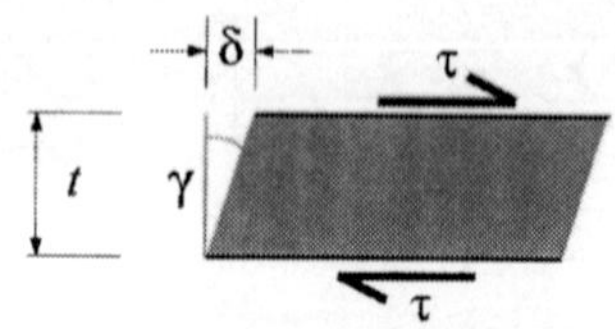

Using these definitions, for small δ, the engineering shear strain of each unit can be written as $\gamma = \tau/G \approx \delta/t$, and so the shear deformation of a single unit of intact rock is $\delta_i = t_i(\tau/G_i)$, with the total shear deformation being given by $\delta_T = \sum_i \delta_i = \sum_i t_i(\tau/G_i)$. Providing δ_T is sufficiently small to allow small angle approximations, then the total shear strain is

$$\varepsilon = \frac{\delta_T}{L} = \frac{\sum_i t_i \frac{\tau}{G_i}}{\sum_i t_i},$$

and as a result the rock mass shear modulus is given by

$$G_m = \frac{\tau}{\varepsilon} = \frac{\tau}{\sum_i t_i \frac{\tau}{G_i} \Big/ \sum_i t_i} = \frac{\tau \sum_i t_i}{\tau \sum_i t_i \frac{1}{G_i}} = \frac{\sum_i t_i}{\sum_i \frac{t_i}{G_i}}.$$

Note that this result is directly analogous to the result obtained in the first part of A8.1.

Fractured strata

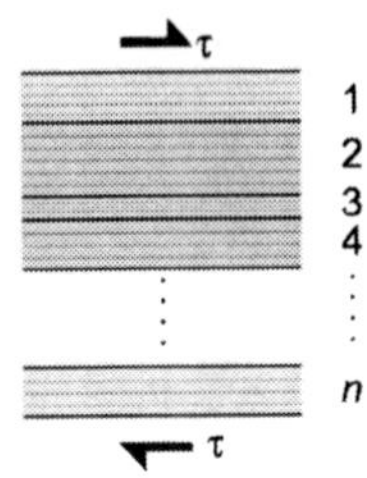

The total thickness of the rock mass is the sum of the thickness of each stratum, $L = \sum_i t_i$. This equation should include a term to account for the thickness of each fracture but, as in Q8.1, we assume that this is negligible compared to the intact rock thickness.

As for the first part of this question, the shear deformation of a single stratum of intact rock is $\delta_i = t_i(\tau/G_i)$ and the shear deformation of any fracture is $\delta_d = \tau/G_d$. For n_i fractures in stratum i, the shear deformation is $\delta_{di} = n_i\delta_d = \lambda_i t_i(\tau/G_{di})$. The total shear deformation is the sum of the shear displacements due to the intact rock and the fractures and is thus given as $\delta_T = \sum_i \delta_i + \sum_i \delta_{di}$, which, upon substitution of the expressions for δ_i and δ_{di}, becomes

$$\delta_T = \sum_i t_i \frac{\tau}{G_i} + \sum_i \lambda_i t_i \frac{\tau}{G_{di}} = \tau\left[\sum_i \frac{t_i}{G_i} + \sum_i \frac{\lambda_i t_i}{G_{di}}\right].$$

Consequently, the total strain is

$$\gamma = \frac{\delta_T}{L} = \frac{\tau\left[\sum_i \frac{t_i}{\sum G_i} + \sum_i \frac{\lambda_i t_i}{G_{di}}\right]}{\sum_i t_i}$$

and so the rock mass shear modulus is

$$G_m = \frac{\tau}{\gamma} = \frac{\sum_i t_i}{\sum_i t_i\left(\frac{1}{G_i} + \frac{\lambda_i}{G_{di}}\right)}.$$

This equation is directly analogous to the result obtained in A8.1, and G_m shows a similar sensitivity to the fracture frequency and fracture stiffness values as illustrated in A8.1 for E_m. Once again, an assessment of the fracture stiffness is essential if the behaviour of the rock mass is to be properly understood.

Q8.3 When the application of stress is not perpendicular to the fractures, as in Q8.1 and Q8.2, it is necessary to transform the stress components in order to establish rock mass deformation moduli using the fracture stiffnesses or compliances. This results in equations for the rock mass modulus, E_m, of the type (Wei and Hudson, 1986[5])

$$\frac{1}{E_m} = \frac{1}{E} + \lambda_1 s^1_{11}\cos^4\alpha + \lambda_1 s^1_{22}\cos^2\alpha\sin^2\alpha + \lambda_2 s^2_{22}\cos^2\alpha\sin^2\alpha + \lambda_2 s^2_{11}\sin^4\alpha$$

[5] This equation is for two orthogonal sets of fractures in 2-D. The method by which the equations for ***n*** sets in 3-D can be developed is given in Wei Z. Q. and Hudson J. A. (1986) The influence of joints on rock modulus. *Proc. Int. Symp. Engineering in Complex Rock Formations* (T. K. Tan, ed.), Pergamon Press, Beijing, pp. 54–62.

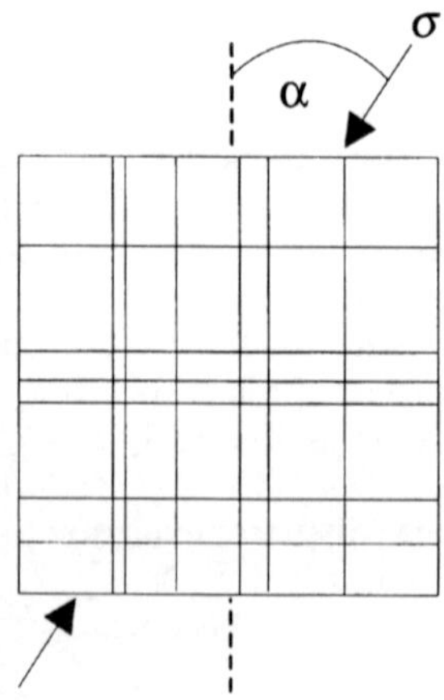

for a 2-D analysis with uniaxial loading.

The equation is for two orthogonal fracture sets, where E is the modulus of the intact rock, λ_1 and λ_2 are the fracture frequencies of the two sets, s^1_{11}, s^1_{22}, s^2_{11}, and s^2_{22} are the normal and shear compliances for the fractures in each set, and α is the angle between the applied normal stress and the normal to the first set.

The equation reduces for one set (i.e. $\lambda_2 = 0$) to

$$\frac{1}{E_m} = \frac{1}{E} + \lambda s^1_{11}\cos^4\alpha + \lambda s^1_{22}\cos^2\alpha\sin^2\alpha$$

and if we put $s^1_{11} = s$ and $s^1_{22} = ks$, and then rearrange, we obtain

$$\frac{1}{E_m} = \frac{1}{E} + \lambda s\cos^2\alpha\left(\cos^2\alpha + k\sin^2\alpha\right)$$

from which we find

$$\frac{E_m}{E} = 1 - E_m\lambda s\cos^2\alpha\left(\cos^2\alpha + k\sin^2\alpha\right).$$

Using $E = 2.75$ GPa, $\lambda = 2$ m^{-1} and $s = 0.05$ m/GPa, the plots shown to the right in Diagram 1 result.

For two sets with $\lambda_1 = \lambda_2 = \lambda$, $s^1_{11} = s^2_{11} = s$ and $s^1_{22} = s^2_{22} = ks$, the basic equation reduces to

$$\frac{E_m}{E} = 1 - E_m\lambda s\left(\cos^4\alpha + 2k\cos^2\alpha\sin^2\alpha + \sin^4\alpha\right).$$

Using $E = 2.75$ GPa, $\lambda = 2$ m^{-1} and $s = 0.05$ m/GPa, the plots in Diagram 2 result.

Explain why terms such as $\cos^4\alpha$ appear in the formula and comment on any general principles that are apparent from these illustrative plots.

A8.3 In Section 3.1, we saw that a $\cos^2$ term is required for resolving stresses — both the force and the area have to be resolved — and this leads to Mohr's circle representation of a 2-D stress state. Terms such as $\cos^4\alpha$ appear in the formula above because four resolutions are now required: two to account for the stress transformation; one to resolve the fracture frequency (see A7.6); and one to resolve the fracture displacements.

Although the plots in Diagrams 1 and 2 have been calculated for specific circumstances, we can note the following general principles.

- The rock mass modulus will always be less than the intact rock modulus (the outer circle in the plots), except when the loading is parallel to the fractures in a rock mass containing only one set.

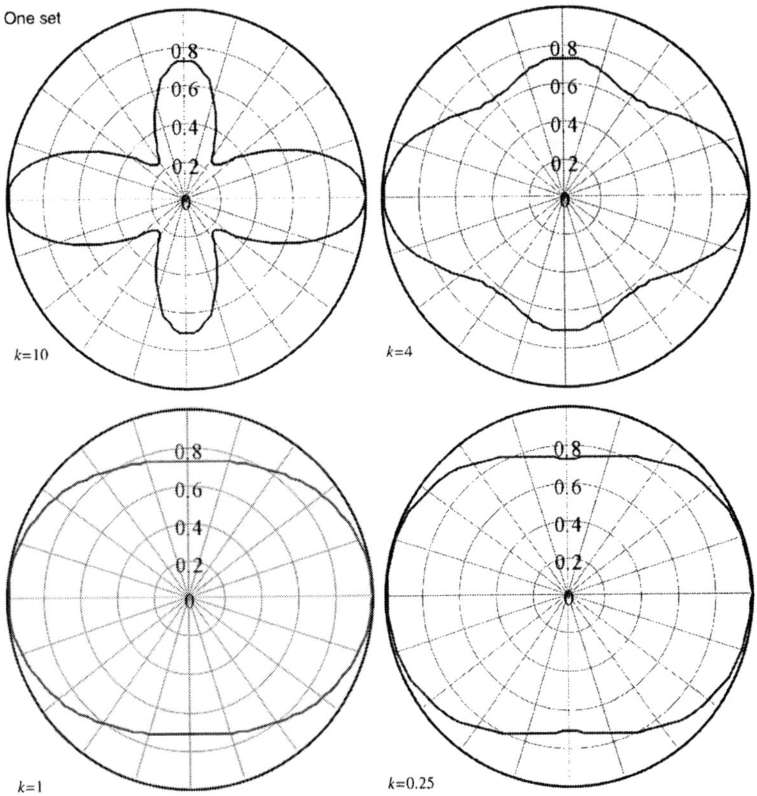

Diagram 1 — one fracture set.

- There will be a variation in the modulus of fractured rock masses as a function of the loading direction (except for the case of a rock mass with two orthogonal sets with equal normal and shear fracture stiffnesses, which has the same modulus regardless of the loading direction, the $k = 1$ plot in Diagram 2 overleaf).
- The directions associated with maximal and minimal rock mass modulus are unlikely to be perpendicular, especially when there is more than one fracture set.

These principles should be remembered when estimating the deformation modulus by any method, or attempting to understand the likely deformation behaviour of a fractured rock mass

Q8.4 A rock mass has the following characteristics: the compressive strength of the intact rock is 80 MPa, the *RMR* (Rock Mass Rating) is 62%, and the *GSI* (Geological Strength Index) is 50. Estimate the *in situ* deformation modulus, E_m.

A8.4 Empirical formulae for estimating E_m are required to answer this question (as found in the references in Footnotes 2 and 4 in this chapter).

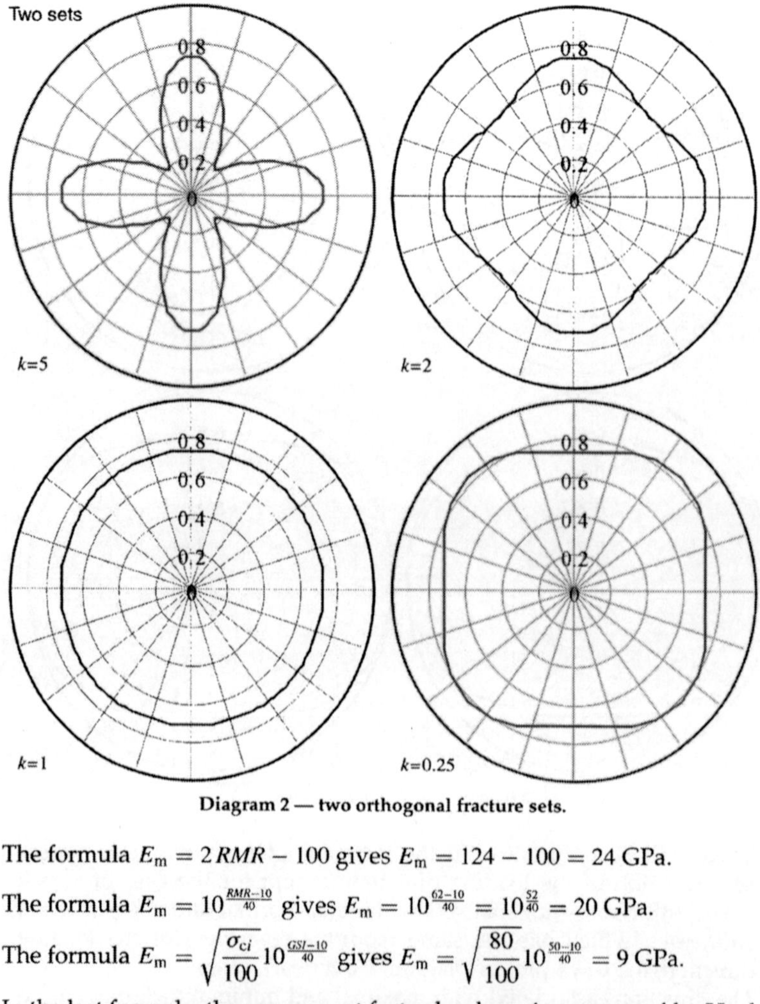

Diagram 2 — two orthogonal fracture sets.

The formula $E_m = 2RMR - 100$ gives $E_m = 124 - 100 = 24$ GPa.

The formula $E_m = 10^{\frac{RMR-10}{40}}$ gives $E_m = 10^{\frac{62-10}{40}} = 10^{\frac{52}{40}} = 20$ GPa.

The formula $E_m = \sqrt{\frac{\sigma_{ci}}{100}} 10^{\frac{GSI-10}{40}}$ gives $E_m = \sqrt{\frac{80}{100}} 10^{\frac{50-10}{40}} = 9$ GPa.

In the last formula, the square root factor has been incorporated by Hoek and Brown (see the reference in Footnote 4) to reflect the increasing role of the intact rock as it becomes weaker.

These empirical methods can provide surprisingly good estimates and they are often the only practical way of estimating the *in situ* deformation modulus, but they should be used with caution. An understanding of the roles of the intact rock modulus and fracture stiffness contributions assists in evaluating the reliability of empirical estimates for given circumstances. We include further aspects of rock mass classification systems in Chapter 12.

Q8.5 To study the effect of a fracture on the rock strength, plot a graph of $(\sigma_1 - \sigma_3)$ vs. β_w using the single plane of weakness formula included in Section 8.1. Assume $\sigma_3 = 10$ MPa, $c = 0$ and $\phi_w = 35°$,

and that the intact rock strength is given by $\sigma_1 = 75 + 5.29\sigma_3$. Also, explain the form of the resulting graph.

A8.5 The plot is shown below. Initially, as β_w is increased from 0°, i.e. from when the fracture is perpendicular to the major principal stress and parallel to the minor principal stress, the fracture does not induce failure and the rock strength is constant, as the strength of the intact rock: 129.7 MPa.

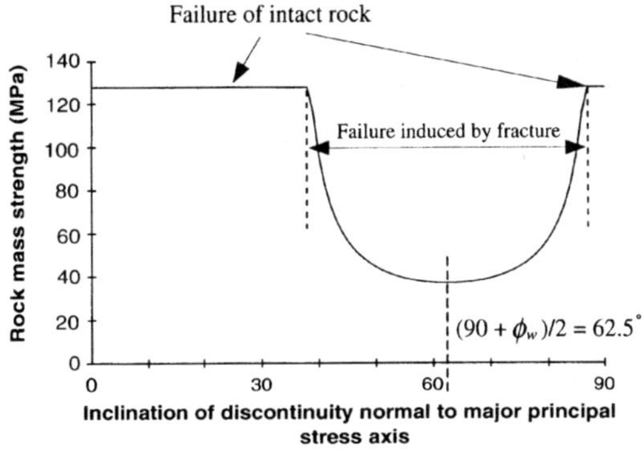

However, as β_w is increased further, and when the normal and shear stresses on the fracture satisfy the Mohr–Coulomb criterion for a value of the major principal stress lower than 129.7 MPa, the fracture causes a reduction in the rock strength; the 'U' shaped curve in the strength locus. Note that this curve has asymptotes at $\phi_w = 35°$ and 90°. The minimal strength is for $\beta_w = (90 + \phi_w)/2 = 62.5°$. Both horizontal portions of the curve occur when the intact rock fails before the fracture induces shear failure.

Q8.6 If a rock mass contains more than one fracture set, we can apply the single plane of weakness theory to each set, and superimpose the results to find a lowest-bound envelope of strength.

(a) Plot the 2-D variation in strength for a rock mass containing two orthogonal sets of fractures, A and B, the strengths of which are $c_A = 100$ kPa, $\phi_A = 20°$ and $c_B = 0°$, $\phi_B = 35°$, when the minor principal stress has the value 10 MPa. The intact rock strength is again given by $\sigma_1 = 75 + 5.29\sigma_3$.

(b) How would this strength variation change if the minor principal stress were reduced to zero?

A8.6 (a) For fracture set A only, the locus of rock strengths (see next page) has been generated using the single plane of weakness formula in Section 8.1.

The strength locus for fracture set B only, is similar, although the minimal strength is higher than that for fracture set A.

To obtain the strength locus for a rock mass containing both sets of fractures, we superimpose the strength loci for the two fracture sets and take the minimal strength at each β_w angle, i.e. the envelope of

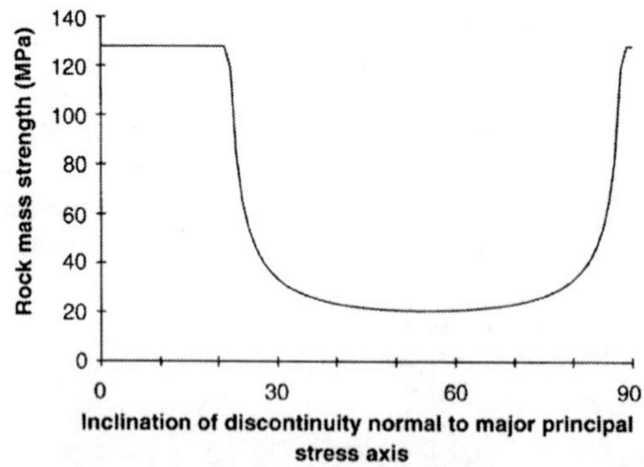

Strength locus of fracture set A

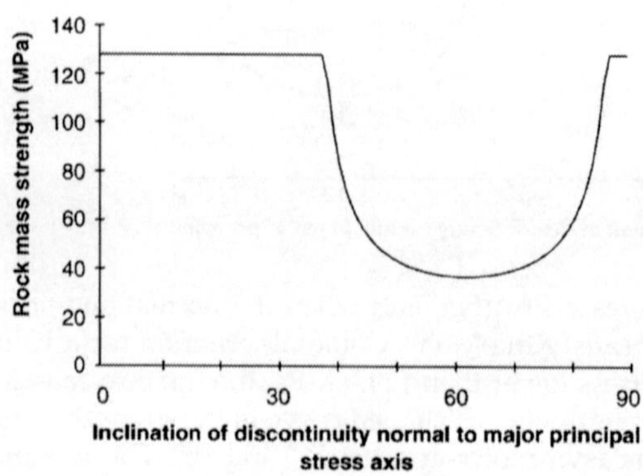

Strength locus of fracture set B

minimal strengths. However, in the rock mass, the two fracture sets are orthogonal, and so, when we superimpose the two curves to determine the lowest strengths of the rock mass, we must apply an offset of + 90° to the values of β_w for fracture set B, as indicated in the sketch below.

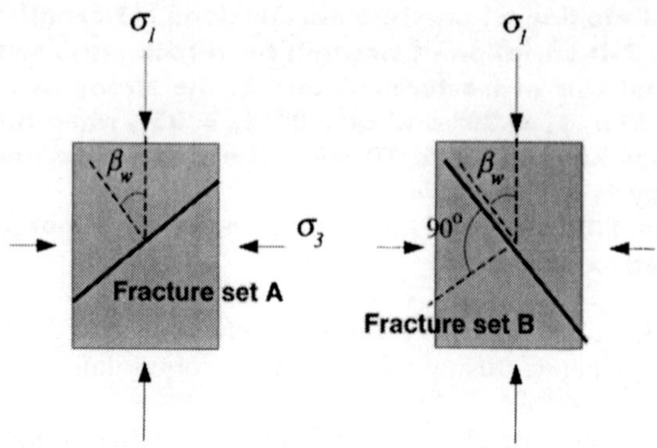

Superimposing the two strength loci in this manner, the strength variation for the rock mass containing the two fracture sets over the full range of β_w, 0° to 360°, is shown below. The lower strength envelope is the heavier line in the 0° to 180° range in the plot below.

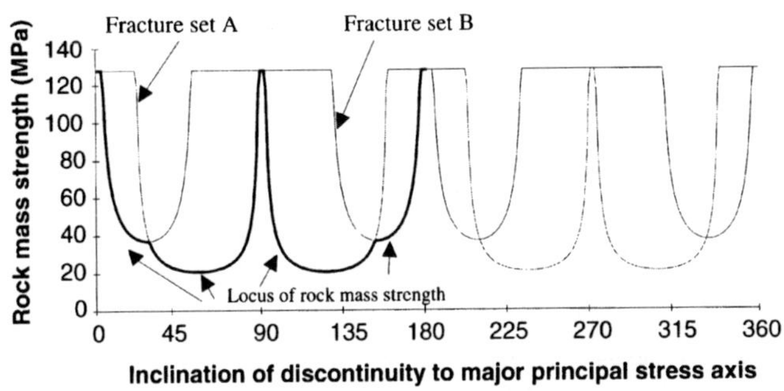

(b) To assess the effect of the minor principal stress, which can be considered as a confining pressure, we repeat the procedure in (a) but with the minor principal stress reduced from 10 MPa to 0 MPa.

For fracture set A, the strength locus (see below) is obtained. Note the dramatic reduction in strength in the region of $\beta_w = (90 + \phi_w)/2 = 55°$. This is because the rock mass, acting as two blocks and without any confining stress, is much more susceptible to the effect of the major principal stress.

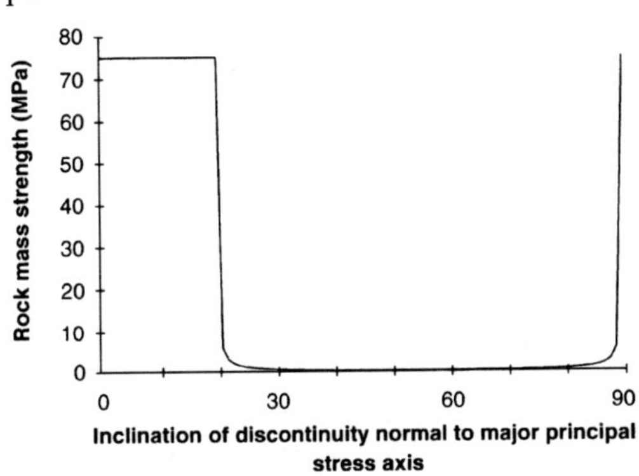

Strength locus of fracture set A

For fracture set B, the strength locus is shown on the next page. The lack of the confining stress and the zero cohesion cause the strength locus to manifest a sudden change as β_w passes the ϕ_w value of 35°.

The superimposition of the two strength loci for fracture sets A and B, together with the lower strength envelope is shown in the plot below. Comparison with the equivalent curves for part (a) of the answer shows that the confining stress has had a significant effect in strengthening the rock mass to being always above 20 MPa strength.

Note that, with the exception of β_w angles close to 0°, 90°, 180° and 270°, the rock mass strength (the heavier line) is virtually zero. Since the peaks at these orientations are too narrow to have any engineering significance, the rock mass strength has been reduced to almost zero by the presence of the two sets of fractures. This example demonstrates the significant loss of strength that occurs when fractures dissect the rock mass.

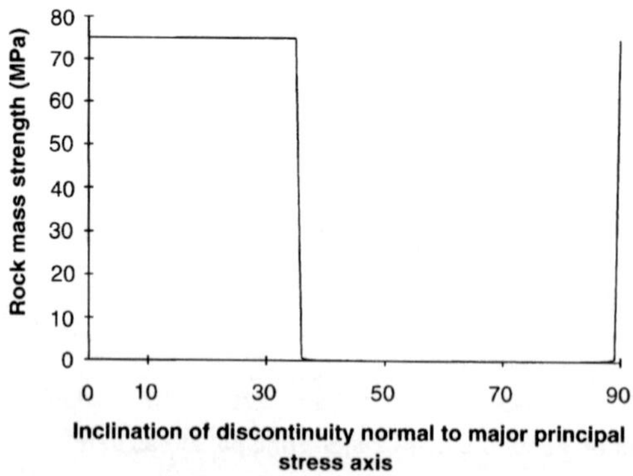

Strength locus of fracture set B

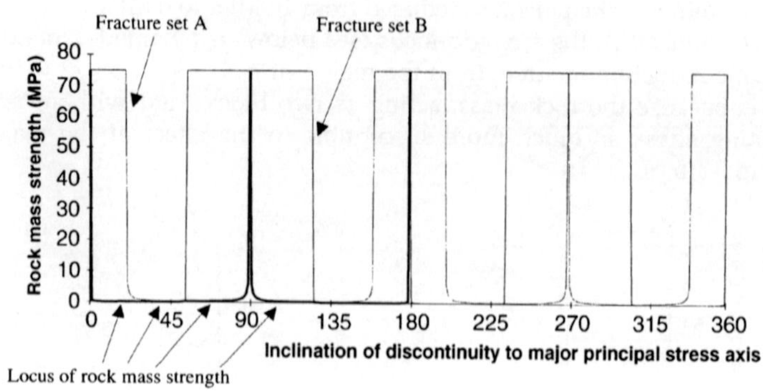

Q8.7 When a rock mass contains a large number of fracture sets and each set has similar strength properties, the rock mass strength can tend to become isotropic, with the multiple plane of weakness theory generating an approximately isotropic strength criterion. Develop such a criterion for the 2-D case of a rock mass that contains four sets of fractures mutually inclined at 45°, the shear strengths of which are given by a linear Mohr–Coulomb criterion with c = 100 kPa and $\phi_w = 30°$. The intact rock strength is given by $\sigma_1 = 75 + 5.29\sigma_3$. Assume that the minor principal stress, σ_3, is 10 MPa.

A8.7 Each fracture set has strength parameters of c = 100 kPa and $\phi_w = 30°$. The rock mass strength variation as a function of maximum principal stress orientation for one such fracture set is shown below.

To investigate the strength of the rock mass, we need to examine the case for the four fracture sets with relative orientations of 0°, ±45° and 90°. Superimposing four of the curves while applying the necessary offsets, and examining the range of β_w for 0–360°, we obtain the follow-

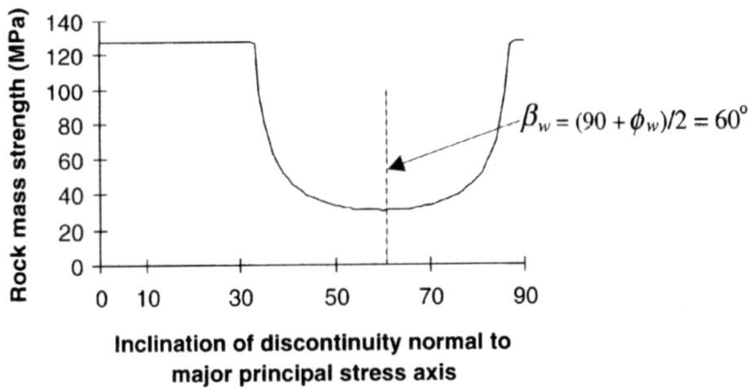

ing diagram, with the heavier line indicating the composite rock mass strength.

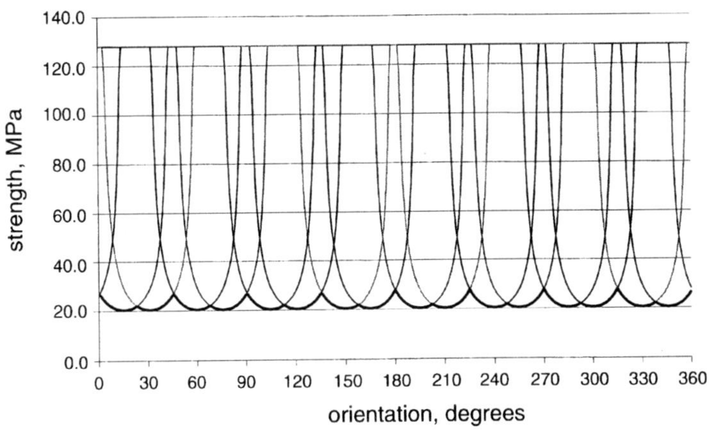

The increasing influence of 1, 2, 3 or 4 such fracture sets being present in the rock mass can be studied by introducing the sets one by one, i.e. as in the polar diagrams on the next page, with fracture sets being introduced progressively with offsets of 0°, 45°, 90° and 135°.

Note that the rock mass strength tends to become more isotropic and that the mean rock mass strength decreases with an increasing number of fracture sets. Here we have assumed that the four fracture sets all have the same values of cohesion and friction angle, but the same tendencies occur for the composite effect of fracture sets with different geometrical and mechanical properties.

Q8.8 How does the significant effect of fractures on the rock mass strength indicate some of the differences between rock mechanics and soil mechanics?

A8.8 Rock is a strong solid material containing fractures which are often of a similar size to the engineering structure. Soil is a weak particulate material with grain sizes much smaller than the engineering structure. Rock failure is often initiated by weaknesses in the rock mass and, as has been seen by the answers to previous questions in this section, the

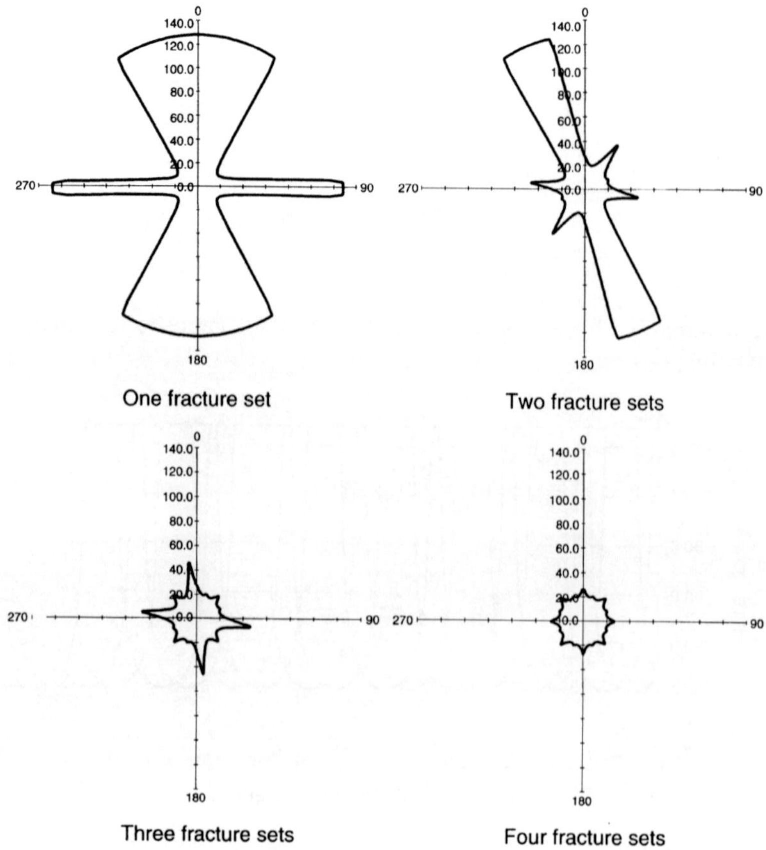

presence of fractures has a significant effect on the rock mass strength. Moreover, the fractures determine the location and orientation of the failure surfaces. In a particulate soil, there are effectively weaknesses at all locations and at all orientations, so the failure location and orientation are often governed not by the soil structure, but by the geometry of the stress field.

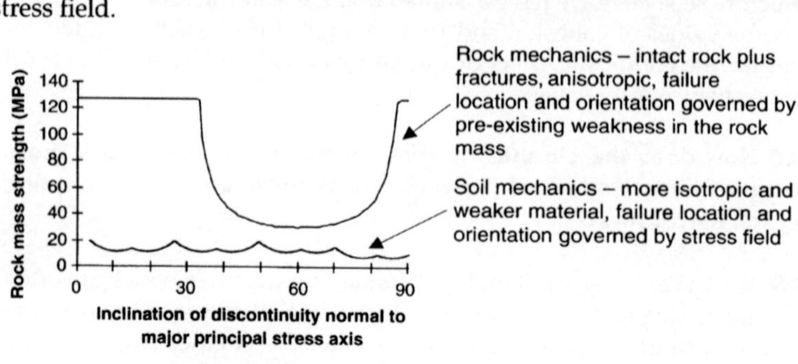

It is useful to bear these ideas in mind when dealing with a weak rock or strong soil, such as shale and chalk, or severely weathered rocks.

These materials may not be amenable to either standard rock mechanics or standard soil mechanics approaches because they exhibit transitional behaviour between the two modes described above.

Q8.9 The peak strength of a poor-quality, closely jointed and weathered granite may be represented by the Hoek–Brown criterion, $\sigma_1 = \sigma_3 + \sqrt{m\sigma_c\sigma_3 + s\sigma_c^2}$, with $m = 1.3$ and $s = 0.00001$.

The uniaxial compressive strength of the intact rock material is estimated as $\sigma_c = 40$ MPa, and the residual strength of the fractures is given by $c_r = 0$, $\phi_r = 15°$.

(a) Plot, in τ–σ space, the expected peak shear strength envelopes for the intact rock material, a fracture at residual strength, and the jointed rock mass for normal stresses up to 10 MPa.

(b) The Hoek–Brown criterion is to be used for the analysis of potential circular slopes cut in the rock mass. Determine values of instantaneous cohesion and friction angle for normal stresses up to 10 MPa.

A8.9 In order to determine the instantaneous values of friction angle and cohesion we must convert the principal stress form of the Hoek–Brown criterion into a shear stress–normal stress space form. During the development of the Hoek–Brown criterion many techniques have been proposed for doing this (Hoek and Brown, 1980, 1988; Hoek, 1990[6]), but for the basic Hoek–Brown criterion of $\sigma_1 = \sigma_3 + \sigma_c\sqrt{m(\sigma_3/\sigma_c) + s}$, the simplest method is to use the following relations (Hoek and Brown, 1988[6])

$$h = 1 + \frac{16}{3}\left(\frac{m\sigma + s\sigma_c}{m^2\sigma_c}\right) \tag{8.1}$$

$$\theta = \frac{1}{3}\left[Q + \arctan\left(\frac{1}{\sqrt{h^3 - 1}}\right)\right] \tag{8.2}$$

$$\phi_i = \arctan\left(\frac{1}{\sqrt{4h\cos^2\theta - 1}}\right) \tag{8.3}$$

$$\tau = \frac{m\sigma_c}{8}\frac{(1 - \sin\phi_i)}{\tan\phi_i} \tag{8.4}$$

In Eq. (8.2), if the angles are evaluated in degrees then $Q = 90°$, and if the angles are evaluated in radians then $Q = \pi/2$.

Finally, substitution of ϕ_i and τ from Eqs. (8.3) and (8.4) into the linear Mohr–Coulomb criterion gives the instantaneous cohesion as

$$c_i = \tau - \sigma\tan\phi_i. \tag{8.5}$$

[6] Hoek E. and Brown E. T. (1980) *Underground Excavations in Rock.* Institution of Mining and Metallurgy, London, 527pp. See also Hoek E. (1990) Estimating Mohr–Coulomb friction and cohesion values from the Hoek–Brown failure criterion. *Int. J. Rock Mech. Min. Sci.*, **27**, 3, 227–229, and Hoek E. and Brown E. T. (1988) Hoek–Brown failure criterion — a 1988 update. *Proc. 15th Canadian Rock Mech. Symp.* University of Toronto, ON, pp. 31–38.

Evaluating these equations for a range of values of normal stress, with values of $\sigma_c = 40$ MPa, $m = 1.3$ and $s = 0.00001$, gives the following results.

σ (MPa)	0.0	0.5	1.0	2.5	5.0	7.5	10.0
h	1.000	1.051	1.103	1.256	1.513	1.769	2.026
θ (rad)	1.04	0.92	0.87	0.79	0.71	0.67	0.64
ϕ_i (rad)	1.43	0.93	0.83	0.68	0.57	0.50	0.45
ϕ_i (deg)	82.1	53.6	47.7	39.2	32.5	28.7	26.0
τ (MPa)	0.009	0.938	1.542	2.938	4.721	6.191	7.479
c_i (MPa)	0.00	0.26	0.44	0.90	1.54	2.09	2.60

In addition to the weathered granite, we can use values of $m = 25$ and $s = 1$ to plot the criterion for the intact granite. The results of this are shown in the following table.

σ (MPa)	0.0	0.5	1.0	2.5	5.0	7.5	10.0
h	1.009	1.011	1.014	1.022	1.035	1.049	1.062
θ (rad)	0.99	0.99	0.98	0.96	0.94	0.92	0.91
ϕ_i (rad)	1.15	1.12	1.10	1.05	0.99	0.94	0.91
ϕ_i (deg)	65.9	64.3	63.0	60.0	56.5	54.0	52.0
τ (MPa)	4.852	5.929	6.938	9.692	13.719	17.319	20.637
c_i (MPa)	4.85	4.89	4.98	5.37	6.16	6.99	7.82

Finally, to plot the criterion for the fractures, we use the Mohr–Coulomb strength parameters given. The results of these calculations are shown in the plots below.

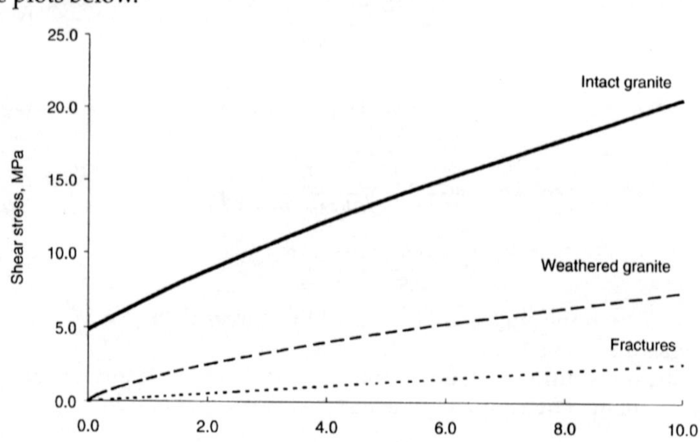

For the more general form of the Hoek–Brown criterion, i.e. $\sigma_1 = \sigma_3 + \sigma_c[m(\sigma_3/\sigma_c) + s]^a$, a more sophisticated analysis is required than that presented here. This makes use of linear regression through pairs of (σ_3, σ_1) data that satisfy the Hoek–Brown criterion to generate equivalent values of (σ, τ) for the Mohr–Coulomb criterion. Further details of this technique are given in Hoek (1990).

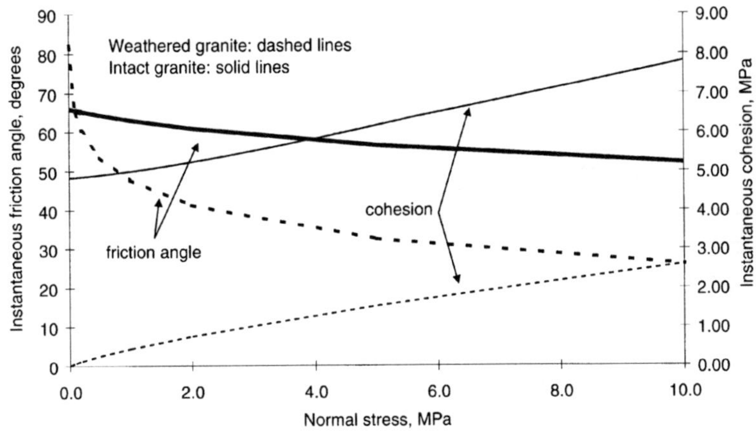

Q8.10 To study the influence of fracture persistence, consider a rock mass containing two fracture sets, A and B, mutually inclined at 45°, as shown to the right. Fracture set A is continuous and contains clay infilling, while fracture set B is rough, clean and intermittent with a 'two-dimensional proportional extent of fracturing' of 0.5. Plot the expected peak strength of this rock mass as a function of the orientation of the principal stresses to the fracture directions, for a minor principal stress of 5 MPa and given the following peak strength characteristics:

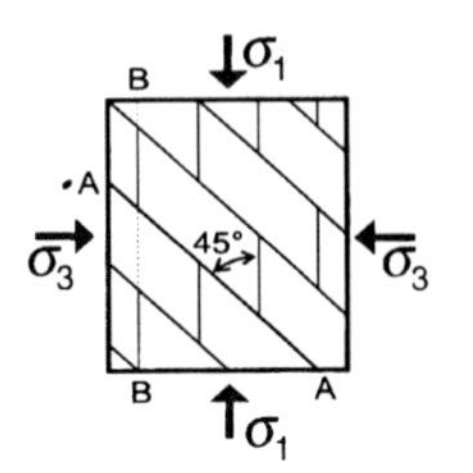

Intact rock: $\sigma_1 = 75 + 5.29\sigma_3$
Fracture set A: c = 100 kPa, $\phi = 20°$,
Fracture set B: c = 0, $\phi = 35°$

A8.10 The key to answering this question is replacing the shear strength of fracture set B with an effective shear strength that is a weighted average of the intact rock and joint parameters

$$\begin{aligned}\tau &= \hat{c} + \sigma_n\left(\tan\hat{\phi}\right)\\ &= \left[\chi c_B + (1-\chi)\,c_i\right] + \sigma_n\left[\chi\tan\phi_B + (1-\chi)\tan\phi_i\right]\end{aligned}$$

where χ is a fracture : intact rock proportionality constant. In this case, we have $\chi = 0.5$.

We require the shear strength parameters of the intact rock in σ–τ space. The strength criterion has been given in principal stress space in the question, but is easily converted using the equation $\sigma_1 = (2c\cos\phi)/(1-\sin\phi) + (1+\sin\phi)/(1-\sin\phi)\sigma_3$.
Comparing this relation with the given strength criterion, we see that $5.29 = (1+\sin\phi_i)/(1-\sin\phi_i) \Rightarrow \sin\phi_i = 4.29/6.29 \Rightarrow \phi_i = 43°$ and hence

$$75 = \frac{2c_i\cos 43}{1-\sin 43} \Rightarrow c_i = \frac{75\,(1-\sin 43)}{2\cos 43} = 16.3\ \text{MPa}$$

To find the strength parameters for the intermittent fracture set, we combine the values obtained for c_i and ϕ_i together with the given values for the strength parameters of fracture set B to find

$$\hat{c} = [0.5 \times 0.0 + (1 - 0.5) \times 16.3] = 8.15\,\text{MPa}$$

and

$$\tan\hat{\phi} = [0.5 \times \tan 35 + (1 - 0.5) \times \tan 43] = 0.8164 \Rightarrow \hat{\phi} = 39.2^\circ$$

Having now found the appropriate strength parameters for the intermittent fracture set, the analysis proceeds as in A8.6 and A8.7 with the results shown below.

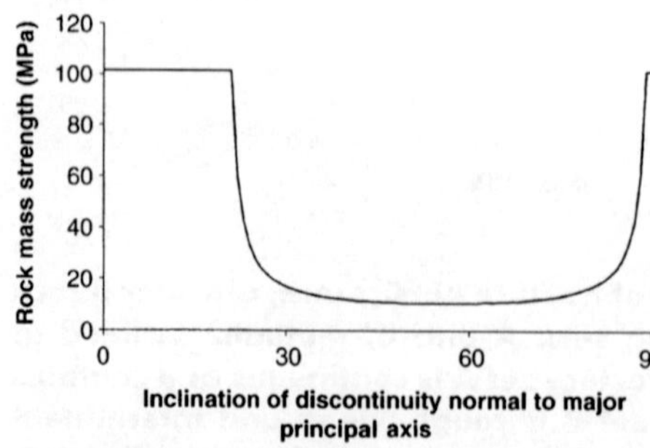

Fracture set A:
$c_A = 100$ kPa and $\phi_A = 20$

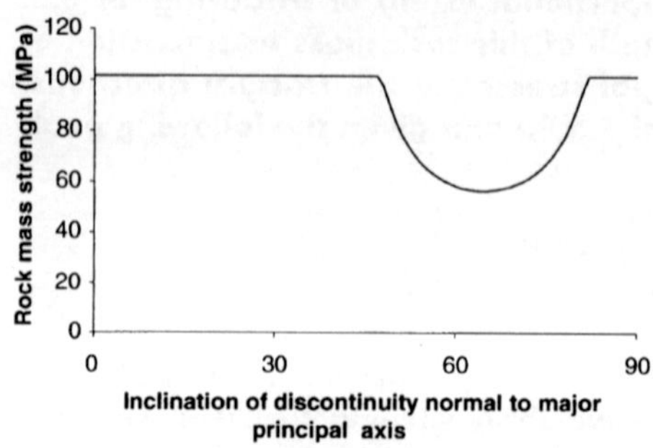

Fracture set B:
$\hat{c} = 8.15$ MPa and $\hat{\phi} = 39.2^\circ$

These diagrams are for each set in isolation. In the rock mass we have $\hat{\beta}_B = 45^\circ$ when $\beta_A = 0^\circ$, and so, using set A as the reference, a -45° shift must be applied to $\hat{\beta}$ to bring the two sets to a common co-ordinate system. The resulting plot is as shown on the next page with the heavier line indicating the composite rock mass strength.

Note that in the second and fourth quadrants, i.e. 90–180° and 270–360°, the strength of the rock mass is only controlled by set A.

8.3 Additional points

We have emphasized that the fractures are the dominant feature governing rock mass behaviour. These fractures dominate the rock mass geometry, deformation modulus, strength, failure behaviour, permeability, and even the local magnitudes and directions of the *in situ* stress field. There is, however, considerable complexity inherent in the geometrical and mechanical properties of rock mass fracture systems, which should be taken into account in direct modelling of the specific fractures. For

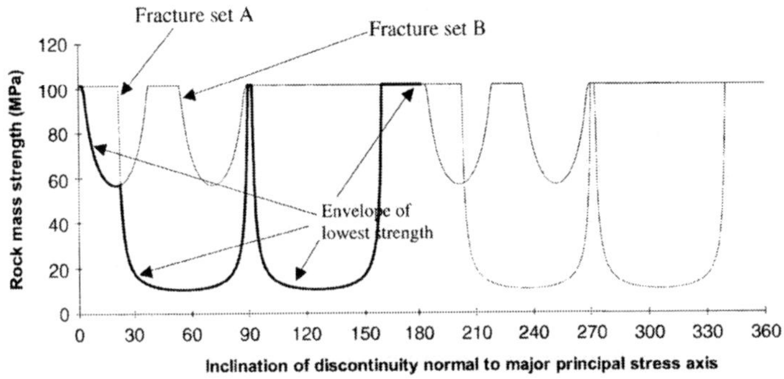

example, the influence of the fractures on rock mass strength which we have highlighted in this chapter has been extended[7] by Amadei (1988) to 3-D and to include the effect of the intermediate principal stress.

When a rock mass is loaded, as in Fig. 8.5, we are interested in the manner in which it will deform, how much it will deform, when failure will occur, and what will happen after failure. Similarly, if an excavation, such as a tunnel, is created in the rock mass, we are interested in the same factors. But, because of the complexity of the rock mass and the

Figure 8.5 Load applied to a rock mass.

[7] Amadei B. (1988) Strength of a regularly jointed rock mass under biaxial and axisymmetric loading. *Int. J. Rock Mech. Min. Sci.*, **25**, 1, 3–13.

difficulty of obtaining detailed information, simplified approaches have been adopted.

The use of rock mass classification methods for evaluating rock mass modulus (Q and A8.4), and the rock mass strength (Q and A8.9) can provide good estimates. However, it is important to understand the factors that govern rock mass deformability and strength in order to be able to assess the validity of these simplified approaches in a given practical situation. We have highlighted this understanding in the questions and answers in this chapter.

A further point is what happens to the rock mass after its strength has been reached? The principles outlined in Section 6.1, A6.1 and A6.6 for intact rock also apply to rock masses. Whether the rock mass failure will be abrupt or progressive depends critically on whether it is being loaded by a 'soft' or 'stiff' system.

9 Permeability

9.1 Permeability of intact rock and rock masses

Permeability is a measure of the ease with which a fluid will flow through rock. Fluids do not flow easily through solid rock, so intact rock has a low permeability. However, rock masses are fractured and, because fractures conduct fluids more readily, it is easier for fluids to flow through them (Fig. 9.1).

Often, the permeability of a rock mass is considered as a scalar quantity, i.e. it can be characterized by one value which is therefore independent of the flow direction. However, in the general case, there will be normal flows (analogous to normal stresses) and there will be cross-flows (analogous to shear stresses). Thus, an improved approach is to consider the **permeability** of a continuum **as a second-order tensor**, analogous to the quantities stress and strain discussed in Chapters 3 and 5. This means that there will be **principal permeability** directions and magnitudes (similar to principal stresses) acting on planes along which there is no cross-flow (similar to zero shear stress on principal stress planes).

The rock mass is not a continuum, but the tensor assumption, with its six independent components, is a significant improvement on a scalar characterization, especially when we know that rock mass properties are directionally dependent because of the fractures.

We noted in Chapter 3 that, to eliminate internal moments in the rock, it is assumed that not all the shear stresses are independent, and this reduces the number of independent stress components from nine to six

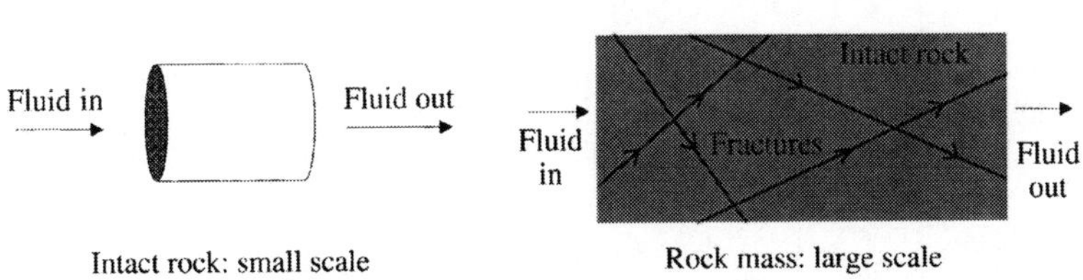

Figure 9.1 Fluid flow through intact rock and fractured rock.

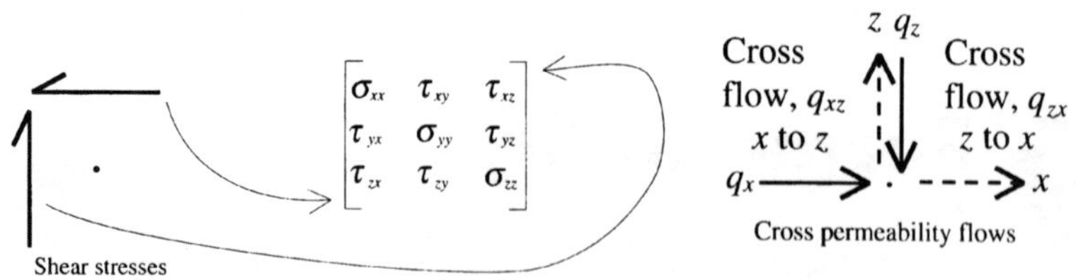

Figure 9.2 Equivalence of the shear stresses and the cross-permeabilities.

(Fig. 9.2). If this assumption is not made, there are nine independent components (for example, as in the Cosserat continuum which does allow for internal moments). In the case of permeability, the equivalent assumption is that the resistances to cross-flow from the x to z axis and from the z to x axis are equal (Fig. 9.2). In intact rock, this assumption seems reasonable, but in a fractured rock mass, the assumption is more questionable.

Thus, **the tensor quantities of stress, strain and permeability are mathematically identical** (see Table 9.1): the same mathematics represents the different physical quantities. As a consequence, we can use Mohr's circle for permeability in the same way as for stress and strain.

The units of permeability, k, are metres squared (L^2). For the case when the fluid is water, the term 'hydraulic conductivity', K, is used; this has units of metre/second (LT^{-1}). The relation between the two is $k = (\mu/\gamma_f)K$ where μ is the water viscosity ($L^{-1}MT^{-1}$) and γ_f is the unit weight of water ($L^{-2}MT^{-2}$).

We distinguish between the permeabilities of the intact rock and the fractured rock mass by using the terms **'primary permeability'** and **'secondary permeability'**, respectively. It is reasonable to assume that the intact rock is continuous and that the primary permeability can be represented by a tensor, with the three orthogonal principal permeabilities. However, for secondary permeability, it should be remembered that the fractures tend to occur in sets and the directions of maximal and min-

Table 9.1 The general and principal components of stress, strain and permeability

	Components of the stress, strain and permeability matrices with reference to known x, y and z axes	The principal components of stress, strain and permeability
Stress	$\begin{bmatrix} \sigma_{xx} & \tau_{xy} & \tau_{xz} \\ & \sigma_{yy} & \tau_{yz} \\ \text{symm.} & & \sigma_{zz} \end{bmatrix}$	$\begin{bmatrix} \sigma_1 & 0 & 0 \\ & \sigma_2 & 0 \\ \text{symm.} & & \sigma_3 \end{bmatrix}$
Strain	$\begin{bmatrix} \varepsilon_{xx} & \varepsilon_{xy} & \varepsilon_{xz} \\ & \varepsilon_{yy} & \varepsilon_{yz} \\ \text{symm.} & & \varepsilon_{zz} \end{bmatrix}$	$\begin{bmatrix} \varepsilon_1 & 0 & 0 \\ & \varepsilon_2 & 0 \\ \text{symm.} & & \varepsilon_3 \end{bmatrix}$
Permeability	$\begin{bmatrix} k_{xx} & k_{xy} & k_{xz} \\ & k_{yy} & k_{yz} \\ \text{symm.} & & k_{zz} \end{bmatrix}$	$\begin{bmatrix} k_1 & 0 & 0 \\ & k_2 & 0 \\ \text{symm.} & & k_3 \end{bmatrix}$

Figure 9.3 Example of water flow through a fractured rock mass. To the right side of the picture, the rock mass is relatively dry. However, in the central portion the rock is wet and there are several jets of water issuing from the rock mass, e.g. at the position indicated by the white arrow.

imal fracture frequencies are generally not orthogonal. The ramifications this has for the concept of principal values and directions for secondary permeability are not clear, but the difficulties of modelling fluid flow in fractured rock masses are evident from the photograph in Fig. 9.3.

Currently, the best way to estimate fracture flow is to use numerical programs that can simulate the fractures directly. From the results, an equivalent tensor representation can be deduced. Initially, these numerical programs were developed for two dimensions and solved for the fracture flow and nodal heads at the fracture intersection points by setting up a series of equations based on Darcy's Law and the fact that the flows into and out of a node are equal (i.e. the continuity equation or Kirchoff's Law). Later, it became possible to solve the problem in three dimensions but, because pressure distributions have to be incorporated along the lines of intersection of the fractures, the extension to 3-D is not simply achieved by 'adding a z component' to the 2-D $x-y$ analysis. It is now also possible to include the primary permeability of the rock blocks between the fractures. In many cases, the contribution of the intact rock will be small; in other cases, such as the sandstone forming petroleum reservoirs, the intact rock contribution is significant.

The quantity of water flowing through a planar rock fracture is theoretically proportional to the cube of the fracture aperture. The consequence of this for modelling and for rock engineering is that the localized nature of water flow through the rock fractures is further concentrated by the effect of the wider fractures (Fig. 9.3). For example, in tunnels in fractured rock, water inflow generally occurs through the wider fractures which are well connected with the surrounding fracture network, rather than equally through all fractures. The idea of the Representative Elemental Volume (REV) (Bear, 1979[1]) was developed because of this type of local influence of the fractures. The REV is the sample size above which a measured rock property becomes essentially constant; any significant property variation caused by the idiosyncrasies in small samples becomes less and less as the sample volume is increased, until a reproducible value is obtained.

Currently, the problem we face is to obtain sufficient information about the fractures to make rock mass simulations realistic. It is only after obtaining good fracture data and using a fracture flow program that one can consider whether the continuum tensor approximation at a given REV level is likely to be useful. This logic is similar to that for the elasticity of fractured rocks: one can only decide whether an isotropic assumption is appropriate once one understands the anisotropic nature of the rock mass.

9.2 Question and answers: permeability

Q9.1 How long does it takes for water, subjected to a 10 m head difference, to pass horizontally through

(a) a 5 m length of intact granite which has an isotropic hydraulic conductivity, *K*, of 1×10^{-12} m/s and

(b) through a 5 m length of fractured limestone with an isotropic hydraulic conductivity of 1×10^{-4} m/s?

A9.1 We use Darcy's law, $Q = KAi$, where Q is the flow rate in m^3/s, K is the hydraulic conductivity in m/s, A is the cross-sectional rock area in m^2, and i is the dimensionless hydraulic gradient, m/m.

(a) The volume of water flowing, V, is Qt, where t is the time. Therefore, $t = V/Q = V/KAi$. For a length of 5 m and a cross-sectional area A, $V = 5A$ and so $t = 5A/KAi$ or $t = 5/Ki$. Substituting $K = 1 \times 10^{-12}$ and $i = 10/5 = 2$, we obtain, $t = 2.5 \times 10^{12}$ s $\approx$ 790,000 years. This rate of water penetration is not significant during engineering works, but it is significant for water flowing through granite masses over geological time: in a hundred million years, the water will have penetrated more than half a kilometre.

(b) For the fractured limestone, $K = 1 \times 10^{-4}$ m/s, and we have $t = 2.5/K = 2.5 \times 10^4$ s $\approx$ 7 h. This result is significant for engineering. It means that for the conditions assumed and with the introduction of water with a 10 m head difference, water would penetrate into the rock mass for 17 m in all directions in a day.

Q9.2 The hydraulic conductivity of an array of parallel fractures (Hoek and Bray, 1977[2]) in the direction parallel to the plane of

[1] The Representative Elemental Volume concept was originally developed for permeability by Bear J. (e.g.) (1979) *Hydraulics of Groundwater*. McGraw-Hill, New York.

[2] Hoek E. and Bray J. W. (1977) *Rock Slope Engineering*. Institution of Mining and Metallurgy, London, 402pp.

the fractures is given by the equation $K = \lambda g e^3/12\nu$ where λ is the fracture frequency, g is gravitational acceleration, e is the fracture aperture, and ν is the kinematic viscosity of the fluid. For a rock mass with a fracture frequency of one fracture per metre and with fracture apertures of 0.01 mm, the hydraulic conductivity is 8.3×10^{-10} m/s.

(a) What is the hydraulic conductivity of a second rock mass which has ten fractures per metre and fracture apertures of 1 mm?

(b) What is the main factor contributing to the difference in the hydraulic conductivity of the two rock masses?

A9.2 (a) We are given K, λ and e, so

$$\frac{g}{12\nu} = \frac{K}{\lambda e^3} = \frac{8.3 \times 10^{-10}}{1 \times (0.01 \times 10^{-3})^3} = 8.3 \times 10^5 \text{ m/s}.$$

Therefore, the hydraulic conductivity of the second rock mass is

$$K = \lambda e^3 \times \left(\frac{g}{12\nu}\right) = 10 \times (1 \times 10^{-3})^3 \times (8.3 \times 10^5) = 8.3 \times 10^{-3} \text{ m/s}.$$

which is a ten million fold increase from the value for the first rock mass.

(b) In the second rock mass, the fracture frequency has increased by a factor of 10 and the aperture by a factor of 100 over the values for the first rock mass. However, the permeability is linearly proportional to the fracture frequency but proportional to the cube of the aperture, giving the ten million fold increase from $10 \times 100 \times 100 \times 100$. So it is the fracture aperture increase that has contributed most to the hydraulic conductivity increase. This illustrates the significant effect of the fractures and their properties on governing the flow of water through a fractured rock mass.

Moreover, we noted in Chapter 3 that the process of rock excavation reduced the normal stress to zero on the excavation surfaces. This enables proximate rock fractures to expand and the hydraulic conductivity of the disturbed zone around the excavation to increase as a function of the cube of the fracture aperture increase.

Q9.3 In question Q9.2, the hydraulic aperture was used. For a fracture with planar and parallel sides, this aperture is the perpendicular distance between the two sides.

(a) Explain the meaning of fracture aperture when the fracture surfaces are rough.

(b) Do you think that the mechanical aperture and the hydraulic aperture of a fracture have the same value? If not, which is greater?

A9.3 (a) When the fracture surfaces are rough, the geometrical aperture will vary all along the fracture. If the aperture geometry were known, the distribution of geometrical fracture apertures could be generated, and a mean and standard deviation calculated. However, the hydraulic aperture will be less than the mean geometrical aperture because the flow will be controlled by regions of smaller aperture, which act as bottlenecks to the flow. In 3-D, it is somewhat different because water flow will occur, along and *across* the fracture aperture, in the

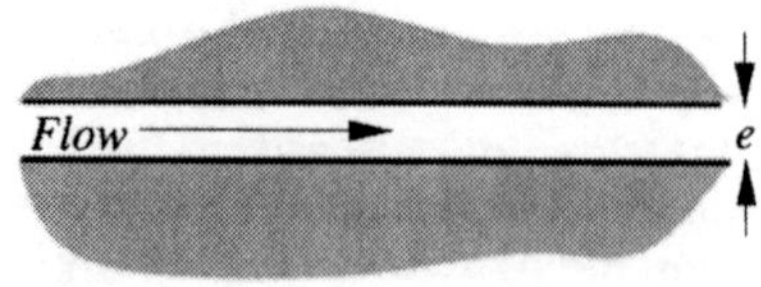

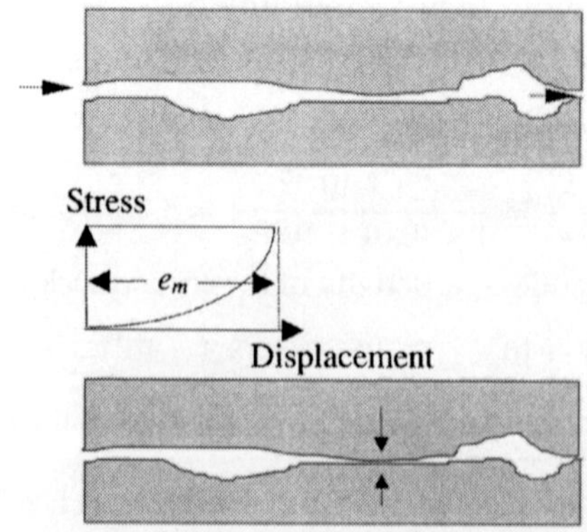

higher-aperture channels formed between the fracture surfaces. Thus, an equivalent hydraulic fracture aperture will be required: the value will be less than the mean geometrical aperture in 2-D, but could be greater in 3-D.

(b) When the fracture is subjected to a compressive normal stress, its aperture reduces. An increasing normal stress is required to continue reducing the aperture, as indicated in the graph above. The mechanical aperture is the amount by which the fracture opens when it is unloaded from a normal stress value equal to the intact rock compressive strength to zero. Thus, the hydraulic aperture and the mechanical aperture are governed by different aspects of the geometry in 3-D: the hydraulic aperture is governed by the channels of aperture spaces, and the mechanical aperture by the regions where there is connecting rock, i.e. where the aperture has zero values. (In 2-D, there can be no fracture flow if the fracture sides are transmitting normal stress, because the aperture will be zero in that region and hence the flow channel will be blocked.) Empirical and back analysis approaches are the only practical methods of establishing equivalent fracture apertures.

Q9.4 What is meant by the term 'transmissivity'?

A9.4 The term 'transmissivity', T, is used for the product of hydraulic conductivity and stratum thickness, t: $T = Kt$. This has a meaning for a continuum, but there are difficulties when the rock mass hydraulic conductivity is governed by fracture flow. As pointed out in the US National Committee for Rock Mechanics publication on rock fractures and fluid flow (Committee on Fracture Characterization and Fluid Flow,

US National Committee for Rock Mechanics, 1996[3]), packer testing at one location in a borehole may encounter a single, highly conductive, extensive, horizontal fracture. When the hydraulic conductivity value has been determined from such test results, it does not make sense to say that a 10 m thickness of the rock stratum is twice as transmissive as a 5 m thickness, when there is only one major fracture conduit present. For this reason, the term 'fracture transmissivity' is receiving increasing usage.

Q9.5 The horizontal section below shows two sets of rock fractures in a 10 m square rock block of unit thickness[4]. On the left-hand side, there is a hydraulic head of 3 m; on the right-hand side, the head is 1 m; and along the top and bottom edges the head linearly decreases from the left to the right.

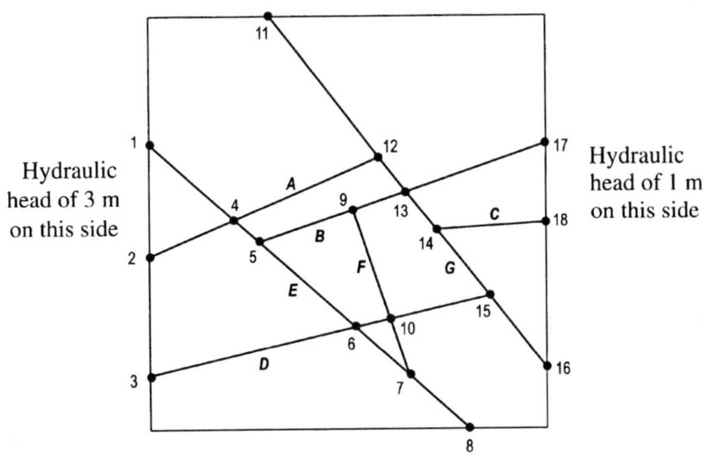

The (x,y) co-ordinates (in m) of the fracture intersections, i.e. the numbered nodes, relative to an origin at the bottom left of the diagram, are given in the table below.

1	2	3	4	5	6	7	8	9	10	11	12	13	14	15	16
0.00	0.00	0.00	2.10	2.74	5.16	6.61	8.06	5.16	6.13	3.06	5.81	6.61	7.42	8.71	10.00
6.94	4.19	1.29	5.16	4.52	2.42	1.13	0.00	5.32	2.58	10.00	6.61	5.65	4.84	3.23	1.45

Assuming that all the fractures have the same aperture, 0.12 mm, and that there is no variation in flow throughout the thickness of the rock block, determine the nodal heads and hence the direction and magnitude of flow in each fracture segment.

[3] Committee on Fracture Characterization and Fluid Flow, US National Committee for Rock Mechanics, (1996) *Rock Fractures and Fluid Flow.* National Academy Press, Washington, DC.

[4] This question is based on Example 11.3 in the book *Discontinuity Analysis for Rock Engineering* by S. D. Priest, with the numerical values slightly changed.

A9.5 To solve such network problems, we have to set up and solve a series of simultaneous equations for the unknown nodal heads. (Because the network is horizontal, we do not need to know the elevation of the network above the datum, i.e. the solution is independent of the elevation head.) We do this by writing an equation for the net flow at each node in terms of the heads at the adjacent nodes and the conductances of the connecting channels, using the fact that the net flow at each node is zero and that the flow in each fracture segment is given by the product of the conductance of the fracture and the head loss along it.

The first step is to establish a matrix of apertures for the network segments linking the various nodes, as follows.

node	1	2	3	4	5	6	7	8	9	10	11	12	13	14	15	16	17	18
1				0.12														
2				0.12														
3						0.12												
4					0.12							0.12						
5						0.12			0.12									
6							0.12			0.12								
7								0.12		0.12								
8																		
9										0.12			0.12					
10															0.12			
11												0.12						
12													0.12					
13														0.12			0.12	
14															0.12			0.12
15																0.12		
16																		
17																		
18																		

In this case, all the apertures are equal, but in the next question, Q9.6, we will be solving the same network for the case when the fracture aperture values are different between segments. Note that this matrix is symmetrical, and we have suppressed the values in the lower left half of the matrix.

Using this matrix, we now compute the associated matrix of conductances, using the formula $c_{i,j} = (ge_{i,j}^3)/(12\nu L_{i,j})$, where $c_{i,j}$, $e_{i,j}$ and $L_{i,j}$ represent the conductance, aperture and length of the fracture segment between node i and node j, and where g and ν are the acceleration due to gravity and the kinematic viscosity of water, respectively. The numerical values of g and ν used here are 9.8 m/s^2 and 1×10^{-6} m^2/s. As the length of the channels is given in m, to simplify the handling of the units, it is convenient to convert the aperture values from mm to m as well. Once this has been done, and remembering that the block has a height of 1 m, it is found that the values are as below with units of conductance in m^2/s.

We now determine an equation for each nodal head in the interior of the network, using the heads of the adjacent nodes. The fundamental

	1	2	3	4	5	6	7	8	9	10	11	12	13	14	15	16	17	18
1				5.0E-7														
2				6.1E-7														
3						2.7E-7												
4					1.7E-6							3.6E-7						
5						4.3E-7			5.6E-7									
6							7.2E-7			1.5E-6								
7								7.8E-7		8.8E-7								
8																		
9										5.0E-7			9.7E-7					
10															5.5E-7			
11												3.2E-7						
12													1.1E-6					
13														1.1E-6			3.9E-7	
14															7.1E-7			5.2E-7
15																6.4E-7		
16																		
17																		
18																		

equation for the head at an interior node is

$$H_j = \frac{\sum_i c_{i,j} H_i}{\sum_i c_{i,j}}$$

(see Section 9.4 of ERM 1, where we show how this equation is derived from the Kirchoff relation for the net flow at a node). Expanding this equation for node 4 results in

$$H_4 = \frac{c_{1,4}H_1 + c_{2,4}H_2 + c_{5,4}H_5 + c_{12,4}H_{12}}{c_{1,4} + c_{2,4} + c_{5,4} + c_{12,4}}$$

Rearranging this to put all of the known heads on the left-hand side of the equation gives

$$-c_{1,4}H_1 - c_{2,4}H_2 = -\left(c_{1,4} + c_{2,4} + c_{5,4} + c_{12,4}\right) H_4 + c_{5,4}H_5 + c_{12,4}H_{12}.$$

Continuing for the remaining nodes gives the following relations

Node 5 $\quad 0 = c_{4,5}H_4 - (c_{4,5} + c_{6,5} + c_{9,5})H_5 + c_{6,5}H_6 + c_{9,5}H_9$

Node 6 $\quad -c_{3,6}H_3 = c_{5,6}H_5 - (c_{3,6} + c_{5,6} + c_{7,6} + c_{10,6})H_6 + c_{7,6}H_7 + c_{10,6}H_{10}$

Node 7 $\quad -c_{8,7}H_8 = c_{6,7}H_6 - (c_{6,7} + c_{8,7} + c_{10,7})H_7 + c_{10,7}H_{10}$

Node 9 $\quad 0 = c_{5,9}H_5 - (c_{5,9} + c_{10,9} + c_{13,9})H_9 + c_{10,9}H_{10} + c_{13,9}H_{13}$

Node 10 $\quad 0 = c_{6,10}H_6 + c_{7,10}H_7 + c_{9,10}H_9 - (c_{6,10} + c_{7,10} + c_{9,10} + c_{15,10})H_{10} + c_{15,10}H_{15}$

Node 12 $\quad -c_{11,12}H_{11} = c_{4,12}H_4 - (c_{4,12} + c_{11,12} + c_{13,12})H_{12} + c_{13,12}H_{13}$

Node 13 $\quad -c_{17,13}H_{17} = c_{9,13}H_9 + c_{12,13}H_{12} - (c_{9,13} + c_{12,13} + c_{14,13} + c_{17,13})H_{13} + c_{14,13}H_{14}$

Node 14 $\quad -c_{18,14}H_{18} = c_{13,14}H_{13} - (c_{13,14} + c_{15,14} + c_{18,14})H_{14} + c_{15,14}H_{15}$

Node 15 $\quad -c_{16,15}H_{16} = c_{10,15}H_{10} + c_{14,15}H_{14} - (c_{10,15} + c_{14,15} + c_{16,15})H_{15}.$

These equations are now assembled into a system of ten simultaneous equations. It is not convenient to solve such a large system by hand, and so a matrix solution using a computer or appropriate hand-held calculator should be sought. With this technique, we build and solve the following matrix equation (note that for the sake of brevity the left-hand matrix below is incomplete, although we have included illustrative rows and columns based on the equations given above).

$$
\begin{bmatrix}
A & c_{5,4} & 0 & 0 & 0 & 0 & c_{12,4} & 0 & 0 & 0 \\
c_{4,5} & \ddots & & 0 & & & & & & 0 \\
0 & & \ddots & c_{7,6} & & & & & & 0 \\
0 & 0 & c_{6,7} & B & 0 & c_{10,7} & 0 & 0 & 0 & 0 \\
0 & & & 0 & \ddots & & & & & 0 \\
0 & & & c_{7,10} & & \ddots & & & & c_{15,10} \\
c_{4,12} & & & 0 & & & \ddots & & & 0 \\
0 & & & 0 & & & & \ddots & & 0 \\
0 & & & 0 & & & & & \ddots & c_{15,14} \\
0 & 0 & 0 & 0 & 0 & c_{10,15} & 0 & 0 & c_{14,15} & C
\end{bmatrix}
\times
\begin{bmatrix} H_4 \\ H_5 \\ H_6 \\ H_7 \\ H_9 \\ H_{10} \\ H_{12} \\ H_{13} \\ H_{14} \\ H_{15} \end{bmatrix}
=
\begin{bmatrix} -c_{1,4}H_1 - c_{2,4}H_2 \\ 0 \\ -c_{3,6}H_3 \\ c_{8,7}H_8 \\ 0 \\ 0 \\ -c_{11,12}H_{11} \\ -c_{17,13}H_{17} \\ -c_{18,14}H_{18} \\ -c_{16,15}H_{16} \end{bmatrix}
$$

where $A = -(c_{1,4} + c_{2,4} + c_{5,4} + c_{12,4})$
$B = -(c_{7,6} + c_{10,7})$, and
$C = -(c_{10,15} + c_{14,15})$

Substituting all the numerical values given above gives the following for the complete matrix equation (note that we have divided by a constant value of 1×10^{-7}).

$$
\begin{vmatrix}
-31.3 & 16.6 & 0.0 & 0.0 & 0.0 & 0.0 & 3.6 & 0.0 & 0.0 & 0.0 \\
16.6 & -26.6 & 4.3 & 0.0 & 5.6 & 0.0 & 0.0 & 0.0 & 0.0 & 0.0 \\
0.0 & 4.3 & -29.1 & 7.2 & 0.0 & 14.9 & 0.0 & 0.0 & 0.0 & 0.0 \\
0.0 & 0.0 & 7.2 & -23.9 & 0.0 & 8.8 & 0.0 & 0.0 & 0.0 & 0.0 \\
0.0 & 5.6 & 0.0 & 0.0 & -20.3 & 5.0 & 0.0 & 9.7 & 0.0 & 0.0 \\
0.0 & 0.0 & 14.9 & 8.8 & 5.0 & -34.2 & 0.0 & 0.0 & 0.0 & 5.5 \\
3.6 & 0.0 & 0.0 & 0.0 & 0.0 & 0.0 & -18.1 & 11.3 & 0.0 & 0.0 \\
0.0 & 0.0 & 0.0 & 0.0 & 9.7 & 0.0 & 11.3 & -35.7 & 10.9 & 0.0 \\
0.0 & 0.0 & 0.0 & 0.0 & 0.0 & 0.0 & 0.0 & 10.9 & -23.1 & 7.1 \\
0.0 & 0.0 & 0.0 & 0.0 & 0.0 & 5.5 & 0.0 & 0.0 & 7.1 & -19.0
\end{vmatrix}
\times
\begin{vmatrix} H_4 \\ H_5 \\ H_6 \\ H_7 \\ H_9 \\ H_{10} \\ H_{12} \\ H_{13} \\ H_{14} \\ H_{15} \end{vmatrix}
=
\begin{vmatrix} -33.3 \\ 0.0 \\ -8.0 \\ -18.8 \\ 0.0 \\ 0.0 \\ -4.5 \\ -3.9 \\ -5.2 \\ -6.4 \end{vmatrix}
$$

With a modern spreadsheet program, the left-hand matrix can be inverted and hence solved for the unknown nodal heads. Doing this leads to the following values (where the units are in m).

H_4	H_5	H_6	H_7	H_9	H_{10}	H_{12}	H_{13}	H_{14}	H_{15}
2.507	2.339	2.229	2.223	1.930	2.062	1.759	1.626	1.438	1.472

Finally, we illustrate these heads in the diagram below, using shading to identify the various flow channels in the network.

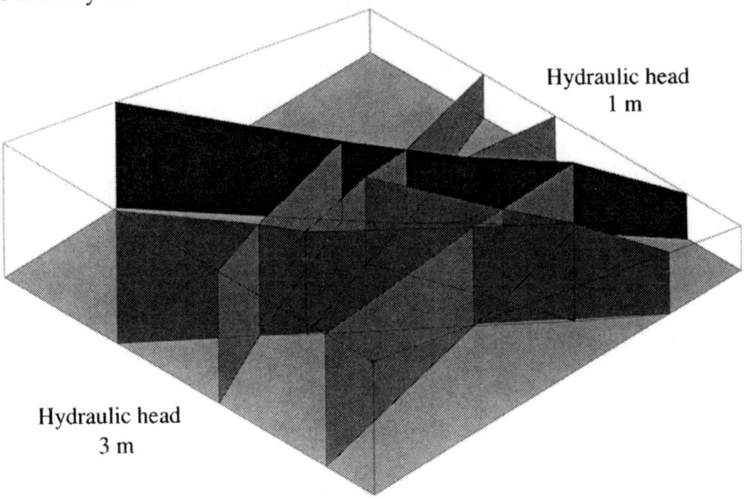

From this diagram, we can see how the head values generally decrease from the left- to the right-hand side of the network, with the water flow always travelling in the direction from the boundary with the highest head to the boundary with the lowest head. However, the circumstances may not be so clear if the apertures of the fracture sets are not the same, as the following question illustrates.

Q9.6 Determine the nodal heads and hence the direction and magnitude of flow in each rock mass fracture segment for the same case as Q9.5, but with different fracture apertures, as given in the matrix below.

	1	2	3	4	5	6	7	8	9	10	11	12	13	14	15	16	17	18
1	■			0.07														
2		■		0.16														
3			■			0.24												
4				■	0.07							0.16						
5					■	0.07			0.18									
6						■	0.07			0.24								
7							■	0.07		0.09								
8								■										
9									■	0.09			0.18					
10										■					0.24			
11											■	0.06						
12												■	0.06					
13													■	0.06			0.18	
14														■	0.06			0.13
15															■	0.06		
16																■		
17																	■	
18																		■

Note that this matrix is symmetric, and we have suppressed the values in the lower left of the matrix. The units are millimetres.

A9.6 We can use the same solution procedure as in A9.5 to determine the nodal head values and hence the flow directions. The aperture values given in the question lead to the following matrix equation.

$$\begin{vmatrix} -27.3 & 3.3 & 0.0 & 0.0 & 0.0 & 0.0 & 8.5 & 0.0 & 0.0 & 0.0 \\ 3.3 & -23.2 & 0.9 & 0.0 & 19.1 & 0.0 & 0.0 & 0.0 & 0.0 & 0.0 \\ 0.0 & 0.9 & -142.4 & 1.4 & 0.0 & 118.8 & 0.0 & 0.0 & 0.0 & 0.0 \\ 0.0 & 0.0 & 1.4 & -6.7 & 0.0 & 3.7 & 0.0 & 0.0 & 0.0 & 0.0 \\ 0.0 & 19.1 & 0.0 & 0.0 & -54.0 & 2.1 & 0.0 & 32.8 & 0.0 & 0.0 \\ 0.0 & 0.0 & 118.8 & 3.7 & 2.1 & -168.9 & 0.0 & 0.0 & 0.0 & 44.3 \\ 8.5 & 0.0 & 0.0 & 0.0 & 0.0 & 0.0 & -10.3 & 1.4 & 0.0 & 0.0 \\ 0.0 & 0.0 & 0.0 & 0.0 & 32.8 & 0.0 & 1.4 & -48.7 & 1.4 & 0.0 \\ 0.0 & 0.0 & 0.0 & 0.0 & 0.0 & 0.0 & 0.0 & 1.4 & -8.9 & 0.9 \\ 0.0 & 0.0 & 0.0 & 0.0 & 0.0 & 44.3 & 0.0 & 0.0 & 0.9 & -46.0 \end{vmatrix} \times \begin{vmatrix} H_4 \\ H_5 \\ H_6 \\ H_7 \\ H_9 \\ H_{10} \\ H_{12} \\ H_{13} \\ H_{14} \\ H_{15} \end{vmatrix} = \begin{vmatrix} -46.6 \\ 0.0 \\ -63.9 \\ -3.7 \\ 0.0 \\ 0.0 \\ -0.6 \\ -13.0 \\ -6.6 \\ -0.8 \end{vmatrix}$$

And from this we find the following values for the unknown nodal heads.

H_4	H_5	H_6	H_7	H_9	H_{10}	H_{12}	H_{13}	H_{14}	H_{15}
2.704	1.864	2.730	2.631	1.680	2.689	2.489	1.509	1.240	2.632

These nodal head values are illustrated in the diagram below.

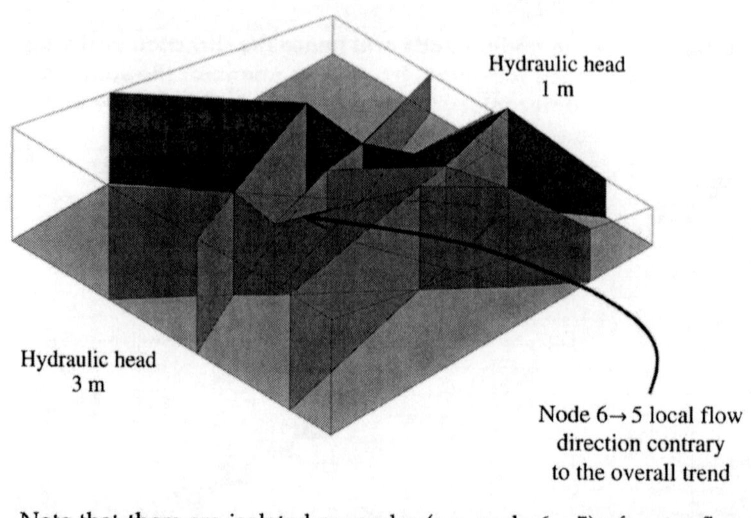

Note that there are isolated examples (e.g. node 6→5) of water flows occurring in a direction contrary to the overall hydraulic head decrease (which is from the left to the right of the diagram). This is in direct contrast to the flow regime determined in A9.5, and illustrates the hazards of assuming regional flow patterns from observations made on a single fracture.

Q9.7 The following rock mass hydraulic conductivity data were obtained from four sets of tests on fractured rock samples with volumes 1 m³, 5 m³, 10 m³, and 15 m³. In each case it was assumed that the hydraulic conductivity was a scalar quantity. The data values are in units of 1 × 10⁻⁶ m/s. As the test volume increased, so did the difficulty and the cost of the tests, so there are fewer data for the tests at larger volumes.

Data for test size of 1 m³

2.122	**1.156**	**3.696**	**1.165**	**1.649**	**2.886**	**1.652**	**2.876**	**1.197**
2.593	**2.114**	**2.771**	**16.214**	**2.529**	**1.700**	**7.658**	**10.928**	**0.627**
9.854	**0.847**	**0.670**	**1.623**	**2.274**	**5.163**	**2.209**		

Data for test size of 5 m³

1.630 1.981 2.436 3.700 1.215 1.767 0.909 0.450 3.512 1.314

Data for test size of 10 m³

1.369 1.188 2.037 1.688

Data for test size of 15 m³

1.487 1.343 1.473 1.738

Using the data above, estimate the REV value[5] for the hydraulic conductivity of the fractured rock mass tested.

A9.7 The rock property data, together with the mean value for each sample volume, are plotted in the diagram below. We can see that there is a larger variation for the smaller sample sizes, and that the mean value progressively decreases as the sample size increases (i.e. from 3.53×10^{-6} m/s, through 1.89×10^{-6} m/s and 1.57×10^{-6} m/s to 1.51×10^{-6} m/s).

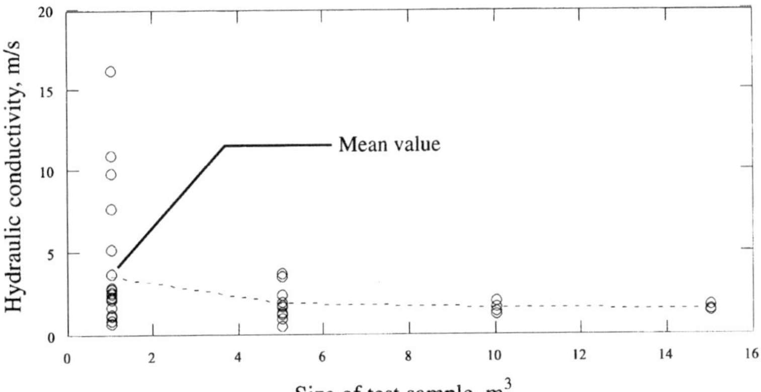

Visual inspection of this plot shows that there does seem to be a stable value for the conductivity at larger volumes, of about 1.5×10^{-6} m/s

[5] The REV definition is given in Section 9.1.

when the REV value is of the order of 8 m^3. More mathematical approaches could be used to evaluate the REV if justified.

We emphasize that the data in this question, and our interpretation, have been presented to illustrate the concept of the REV. In another case, the mean might increase, decrease or oscillate as the tested volume increases. Also, if the rock mass is heterogeneous, and hence its measured value changes with location, it is possible that no REV will be identified because the rock properties could be changing with location faster than the REV is being reached by increasing the sampling size. Numerical simulation of fracture patterns and the associated fracture connectivity and REVs was pioneered by Long (1983)[6].

Q9.8 The values and directions for the principal hydraulic conductivities in a dolomitic rock, shown below, were obtained from back analysis of induced drainage discharges at the Morro da Usina Mine, Vazante District, State of Minas Gerais, Brazil[7]. These are the mean principal hydraulic conductivities for a rock mass volume greater than 3 km^3. What geological circumstances could cause these values and orientations?

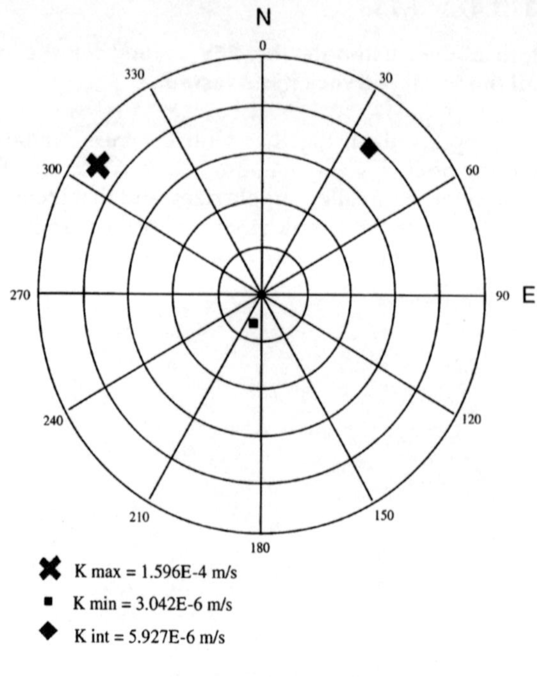

[6] Long J. C. S. (1983) *Investigation of equivalent porous medium permeability in networks of discontinuous fractures.* PhD Thesis, University of California, Berkeley.

[7] The authors are grateful to Prof. E. Quadros and Dr. F. O. Franciss for permission to use this example.

A9.8 We note that the maximum principal hydraulic conductivity has a value of 1.596×10^{-4} m/s in a sub-horizonal orientation to the northwest. The intermediate value is 5.927×10^{-6} m/s in a sub-horizontal direction to the northeast, and the minimum value is 3.042×10^{-6} m/s in a sub-vertical direction. The maximum value is about 52 times greater than the minimum value, but the intermediate value is only about twice the minimum value.

Now consider the geological setting. Dolomitic rock has bedding planes and often has two sets of joints orthogonal to each other and to the bedding. The principal hydraulic conductivity values are sub-horizontal and sub-vertical, indicating that the strata are lying horizontally. Thus, with no further information, we would predict that there is a hydraulically significant set of vertical fractures striking southeast to northwest with a less significant vertical set striking northeast to southwest. In fact, this was the case: the transmissivities of the major northwest fractures and northeast sheared contacts varied between 0.01 and 0.001 m^2/s.

Q9.9 (a) Explain the term 'effective stress'.

(b) If a fracture contains water under pressure and the stresses at the fracture surfaces are being considered, what would be the effect on the normal stress and on the shear stress in the rock of steadily increasing the water pressure?

A9.9 (a) The term 'effective stress', σ', was developed by Terzaghi as the normal stress component, σ, minus the water pressure, u: $\sigma' = \sigma - u$. Goodman (1998)[8] explains that, in 1923, Terzaghi "fully understood now that the addition of an increment of external pressure to a clay resulted in a temporary increment of water pressure of equal magnitude . . . when the water pressure in the soil's pores had a value u, and the external pressure had a magnitude p, only the value $p-u$ was effective in causing force between the grains."

The water pressure is subtracted only from the normal stress components because no shear stresses can be sustained in a fluid. In a non-particulate intact rock, the validity of the use of effective stresses is not clear, especially given the long periods of time that water pressures require for dissipation in intact rock, as we saw in A9.1. The concept is more likely to be valid in fractured rock masses, where the water pressure in the fractures can exert a definite overall pressure, as shown in the sketch below.

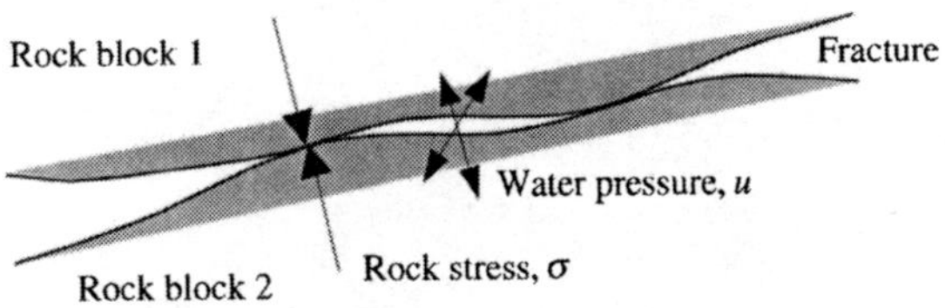

[8] See the biography of K. Terzaghi by Goodman R. E. (1998) *The Engineer as Artist*. ASCE Press, Reston, VA, 340pp.

One can imagine that the normal stress component across the fracture is indeed reduced by the jacking effect of the water.

(b) Unfortunately, the situation is not as simple as may be thought because the fracture has already disturbed the *in situ* stress field, rotating and altering the principal stress components. Also, as the water pressure is steadily increased, the ratio between the amount of shear stress and the amount of normal stress that can be transmitted across the fracture changes, until eventually no shear stress will be transmitted when the fracture surfaces are separated, and only normal stress can be transmitted across the fracture, now the same value, u, along the whole fracture.

In A8.8, we explained that the main difference between soil mechanics and rock mechanics is that soil is a particulate medium in which the grains are much smaller than the engineered structure. The effective stress concept holds in rock mechanics, providing a sufficiently large, strongly fractured rock mass is being considered, but often the dimensions of the engineered structure are of the same order as the sketch above and it is the local, not global, conditions that govern engineering success or failure in rock mechanics.

Q9.10 In the photograph below, the water pressure in a borehole in marble has been steadily increased. In addition to the normal pressure created at the rock surface, the water pressure in this configuration also creates a circumferential tensile stress of the same magnitude in the rock, as in Q4.6. A crack developed and was propagated under servo-controlled conditions from A to B. The water not only penetrated the crack but also the intact marble, forming the dark inverted V visible in the photograph. How do we deal with the effective stress under these circumstances?

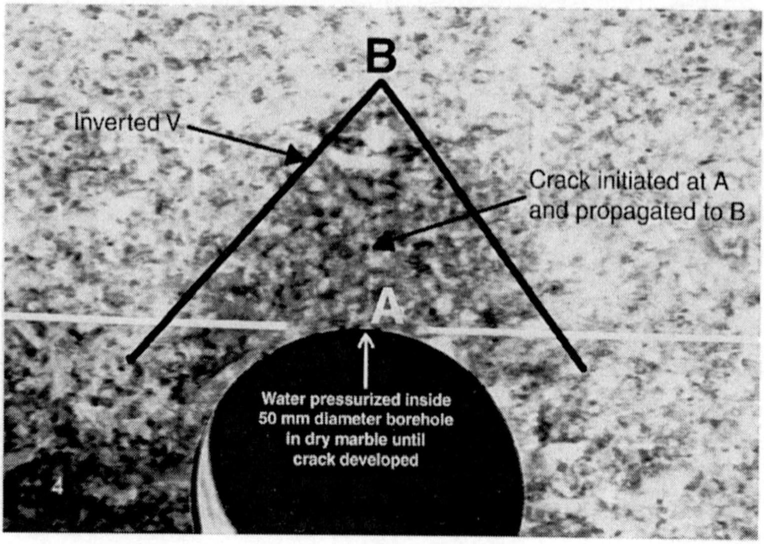

A9.10 In view of the complications described in A9.9, plus in this case the limited crack length, a semi-empirical approach to the inclusion of the water pressure is the most prudent approach. Haimson (1987)[9] uses the poroelastic equation $P_c - P_o = (T + 3\sigma_h - \sigma_H - 2P_o)/K$, where P_c is the critical pressure, P_o is the pore pressure, $K = 2 - \alpha(1 - 2\nu)/(1 - \nu)$ where α is the Biot 'constant', and $1 \geq \alpha \geq 0$, hence $2 \geq K \geq 1$. Readers are referred to the publications of Haimson for more detailed discussion of hydraulic fracturing and to Cheng (1998)[10] for an explanation of poroelasticity.

9.3 Additional points

From the information in Section 9.1 and the answers in Section 9.2, we have highlighted the fact that mechanisms in engineering rock mechanics can be complex, on the small and large scales. It is easy to be critical of semi-empirical approaches and rock mass classification methods, but of necessity there are always some types of simplification and approximation being made in engineering rock mechanics. Sometimes these approximations are not obvious, as in the isotropic assumptions made in the theory of elasticity which were discussed in Chapter 5; sometimes they are obvious, as for the case of water flow as discussed in the current chapter.

The subject of water flow in intact rock and rock masses is one of the most intractable subjects in engineering rock mechanics. For this reason, there are many aspects of the subject requiring further research studies. In the US National Committee on Rock Mechanics book (see Footnote 3 in this chapter) on 'Rock Fractures and Fluid Flow', eight research recommendations are made. These are that the following should be developed:

- additional *in situ* research facilities;
- improved conceptual models for fluid flow and transport;
- improved understanding of the origin and development of fracture systems;
- improved fracture detection methods;
- realistic numerical models;
- understanding of stress, flow, temperature and chemistry coupling;
- further research on waste isolation in fractured rock.

The last two recommendations are included because the subject of water flow in fractured rocks is particularly important for radioactive waste disposal. By definition, the waste will have been successfully isolated in an underground repository only if radionuclides do not migrate in the groundwater from the repository to the biosphere. However, to make an engineering prediction for a long period into the future requires ad-

---

[9] Haimson B. C. (1987) Measurement of *in situ* stress, in *Geophysics, Methods of Experimental Physics, Vol. 24B* (C. G. Sammis and T. L. Henyey (eds), Academic Press, New York, pp. 377–408.

[10] Cheng A. H.-D. (1998) On generalized plane strain poroelasticity. *Int. J. Rock Mech. Min. Sci.*, **35**, 2, 183–193.

vanced modelling and approximating many thermo-hydro-mechanical processes (Stephansson et al., 1999 [11]).

[11] Stephansson O. Hudson J. A. Tsang C.-F. Jing L. and Andersson J. (1999) *DECOVALEX II Project. Coupled THM issues related to repository design and performance, task 4.* SKI Report 99:7, Stockholm.

10 Anisotropy and inhomogeneity

10.1 Rock masses: order and disorder

"The use of laboratory tests to predict the reaction of a rock massif encounters three major and inter-related obstacles. The first of these is the frequent reluctance of the user to recognize and accept the fact that rock is a heterogeneous material. It seldom is isotropic; it very often acts as an anelastic medium; it rarely occurs as a continuous medium. And, all too frequently the rock *in situ* may have natural stresses that cannot be quantitatively examined and thus cannot be simulated in the laboratory. The second major obstacle is that as yet we have no way to quantitatively assess the influence of gross geological defects such as faults, joints, fractures, folds, etc. Thus we are unable to simulate their effects in laboratory testing."

From the information and philosophy presented so far in this book, the text in the paragraph above follows naturally and might have been written recently. In fact, it was the beginning of an article (Judd, 1965[1]) by William Judd published in 1965. It is evident that the problems in engineering rock mechanics have been recognized for a long time and we hope that the same paragraph will not still be appropriate in, say, 2065.

We have seen in the previous chapters that a rock mass is a complex geometrical and mechanical assemblage. Although specific physical mechanisms have created the rock mass in the form we see today, it is difficult to sample and characterize the geometrical and mechanical complexity resulting from the superimposition of a sequence of geological processes. From the engineering point of view, there are two crucial aspects:

(a) we should be aware of this rock mass complexity; and
(b) we should find ways to characterize the rock adequately for engineering purposes.

Generally, it is not necessary to interpret the nature of the present-day rock mass in order to understand exactly which mechanisms left their

[1] Judd W. R. (1965) Some rock mechanics problems in correlating laboratory results with prototype reactions. *Int. J. Rock Mech. Min. Sci.*, **2**, 197–218.

overlapping footprints in the geometry and mechanical properties that we see today; this is the realm of structural geology analyses. However, we do wish to capture the geometrical and mechanical essence of the rock mass in so far as it is required for engineering. As highlighted by both William Judd and previous chapters in this book, an understanding of the geological setting, the *in situ* stress, strain, the intact rock, the fractures, the nature of rock masses, and permeability is essential for engineering analysis.

In this chapter, we now consider the consistency of the rock mass properties. Is the rock deformability and strength the same everywhere? Is the rock strength the same when measured in different directions? Are the fractures in sets of parallel fractures? In short, what is the orientational and spatial consistency of the rock mass in question, and how do we characterize any variations that occur? To emphasize the fact that the rock is not a perfect material for modelling, we use two acronyms, as introduced in the Frontispiece.

CHILE: Continuous, Homogeneous, Isotropic, Linearly Elastic (a model material).
DIANE: Discontinuous, Inhomogeneous, Anisotropic, and Not Elastic (the rock reality).

The terms in these acronyms have the following meanings.

Continuous is mechanically continuous; there can be variations in the mechanical property values but there are no mechanical breaks.
Discontinuous does contain mechanical breaks having effectively zero tensile strength.
Homogeneous[2] has the same property values at all locations.
Inhomogeneous (or heterogeneous) has different property values at different locations.
Isotropic[2] has the same property values in different directions.
Anisotropic has different property values in different directions.
Linearly Elastic the stress–strain curve is a line with a constant slope, all strains are instantaneous, and all energy can be recovered.
Not Elastic on unloading not all energy input can be recovered and strains may be time dependent.

In the past, the rock mass was often modelled as a CHILE material. Nowadays, there is recognition that the rock reality is DIANE and that numerical modelling should be able to incorporate all the DIANE aspects as required by the rock engineering design problem in hand. Often, the pragmatic approach to characterizing a rock mass is achieved by dividing the rock mass into structural domains, each having different property values. These domains can be the lithological domains, or they can be more detailed and be directly based on geotechnical changes.

[2] Greek *homos* ≡ same, *genos* ≡ kind, *isos* ≡ equal, *tropos* ≡ way.

The understanding of the correct use of the statistical terms 'resolution', 'accuracy', 'bias' and 'precision' is helpful for this procedure. We give examples of their correct use here.

Resolution the number of digits used to specify a value, e.g. the value of π can be given as 3, 3.1, 3.14, 3.141, etc.

Accuracy the ability to obtain the correct answer on the average, regardless of the spread of results.

Bias the difference between the sample mean and the true population mean.

Precision the spread of results, regardless of whether they are accurate or not.

The questions in this chapter have been included to emphasize the anisotropy and inhomogeneity problem and to indicate procedures for characterizing these features of the rock mass. We emphasize that the DIANE nature of the rock should not be ignored by simply hoping that a CHILE model is adequate.

10.2 Questions and answers: anisotropy and inhomogeneity

***Q10.1* On the rock mass photograph below, identify example features that indicate that the rock mass is more DIANE than CHILE.**

A10.1 In the picture, the limestone beds have been deformed to a vertical position. The bedding planes form fractures and there is a variety of other fractures visible, making the rock mass Discontinuous. Note that one weak stratum where the engineer is standing has been eroded by the action of rainwater. Thus, the bedding planes, shear planes, joints and erosion features form the fractures. The rock mass is Inhomogeneous because the strata have different property values at different locations (the eroded material was the weakest).

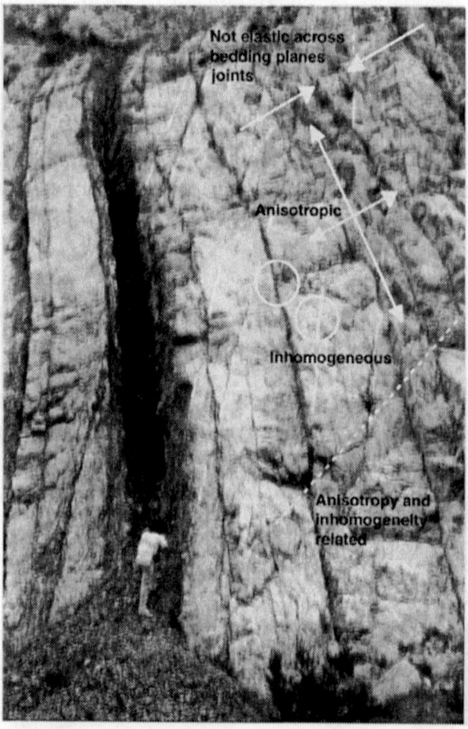

The rock mass is Anisotropic because the fracture frequency and other properties in different directions are significantly different. Such clearly defined fractures will also have a strongly 'Not Elastic' character.

Q10.2 With reference to the photograph in Q10.1, explain why inhomogeneity and anisotropy can be related.

A10.2 The diagram to the right shows four strata with different properties: there is inhomogeneity within and between the strata.

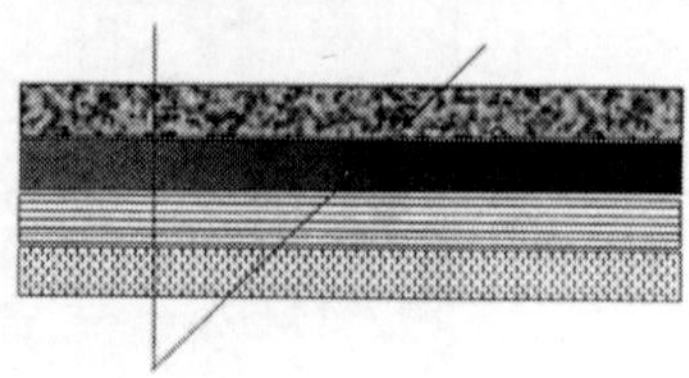
1
2
3
4

Assume strata 1 and 4 are internally homogeneous, but strata 2 and 3 are internally inhomogeneous. A scanline perpendicular to all the strata will therefore give different rock property values depending on where it is located. In any given region, the property values will also depend on scanline direction because different amounts of each scanline will occur per metre of an inclined scanline. Moreover, different types of anisotropy will be experienced at different locations in the different directions because the strata are inhomogeneous.

Q10.3 The following data are pairs of point load strength (*PL*) and uniaxial compressive strength (*UCS*) values for a particular rock type.

***UCS* (MPa)**	**52.4**	**60.7**	**44.6**	**66.6**	**47.6**	**56.3**	**61.5**	**48.2**	**49.6**	**47.2**	**56.7**	**48.4**	**61.5**	**52.9**
***PL* (MPa)**	**2.51**	**2.87**	**2.14**	**3.04**	**2.31**	**2.64**	**2.90**	**2.21**	**2.25**	**2.18**	**2.55**	**2.23**	**2.75**	**2.55**

We wish to correlate these strength values, and can do so either in the form $PL = a\sigma_c + b$ or in the form $\sigma_c = cPL + d$. On the basis of the best independent variable, which of these forms is appropriate?

Determine values for the appropriate constants (i.e. either *a* and *b*, or *c* and *d*).

A10.3 When a straight line of the form $y = mx + c$ (or any other curve) is fitted to a set of observations, we are assuming that the x-values represent the independent variable (i.e. are the known controlled variable), and that the y-values represent the dependent variable. In the case of the two variables, Point Load strength and Uniaxial Compressive Strength, we know that the *PL* values — from an index test — exhibit a wide spread; whereas the *UCS* values — from a relatively precise and direct laboratory procedure — have a much reduced spread. As a result, and for this case, we can take uniaxial compressive strength as the independent variable and fit the straight line $PL = a\sigma_c + b$ to the data.

Fitting a straight line to the data (taking care to make sure that the line is constrained to pass through the origin), gives a value of $a = 0.0465$. We then rearrange the equation to give the relation

$$\sigma_c = \frac{1}{0.046532} PL \text{ or } \sigma_c = 21.49\, PL.$$

Had we tried to fit an equation of the form $\sigma_c = cPL + d$ to the data, we would have found that $c = 21.48$. In this case the difference between the two coefficients is negligible, and not significant for engineering considerations, but the second method is incorrect and in other cases could introduce a much greater difference having engineering significance. It is important to carefully assess which of the variables is the independent variable and use it appropriately.

Q10.4 Imagine that a new index test for determining the tensile strength of specimens of intact rock is under development in the Rock Mechanics Laboratory at Imperial College. This test involves bonding a steel rod to the surface of a specimen with high strength adhesive, and then measuring the tensile load required to pull the rod together with a small piece of rock away from the main block of rock. Four test configurations are under consideration, and for each of these a theoretical relation between rock strength and pull-off force has been developed. Test results for the four configurations,

together with the appropriate theoretical relation, are given below. The first row in each table is the load measured in the new test. The second row in each table is the tensile strength of the rock as measured by a standard method.

Classify each configuration in terms of accuracy and precision, and hence recommend which configuration(s) should be retained for further development.

***Configuration 1*: strength = 0.049 × load**

Load (*N*)	67.3	76.8	83.9	104.8	153.7	168.9	191.2	194.7	237.5	258.3
Strength (MPa)	4.2	4.8	5.2	6.5	9.6	10.6	11.7	12.2	14.6	16.1

***Configuration 2*: strength = 0.066 × load**

Load (*N*)	68.9	105.3	106.2	120.1	148.5	164.8	197.4	220.5	232.8	236.9
Strength (MPa)	4.7	2.4	3.3	7.4	6.7	10.9	6.3	7.8	8.2	7.6

***Configuration 3*: strength = 0.074 × load**

Load (*N*)	83.5	95.0	111.7	151.6	170.0	189.5	190.2	193.9	201.1	205.3
Strength (MPa)	6.1	7.1	8.3	11.4	12.7	13.8	14.3	14.3	14.9	15.3

***Configuration 4*: strength = 0.094 × load**

Load (*N*)	68.9	105.3	106.2	120.1	148.5	164.8	197.4	220.5	232.8	236.9
Strength (MPa)	5.5	10.1	10.5	11.6	14.4	12.3	20.1	22.9	20.9	21.8

A10.4 In each of the test configurations, we should take tensile strength (as listed in the second row of each table) as the independent variable, because the load is being measured in the index test and we fit a line of the form 'load = m × strength' to the data. If the fitted curve has a different gradient to the theoretical relation, then the test is inaccurate: there is bias in the results. If the fitted curve has a low coefficient of determination, r^2, there will be considerable spread about the fitted curve, and so the test is imprecise. Thus, to classify each configuration a straight line is fitted to the data, the slope of the line is compared to the theoretical relation to determine the accuracy and the spread of results is studied to determine the precision.

The test results are shown graphically below. In these plots the dots represent the test results and the lines represent the theoretical relations.

Configuration 1: strength = 0.049 × load, or load = 20.41 × strength.
The best fit gives *load* = 16.10 × *strength* and $r^2 = 0.999$. The significant difference between the gradients indicates that the test is inaccurate whereas the large value of r^2 indicates that it is precise.

Configuration 2: strength = 0.066 × load, or load = 15.15 × strength.
The best fit gives *load* = 23.37 × *strength* and $r^2 = 0.221$. The test is inaccurate and imprecise.

Configuration 3: strength = 0.074 × load, or load = 13.51 × strength.
The best fit gives *load* = 13.47 × *strength* and $r^2 = 0.999$. The test is accurate and precise.

Configuration 4: strength = 0.094 × load, or load = 10.64 × strength.
The best fit gives *load* = 10.54 × *strength* and $r^2 = 0.930$. The test is accurate but not very precise.

Configuration 3 is the best because it is accurate and precise. Configuration 1 is precise, but its inaccuracy means that an adjustment will have to

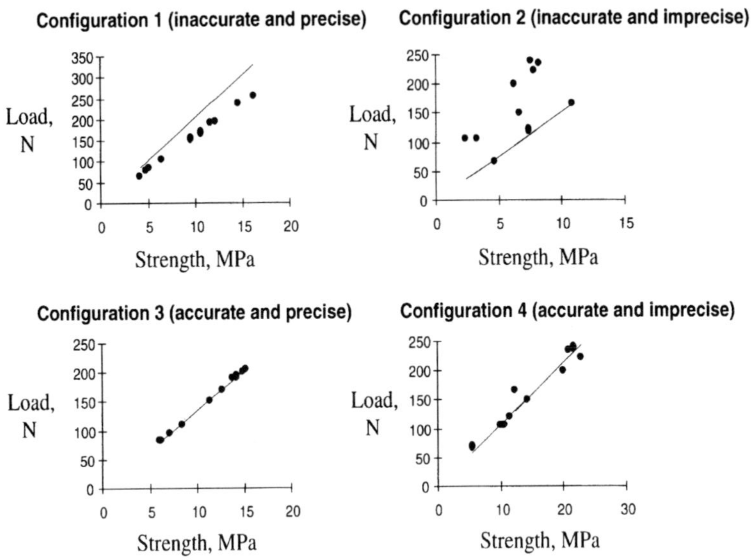

be made for the bias. Configuration 2 has hopeless precision and should be abandoned. Configuration 4 is better, but the imprecision is probably too large to be worthwhile in comparison to configurations 1 and 3.

***Q10.5* A vertical site investigation borehole intersects a stratum of sandstone which is dipping at 17°. A length of intact core from this borehole was taken to the laboratory for hydraulic conductivity testing, and two small-diameter plugs drilled from it: one axially and one radially. These plugs were then tested in a permeameter, and the values of the hydraulic conductivity were found to be 1.728×10^{-7} m/s and 1.557×10^{-6} m/s in the axial and the radial directions, respectively. Stating any assumptions that you make, compute estimates of the hydraulic conductivity normal and parallel to the bedding.**

A10.5 We have been given hydraulic conductivity values in specific directions. Assuming that hydraulic conductivity is a tensor quantity, we can determine the components in any directions and the principal values using the transformation equations. We know the dip of the bedding relative to the tested plug directions and we assume that the anisotropy in conductivity is structurally related (i.e. the principal directions are normal and parallel to the bedding).

For the axial plug we have $k_a = 1.728 \times 10^{-7}$ m/s, and an orientation of $\theta_a = 90 + 17 = 107°$. For the radial plug, the orientation is $\theta_r = 17°$, with $k_r = 1.557 \times 10^{-6}$ m/s. Writing down the transformation equations[3] for these conditions, in terms of the principal values k_n (normal to bedding)

---

[3] Angles: anticlockwise positive, zero being horizontal to the right.

and k_p (parallel to bedding) we obtain

$$k_a = k_p \cdot \cos^2\theta_a + k_n \cdot \sin^2\theta_a \qquad k_r = k_p \cdot \cos^2\theta_r + k_n \cdot \sin^2\theta_r$$

which, in matrix form, are

$$\begin{bmatrix} k_a \\ k_r \end{bmatrix} = \begin{bmatrix} \cos^2\theta_a & \sin^2\theta_a \\ \cos^2\theta_r & \sin^2\theta_r \end{bmatrix} \cdot \begin{bmatrix} k_p \\ k_n \end{bmatrix}.$$

Solving for k_n and k_p as the principal values gives $k_p = 1.7 \times 10^{-6}$ m/s and $k_n = 3.008 \times 10^{-8}$ m/s.

Note that $k_p > k_n$, a feature which is common in such rocks. Defining an anisotropy index in terms of the ratio of these principal values, we have $\rho = k_p/k_n$ and hence $\rho = 56.5$. The polar plot equations can be used to produce a polar plot of the hydraulic conductivity in a section perpendicular to the bedding, and this is shown below.

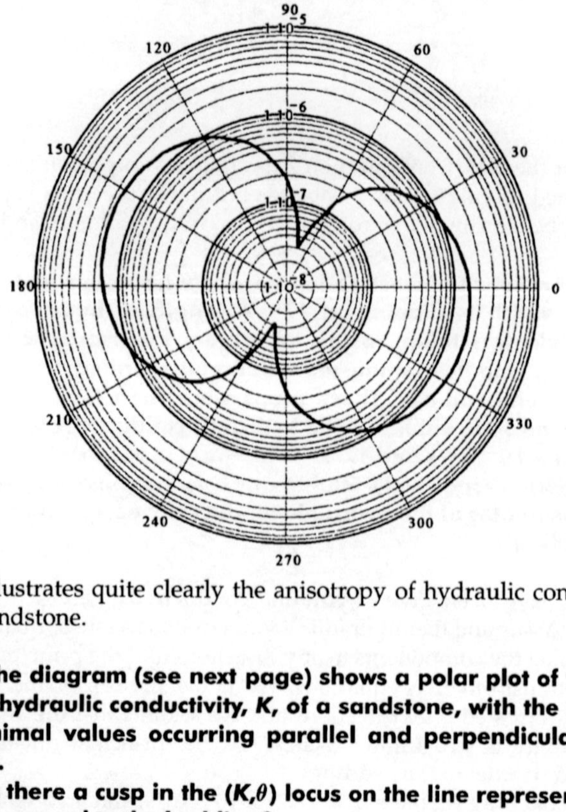

This illustrates quite clearly the anisotropy of hydraulic conductivity in this sandstone.

Q10.6 The diagram (see next page) shows a polar plot of the variation in hydraulic conductivity, *K*, of a sandstone, with the maximal and minimal values occurring parallel and perpendicular to the bedding.

Why is there a cusp in the (*K*,θ) locus on the line representing the direction normal to the bedding?

A10.6 The occurrence of cusps in rock mechanics loci is not unexpected. In the fracture frequency formula, $\lambda_s = \sum_{i=1}^{n} \lambda_i |\cos\theta_i|$, used in A7.6, there are cusps caused by the absolute value function, i.e. $|\cos\theta_i|$. However, that is not the case here.

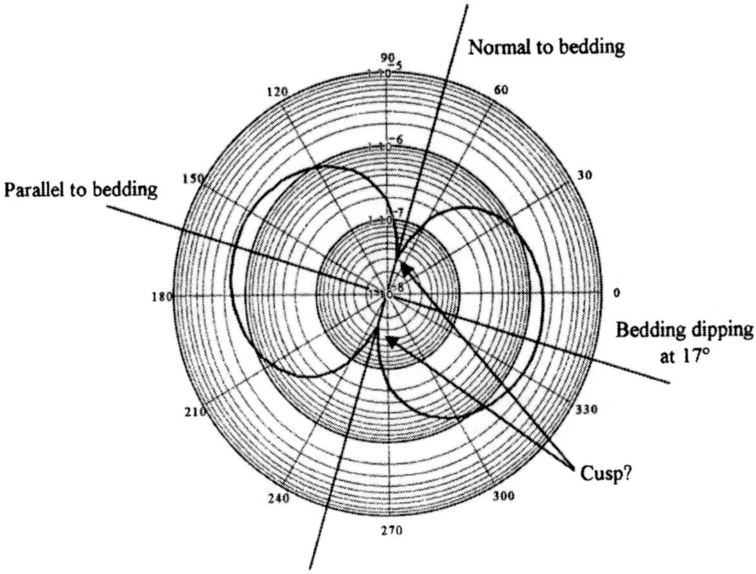

Question 10.6 is included to test your powers of conviction. A tensor component can only vary according to the transformation equations, which are in the form $a\cos^2\theta + b\sin^2\theta$. Therefore, the maximal and minimal values, in this case parallel and perpendicular to the bedding, must have the same form of mathematical variation. The maximum is not on a cusp, so the minimum cannot be. Therefore, there must be something about the way the variation is plotted that causes the cusp, and indeed there is.

In the left-hand diagram below, the variation is shown for the three cases, R_2, R_4, and R_8, in which the maximal and minimal components of the hydraulic conductivity tensor are in the ratios 2:1, 4:1, and 8:1, respectively. It is especially clear in the R_2 case that there is no cusp at

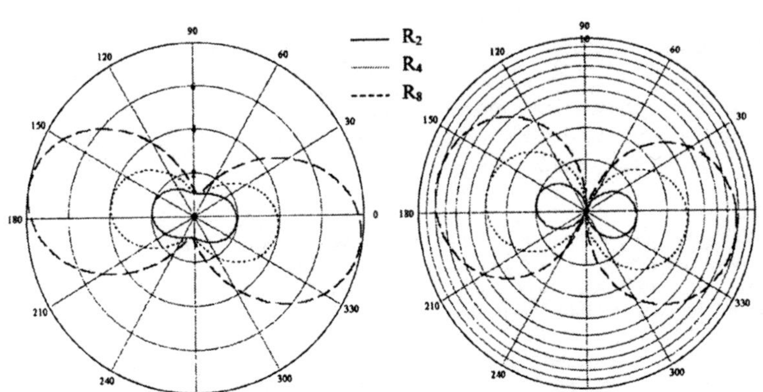

Linear radial scale

Logarithmic radial scale

the location of the minimum value. Moreover, the plot in the question is a ($\log k, \theta$) plot, which accentuates the apparent cusp, as indicated in the lower-right diagram in which the same information (as given in the left-hand diagram) is plotted.

Q10.7 What does the term 'structural domain' mean in the context of engineering rock mechanics?

A10.7 Because of the complexity of the inhomogeneity and anisotropy encountered in rock masses, as conceptually illustrated in A10.2, some kind of simplification is required for engineering modelling. One approach is to divide the rock mass into volumetric domains, within each of which the same rock mechanics properties are essentially constant. The word 'essentially' means that although the properties will be varying, they can be assumed to be constant for the rock mass and modelling in hand. Since all rock masses are different, and the purpose and type of modelling will vary with the project objective, there can be no absolute rules for choosing structural domains. Often they are chosen according to the lithological boundaries because the rock mechanics properties can be strongly related to the geology; but this is not always the case and the choice can also be made on the basis of site investigation sampling schemes supported by geostatistical[4] analyses.

Q10.8 During a site investigation for quarry development, geometrical properties of the fractures on a large surface rock exposure were measured. Eight sampling squares, each 100 m × 100 m, were established on the rock exposure. One of the set of statistics produced during the site investigation was the number of fractures with a mean aperture exceeding 10 mm in each sampling square. These results were as follows:

Square:	**1**	**2**	**3**	**4**	**5**	**6**	**7**	**8**
	461	**397**	**453**	**362**	**389**	**421**	**382**	**423**

On the basis of these data alone, is there sufficient evidence to conclude that the sampling squares should be regarded as different structural domains?

A10.8 The purpose of this question is to highlight the issues and hazards in choosing structural domains for rock engineering design purposes. The key issues[5] are the following:

[4] The subject of geostatistics is not just the statistical analysis of geological data, but a form of statistics in which not only the sample value is used but also the location of the sample. Geostatistical trend analyses are powerful because they can indicate when different domains should be used. A good introduction to this concept is found in Davis (1973), see Footnote 6.

[5] The question of whether such variations in data do reflect underlying causes or are a result of the sampling process pervades much of science and daily life. In 1999, a crime prevention officer in the Avon and Somerset police in the UK claimed that the sign of the zodiac under which a person was born indicates how likely their car is to be stolen. This

(a) establishing the purpose behind identifying such zones;
(b) have the appropriate rock mechanics properties been selected in conjunction with, say, structural geology information to identify these domains? and
(c) given the criteria for zoning the rock mass, how do we distinguish sampling artifacts from real differences in different domains?

To answer the question, it would be necessary to have further information on the exact purpose of the site investigation and the other properties measured, but dividing the rock mass into different structural domains can be useful, e.g. different domains could have different susceptibilities to slope failure.

Based just on the data given, the question is whether the values indicate a real rock structure difference or whether the variation in the values could have arisen by the sampling process. If we thought it justified, we could perform statistical tests to examine whether statistical deficiencies exist. However, this would not indicate the existence of structural domains; it could be highlighting some attribute of the sampling process. Moreover, we have not been told the purpose of the project and which rock mechanics properties will be required in the modelling. Thus, there is insufficient evidence to conclude from these data alone that structural domains are required.

Q10.9 The following list of fracture locations, quoted in metres, is taken from the fracture log of a borehole core which transects a well known stratigraphic boundary between two units of limestone. Evidently, this boundary is clear to sedimentologists, but not to geotechnical engineers.

5.780, 6.391, 6.761, 7.105, 7.180, 7.401, 7.478, 8.142, 8.455, 9.139, 10.072, 10.264, 10.470, 10.539, 10.678, 11.421, 11.541, 12.178, 12.596, 12.620, 12.736, 12.936, 13.134, 13.325, 13.430

Use the concept of moving averages to help locate this boundary.

A10.9 To attempt to find this boundary using moving averages, we firstly compute the spacing values and then, taking these spacing data in groups, compute the mean spacing and mid-point distance of the group. The mean spacing is plotted against mid-point distance, and the

---

was based on the records of 8698 people who had their car stolen in the area from April 1998 to March 1999. The number of cars stolen versus the star sign data were

May 21–Jun 21 Gemini 811	Apr 21–May 20 Taurus 794	Jun 22–Jul 23 Cancer 785	Jul 24–Aug 23 Leo 756	Mar 21–Apr 20 Aries 754	Feb 20–Mar 20 Pisces 730
Aug 24–Sep 23 Virgo 719	Dec 23–Jan 20 Capricorn 696	Sep 24–Oct 22 Libra 686	Jan 21–Feb 19 Aquarius 671	Oct 23–Nov 22 Scorpio 657	Nov 23–Dec 22 Sagittarius 639

Interpretations of these data are, for example, that Geminis are inattentive and leave their keys in the car, whereas Sagittarians buy their dream car and then ensure that it is secure. What do you think? Is the adoption of such a predictive model valid, based on the data alone?

process repeated for different size groups of data. The 'step' in these plots indicates the approximate location of the boundary, and possibly a structural domain boundary. The results of groups of 9, 10, 11, 12, 13, and 14 values are shown below.

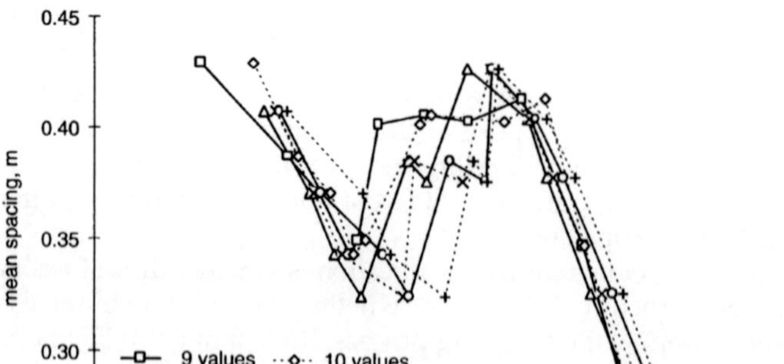

It is evident from this graph that there is a definite change in the mean spacing value at about 11 m: before 11 m, the mean value is about 0.38 m; after 11 m, the mean value reduces to about 0.26 m. As a result, we can say that there is some change in the underlying spacing data at about the 11 m position.

One of the features of presenting data smoothed through the use of moving averages is that there may be a shifting in the peaks and troughs of the smoothed curves. As Davis (1973)[6] has noted, "A moving average will 'lead' an upward run in the data; that is, the smoothed curve will rise at a greater rate than the data themselves. Likewise, the smoothed curve will 'trail' a downward run." Thus, in the curves above we would anticipate that the stratigraphic boundary will be at a position slightly less than 11 m. Davis goes on to explain that when such smoothing techniques are applied to data such as seismic records, the process is called 'filtering' and the smoothed logs are often easier to correlate with geological boundaries. The major, or long-term, features of the record are emphasized at the expense of short-term variations, known as 'loss of high frequencies'. The smoothing is called a 'low-pass filter'.

For the current question, the identification of a stratigraphic boundary between two units of limestone, the smoothing process as a low-pass filter is precisely what is required, and enables the clearer identification of the boundary, as illustrated. Note that in this question, the spatial location of the fracture events has been taken into account because we have explicitly used the position of the fractures. In classical statistical analysis, such location is not generally taken into account. Because the spatial information is critical, e.g. for deciding on the location of

[6] Davis J. C. (1973) *Statistics and Data Analysis in Geology*. Wiley, New York, 550pp.

structural domains, it should be utilized where possible, and is the basis of the next question.

Q10.10 The subject of geostatistics deals with the variation of property values in space, and so anisotropy and inhomogeneity can be quantitatively studied. The basic device of geostatistics is the semi-variogram, defined as

$$\gamma(h) = \left(\frac{1}{2n}\right) \sum_{i=1}^{n} [p(x) - p(x+h)]^2$$

where $\gamma(h)$ is the semi-variogram statistic for samples distance h apart, n is the number of sample pairs, $p(x)$ is the rock property value at location x, and $p(x+h)$ is the value at location $x+h$.

Using this statistic, a graph can be constructed of $\gamma(h)$ *versus* h to indicate the variation in rock property values as a function of distance between the observations.

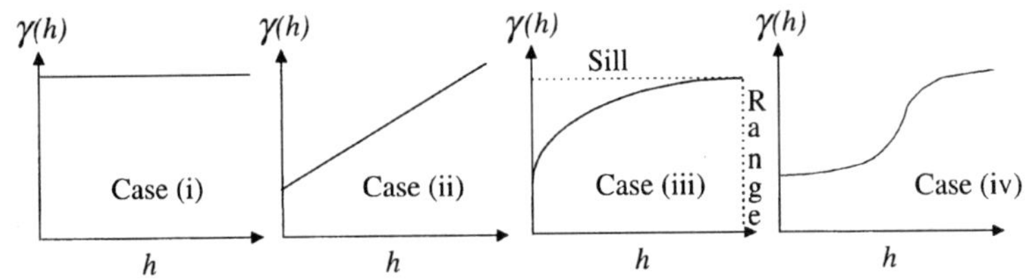

In the diagram above, there are four examples of such semi-variogram curves for a rock mass. In each case, the length of the h axis represents 50 m.

(a) Explain what type of variation in rock property values, or inhomogeneity, each type of semi-variogram represents.

(b) For each type of semi-variogram behaviour, how far away from a borehole would you feel confident in extrapolating results obtained from a borehole core?[7]

A10.10 (a) If the rock mass were homogeneous, $p(x)$ and $p(x+h)$ would always have the same value for all values of x and h, and the semi-variogram statistic $\gamma(h)$ would be zero for all values of x and h. However, given the form of the equation for $\gamma(h)$, the statistic will take on positive values when there are differences between the rock property values at different sample locations, i.e. when there are differences between $p(x)$, the rock property value at location x, and $p(x+h)$, the value at location $x+h$. Thus, we might expect that as h is increased, i.e. the distance between the samples is increased, that $\gamma(h)$ will increase because the rock properties will have changed by a greater amount over larger h distances.

The magnitude of $\gamma(h)$ and the way in which $\gamma(h)$ increases is thus an indication of the type of heterogeneity present in the rock mass.

[7] Borehole information represents specific information from within the rock mass; we have to interpolate between the boreholes to estimate the properties between them.

Moreover, if the semi-variogram statistic is obtained for different directions, the anisotropy of the rock mass inhomogeneity can also be studied.

Case (i): nugget effect. No correlation between sample values. Consistent inhomogeneity across the site. Independent isolated values, like nuggets. This is the model assumed in conventional statistics.

Case (ii): linear effect. There is some basic variation with repeated samples (i.e. as evidenced at $h = 0$). Correlation then linearly decreases with sample separation distance.

Case (iii): 'spherical' semi-variogram. At sample separations less than the range (the range is the sample separation distance, h, for which there is a correlation), there is some correlation between the sample values: this is the region of correlation. The correlation reduces rapidly as h increases.

Case (iv): 'Gaussian' semi-variogram. Similar to Case (iii) but the region of better correlation extends to about half the range.

(b) *Case (i)*: not at all; *Case (ii)*: about 25 m; *Case (iii)*: about 10 m; *Case (iv)*: about 25 m.

The semi-variogram methodology using data from known locations and the associated geostatistical techniques such as kriging to predict rock property values in other locations was originally developed for characterizing and predicting ore grades for mining. This question and answer have been included to illustrate the potential value of geostatistical techniques for characterizing rock mass property data: it is a tool ideally suited for rock characterization but currently is under-used.

10.3 Additional points

In terms of the **DIANE** nature of real rock masses, almost all rock masses are fractured and hence **D**iscontinuous. The fractures are critical because mechanical failure usually occurs through the presence of a major low-strength feature, such as reactivation of movement on a fracture, rock blocks moving, or the influence of water in the fractures. We should always assume that the rock mass is fractured.

Similarly, it should be assumed that the rock mass is **I**nhomogeneous, unless there is some evidence that the degree of inhomogeneity is not significant for the rock engineering design study underway. The best method of characterizing and simplifying inhomogeneity is to divide the rock mass into structural domains such that the property values can be assumed essentially constant within a domain, and the best method for the choosing of structural domains is commonsense geology supported by sampling and geostatistics.

Remember also that **A**nisotropy is often 'built into' the parameters being used. We saw that stress and strain, represented by second-order tensors, have principal axes. Thus, the possibility of an anisotropic stress state is automatically included in the definition of stress with the three principal stresses acting in mutually orthogonal directions. Similarly, in

Chapter 5, we saw how the elastic strain compliance matrix can represent differing types of anisotropy, and especially transverse isotropy and orthotropy. In Chapter 9, we discussed permeability and hydraulic conductivity in terms of the same tensorial representation and in this chapter we evaluated a directional hydraulic conductivity problem. When a parameter does not directly incorporate a directional dependency, e.g. the uniaxial compressive strength or the fracture frequency, it is tempting to ignore any anisotropy that might be present. This temptation should be resisted: isotropy should only be assumed once there is evidence that the degree of anisotropy is not significant.

The rock mass will always be 'Not Elastic' because there will always be some time-dependent component to the induced strains, and loading/unloading curves will always exhibit hysteresis. So, even though the elastic stress and strain distributions can often be helpful in understanding the mechanics of a rock mass (e.g. in highly stressed deep South African gold mines), it should always be remembered that the rock mass response to engineering perturbations is Not Elastic. Establishing to what extent the inelastic rock mass response can be adequately modelled for engineering purposes by an elastic model is part of the art of engineering rock mechanics.

11 Testing techniques

11.1 Rock properties

The term 'rock properties' refers here to those intact rock or rock mass properties that are needed for engineering design purposes. For example, the rock properties may be used to

- obtain a general impression of the mechanical nature of the rock mass, e.g. the rock is strong because it has a compressive strength of 300 MPa,
- compare the rock properties with a previous project where the rock properties were also obtained, e.g. this rock is stronger than the one we had at the Golconda Mine,
- generate a rock mass classification scheme value, e.g. the RQD is needed for the Rock Mass Rating (RMR) scheme, see Chapter 12, or
- support numerical modelling, e.g. the shear stiffness of fractures is required for a distinct element numerical code.

Many of the required rock properties can be categorized according to the subjects of the earlier chapters, as shown in Table 11.1.

Strictly speaking, *in situ* stress is a site property rather than a rock property, but testing techniques are required to determine the *in situ* stress and so it is one of the categories below. The 'permeability' could be included as a separate item under each of the 'intact rock', 'fractures' and 'rock mass', but we prefer to consider the property in a separate category because the subject involves the connectivity of the rock mass fractures. In each case, there should be information about any variation in these properties across the site, which was the theme of Chapter 10.

Table 11.1 Examples of rock properties measured in a site investigation programme

In situ stress	Intact rock	Fractures	Rock mass	Permeability
Magnitudes and directions of the three principal stresses	Deformation, strength and failure properties	Geometrical occurrence and mechanical properties	Deformation, strength and failure properties	Nature of any flow through the intact rock and rock mass

The structural geology and hydrogeological setting information will be strategically helpful for this purpose.

The rock properties can be measured directly or indirectly. For example, in Q10.3, the uniaxial compression test and point load test values were compared. Because the uniaxial compression test provides a direct value for the compressive strength, it is a direct test. On the other hand, because the point load test gives an index value which is used to indicate the uniaxial compressive strength via a correlation factor, the point load test is an index test. There are many possibilities for such indirect tests in rock mechanics and an advantage of them is that they can provide many more results than direct tests, more rapidly and more cheaply. Their disadvantage is a possible lack of precision and knowing whether or not there is any bias in the values. To make decisions about which type of test to use, one has to recall why the rock properties are required and the resources available, and hence whether direct tests, indirect tests, or a mix of the two types are best suited to the project in hand.

In the questions that follow in Section 11.2, we provide a flavour of the nature of site investigation and how some of the testing problems are solved. This chapter is the first where we link the rock mechanics with the rock engineering. It is important when practising rock engineering to understand the rock mechanics concepts first — which has been our aim in Chapters 1–10. Now, we highlight the engineering thinking that is required to assess and measure the rock properties.

11.2 Questions and answers: testing techniques

***Q11.1* The section of site investigation borehole core shown in the photograph on the next page is from a vertical borehole and contains three stress-induced fractures. The top of the core is a stress fracture of the same kind. The bottom end is a drilling break. Assuming that the strip of translucent tape (adjacent to the numbers written on the core in the photograph) is on the northern side of the core, in which horizontal directions do you think the major and minor principal stresses act?**

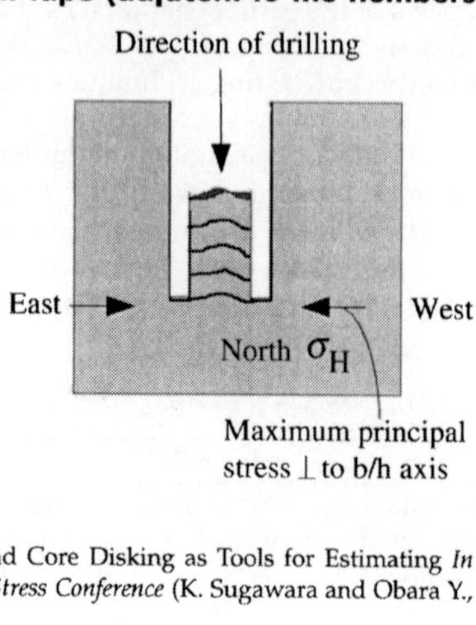

A11.1 The three central fractures and the top end of the core section are all fractures caused by the *in situ* stresses being concentrated at the end of the borehole during core drilling with a hollow drill bit, see diagram to right (Haimson, 1997[1];

[1] Haimson B. (1997) Borehole Breakouts and Core Disking as Tools for Estimating *In Situ* Stress in Deep Holes, in *Proc. of the Rock Stress Conference* (K. Sugawara and Obara Y., eds). Balkema, Rotterdam, pp. 35–42.

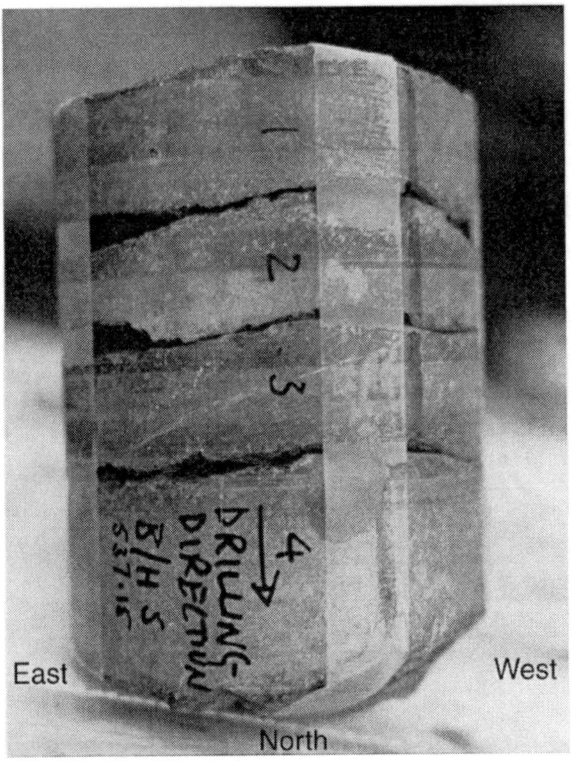

Bankwitz and Bankwitz, 1995[2]). The direction of the trough in the disc indicates the E–W direction of the σ_H major horizontal principal stress because the discs form in a similar way to the Brazilian tensile test failure — initiating on the diameter corresponding to the primary loading direction and propagating perpendicular to the least stress (which is vertical). The saddle shape occurs because the failure propagation follows the

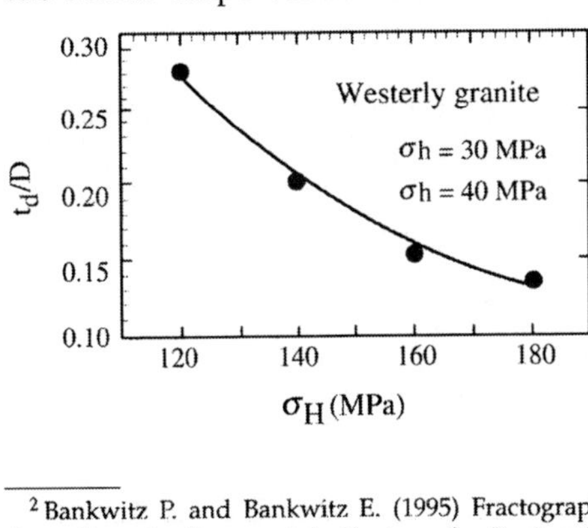

Example of the relation between core disk thickness (normalized by the core diameter) and σ_H for given σ_h and σ_v, from Haimson, 1997.

[2] Bankwitz P. and Bankwitz E. (1995) Fractographic Features on Joints of KTB Drill Cores (Bavaria, Germany), in *Fractography, Fracture Topography as a Tool in Fracture Mechanics and Stress Analysis* (M. S. Ameen, ed.). Geological Society Special Publication No. 92, pp. 39–58.

line of least resistance, rising towards the lower stressed region in the drill core. Thus, the major principal stress, σ_H, acts E–W and the minor principal stress, σ_h, acts N–S. The ratio of disc thickness to diameter indicates (according to the diagram included above from Haimson, 1997) that the local σ_H could be about 120 MPa.

Q11.2 With reference to fracture property measurements made during a site investigation
– on borehole rock core,
– on the borehole walls, and
– on rock exposures.
Complete the table below indicating your opinion of how well you think that the listed fracture properties can be measured or estimated. The first column of the table represents the ten fracture measurements recommended by the ISRM, as in Fig. 7.7. Use the letters G for Good, M for Medium, and P for Poor.

Characteristic	Measurement method	Core	B/H wall	Exposure
Orientation	Compass-clinometer			
Spacing	Measuring tape			
Persistence	Measuring tape			
Roughness	Against reference chart			
Wall strength	Schmidt hammer			
Aperture	Feeler gauge			
Filling	Visual			
Seepage	Timed observations			
Number of sets	Stereographic projection			
Block size	3-D fracture frequency			

A11.2 The completed table follows. These measurements cover a wide range of attributes — rock mass geometry, intact rock strength and

Characteristic	Measurement method	Core	B/H wall	Exposure
Orientation	Compass-clinometer	M	G	G
Spacing	Measuring tape	G	G	G
Persistence	Measuring tape	P	P	G/M
Roughness	Against reference chart	M	P	G
Wall strength	Schmidt hammer	M	P	G
Aperture	Feeler gauge	P	M	G
Filling	Visual	P	P	G
Seepage	Timed observations	P	P/M	G
Number of sets	Stereographic projection	M	G	G
Block size	3-D fracture frequency	P	P	G

hydrogeological properties. Also, some of the measurements are more easily performed during one type of site investigation than another. Although there is, therefore, some subjectivity in the G, M, P quality coding, the completed table should be of this general form.

Q11.3 The results of a series of scanline surveys at a particular site are as follows:

Scanline trend (°)	**000**	**355**	**085**	**153**	**216**	**271**
Scanline plunge (°)	**90**	**35**	**28**	**51**	**05**	**12**
Fracture frequency (m^{-1})	**5.54**	**7.93**	**6.02**	**7.00**	**6.99**	**7.65**

Analysis of the fractures intersected by the scanlines has shown that the rock mass contains four fracture sets, with orientations 145/08, 148/88, 021/76 and 087/69 (given as dip direction/dip angle). What is the best estimate of the frequency of each fracture set?

A11.3 For the situation when we know the orientation and fracture frequency of each of the four fracture sets, then the fracture frequency along a scanline in any particular direction, Λ_s, is given by (see A7.6)

$$\Lambda_s = \lambda_1|\cos\theta_{s1}| + \lambda_2|\cos\theta_{s2}| + \lambda_3|\cos\theta_{s3}| + \lambda_4|\cos\theta_{s4}|$$

where θ_{s1} is the angle between the normal to Set 1 and the direction in which we are interested, and the other angles are similarly defined.

If we write out this equation for each scanline, and arrange the results in matrix form, we obtain

$$\begin{bmatrix} \Lambda_1 \\ \Lambda_2 \\ \vdots \\ \Lambda_6 \end{bmatrix} = \begin{bmatrix} |\cos\theta_{11}| & |\cos\theta_{12}| & \cdots & |\cos\theta_{14}| \\ |\cos\theta_{21}| & |\cos\theta_{22}| & \cdots & |\cos\theta_{24}| \\ \vdots & \vdots & \vdots & \ddots \\ |\cos\theta_{61}| & |\cos\theta_{62}| & \cdots & |\cos\theta_{64}| \end{bmatrix} \begin{bmatrix} \lambda_1 \\ \lambda_2 \\ \vdots \\ \lambda_4 \end{bmatrix} \quad \text{or } \boldsymbol{\Lambda} = \boldsymbol{\Omega}\cdot\boldsymbol{\lambda}$$

Hence, for the case when the vector $\boldsymbol{\Lambda}$ is known (i.e. the scanline results) and the vector $\boldsymbol{\lambda}$ is unknown (i.e. the fracture set frequencies), we need to solve this matrix equation for $\boldsymbol{\lambda}$. However, as the matrix $\boldsymbol{\Omega}$ has more rows than columns, it cannot be simply inverted, and so the best estimate for $\boldsymbol{\lambda}$ is given by the least-squares solution,

$$\boldsymbol{\lambda} = (\boldsymbol{\Omega}^T\cdot\boldsymbol{\Omega})^{-1}\cdot(\boldsymbol{\Omega}^T\cdot\boldsymbol{\Lambda}).$$

In order to evaluate this equation, we firstly determine the angles between the individual scanlines and the normals to the fracture sets. These angles can either be computed using vector methods, or measured on the hemispherical projection. Using the former method leads to the following results for the angles:

	Set 1	Set 2	Set 3	Set 4
Scanline 1	8.0	88.0	76.0	69.0
Scanline 2	48.2	41.5	125.1	76.6
Scanline 3	66.2	112.6	105.2	131.0
Scanline 4	46.9	126.8	53.4	87.7
Scanline 5	87.6	111.7	17.3	51.9
Scanline 6	73.4	57.3	68.0	9.8

and from this we obtain the matrix of $|\cos\theta_{si}|$ values,

$$\Omega = \begin{bmatrix} 0.990 & 0.035 & 0.242 & 0.358 \\ 0.667 & 0.749 & 0.576 & 0.232 \\ 0.403 & 0.384 & 0.262 & 0.656 \\ 0.683 & 0.599 & 0.597 & 0.040 \\ 0.041 & 0.370 & 0.955 & 0.617 \\ 0.286 & 0.540 & 0.375 & 0.985 \end{bmatrix}.$$

As a result, the frequencies of the fracture sets are found to be

$$\lambda = \left(\Omega^{\mathrm{T}} \cdot \Omega\right)^{-1} \cdot \left(\Omega^{\mathrm{T}} \cdot \Lambda\right) =$$

$$\begin{bmatrix} 2.138 & 1.268 & 1.283 & 1.108 \\ 1.268 & 1.498 & 1.454 & 1.222 \\ 1.283 & 1.454 & 1.867 & 1.374 \\ 1.108 & 1.222 & 1.374 & 1.965 \end{bmatrix}^{-1} \times \begin{bmatrix} 20.457 \\ 19.360 \\ 21.200 \\ 19.899 \end{bmatrix} = \begin{bmatrix} 3.42 \\ 3.91 \\ 3.54 \\ 3.29 \end{bmatrix} \mathrm{m}^{-1}.$$

Knowing the fracture frequencies of the sets enables the calculation of the fracture frequencies along lines of any orientation in a rock mass, as required in the next question in which the fracture frequencies have been slightly changed.

Q11.4 As part of a site investigation study, a rock mass was found to contain four fracture sets with dip/dip direction and frequencies as follows (see figure on next page):

Set 1: 08/145, 3.48/m
Set 2: 88/148, 3.91/m
Set 3: 76/021, 3.58/m
Set 4: 69/087, 3.26/m

In order to establish in which directions through the rock mass an excavation will encounter the minimal and maximal numbers of fractures, the fracture frequency in different directions through the rock mass, λ_s, was calculated using the formula $\lambda_s = \sum_{i=1}^{n} |\lambda_i \cos\theta_i|$ (see Q7.6 and A7.6). The results are presented below on a hemispherical projection, with the contouring corresponding to the fracture frequency values in the different directions.

Explain from first principles why the directions of the minimal and maximal frequencies occur where they do.

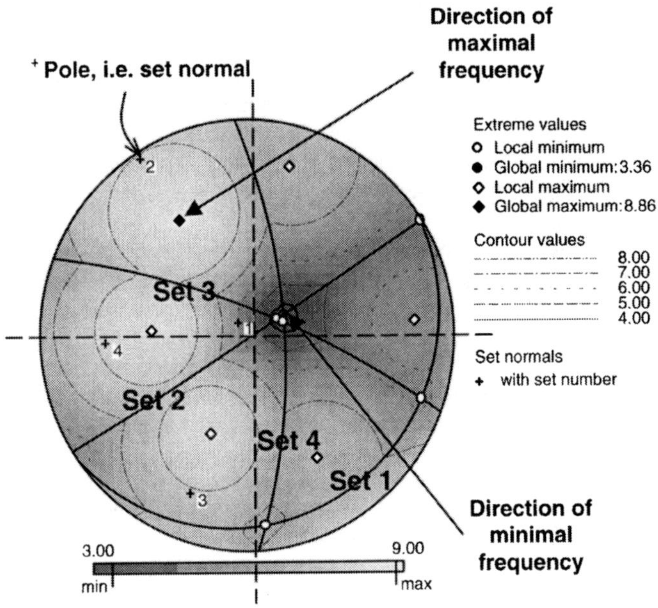

A11.4 *Minimal value.* In any given rock mass, the minimum fracture frequency lies along a direction that is formed by the intersection of two fracture sets. This is because no fractures from the relevant fracture sets will be intersected along such a direction. To identify the global frequency minimum we therefore need to determine the direction of the intersection of each pair of fractures, and then compute the fracture frequency in that direction. The directions of the various intersections, and the frequency computed along those directions, are as follows:

Sets 1 and 2	Sets 1 and 3	Sets 1 and 4	Sets 2 and 3	Sets 2 and 4	Sets 3 and 4
058/00	109/07	174/07	064/71	063/67	072/68
5.42 m^{-1}	5.68 m^{-1}	6.65 m^{-1}	3.44 m^{-1}	3.36 m^{-1}	3.38 m^{-1}

The global minimum is therefore in the direction of the intersection of Sets 2 and 4, and has a magnitude of 3.36 m^{-1}. Notice that the minima in the directions of the intersections of Sets 2 and 3, Sets 2 and 4, and Sets 3 and 4 are all similar in magnitude. This is because the directions of these intersections are all similar.

Maximal value. Although the directions of the various minima coincide with the directions of fracture set intersections, the directions of the various maxima are not so well defined and can only be found by rigorous computation. The maximal value occurs where the sum of the fracture contributions from all intersected sets is maximized, and an approximate value could be found by simply scanning all the fracture frequency values used to generate the contoured hemispherical projection to find the maximum value.

A more elegant method utilizes the vector-like nature of fracture frequency (i.e. it has both a magnitude and a direction): the maximal fracture frequency is found as the resultant of the individual set frequencies. However, because the maximal frequency for each fracture set occurs in two opposite directions (e.g. for a horizontal set the maximal frequency occurs in both a vertical upward and a vertical downward direction), the resultant formed from either of these directions must be considered.

Clearly, therefore, a number of candidates for the overall maximum will be found using this procedure. In fact, these will either be the various local maxima that can be seen in each zone of the projection bounded by great circles, or maxima that — whilst mathematically valid — do not physically exist. To identify which is which, the actual frequency in the direction of a mathematical maximum is computed and the two results compared; they are equal for maxima that physically exist.

In order to investigate the various maxima shown on the projection, we start by computing the Cartesian components of the normal to each fracture set. For a right-handed set of axes with x directed east, y directed north and z directed upwards, these components are given by $n_x = \sin(\alpha_n)\cos(\beta_n)$, $n_y = \cos(\alpha_n)\cos(\beta_n)$ and $n_z = -\sin(\beta_n)$, where α_n and β_n are the trend and plunge of the normal. For the fracture sets used here, these are as follows:

α	β	α_n	β_n	n_x	n_y	n_z
145	08	325	82	−0.080	0.114	−0.990
148	88	328	02	−0.530	0.848	−0.035
021	76	201	14	−0.348	−0.906	−0.242
087	69	267	21	−0.932	−0.049	−0.358

These values represent downward-directed normals, and to convert them to upwards-directed normals we simply multiply each component by −1. The components of each of the various resultants are then given as $r_x = \sum_i s_i n_{x_i}$, $r_y = \sum_i s_i n_{y_i}$ and $r_z = \sum_i s_i n_{z_i}$, where each s_i takes the value ±1 in order to cycle through all candidate resultants. Applying these to the data used here results in the table given below.

Candidate	s_1	s_2	s_3	s_4	$r_x = \sum_i s_i n_{x_i}$	$r_y = \sum_i s_i n_{y_i}$	$r_z = \sum_i s_i n_{z_i}$	Resultant
1	1	1	1	1	−6.633	0.308	−5.617	8.697
2	1	1	1	−1	−0.554	0.627	−3.280	3.385
3	1	1	−1	1	−4.143	6.794	−3.885	8.855
4	1	1	−1	−1	1.936	7.113	−1.548	7.532
5	1	−1	1	1	−2.491	−6.319	−5.344	8.643
6	1	−1	1	−1	3.587	−6.001	−3.007	7.611
7	1	−1	−1	1	−0.002	0.167	−3.612	3.616
8	1	−1	−1	−1	6.077	0.485	−1.275	6.228

The Cartesian components of each candidate are found by normalizing each resultant in the table above to a magnitude of unity. If we represent these Cartesian components by (m_x, m_y, m_z), they are used in the usual formula of

$$\lambda_s = \sum_i \lambda_i |\cos(\theta_i)| = \sum_i \lambda_i |m_x n_{x_i} + m_y n_{y_i} + m_z n_{z_i}|$$

to compute the actual frequency in the direction they represent. This computation, together with the magnitude of the resultant shown above, is given below.

Candidate	m_x	m_y	m_z	$\lambda_1 \lvert \cos(\theta_1) \rvert$	$\lambda_2 \lvert \cos(\theta_2) \rvert$	$\lambda_3 \lvert \cos(\theta_3) \rvert$	$\lambda_4 \lvert \cos(\theta_4) \rvert$	λ_s, m^{-1}	Resultant
1	−0.763	0.035	−0.646	2.452	1.785	1.394	3.067	8.697	8.697
2	−0.164	0.185	−0.969	3.458	1.085	0.442	1.600	6.585	3.385
3	−0.468	0.767	−0.439	1.946	3.571	1.526	1.812	8.855	8.855
4	0.257	0.944	−0.206	1.012	2.625	3.204	0.691	7.532	7.532
5	−0.288	−0.731	−0.618	1.921	1.742	3.265	1.715	8.643	8.643
6	0.471	−0.788	−0.395	0.918	3.535	2.312	0.845	7.611	7.611
7	0.000	0.046	−0.999	3.461	0.290	0.716	1.161	5.628	3.616
8	0.976	0.078	−0.205	0.465	1.734	1.290	2.739	6.228	6.228

This shows that candidates 2 and 7 do not physically exist. Of the remaining candidates, their orientations are found from their Cartesian components to be 273/40, 329/26, 015/12, 202/38, 149/23, and 085/12. Of these, the global maximum is at 329/26 with a magnitude of 8.855 m^{-1}.

This answer illustrates an important point: **the directions corresponding to the minimal and maximal numbers of intersected fractures are not perpendicular**. The reason why this is important is that it raises questions about the validity of tensor representations of rock mass properties, in which the principal directions corresponding to the extreme values are orthogonal, as is the case for stress, strain and permeability. For example, we saw in Q8.3 and A8.3 that the directions of minimal and maximal rock mass deformability modulus may not be perpendicular.

Q11.5 When cyclic deformability tests are conducted on rock masses, the typical force–displacement curve is as shown to the right (Schneider, 1967[3]; Goodman, 1989).

Explain why, with cycles of repeated unloading and reloading, the curve manifests different unloading and reloading moduli, permanent deformations and hysteresis.

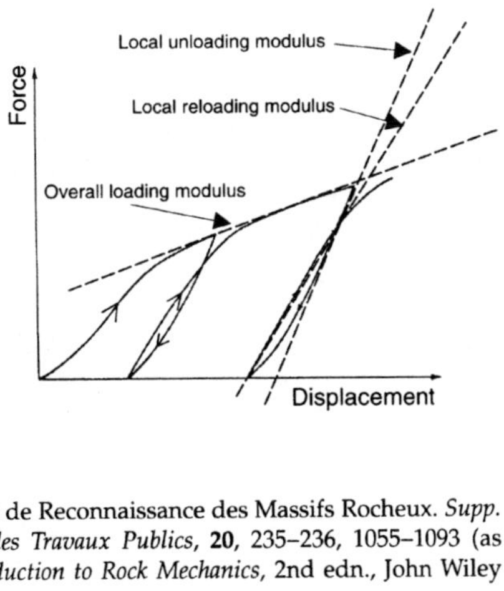

[3] Schneider B. (1967) Moyens Nouveaux de Reconnaissance des Massifs Rocheux. *Supp. to Annales de l'Inst. Tech. de Batiment et des Travaux Publics*, **20**, 235–236, 1055–1093 (as illustrated in Goodman R. E. (1989) *Introduction to Rock Mechanics*, 2nd edn., John Wiley and Sons, New York, 562pp).

A11.5 The mechanical behaviour of a rock mass is dominated by the fractures, which significantly reduce the modulus from the intact rock value. The deformation modulus of a rock mass is often only about one tenth that of the intact rock modulus. Thus, we would expect that the idiosyncrasies of the force–displacement curve above are caused by the fractures. Indeed, when fractures are compressed, they have a non-linear behaviour as the asperities are deformed and crushed. However, when the fractures are subsequently unloaded, the behaviour will be more linearly elastic and stiffer, because the asperities have been crushed. This is also the reason for the permanent deformation and the hysteresis that occurs on initial loading and unloading. After one or two cycles of such loading and unloading, the fracture surfaces in the rock mass have been sufficiently disturbed to make the rock mass modulus higher and it responds in a more reproducible manner. The site investigation question is which modulus the design calculations require — and this will depend on whether the rock mass will be repeatedly loaded during engineering operations.

Q11.6 **The tensile strength of an architectural granite was measured to ensure that the granite would be strong enough to form the structural elements of a pedestrian bridge in a shopping mall. Ten specimens were tested in each of four test configurations (illustrated in A6.9), and the values obtained were as follows.**

Type of test	Mean value (MPa)	Standard deviation (MPa)
Direct tension test	8.4	3.2
Point load test	9.6	3.8
Beam test	10.4	4.5
Ring test	12.9	6.7

Are these results consistent with what you know about tensile strength variation and which value would you use for the structural calculations?

A11.6 We expect tensile strength variation (see A6.9) both within and between tests. The tensile strength increases with (a) a lower test volume subjected to the high stress, and (b) a higher tensile stress gradient. We note that the different tensile strength test configurations have the following characteristics.

Direct tension test. The whole test volume is subject to the same high tensile stress and there is no tensile stress gradient. This indicates a low measured strength.

Point load test. The region of the specimen between the loading platens is subjected to a high tensile stress, and there is no significant tensile stress gradient.

Beam test. It is only the region at the opposite face of the beam from the loading point that is subjected to the high tensile stress, where there is a linear tensile stress gradient.

Ring test. Only the regions on the inner surface in line with the loading are subjected to the high tensile stress, where there is a steep tensile stress gradient. This results in a high measured strength.

Thus, the mean values and standard deviations of the test results are entirely consistent with our understanding of tensile strength variation. For the structural calculations, we should use the value obtained in a similar configuration to the design configuration, i.e. the beam test. We should also obtain an estimate of any scale effect in moving up to the size of the bridge elements, and apply a safety factor.

Q11.7 **The following table shows data obtained from a single-stage triaxial compression test on a cylindrical rock sample, conducted with closed-loop servo-control, at a confining stress of 10.0 MPa, and at zero pore pressure.**

Total axial load (kN)	Sample height (mm)	Sample diameter (mm)
0.00	100.84	50.20
19.89	100.80	50.20
39.60	100.77	50.20
63.40	100.74	50.20
88.67	100.71	50.21
116.18	100.68	50.21
144.68	100.65	50.22
162.38	100.63	50.22
185.23	100.58	50.24
190.62	100.56	50.25
191.99	100.54	50.25
180.22	100.52	50.26
137.56	100.49	50.26
115.79	100.46	50.27
101.93	100.43	50.28
97.97	100.40	50.28
96.98	100.37	50.28

Estimate values for the following:

(i) yield strength σ_y;
(ii) peak strength σ_{max};
(iii) residual strength σ_r;
(iv) tangent Young's modulus E_{tan} at 50% peak axial stress; and
(v) tangent Poisson's ratio ν_{tan} at 50% peak axial stress.

A11.7 From the test data recorded, we compute the axial stress, axial strain and diametral strain in order to plot the axial stress–axial strain and diametral strain–axial strain curves and to visually estimate the various strength parameters. The stress and strain values are given in the table below. Note that the axial strain values are positive (because the convention of contraction positive is being used), but the diametral strain values are negative (because the specimen expands circumferentially during the compression test).

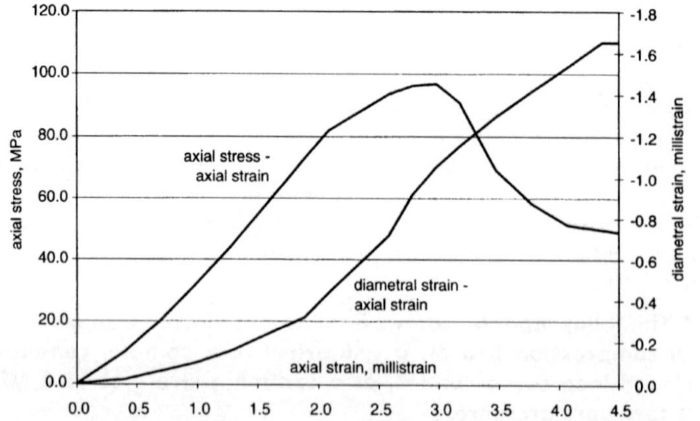

Axial stress	Axial strain	Diametral strain
0.00	0.000	0.000
10.05	0.397	−0.020
20.01	0.694	−0.060
32.03	0.992	−0.100
44.79	1.289	−0.159
58.67	1.587	−0.239
73.05	1.884	−0.319
81.97	2.083	−0.438
93.45	2.578	−0.717
96.13	2.777	−0.916
96.80	2.975	−1.056
90.85	3.173	−1.155
69.32	3.471	−1.295
58.34	3.768	−1.414
51.34	4.066	−1.534
49.34	4.363	−1.653
48.84	4.661	−1.653

The curves associated with these results are shown above. The strength parameters are read from these curves, and have the following values:

(i) yield strength $\sigma_y \approx 83$ MPa (this is where the axial stress–axial strain curve becomes visibly non-linear);

(ii) peak strength $\sigma_{max} \approx 97$ MPa (this is the maximum axial stress the specimen sustained);

(iii) residual strength $\sigma_r \approx 49$ MPa (this is the final stress value given, which may have reduced further had the test been continued);

(iv) tangent Young's modulus E_{tan} at 50% peak axial stress is computed from the axial stress and axial strain values immediately above and below 50% of the peak axial stress. The value is given by

$$\frac{58.70 - 44.80}{1.171 - 0.893} = \frac{13.9}{0.278} = 50.00 \text{ GPa}.$$

In practice, we would compute this value from several ranges of values to estimate any variation.

(v) tangent Poisson's ratio ν_{tan} at 50% peak axial stress is similarly computed from the diametral and axial strain values immediately above and below 50% of the peak axial stress. It is given by

$$\frac{0.159 - 0.100}{1.171 - 0.893} = \frac{0.059}{0.278} = 0.21.$$

***Q11.8* Explain why the introduction of servo-controlled testing machines in the early 1970s revolutionized rock mechanics laboratory testing, and why we are now able to test rocks under virtually any loading conditions.**

A11.8 The principle of a servo-controlled testing machine is that a feedback signal representing some experimental value is continuously

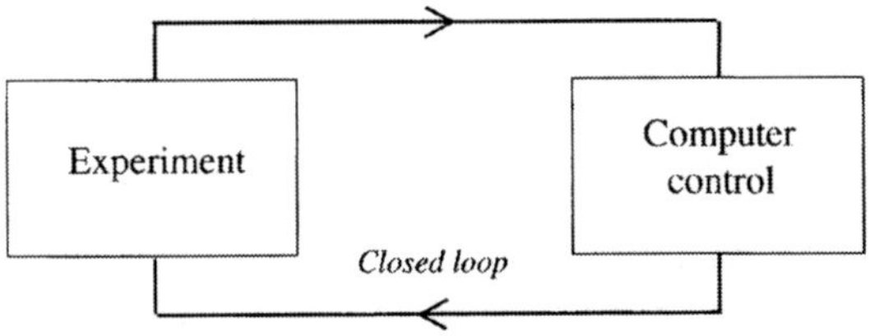

compared with a program signal representing the desired value: any discrepancy is corrected with only a 5 millisecond response time. This is known as a closed-loop control system.

Two complete stress–strain curves for rock are shown in the plot to the right. The pre-peak portion is the region OA. The two types of curve are categorized in terms of the characteristic of the post-peak region: either the curve monotonically increases in axial strain (the thick curve) or it does not (the thinner curve). The former, is termed a Class I curve; the latter is termed a Class II curve (Fairhurst and Hudson, 1999 [4]).

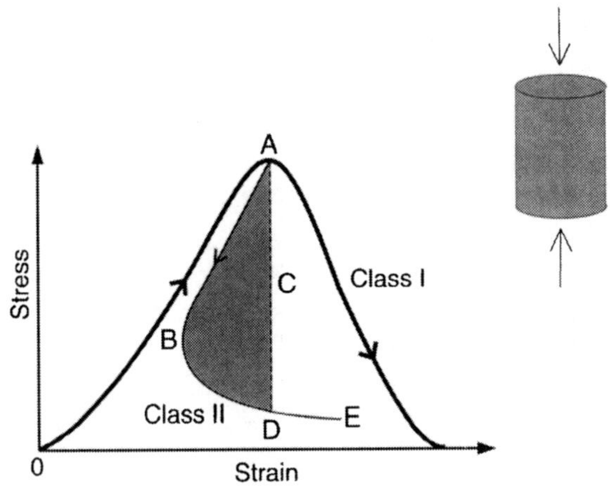

It is important to understand these two types of curve in order to optimize the control of rock failure. Cylindrical specimens that exhibit Class I behaviour tend to be somewhat ductile in nature when loaded axially, whereas specimens that exhibit Class II behaviour respond in a brittle fashion to axial loading. A test conducted in axial strain control is generally sufficient to obtain the complete stress–strain curve of specimens exhibiting Class I behaviour, but alternative control techniques, such as using circumferential strain as the independent (or control) variable, are necessary when testing specimens that exhibit Class II behaviour because the stress–strain curve does not then monotonically increase in axial strain. Note that the shaded area ABDCA is the surplus energy which would be supplied by a rigid machine (one with infinite

[4] Fairhurst C. E. and Hudson J. A. (1999) Draft ISRM suggested method for the complete stress–strain curve for intact rock in uniaxial compression. *Int. J. Rock Mech. Min. Sci.*, **36**, 3, 279–289.

stiffness indicated by the line AD) or a servo-controlled machine with axial strain control — both leading to uncontrolled failure — and so energy has to be withdrawn from the specimen to sustain continued controlled failure. It is an awesome experience to stand next to a granite specimen being tested under such controlled conditions and watch it quietly change from solid rock to fragmented grains and dust particles.

In a uniaxial compression test, one could, for example, control the stress rate, the strain rate, the energy input rate, the pore pressure, or the acoustic emission output rate. Servo-control can be used in any testing configuration and this type of testing machine is limited only by one's imagination. That is why such machines have revolutionized rock testing and enable virtually any test to be servo-controlled. Also, more realistic loading conditions can be applied, so that the rock can be tested in the same way that it is loaded in the engineering scheme. We anticipate that the next phase of development will be the use of servo-control in field tests.

Q11.9 **(a) The results in the table below represent shear displacement and shear stress recorded during a direct shear test on a fracture in slate. The shear displacement range was from 0 to 15 mm as shown in the table below. The normal stress during the test was 0.34 MPa.**

Shear displacement (mm)	**0.0**	**0.5**	**1.0**	**1.5**	**2.0**	**2.5**	**3.0**	**3.5**
Shear stress (kPa)	**0**	**281**	**344**	**344**	**328**	**281**	**281**	**297**
Shear displacement (mm)	**4.0**	**4.5**	**5.0**	**5.5**	**6.0**	**6.5**	**7.0**	**7.5**
Shear stress (kPa)	**281**	**281**	**266**	**266**	**266**	**281**	**281**	**281**
Shear displacement (mm)	**8.0**	**8.5**	**9.0**	**9.5**	**10.0**	**10.5**	**11.0**	**11.5**
Shear stress (kPa)	**297**	**297**	**297**	**313**	**313**	**313**	**313**	**313**
Shear displacement (mm)	**12.0**	**12.5**	**13.0**	**13.5**	**14.0**	**14.5**	**15.0**	
Shear stress (kPa)	**313**	**313**	**313**	**313**	**313**	**313**	**313**	

Use these results to determine the residual shear strength of the fracture.

(b) A series of direct shear tests was undertaken at different normal stress values on samples from the fracture, and the peak shear stress encountered during each test was recorded, as shown in the table below.

Normal stress (kPa)	**336**	**648**	**961**	**1273**	**1586**
Peak shear stress (kPa)	**344**	**516**	**719**	**953**	**1156**

Use these results to determine the basic friction angle, ϕ, and the asperity angle, *i*, for the fracture. Also comment on the validity of the bi-linear approximation for the failure locus.

A11.9 This question has been included to illustrate typical laboratory testing results and their interpretation. Such laboratory testing is invariably conducted if specific properties are required for an analytical or numerical model. It is therefore essential that we know how to extract such values from test results — which is sometimes straightforward, and sometimes open to interpretation.

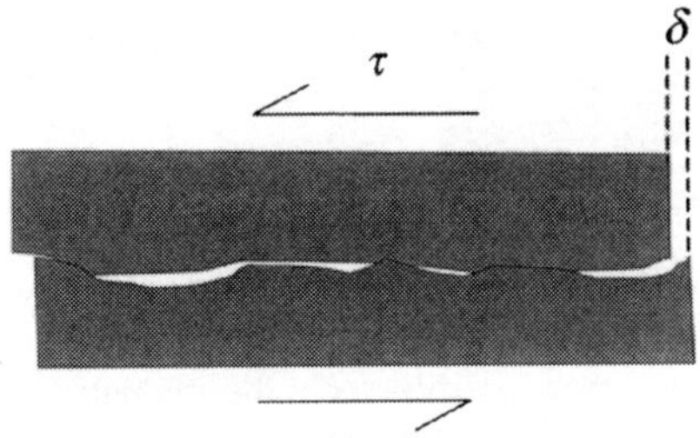

Schematic cross-section of the direct shear test

(a) Plotting the displacement and stress results for the shear test, illustrated above, generates the diagram shown below. There is a well-defined peak strength, followed by a poorly defined residual strength. By poorly defined, we mean that a residual strength is encountered at a displacement of about 6 mm, but should the residual strength be this or the one reached at a larger maximum displacement? Although, by definition, we regard the residual strength as that reached at large displacements, this may not be appropriate if the fracture has failed at some intermediate lower strength — before reaching the higher strength attained at a larger displacement.

By analogy with the complete stress–strain curve in compression, the *in situ* stability of the fracture depends on the stiffness of the loading system (i.e. the engineering structure for a field project), and the appropriate value for the residual strength can only be chosen given the engineering circumstances. Thus, it may be that the appropriate residual shear strength is either about 260 kPa or about 310 kPa. If a shear displacement of more than 8 mm is sufficient to cause catastrophic collapse of a structure, it is of no value to know what the strength would be at subsequent, higher, shear displacement values.

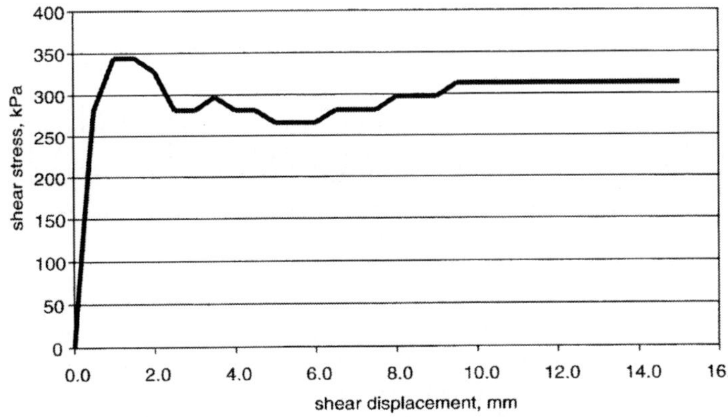

(b) Plotting the normal stress and peak shear stress values provides a failure locus, and the values lie close to a straight line, as shown below. The basic friction angle is given by the slope of this straight line, 33.4°.

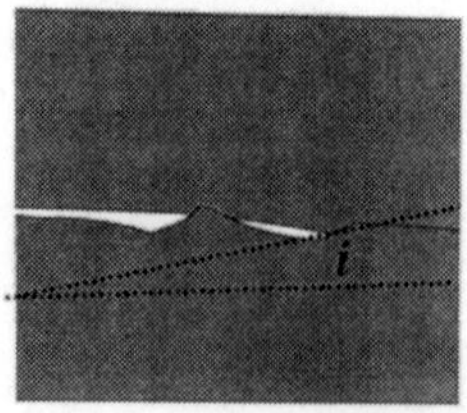

The asperity angle *i*

The bi-linear approximation to the failure locus requires that the initial portion of the locus emanates from the origin. However, there is no way of determining where this initial portion intersects the measured failure locus — we have assumed it to be where the locus intersects the normal stress used in the first test. The dashed extension of the measured failure locus is fictitious, and is sometimes used to determine an 'apparent cohesion'. To compute the asperity angle, i, we take the gradient of the initial segment of the locus, $\phi + i$, which has a value of 44°, and subtract from it the basic friction angle ϕ, to give an asperity angle of $44° - 33.4° = 10.6°$.

All five of the test results lie close to a straight line, and hence indicate that any bi-linear behaviour would occur at a normal stress value lower than the smallest one used here. We have assumed that the initial part of the bi-linear approximation extends between the origin and the first test result, but there is no means of establishing this on the basis of the results given.

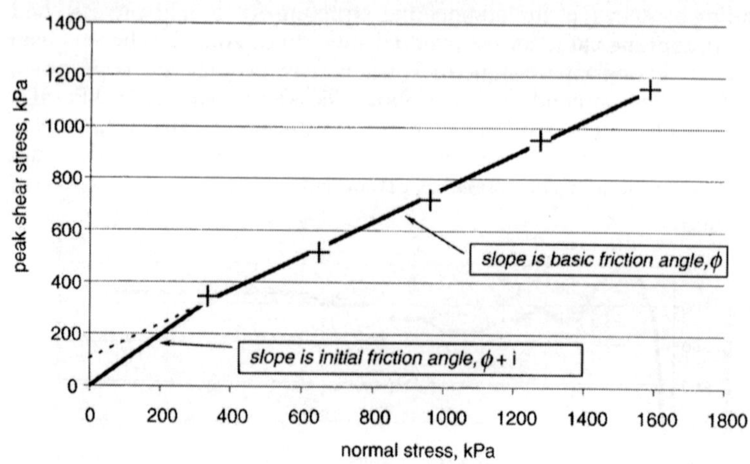

Q11.10 The diagram below shows example results from using a numerical modelling code for predicting the elastic displacements (indicated by the arrows) of a 2-D assemblage of distinct rock blocks through which a tunnel has been excavated. The plot shows the displacement vectors.

(a) Write down a list of rock properties that you think would be required as input to such a modelling exercise.

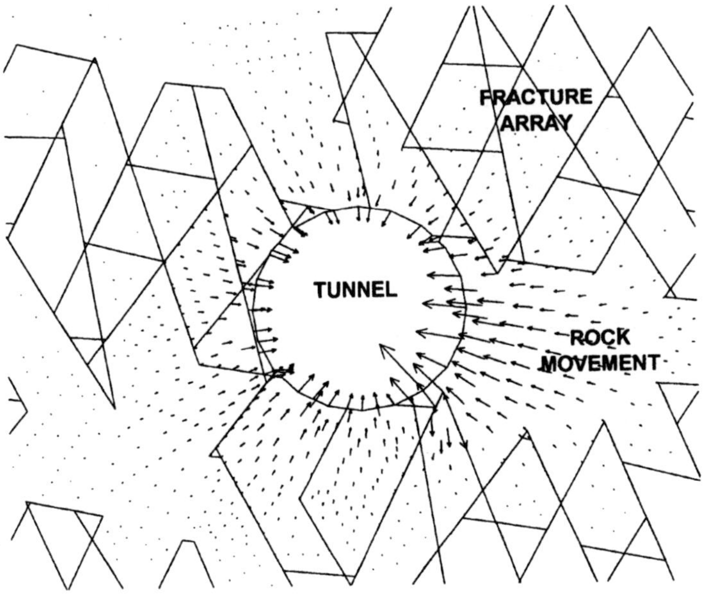

(b) Indicate which of these are likely to be practicably measureable.

A11.10 (a) There are many possible answers to this question, but the rock parameters actually used by the UDEC code which generated the results are:

- the magnitudes and directions of the *in situ* principal stresses;
- Young's modulus and Poisson's ratio of the intact rock;
- the number of fracture sets and;
- for each fracture set, the orientation, frequency, persistence, cohesion, angle of friction, normal stiffness, shear stiffness, and dilation angle.

(b) The *in situ* stress state and intact rock parameters can usually be established fairly well, assuming homogeneity across the region of interest. The number of fracture sets is sometimes clear, but sometimes not in complex circumstances. Given that fracture sets have in fact been established, the orientation and frequency are relatively easy to specify, but the persistence is impossible to measure completely and hence to specify. The cohesion and angle of friction are also relatively easy to measure or estimate, but establishing values for the normal stiffness, shear stiffness and dilation angle that represent the *in situ* fractures is much more difficult.

Thus we see that a disadvantage of such numerical modelling is that a large proportion of the required input parameters may not be obtainable. However, the advantage of the numerical modelling approach is that the sensitivity of the instability mechanisms to the input parameters can be studied in detail. Thus, it is better to use numerical modelling for a parametric study of the overall rock mass behaviour, rather than trying to establish specific values at specific points in the rock mass.

11.3 Additional points

We emphasize that it is not possible to specify the components of a site investigation and the associated testing programme that will be universally appropriate for all rock engineering projects. This is because different projects have different objectives (e.g. the objective of mining engineering is to obtain the rock, whereas the objective of civil engineering is to utilize the space created). Thus, it is necessary to understand the complexities of the rock mass geometry and mechanical properties in order to make sensible decisions on the optimal site investigation given a specific engineering objective.

Sometimes, even the most basic parameters cannot be measured directly. For example, assume that you have been contracted to measure the rock stresses during a site investigation. The client, who takes an interest in rock mechanics, visits your measurement site. He is watching you using the flatjack technique — measuring the normal rock stress components. You have installed two pins in a rock wall, measured the distance between them, cut a slot in the rock wall between the pins, cemented a 'flatjack' into the slot, and hydraulically inflated the jack until the original distance between the pins is re-established, giving the normal stress component perpendicular to the flatjack. The client asks you how many measurements you are going to make. You explain that there will be six such normal rock stress measurements at each test location, using six flatjacks at six orientations, so that the six normal stresses can be used to establish the three normal stresses and three shear stresses of the rock stress matrix. Suddenly, the client has an inspiration, "Surely, it would be better and more elegant to measure the three normal stresses and the three shear stresses directly, rather than measuring six normal stresses?" The client's inspiration is a good idea but, unfortunately, no one has yet found a way to measure a shear stress directly.

For standardization, it is necessary to have guidance on how to establish specific rock parameters. The International Society for Rock Mechanics (ISRM) and the American Society for Testing Materials (ASTM) publish recommended procedures for establishing a wide range of parameters. We gave lists of these in Chapter 11 of ERM 1. Starting in 1998, a second series of ISRM Suggested Methods has been under development; these can be found in the issues of the International Journal of Rock Mechanics and Mining Sciences from 1999 onwards (e.g. see Fairhurst and Hudson, 1999).

12 Rock mass classification

12.1 Rock parameters and classification schemes

The purpose of **rock mass classification** is to establish the quality of a particular rock mass (or part of a rock mass) by assigning rating values to a **set of rock parameters**. Webster's dictionary defines 'classification' as "the act of classifying or forming into a class or classes, so as to bring together those beings or things which most resemble each other, and to separate those that differ". This definition immediately highlights two main issues in rock mass classification: the purpose of the classification has to be established and the method of classification has to be commensurate with the purpose.

For example, if we only used the uniaxial compressive strength of the intact rock and the fracture frequency of the rock mass, we could generate a rock mass classification scheme for characterizing sections of rock in a tunnel as shown in Table 12.1.

On the basis of this scheme, all rock masses must then be one of the categories, *A1, A2, B1, B2*. We could call this a **Rock Index** and assign the words 'Good' to *A1*, 'Fair' to *A2* and *B1*, and 'Poor' to *B2*. But what is the purpose of this classification? Perhaps, the Rock Index would indicate the excavatability and stability of the rock masses in each category. If so, is the classification the best one for that purpose?

There are **four main steps** in the development of any rock mass classification scheme:

- decide on the objective of the rock mass classification scheme;
- decide on the parameters to be used, their ranges and ratings;
- decide on the algebra to be used for the rock index (e.g. do we simply

Table 12.1 Illustrative simple rock mass classification scheme

Parameter	Ratings, R	
Uniaxial compressive strength, σ_c	If $\sigma_c \geq 100$ MPa, $R = A$	If $\sigma_c < 100$ MPa, $R = B$
Fracture frequency, λ	If $\lambda \leq 4$/m, $R = 1$	If $\lambda > 4$/m, $R = 2$

select values from a table, do we add rating values together, do we multiply ratings together, or something else?); and
- calibrate the rock index value against the objective.

The **advantage** of using a rock mass classification scheme is that it is a simple and effective way of representing rock mass quality and of encapsulating precedent practice. The **disadvantage** is that one cannot use it for a different objective or in significantly new circumstances.

The rock mass classifications that have been developed to date follow this basic approach, but include more parameters and use a greater number of classes than the simple 'good', 'fair', 'poor' example we gave above. For example, by adding a third parameter to the classification given in Table 12.1, 'thickness of the layers', and using more rating values (Vervoort and de Wit, 1997[1]), a useful rock index for rock dredging has been developed. By judicious choice of the relevant parameters, such rock mass classification schemes can be a powerful tool for rock engineering.

The two main classification systems, Rock Mass Rating and Tunnelling Quality Index (***RMR* and *Q***), have both been widely applied and there is now a large database of projects where they have been used as the main indicator of rock stabilization requirements in rock tunnelling. The systems provide a coherent method of using precedent practice experience and can now be linked to numerical analysis approaches.

With all schemes, the key issues are the objective of the classification system, choice of the optimal parameters, assigning numerical ratings to parameter values, the algebraic manipulation of the parameter ratings, and drawing conclusions from the mean and variation of the overall rock quality index values.

12.2 Questions and answers: rock mass classification

Rock mass classification schemes are designed to be used in the field, but it is possible to apply them in the office, given a description of the rock mass. Indeed, they are often used in this way, especially as they can be a means of translating a site investigation report into an input for design. In the following questions, you are asked to apply rock mass classification schemes given a description of two rock masses[2] (Bieniawski, 1989).

Q12.1 **A mudstone rock mass at a depth of 200 m contains three fracture sets. One set comprises bedding planes; these are highly weathered, slightly rough surfaces, and are continuous with an orientation of 180/10. Another set is jointing; these joints are slightly weathered, slightly rough, and have an orientation of 185/75. The**

[1] Vervoort A. and de Wit K. (1997) Use of rock mass classifications for dredging. *Int. J. Rock Mech. Min. Sci.*, **34**, 5, 859–864.

[2] To answer Q12.1–12.5, it is necessary to use the *RMR* and *Q* rock mass classification tables in Appendix 3 of this book. Further explanation of these tables is presented in Bieniawski Z. T. (1989) *Engineering Rock Mass Classifications*. Wiley, New York, 251pp.

third set is also jointing; again, the joints are slightly weathered and slightly rough, and have an orientation of 090/80. The strength of the intact rock has been assessed as 55 MPa, and values for the *RQD* and mean fracture spacing are reported as 60% and 0.4 m, respectively.

Use the *RMR* system to classify this rock mass, and assess the stability of a 10 m wide excavation being driven from east to west.

A12.1 In order to apply the *RMR* system, five principal parameters are assessed: intact rock strength, groundwater conditions, *RQD*, fracture spacing, and fracture condition. This provides the basic *RMR* value for the rock mass. The orientation of the fractures is then accounted for through the use of rating adjustment factors to determine the final *RMR* value.

In the example here we have three sets of fractures, and so we will need to apply the *RMR* system to each set in turn, and hence identify the set that is most critical for this particular excavation. Of course, the classification parameters of intact rock strength and groundwater conditions relate to the rock mass as a whole, rather than to specific fracture sets, and are applied identically to each set. Here, the values for *RQD* and mean fracture spacing are reported generally, rather than for specific sets, and so these will be applied equally. We will assess all of these parameters first, and then move on to the parameters that are specific to the fracture sets.

Overall rock mass

The strength of the intact rock has been assessed as 55 MPa, and so a conservative value for the strength rating is 6. The excavation is situated at a depth of 200 m, and the rock mass is a mudstone. At a depth of 200 m, the vertical stress component will be in the region of 5 MPa (assuming a unit weight of rock of 25 kN/m^3), and this stress will probably be sufficient to keep the fractures tightly closed. Taken together with the fact that mudstones have very low primary and secondary permeability, we can infer that the groundwater conditions will probably be in the range of damp to wet, with rating values ranging from 7 to 10.

For an average *RQD* value of 60%, the rating value is about 12, and for a mean fracture spacing of 0.4 m the rating value is about 10. Adding these four ratings together gives a total rating value of
$6 + (7 \text{ to } 10) + 12 + 10 = 35 \text{ to } 38$.

Classification using Set 1

These bedding planes are highly weathered, with slightly rough surfaces, and are continuous. The rating values for these specific attributes are 1, 3 and 0, respectively. An assessment of aperture can be made on the basis of the *in situ* stress state at the location of the excavation: a stress of 5 MPa will mean that the aperture is low, and so a reasonable rating value is 5. We have no information regarding infilling but, as there is no evidence to suggest the presence of any, we can assume a rating value of 6. The total rating value for these fractures is then $1 + 3 + 0 + 5 + 6 = 15$. The dip direction of the bedding is 180°, and so the excavation is being driven along the strike. This is considered to result in 'fair' conditions, with a corresponding rating adjustment value of −5.

Thus, the overall *RMR* value based on Set 1 is (35 to 38) + 15 = 50 to 53, which is classified as 'fair rock'. Taking into account the orientation rating adjustment of −5, the *RMR* value is reduced to 45 to 48, but this does not alter the classification.

Classification using Set 2

We know that this jointing is slightly weathered, slightly rough, and has an orientation of 185/75. We can therefore assign rating values of 5, 3 and −12 to these attributes. Knowing that the fractures are joints in a mudstone, their persistence will probably be in the range of 1 m to 2 m, and an appropriate rating value for this is 2. As with Set 1, we can assign rating values for the aperture and infilling of 5 and 6, respectively. The total rating value for these fractures is then 5 + 3 + 2 + 5 + 6 = 21, giving an overall *RMR* value of (35 to 38) + 21 = 56 to 59. Again, this is classified as 'fair rock'. Taking the orientation rating adjustment of −12 into account reduces the *RMR* value to 44 to 47.

Classification using Set 3

The jointing representing Set 3 has identical mechanical characteristics to Set 2, and so has a rating value of 21, giving an overall *RMR* value of 56 to 59 and a classification of 'fair rock'. The orientation rating adjustment is now −5 (the strike is perpendicular to the excavation axis, and we are driving against the dip which is classed as 'fair'), which reduces the *RMR* value to 51 to 54.

Overall assessment

We can see that Set 2 leads to the most critical classification, with a range of probable *RMR* values of 44 to 47. Using a chart linking *RMR* and excavation span to stand-up time shows that an excavation 10 m wide in such a rock mass would suffer from immediate collapse, and so we can see that the engineering design will need to incorporate rock stabilization measures (i.e. support or reinforcement). In addition, some form of staged excavation may also be necessary, whereby a small pilot excavation is formed and then systematically opened out to the full size as the engineering behaviour of the rock mass is steadily improved as the stabilization measures are applied.

Q12.2 A 7-m-diameter tunnel is to be driven through a sequence of shale and basalt rock at a maximum depth of 61 m. The shales dip towards the east, and the basalts form sub-vertical dykes. The bedding dips between 15° and 20°, the joints dip between 70° and 90°. The joints in the shale are rough, and most of them are thin and healed with calcite, but overall the rock is described as 'blocky and seamy'. The groundwater level is about 50 m above the invert of the tunnel. The average uniaxial compressive strength of the shale is 53 MPa, of the basalt it is 71 MPa. The vertical stress is about 1.0 MPa, and the horizontal stress is about 3.4 MPa. The snaking nature of

the tunnel's route means that at some place along its length it will head in all directions between 090° and 180°.

Use the *RMR* system to predict how the rock will behave in the excavation.

A12.2 This is an example of a commonly occurring problem where, at first sight, much useful information is given but, when we investigate further, we find that there is little on which to base a rock mass classification. Often we are not able to obtain further information, and so it is necessary to carefully consider what are the appropriate rating values. Also, there may be dispute about the rating values we select, and so it is prudent to investigate the sensitivity of the rating value assessments to these values.

Overall environmental conditions

As the groundwater level is 50 m above the tunnel invert, a water pressure of 0.5 MPa will be induced at the tunnel level. The major principal stress is 3.4 MPa acting horizontally, and so the ratio of water pressure to major principal stress is $0.5/3.4 \approx 0.15$, which is regarded as 'wet' and therefore attracts a rating value of 7.

Shale

The shale is described as 'blocky and seamy' and, although descriptions do not enter into the *RMR* assessment directly, they can be used to help assess the *RQD* and fracture spacing values. Shales are sedimentary rocks, and tend to form distinct beds. The description 'blocky and seamy' allows us to picture a rock mass that splits easily along the bedding, but is broken into blocks by the joints. Thus, it is likely that both the *RQD* and the mean fracture spacing values will be low for such a rock, and so we may select rating values of, say, 10 and 8 for these two parameters.

We are told that the 'joints in the shale are rough, and most of them are thin and healed with calcite'. We do not know whether this description is just for the joints, or for the joints and bedding planes. If we assume it is for the joints only, then how do we assess the bedding planes? Bedding planes tend to be extensive, and in a shale will be smooth. Joints on the other hand will probably have a persistence of no more than a few metres. We can use this understanding to make preliminary rating assessments. The fact that the joints are 'thin and healed with calcite' allows us to assess the aperture rating as being about 5, and the infilling as about 5 as well. Finally, the tunnel is to be excavated at quite a shallow depth — no more than about 60 m — and on this basis it would be prudent to assume that some weathering will have taken place. A rating value of 4 for weathering is therefore suggested.

We can now assess these fracture sets as follows:

	Persistence	Aperture	Roughness	Infilling	Weathering	Total
Bedding	0	5	1	5	4	15
Joints	5	5	5	5	4	24

This shows that the rating value for the bedding is much lower than that of the jointing, and so it is likely that the bedding will be the most critical fracture set for determining tunnel stability. As the dip of the bedding is no more than 20°, this is considered to result in 'fair' conditions, and so the rating adjustment for orientation will be −5. Note that if we consider the joints, then when the tunnel is heading south these features will give rise to 'very unfavourable' conditions, with a rating adjustment of −12. Combining the fracture assessment and the rating adjustment for the bedding gives a value of 15 − 5 = 10, and for the joints gives a value of 24 − 12 = 12. This shows that the bedding is, as we surmised above, the most critical feature, but there is little difference between the joints and the bedding.

Finally, we know the strength of the shale is 53 MPa, and this gives a rating value of about 5. The basic *RMR* value for the shale is then as follows:

Strength	Groundwater	*RQD*	Spacing	Fracture Condition	Total
5	7	10	8	15	45

Taking the rating adjustment for orientation into account reduces this to 40.

Basalt dykes

We are given no definite geomechanical data regarding the basalt dykes, other than that they are sub-vertical features. We need to turn to our geological knowledge in order to make an assessment of this rock type. Firstly, we can assume that these dykes will be of limited thickness, say, no more than 5 m. As the cooling joints in a dyke run across the plane of the dyke — rather than parallel to it — this will set a maximum fracture persistence of 5 m, giving a rating value of 2. The aperture of these cooling joints could be as large as 1 mm, giving a rating value of about 3. The joints are liable to be 'slightly rough', with no infilling and only slight weathering. The rating values for these attributes are 3, 6 and 5, respectively. The overall assessment for the joints in the dykes is then 2+3+3+6+5 = 19. The rating adjustment for the orientation of these joints is difficult to assess but, given the limited extent of the joints, we can perhaps assign an effect of 'fair' and a corresponding rating value of −5.

The *RQD* of the dykes will probably be high, and a rating value of around 15 will be suitable. The mean fracture spacing may be of the order of 0.5 m, and the rating value for this is 10. Finally, the strength of the basalt is 71 MPa, and the corresponding rating value is about 7.

The basic *RMR* value for the dykes is then as follows:

Strength	Groundwater	*RQD*	Spacing	Fracture condition	Total
7	7	15	10	19	58

Taking the rating adjustment for orientation into account reduces this to 53.

We must bear in mind that the dykes may have been acting as conduits for groundwater flow for a long time, and may be highly weathered.

If there is any evidence for such weathering, then another assessment should be made taking this into account.

Overall assessment

The overall *RMR* values for the two rock types are 40 for the shale and 53 for the basalt dykes. For a 7-m-diameter tunnel, we find that the shale will suffer immediate collapse, whereas the basalt will be able to stand unsupported for around 1 week. We can now suggest that the tunnel excavation should be carried out using some form of shield or tunnel boring machine to offer continuous support, followed by the installation of an immediate support system such as a pre-cast concrete lining. This overall system may not work well when a dyke is encountered (the rock may be too strong to excavate mechanically), but the additional stand-up time available in this material will allow the use of a different support or reinforcement system, say, shotcrete or rockbolts.

Q12.3 Use the Q system to assess the stability of the rock mass as described in Q12.1.

A12.3 In order to determine the *Q* value for a rock mass, we need to determine rating values for each of six parameters: *RQD*, joint set number, joint roughness number, joint alteration number, joint water reduction factor, and stress reduction factor.

***RQD* rating**

The average *RQD* value is 60%, and hence the rating value is 60.

Joint set number (J_n)

As there are three sets of fractures, the appropriate value for this parameter is 9.

Joint roughness number (J_r)

The bedding planes are slightly rough, continuous surfaces. No single rating entry fits this description exactly, but that of 'rough or irregular, planar' seems most appropriate for such a large-scale feature as a bedding plane. The rating value is then 1.5.

The jointing is slightly weathered and slightly rough. Joints are likely to be relatively small-scale features (persistence of the order of 1 m to 2 m) and so a realistic rating value is 3, appropriate for 'rough or irregular, undulating' features.

Although we could perform two calculations and use both of these values in turn, it is only appropriate to use the one that represents the most critical fracture set. Thus, as the rock is a mudstone, we should take particular account of the highly continuous bedding planes, and hence use the value of 1.5.

Joint alteration number (J_a)

Although we know that the bedding planes are highly weathered and that the jointing is slightly weathered, we have no information regarding fracture infilling materials. However, the general description of the geological environment would suggest that the fracture surfaces are in contact. This, taken in conjunction with the fact that the rock type is a mudstone, leads to the selection of '...low-friction clay mineral coatings...' as the most appropriate entry for joint alteration, giving a rating value of 4.0. It is worthwhile noting that, had we selected entries for 'rock wall contact before 10 cm shear', we may have felt that the entry for 'clay mineral fillings (continuous, <5 mm in thickness)' was the most appropriate, giving a rating value of 8.0. We should bear this in mind and examine the final classification for sensitivity to this rating value.

Joint water reduction factor (J_w)
The presence of water has not been noted in the description and, given the low primary and secondary permeability of mudstone, it is reasonable to select 'dry excavations or minor inflow, e.g. 5 l/min locally' for this parameter, and set the rating value to 1.0.

Stress reduction factor (*SRF*)
The tunnel is to be excavated at a depth of 200 m in a rock with a compressive strength of 55 MPa. The vertical stress at this depth will be in the order of 5 MPa (assuming a unit weight of rock of 25 kN/m^3), and assuming that this will be the major principal stress leading to a strength/stress ratio of $55/5 = 11$. However, at this depth it is possible that the major principal stress could be horizontal with a magnitude twice that of the vertical stress, which will lead to a strength/stress ratio of 5.5. Taking these two results together indicates that we should regard this environment as 'high-stress, very tight structure (usually favourable to stability, may be unfavourable for wall stability)' and take the rating for SRF to lie in the range of 0.5 to 2.0. An initial assessment can be made using a value of 1.0, but we should be prepared to investigate the effect of varying this rating.

***Q* value and assessment**
The Q value for the rock mass is now computed as

$$Q = \frac{RQD}{J_n} \times \frac{J_r}{J_a} \times \frac{J_w}{\mathrm{SRF}} = \frac{60}{9} \times \frac{1.5}{4.0} \times \frac{1.0}{1.0} \approx 2.5$$

for which the classification is 'poor'. If, as was noted above, we investigate the sensitivity of Q to our uncertainty by increasing the joint alteration number to 8.0 and increasing the stress reduction factor to 2.0, then we obtain a value of

$$Q = \frac{RQD}{J_n} \times \frac{J_r}{J_a} \times \frac{J_w}{\mathrm{SRF}} = \frac{60}{9} \times \frac{1.5}{8.0} \times \frac{1.0}{2.0} \approx 0.6$$

which is regarded as 'very poor'.

In order to determine the engineering ramifications of these assessments, we now need to determine the 'equivalent dimension' of the excavation. This is the actual size of the excavation scaled to account for the degree of security we require in our assessment (i.e. reducing the size gives us less required security, and is appropriate for temporary mine openings; increasing the size increases the required security and is appropriate for openings to which the general public have access). Here, we have no information on which to compute the equivalent dimension, and so we take it as the actual size, i.e. 10 m. Charts and tables are available that show how the reinforcement and support requirements vary for various combinations of Q and equivalent dimension, and using such aids leads to the following assessment for a 10 m span:

Poor rock Untensioned rockbolts, at 1 m to 1.5 m spacings, together with mesh-reinforced shotcrete applied to a thickness of 5 cm to 10 cm.

Very poor rock Untensioned rockbolts, at 1 m spacings, together with mesh-reinforced shotcrete applied to a thickness of 5 cm to 7.5 cm.

The similarity between these schemes would allow us to develop a flexible system for application underground, such that the inevitable variations in rock mass quality encountered during construction could be dealt with easily. It is interesting to see that the shotcrete thickness is lower for the very poor rock than for the poor rock. This is because the rockbolt spacing is also lower, and so the shotcrete spans smaller distances.

Q12.4 Use the Q system to assess the stability of the rock mass described in Q12.2.

A12.4 In this example we will need to assess the two principal rock types — shale and basalt dykes — separately. The lack of geomechanical data means that we will need to apply a good deal of judgement in order to generate a classification for the rock mass.

In order to determine the Q value for a rock mass, we need to determine rating values for each of six parameters: *RQD*, joint set number, joint roughness number, joint alteration number, joint water reduction factor, and stress reduction factor.

RQD rating

No values for *RQD* are given, but using the values determined as part of our assessment using *RMR* gives a rating of 50 for the shale and 75 for the basalt dykes.

Joint set number

It is appropriate to assume that the shale contains three fracture sets, for which the joint set number is 9, and to assume that the cooling joints in the dykes can best be described as '...four or more joint sets, random, heavily jointed...', for which the rating value is 15.

Joint roughness number

The bedding in the shale is likely to be a particularly extensive feature, and may therefore be the most critical feature. For a large-scale feature such as a bedding plane, the description that best fits this is 'rough or irregular, planar', and hence the rating value is 1.5.

In the basalt dykes the joints will tend to be discontinuous, small-scale features, and the appropriate rating value for them will be 4, although the slightly lower value of 3 may be suitable.

Joint alteration number

The joints in the shale are quoted as 'thin and healed with calcite', and so an appropriate classification for these is 'tightly healed, hard, non-softening, impermeable filling, i.e. quartz or epidote' which gives a rating value of 0.75. However, it is possible that the shale may degrade readily, and so we should consider the classification 'softening or low-friction clay mineral coatings', for which the rating value is 4.0.

For the basalt, a prudent classification would be 'slightly altered joint walls', giving a rating value of 2.0.

Joint water reduction factor

The groundwater level is about 50 m above the tunnel invert, and so this could lead to a water pressure of 5 kg/cm^2, for which the rating value is 0.5.

Stress reduction factor

The average uniaxial compressive strength of the shale is 53 MPa, and of the basalt it is 71 MPa. The vertical stress is about 1.0 MPa, and the horizontal stress is about 3.4 MPa. The major principal stress is horizontal with a magnitude of 3.4 MPa, and the compressive strength of the rock types is 53 MPa for the shale and 71 MPa for the basalt. The strength/stress ratio for these two cases is then $53/3.4 = 15.6$ and $71/3.4 = 20.9$, respectively, and so for both of them the stress reduction factor is 1.0.

***Q* value and assessment**

The Q value for the shale is now computed as

$$Q = \frac{RQD}{J_n} \times \frac{J_r}{J_a} \times \frac{J_w}{\mathrm{SRF}} = \frac{50}{9} \times \frac{1.5}{0.75} \times \frac{0.5}{1.0} \approx 5.6$$

for which the classification is 'fair', and the Q value for the basalt is

$$Q = \frac{RQD}{J_n} \times \frac{J_r}{J_a} \times \frac{J_w}{\mathrm{SRF}} = \frac{75}{15} \times \frac{3}{2} \times \frac{0.5}{1.0} \approx 3.8$$

for which the classification is 'poor'. We noted above that for the shale a value for J_a of 4.0 may be more suitable than 0.75, and adopting this value reduces Q for the shale to 1.0, which is on the boundary of 'very poor' and 'poor' rock.

If we take the 'equivalent dimension' of the excavation to be its actual dimension, i.e. 7 m, then by reference to charts and tables of reinforcement and support requirements we find that the following are appropriate:

Fair rock Untensioned rockbolts at 1 m to 1.5 m spacings, together with mesh.

Poor rock Shotcrete applied to a thickness of 2.5 cm to 7.5 cm, or untensioned rockbolts at 1 m to 1.5 m spacings, together with mesh-reinforced shotcrete applied to a thickness of 5 cm to 10 cm.

Very poor rock Untensioned rockbolts, at 1 m spacings, together with mesh-reinforced shotcrete applied to a thickness of 5 cm to 7.5 cm.

Note how the stabilization systems become heavier as the quality of the rock reduces. Once again, we will be able to exploit the similarity between these schemes to develop a flexible system for use in the tunnel, such that the inevitable variations in rock mass quality encountered during construction could be dealt with easily.

Q12.5 Using your assessments of *RMR* and Q for questions Q12.1, Q12.2, Q12.3 and Q12.4, investigate the relation between the Q and *RMR* values. Do your results correspond with a generally accepted relation, *RMR* = 9 ln Q + 44?

A12.5 From answers A12.1–A12.4, we now have the following information:

Project	*RMR*	*Q*
Excavation in mudstone	44–47	0.6–2.5
Tunnel in shale	40	1–5.6
Tunnel in basalt dykes	53	3.8

These results, together with the relation $RMR = 9 \ln(Q) + 44$, are plotted below:

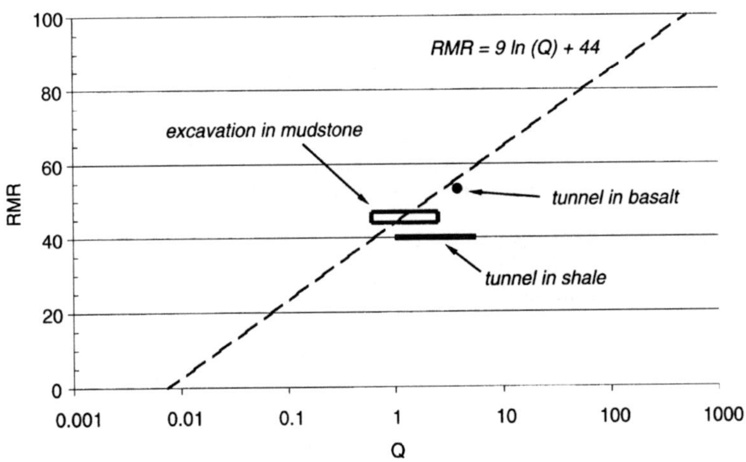

Notice that a box has been plotted for the excavation in mudstone, as we have a range of values for both *Q* and *RMR*. For the tunnel in shale, we have a range of values for *Q* and a single value for *RMR* and so a line has been plotted, and for the tunnel in basalt we have a single value for both Q and RMR, which means that only a single point is plotted.

This plot shows that the values we have determined in Q12.1–Q12.4 plot close to the line representing the empirical relation between Q and RMR, despite the fact that we had to use judgement to determine many of the various rating values. This highlights one of the strengths of rock mass classification systems: they are really quite robust in application. However, it is important not to attempt to be too precise when using them. For example, trying to distinguish between Q values of, say, 1.2 and 1.3 is not a useful exercise.

Q12.6 The diagram below (Singh and Goel, 1999[3]) shows *RMR–Q* correlations for case studies in India, Scandinavia, UK and USA.

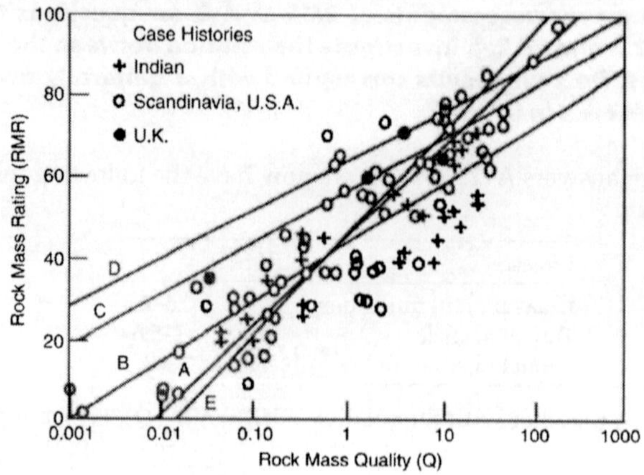

The suggested *RMR–Q* correlation lines shown on the diagram are

A	$RMR = 9 \ln Q + 44$
B	$RMR = 5.9 \ln Q + 43$
C	$RMR = 5.4 \ln Q + 55.2$
D	$RMR = 5 \ln Q + 60.8$
E	$RMR = 10.5 \ln Q + 41.8$

For a rock engineering design project where a correlation between *RMR* and Q is required to support the design, which of the correlations would you choose?

A12.6 In A12.1–A12.4, it is evident that the assessments of RMR and Q require some experience and engineering judgement. Moreover, the ratings apply for the specific rock mass at the site and to the project in hand. Therefore, it is better to try to extract more information from the site for direct assessment of RMR and Q than to use the correlation lines. However, if there are reasons for using a correlation

[3] The information in this question is from pp. 93–94 in Singh B. and Goel R. K. (1999) *Rock Mass Classification*. Elsevier, Oxford, 267pp.

line, then RMR = 9 ln Q + 44 is the most well known line, although RMR = 5.9 ln Q + 43 has the highest correlation coefficient for the 115 case studies used by Singh and Goel which include 34 from India. Singh and Goel (1999) also indicate how to improve the correlation by using a rock condition rating and a rock mass number.

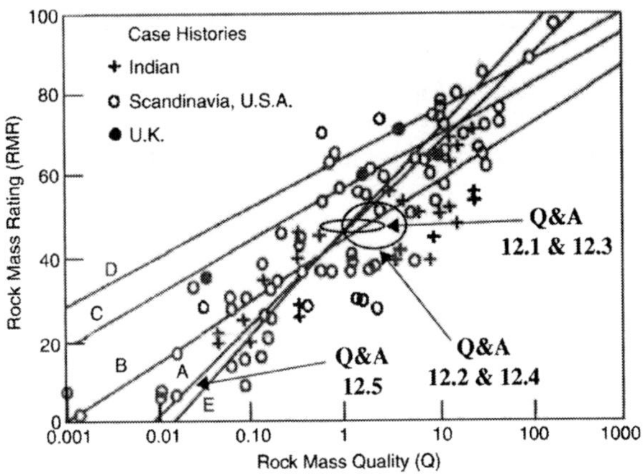

The *RMR* and Q values given in A12.1–A12.4 are plotted on the diagram above. Line A is the one plotted in A12.5.

Q12.7 Imagine that a rock mass classification system is required to assess the instability of natural slopes in the Italian Alps. List 15–25 parameters that you think would be most useful for such a classification scheme.

A12.7 A scheme that was developed for this purpose (Mazzoccola and Hudson, 1996[4]) included the following nineteen parameters.

Geology	Folds	Faults	Rainfall	Freeze/thaw
Previous instability	Fracture intact wall strength	Weathering	Number of fracture sets	Fracture orientation
Fracture aperture	Fracture persistence	Fracture spacing	Mechanical properties	Rock mass strength
Hydraulic conditions	Slope orientation	Slope dimensions	*In situ* stress	

When a standard scheme, such as the *RMR* or Q system, is being used, the results can be compared with extensive previous experience (Q and A 12.5 and 12.6) and conclusions drawn about engineering design. In the case of a new classification for a new purpose — in this case a natural slope instability classification — this is not possible, although there is the advantage that we can include all the paramet-

[4] Mazzoccola D. F. and Hudson J. A. (1996). A comprehensive method of rock mass characterization for indicating natural slope instability. *Q. J. Eng. Geol.*, **29**, 37–56.

ers that we consider essential. However, there is the disadvantage that *a priori* assignments of ratings for the parameter values encountered on site are not straightforward (because the values have not yet been measured) and questions about the range and sensitivity of the parameter ratings can only be answered after the measurements have been made.

Nevertheless, the exercise of creating a new scheme is worthwhile because it will indicate the relative nature or quality of different sites. In the case of the Italian Alps work, it was possible to establish the relative instability of twenty natural slopes, which directly indicates which slopes are the most hazardous. The paper by Mazzoccola and Hudson (1996) also demonstrates how to establish weighting factors for the parameter ratings for any given rock mechanics or rock engineering objective.

Q12.8 A rock mass classification system is required for assessing the suitability of different rock formations for storing compressed domestic gas in unlined rock caverns along the route of a main gas transmission line. List the rock parameters that you would use in a rock mass–rock engineering classification scheme for this objective.

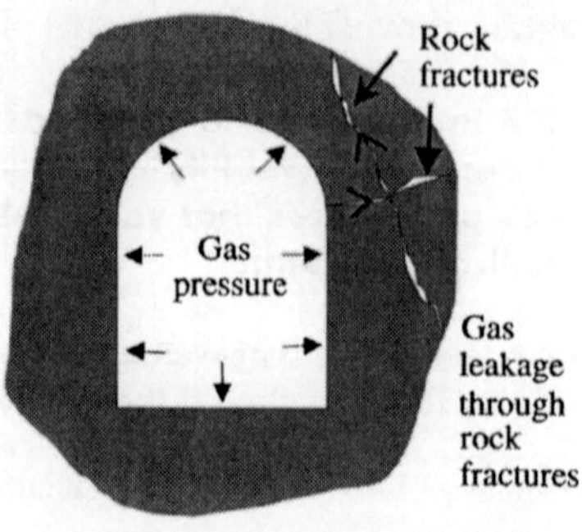

A12.8 The principal rock engineering objective is to reduce potential gas leakage to an acceptably low level. The primary gas leakage pathway is via rock fractures, and possibly also the intact rock. Therefore, the following parameters should be included in a rock mass classification scheme tailored to this engineering objective.

— Minimum principal stress value, σ_3, because the gas pressure can open fractures.

— Compressive strength of the rock, σ_c, because the cavern should be stable and the compressive strength indicates the intact rock quality.

— Groundwater pressure at cavern crown level (before cavern excavation), p_w, because the gas has to pass through the water-filled rock fractures.

— Stress anisotropy ratio, $(\sigma_1/\sigma_3) = \sigma_R$, because anisotropic stress conditions are destabilizing to an excavation.

— Maximal stored gas pressure, p_g, because the 'rock mass–rock engineering' composite conditions are a function of this engineering variable.

— Fracture frequency, λ, because the number of fractures will affect leakage.

— Fracture aperture, e, because the aperture will also affect leakage potential.

The next step is to establish whether an increase in each of these seven parameters is good or bad in the context of a classification index in which higher values indicate improved engineering circumstances.

Minimum principal stress value	σ_3	an increase is good
Compressive strength of the rock	σ_c	an increase is good
Groundwater pressure at cavern crown level (before cavern excavation)	p_w	an increase is good
Stress anisotropy ratio	$(\sigma_1/\sigma_3) = \sigma_R$	an increase is bad
Maximal stored gas pressure	p_g	an increase is bad
Fracture frequency	λ	an increase is bad
Fracture aperture	e	an increase is bad

To develop a 'Gas Cavern Tightness Index' (GCTI), the parameters or parameter ratings have to be algebraically arranged so that variations in the parameter values are correctly reflected in the GCTI index according to the table above. If we use the same approach as the *Q* system, in which the quotients have an interpretation, a suitable GCTI would be

$$\text{GCTI} = \frac{\sigma_c}{\sigma_R} \times \frac{\sigma_3}{\lambda e^3} \times \frac{p_w}{p_g}.$$

The individual quotients can then be interpreted as follows:

$\dfrac{\sigma_c}{\sigma_R}$ is related to the stability of the intact rock around the cavern;

$\dfrac{\sigma_3}{\lambda e^3}$ is related to the resistance to gas leakage; and

$\dfrac{p_w}{p_g}$ is related to the effectiveness of the groundwater confinement.

The reader can no doubt think of improvements to this initial index. Our intention here is only to provide an example of the principles used in developing new rock mass classification indices.

Q12.9 The following parameter values have been determined for three rock mass types found along the route of a major new highway tunnel that passes at a high level through the flank of a mountain range:

	Strength (MPa)	RQD (%)	Mean fracture spacing (m)
Sandstone	80	45	0.4
Mudstone	20	75	0.3
Syenite intrusions	250	10	0.2

The fractures within each rock mass type have the properties shown in the following table:

	Persistence (m)	Aperture (mm)	Roughness	Infilling	Weathering
Sandstone	5–8	~1.5	rough	none	none
Mudstone	1.5–2.5	~0.5	slight	none	slight
Syenite	2	~6	very	none	none

Write down a description for each of these three rock mass types, and describe their likely engineering behaviour.

Then apply the *RMR* system to these rock mass types, and compare the assessment of their engineering behaviour made in this way with the description you wrote down earlier.

What do you conclude from this exercise about the ability of *RMR* to discriminate between the engineering behaviour of these particular rock mass types?

A12.9 Descriptions of the rock mass types.

On the basis of the rating summaries given we can describe each rock type as follows.

Sandstone

A strong rock, probably with extensive bedding planes that cause the rock to break into beds that are on average 0.4 m thick. The moderately low *RQD* indicates that a large number of these units will be thinner than 0.1 m, and so a slabby structure should be expected. The roughness of the fractures, together with the appreciable apertures, indicates that the rock mass may be loose.

Mudstone

Very weak, but no evidence of extensive bedding planes. The fractures are tight and only slightly rough, and taken together with the *RQD* of 75% this indicates that the rock mass is not highly fractured — it is probably blocky, with most of the blocks being around 0.3 m in size.

Syenite intrusions

A very strong rock, but with very low *RQD* and mean spacing values. The fractures are very rough but have large aperture. The rock mass probably has a 'sugar cube' type appearance, but may have a high degree of mechanical interlock.

In summary, we have one rock that is very strong but highly broken (syenite), another that is weak but comprising reasonably large blocks (mudstone), with the third somewhere in between (sandstone).

***RMR* ratings**

The *RMR* rating values associated with the fracture summaries for each of the three rock types in the Question are presented in the table below (rating values are given between parentheses).

	Persistence (m)	Aperture (mm)	Roughness	Infilling	Weathering	Total
Sandstone	5–8 (2)	~1.5 (3)	rough (5)	none (6)	none (6)	(22)
Mudstone	1.5–2.5 (4)	~0.5 (4)	slight (3)	none (6)	slight (5)	(22)
Syenite	2 (4)	~6 (0)	very (6)	none (6)	none (6)	(22)

As we can see, the very different fracture descriptions have led to an identical rating value of 22 across all rock types.

Before continuing to determine the basic *RMR* value for each rock type, we need to assess the groundwater conditions. We know that the tunnel passes high through the flank of a mountain range, and from this we can infer that it will probably be located above the groundwater level. Accordingly, we can assess the conditions as 'damp' and assign a rating value of 10 to the groundwater.

Taking the information we have been given about the strength and degree of fracturing, together with the fracture condition and groundwater rating values, we can now obtain the basic *RMR* for each rock type (rating values are given between parentheses):

	Strength (MPa)	*RQD* (%)	Fracture spacing (m)	Fracture condition	Ground-water	Total rating
Sandstone	80 (8)	45 (9)	0.4 (10)	(22)	(10)	(59)
Mudstone	20 (3)	75 (15)	0.3 (9)	(22)	(10)	(59)
Syenite	250 (15)	10 (4)	0.2 (8)	(22)	(10)	(59)

The basic *RMR* rating of 59 is the same for all three rock types. This value of 59 classifies each rock type as being on the boundary of 'fair rock' and 'good rock'.

Rock type comparisons

These three rock types which, on the basis of their descriptions, should display very different engineering behaviour, all score the same *RMR* value. It appears therefore that *RMR* is not a good discriminator of engineering behaviour for these rock types.

This is not a deficiency with the *RMR* system in itself, but is more to do with the use of addition to compute a single overall value of *RMR*. This is demonstrated if we examine the ratings associated with strength, *RQD* and fracture spacing for these rock types. It is reasonable to assume that *RQD* and fracture spacing are related to each other, and that they are essentially independent of rock strength. On this basis we can produce the table given below, and then plot the results as shown in the figure.

	Strength	*RQD* + fracture spacing	Partial rating
Sandstone	8	19	27
Mudstone	3	24	27
Syenite	15	12	27

Geometrically, the use of addition to combine values induces what is known as the 'city block metric', so called because it reflects the lengths

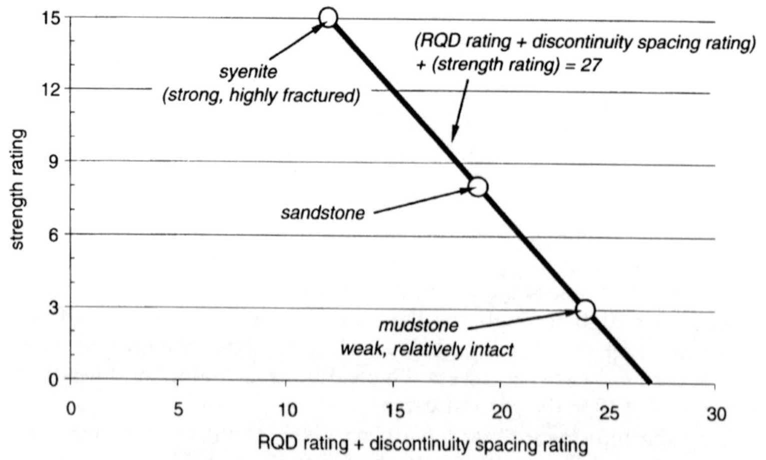

of the routes one can walk through a city that is built on a rectangular grid. In the example here, for the partial *RMR* rating value of 27, i.e. a path length of 27 units, there are many paths of this length formed from a combination of a horizontal distance (i.e. combined *RQD* and fracture spacing rating) and a vertical distance (i.e. strength rating), represented by the solid line in the figure. Within the limit of acceptable values for the strength rating (0 to 15), the end points of these paths are shown as the solid line.

The figure shows that the rating value locus of 27 extends from extremely strong and highly fractured rock through to very weak and unfractured rock and, as a result, it is not possible to discriminate between these rock types using *RMR*. In fact, this is true for all classification schemes that depend on a single classification value computed using simple arithmetic; we have chosen *RMR* as the example here.

A technique to overcome this difficulty is to quote classification values as a vector: for *RMR*, there are five parameters and so it would be a five-dimensional vector, and for *Q* it would be a six-dimensional vector. However, one of the problems of adopting such a technique is that it would require the rock engineering community to reinterpret the large database of projects on which these schemes had been used.

Q12.10 The following measurements of mean fracture spacing (in metres) have been made on core from 12 boreholes as part of a site investigation project:

0.259	**0.304**	**0.875**	**0.292**	**0.467**	**0.412**	**0.350**	**0.368**	**0.438**	**0.389**	**0.280**	**0.318**

As the rock mass is to be characterized using the *Q* system, the following parameters have also been determined: $J_n = 9$; $J_r = 1.5$; $J_a = 2$; *SRF* = 2.5; and $J_w = 1$.

(a) Using the frequency measurements to determine *RQD* values and thence *Q* values with the additional parameters given,

comment on the inhomogeneity of the rock mass in terms of (i) fracture frequency and (ii) Q.

(b) A technique for increasing the range of *RQD* values in a given rock mass is to adopt a different *RQD* threshold value (from the usual value of 0.1 m) computed using

$t^* = 2\ln(\lambda_{max}/\lambda_{min})/(\lambda_{max} - \lambda_{min})$,

where λ_{max} and λ_{min} are the extreme values of the fracture frequency occurring in the rock mass. Use this technique to compute new values of Q, and compare the results with those found in Part (a).

A12.10 (a) In order to compute Q we use the relation $Q = (\text{RQD}/J_n) \times (J_r/J_a) \times (\text{SRF}/J_w)$, and to determine values for RQD we use the relation $\text{RQD} = 100(\lambda t + 1)\exp(-\lambda t)$, where λ, the fracture frequency, is given by the reciprocal of the mean spacing. In this relation for RQD, the threshold value t is taken to be 0.1 m. The results of the calculations are given in the following table and plotted in the figure below.

Borehole	1	2	3	4	5	6	7	8	9	10	11	12
Spacing (m)	0.259	0.304	0.875	0.292	0.467	0.412	0.350	0.368	0.438	0.389	0.280	0.318
Frequency (m^{-1})	3.861	3.289	1.143	3.425	2.141	2.427	2.857	2.717	2.283	2.571	3.571	3.145
RQD (%)	94.2	95.6	99.4	95.3	98.0	97.5	96.6	96.9	97.8	97.2	95.0	96.0
Q	19.6	19.9	20.7	19.9	20.4	20.3	20.1	20.2	20.4	20.3	19.8	20.0

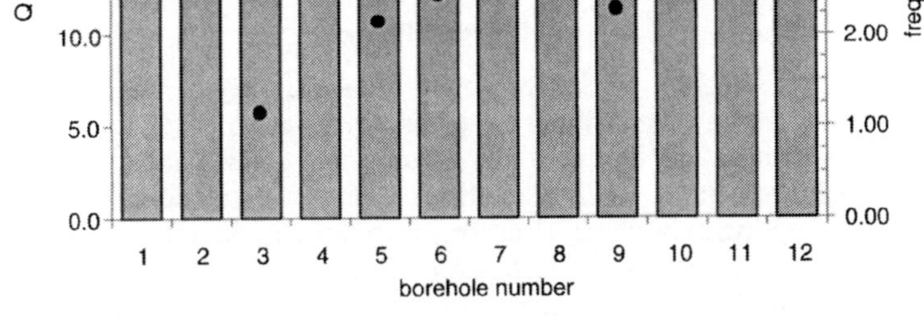

We can see immediately from the plot that the rock mass is significantly inhomogeneous in terms of fracture frequency (for these data, the ratio of maximal frequency to minimal frequency is 3.38), and that this information is lost when we consider Q values (the ratio of maximal Q to minimal Q is only 1.06). Clearly, Q is not sensitive to changes in fracture frequency (or its reciprocal fracture spacing).

(b) On the basis of the site investigation results, the minimal and maximal values of fracture frequency in the rock mass are 1.143 m^{-1} and 3.861 m^{-1}, respectively, and from these we find that the appropriate value of *RQD* threshold is

$$t^* = \frac{2}{\lambda_{max} - \lambda_{min}} \ln\left(\frac{\lambda_{max}}{\lambda_{min}}\right) = \frac{2}{3.861 - 1.143} \ln\left(\frac{3.861}{1.143}\right) = 0.90\,\text{m}.$$

The new values of *RQD* and *Q* — called *RQD** and *Q** to distinguish them from the customary formulations — are as shown in the following table.

Borehole	1	2	3	4	5	6	7	8	9	10	11	12
Spacing (m)	0.259	0.304	0.875	0.292	0.467	0.412	0.350	0.368	0.438	0.389	0.280	0.318
Frequency (m^{-1})	3.861	3.289	1.143	3.425	2.141	2.427	2.857	2.717	2.283	2.571	3.571	3.145
*RQD** (%)	14.0	20.7	72.7	18.9	42.9	36.1	27.5	30.1	39.4	33.0	17.1	22.8
*Q**	2.9	4.3	15.1	3.9	8.9	7.5	5.7	6.3	8.2	6.9	3.6	4.8

These results for *Q**, together with the results from the earlier calculation, are plotted in the figure below. We can see from this that *Q** has a greater discrimination than *Q* with regard to the inhomogeneity, at the expense of much reduced values.

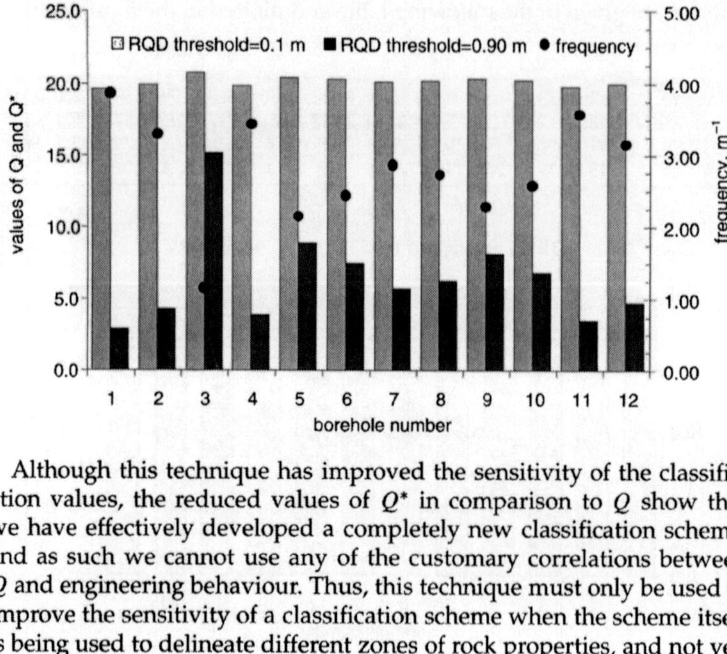

Although this technique has improved the sensitivity of the classification values, the reduced values of *Q** in comparison to *Q* show that we have effectively developed a completely new classification scheme, and as such we cannot use any of the customary correlations between *Q* and engineering behaviour. Thus, this technique must only be used to improve the sensitivity of a classification scheme when the scheme itself is being used to delineate different zones of rock properties, and not yet, for example, for selection of support and reinforcement.

12.3 Additional points

A substantial extension to the *RMR* system has been made by Romana (1993)[5] with his *SMR* system for assessing slope stability. The *SMR* value

[5] Romana M. R. (1993) A geomechanical classification for slopes: slope mass rating, in *Comprehensive Rock Engineering*, Vol. 3, Ch. 23 (J. A. Hudson, ed.), Pergamon Press, Oxford, pp. 575–599.

can be written in the form

$$SMR = RMR - F_g + F_e,$$

where F_g is a factor representing the geometry of the potential instability present in a rock slope, and F_e is a factor corresponding to the excavation method. Geometries that are intrinsically more unstable have higher values of F_g (and hence reduce *SMR*), and excavation methods that induce little perturbation in the rock mass have high values of F_e (and hence increase *SMR*).

Additional work on applying classification systems to slope stability has been reported by Sonmez and Ulusay (1999)[6]. In this paper, it is noted that the rock mass classification should refer not to the rock mass in its undisturbed condition but to the excavation-disturbed rock mass — which is the one that hosts the rock engineering structure. The authors suggest methods for assessing the excavation disturbance effect on rock mass classification values.

The **advantages** of using rock mass classification systems are that

- the rock mass quality can be assessed simply, rapidly and continuously,
- the classification values can be established by trained site personnel (i.e. a high level of general engineering expertise is not required),
- continuous rock mass assessment using logging sheets will alert contractors and consulting engineers to significant changes in rock conditions, and
- engineering design is coherently based on previous experience.

The **disadvantages** of using a rock mass classification system are that

- the systems currently in use are historical and idiosyncratic,
- the algebra and ratings values of the systems have not been scientifically considered, and
- they cannot be used for the full range of engineering objectives.

[6] Sonmez H. and Ulusay R. (1999) Modifications to the geological strength index (GSI) and their applicability to stability of Slopes. *Int. J. Rock Mech. Min. Sci.*, **36**, 743–760.

13 Rock dynamics and time dependency

13.1 Strain rates

Rock behaviour mechanisms can occur at significantly different rates: during the excavation of rock using explosives, rock failure occurs in milliseconds; a rock block can slip out of a cavern roof in a second; a shaft might take a day to fill up with water; a chalk or mudstone face can deteriorate in a few days. It may take months or years for water to move through a granitic rock mass and, during the 120 years design life of an unlined rock cavern for civil engineering purposes, creep processes could lead to roof collapse. Some geological processes occur over millions of years.

In view of this time dependency over a wide time range, it is convenient to consider the rate at which such processes occur in terms of strain rate. Consider a rock cylinder subjected to uniaxial compression along its axis and that the rock's compressive strength is reached at 0.5% strain, i.e. 0.005 or 5×10^{-3} strain. At a **strain rate** of $1 \times 10^{-3}\ s^{-1}$, the rock specimen will fail in 5 s. At a strain rate of $1 \times 10^{-4}\ s^{-1}$, the rock specimen will fail in 50 s. For explosive failure in 2 ms, the strain rate would be $5 \times 10^{-3}/2 \times 10^{-3} = 2.5$ or $2.5 \times 10^{0}\ s^{-1}$. Note that there are four orders of magnitude difference between the slowest and fastest of these examples.

A range of strain rates is shown in Fig. 13.1. These are the strain rates per second, and above the strain rate scale some failure times for the rock specimen are shown.

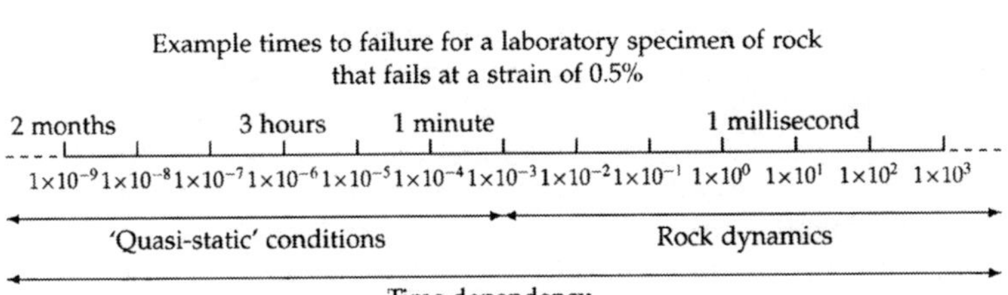

Figure 13.1 Range of strain rates

The distinction between 'quasi-static' and 'dynamic' shown in Fig. 13.1 is arbitrary, although the realm of rock dynamics is usually regarded as being those strain rates at which '**quasi-static**' laboratory tests are impractical and where vibrations will occur, above, say, $1 \times 10^{-3}\ \mathrm{s}^{-1}$. Note, however, that the phenomenon of time dependency occurs over the full spectrum of strain rates.

We emphasized in ERM 1, and in the questions in earlier chapters of this book, that there is no time component in elasticity: the relations between stress and strain are not a function of time. Hence, rather than say that elasticity occurs at an infinite strain rate, it is better to say that elasticity is independent of time. However, for the high strain rates, we can discuss the velocities of **propagation of stress waves** in elastic rocks, which can be determined as a consequence of Hooke's Law and the equations of motion. In A3.11 we showed that the sum of the rates of changes of the stress components in a given direction is zero when no force is applied, e.g. $\frac{\partial \sigma_{xx}}{\partial x} + \frac{\partial \tau_{yx}}{\partial x} + \frac{\partial \tau_{zx}}{\partial x} = 0$. When the sum of these rates of change is not zero (meaning that a net force is applied), then force = mass × acceleration, or $\frac{\partial \sigma_{xx}}{\partial x} + \frac{\partial \tau_{yx}}{\partial x} + \frac{\partial \tau_{zx}}{\partial x} = \rho \frac{\partial^2 u}{\partial t^2}$. This is for equilibrium in the x-direction, where ρ is the rock density and u is displacement in the x-direction.

Considering a one-dimensional situation, for example, stress waves travelling along a rod, this equation of motion reduces to $\frac{\partial \sigma_{xx}}{\partial x} = \rho \frac{\partial^2 u}{\partial t^2}$. Because $E = \frac{\text{stress}}{\text{strain}} = \sigma_{xx} / \left(\frac{\partial u}{\partial x}\right)$, $\sigma_{xx} = E\frac{\partial u}{\partial x}$, and hence $E\frac{\partial^2 u}{\partial x^2} = \rho \frac{\partial^2 u}{\partial t^2}$. Rearranging this gives $\frac{E}{\rho} = \frac{\partial^2 u}{\partial t^2} \cdot \frac{\partial^2 x}{\partial u^2} = \rho \frac{\partial^2 u}{\partial t^2}$, which shows that this relation corresponds to a **longitudinal stress wave** velocity of $V_\mathrm{p} = \sqrt{\frac{E}{\rho}}$.

In the two-dimensional case for an isotropic material, the longitudinal velocity in a plate is $V_\mathrm{p} = \sqrt{\frac{E}{\rho(1-\nu^2)}}$, and in three dimensions, $V_\mathrm{p} = \sqrt{\frac{E(1-\nu)}{\rho(1-2\nu)(1+\nu)}}$. The longitudinal stress waves are also referred to as dilatational, compressive, or primary waves.

The **transverse stress wave**, which has a different velocity from the longitudinal stress wave, can also be referred to as a distortional, shear, or secondary wave. There are also longitudinal surface (or Rayleigh) waves, transverse surface (or Love) waves, and Stoneley waves which occur at the boundary of two connected elastic rock strata.

There is a variety of **time-dependent effects** at lower strain rates. **Creep** occurs when the stress is held constant and the strain changes. **Relaxation** occurs when the strain is held constant and the stress changes. **Fatigue** occurs when there is increasing strain during stress cycling. When the complete stress–strain curve is obtained in uniaxial compression at strain rates of, for example, 1×10^{-3} s^{-1}, 1×10^{-4} s^{-1}, and 1×10^{-5} s^{-1}, the results will be strain rate-dependent: the compressive strength might be 20% higher at a strain rate of 1×10^{-3} s^{-1} compared to the compressive strength obtained at a strain rate of 1×10^{-5} s^{-1}; and the shape of the post-peak curve will be influenced by time-dependent processes.

Thus, time-dependent phenomena are important for both the discipline of rock mechanics and the rock engineering applications. Many rocks exhibit significant time dependency, yet we do not have such comprehensive methods of characterizing and predicting time-dependent behaviour as compared to either elastic or plastic behaviour.

13.2 Questions and answers: rock dynamics and time dependency

***Q13.1* There is no time component in the theory of elasticity. Why then does Young's modulus, expressed in units of stress, have time in its dimensions: $L^{-1}MT^{-2}$?**

A13.1 Although it is correct that the relation between stress and strain is independent of time, the units contain the time dimension because stress is defined as force/area and force is defined using Newton's second law which contains acceleration. Young's modulus is expressed as E = stress/strain = (F/A)/strain = (kg m s^{-2}/m^2)/strain = kg m^{-1} s^{-2}/strain, and hence we see that Young's modulus involves time. The dimensions of Young's modulus are $L^{-1}MT^{-2}$, as given in the units section at the beginning of this book.

The SI force unit, the newton, is defined as the force required to accelerate a mass of one kilogram at a rate of one metre per second per second — hence involving time. Note that all derived SI units are developed via some physical relation using the three base units of metre, kilogram and second.

***Q13.2* A 10-mm-diameter core of intact marble is carefully drilled out to a length of 1 m. The core is suspended horizontally by steel wires and then struck gently at one end to produce a longitudinal stress wave through the bar, as shown below. This is known as the Hopkinson bar experiment, used to study the transmission of stress waves.**

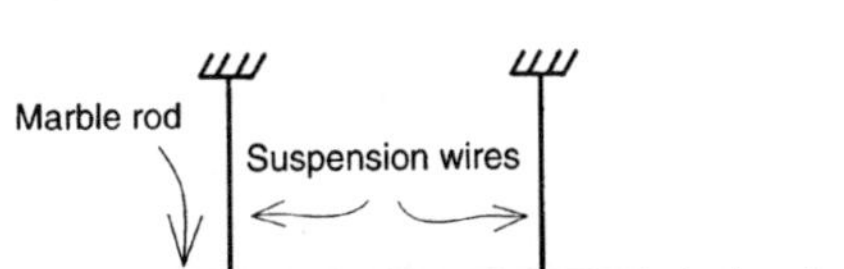

(a) If Young's modulus of the marble is 50 GPa and the unit weight is 27 kN/m^3, estimate the time taken for the longitudinal stress wave to travel from one end of the core to the other.

(b) Given that marble has a sufficiently high compressive strength to sustain the compressive wave but has a low tensile strength, where will the bar break, and why?

A13.2 (a) We assume that the longitudinal velocity in the core is given by $V_p = \sqrt{E/\rho}$, where E is Young's modulus and ρ is the density. For a unit weight of 27 kN/m^3 and an acceleration due to gravity of 10 m/s^2, the density is $27 \times 10^3/10 = 2700$ kg/m^3. Thus, we find $V_p = \sqrt{50 \times 10^9/2700} = 4300$ m/s. As the bar has a length of 1 m, the wave will therefore take $1/4300 = 2.32 \times 10^{-4}$ s, or 232 μs to travel along it.

(b) The compressive wave travels down the core but is reflected at the free end as a tensile wave.

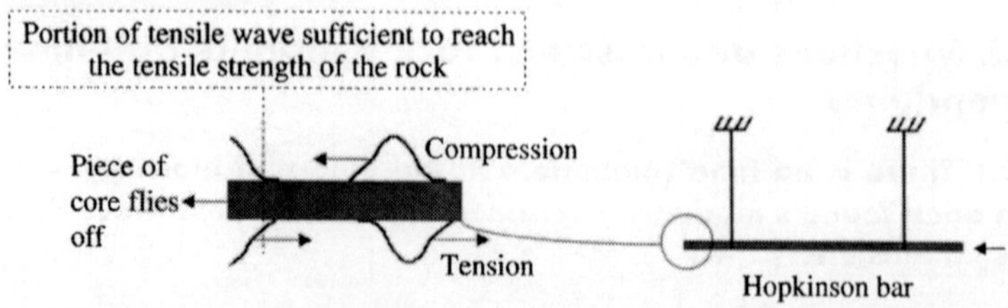

Once the absolute amplitude of the reflected tensile wave is sufficiently greater than the absolute amplitude of the incident compressive wave, the tensile strength will be reached and a piece will fly off the end of the length of core.

Q13.3 **What is the ratio V_p^2/V_s^2 in terms of the elastic rock constants and what is the specific value of the expression for a rock with $\nu = 0.27$?**

A13.3 We have $V_p = \sqrt{E(1-\nu)/\rho(1-2\nu)(1+\nu)}$ and $V_s = \sqrt{E/2\rho(1+\nu)}$, and so $V_p^2/V_s^2 = 2(1-\nu)/(1-2\nu)$. This is a useful relation for seismic rock mass investigations enabling easier evaluation of the dynamic elastic constants. For a rock with $\nu = 0.27$, $V_p^2/V_s^2 = 2(1-0.27)/(1-0.54) = 3.174$. Thus, we see that longitudinal waves propagate faster than shear waves.

If the density is known, Poisson's ratio can be estimated from V_p^2/V_s^2 and then the dynamic Young modulus estimated from V_p. Also, the relations between V_p, V_s and the elastic constants mean that rock masses can be classified using the values of V_p and V_s.

Q13.4 **A 100-mm-long rock specimen is to be tested in uniaxial compression using strain control in a servo-controlled testing machine. The Young's modulus of the rock is 60 GPa and the compressive strength is 200 MPa. We should like to reach the compressive strength in the test in about 10 minutes. What displacement rate should be used for the testing machine program, and what is the corresponding rock strain rate?**

A13.4 Firstly, we have to calculate the rock specimen displacement at the compressive strength. The strain at the compressive strength is stress/E = 200/60000 = 0.0033. For a specimen 100 mm long, this is 100 × 0.0033 = 0.33 mm displacement. To reach this displacement in 10 minutes requires a displacement rate of 0.33/600 mm/s = 0.00055 mm/s or 5.5×10^{-7} m/s. This is equivalent to a strain rate of 0.00055/100 per s = 0.0000055 s^{-1} = 5.5×10^{-6} s^{-1}.

The displacement rate used in practice would have to be slightly faster than the calculated value of 5.5×10^{-7} m/s because the stress–strain curve will have a lower slope than 60 GPa both in the initial portion, when 'bedding down' occurs, and near the peak stress, when cracking occurs.

***Q13.5* The results in the table below show the axial displacement and radial strain induced in a cylindrical specimen of weak chalk during a uniaxial creep test. In this test, the specimen was initially 250 mm high and was subjected to an axial stress of 55 MPa. After 3 hours the test was stopped, at which stage creep had ceased and the displacement had become constant at 0.4545 mm.**

Time (min):	**0**	**1**	**2**	**3**	**4**	**5**	**6**	**7**
Axial displacement (mm):	**0.409**	**0.414**	**0.419**	**0.423**	**0.427**	**0.430**	**0.433**	**0.435**
Radial strain ($\times 10^{-6}$):	**−451**	**−461**	**−471**	**−479**	**−487**	**−493**	**−499**	**−504**
Time (min):	**8**	**9**	**10**	**11**	**12**	**13**	**14**	**15**
Axial displacement (mm):	**0.438**	**0.440**	**0.441**	**0.443**	**0.444**	**0.445**	**0.447**	**0.447**
Radial strain ($\times 10^{-6}$):	**−509**	**−513**	**−516**	**−519**	**−522**	**−524**	**−526**	**−528**

On the basis of these results, select a simple viscoelastic model for the rock, and determine values for the various viscoelastic constants.

A13.5 Simple viscoelastic models are built up from elastic (spring) and viscous (dashpot) elements connected in series, parallel or a combination of both. In order to decide on a likely model, we start by computing the axial strain from these data, and then plotting the results (see next page) in order to visualize the behaviour of the specimen.

Time (min):	0	1	2	3	4	5	6	7	
Axial strain ($\times 10^{-3}$):	1.635	1.656	1.675	1.692	1.707	1.720	1.731	1.742	
Time (min):	8	9	10	11	12	13	14	15	180
Axial strain ($\times 10^{-3}$):	1.751	1.759	1.766	1.772	1.777	1.782	1.786	1.790	1.818

From this we see that the specimen displays an immediate elastic response when the axial stress is applied, indicating that the viscoelastic model we choose must have an elastic element directly in series with the applied stress. Also, because the viscous creep of the material stops after a definite time, then the viscous element must be connected in parallel

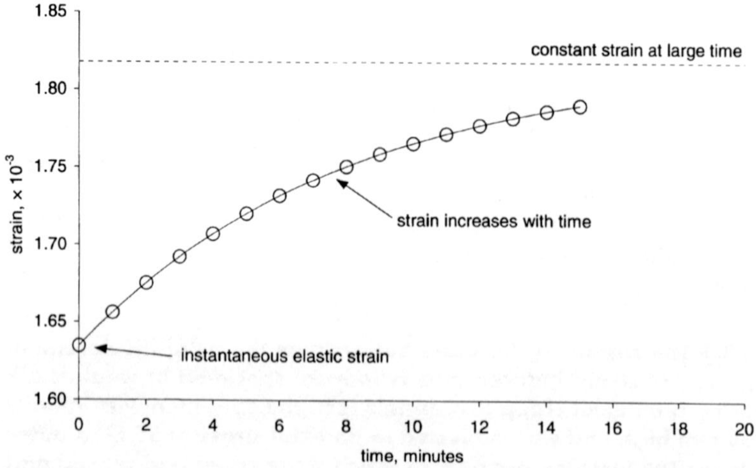

with another spring element. Assembling these elements results in a generalized Kelvin material, which is shown schematically below.

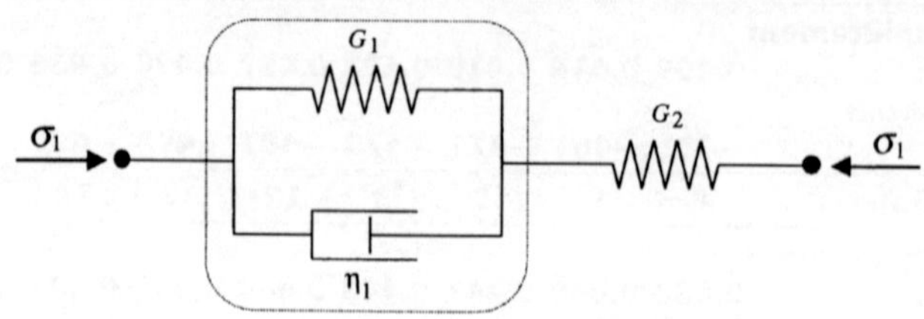

In this model, the elastic element with stiffness G_2 is responsible for the immediate elastic response. If we consider the part of the model within the dotted line, then immediately after the loading is applied the stress is borne by the viscous element, but as time passes the stress is eventually borne entirely by the elastic element. Thus the viscous element, with viscosity η_1, is responsible for the creep and the elastic element, with stiffness G_1, is responsible for causing the creep to cease.

Viscoelastic materials are usually assumed to act elastically when subjected to a hydrostatic stress state (if this were not true, viscoelastic materials would creep when submerged in water), and creep only when subjected to a stress state that causes distortion. For a material that acts as a generalized Kelvin material in distortion and as an elastic material in hydrostatic compression, the constitutive equation for a uniaxial stress state is (Jaeger and Cook, 1979; Goodman, 1989)[1]

$$\varepsilon_1 = \frac{\sigma_1}{3}\left\{\frac{2}{3K} + \frac{1}{G_2} + \frac{1}{G_1}\left[1 - \exp\left(-\frac{G_1 t}{\eta_1}\right)\right]\right\}.$$

[1] Techniques for deriving such constitutive relations are given by Jaeger J. C. and Cook N. G. W. (1979) *Fundamentals of Rock Mechanics*. Chapman and Hall, London, 3rd edn., 593pp. and Goodman R. E. (1989) *Introduction to Rock Mechanics*. Wiley, Chichester, 2nd edn., 562pp.

At $t = \infty$ this reduces to

$$\varepsilon_1(\infty) = \frac{\sigma_1}{3}\left\{\frac{2}{3K} + \frac{1}{G_2} + \frac{1}{G_1}\right\}$$

and so the difference between the long-term strain and the strain developed at time t is

$$\varepsilon_1(\infty) - \varepsilon_1(t) = \Delta\varepsilon = \frac{\sigma_1}{3}\left\{\frac{1}{G_1}\exp\left(-\frac{G_1 t}{\eta_1}\right)\right\}.$$

Rearranging this leads to

$$\ln(\Delta\varepsilon) = -\frac{G_1}{\eta_1}t + \ln\left(\frac{\sigma_1}{3G_1}\right),$$

which is the equation of a straight line, and shows how we can obtain values for G_1 and η_1.

The table below shows the data required to plot this straight line, and the plot is given below.

t (s):	0	60	120	180	240	300	360	420
$\ln(\Delta\varepsilon)$	−25.33	−25.45	−25.58	−25.70	−25.83	−25.95	−26.08	−26.20
t (s):	480	540	600	660	720	780	840	900
$\ln(\Delta\varepsilon)$	−26.33	−26.45	−26.58	−26.70	−26.83	−26.95	−27.08	−27.20

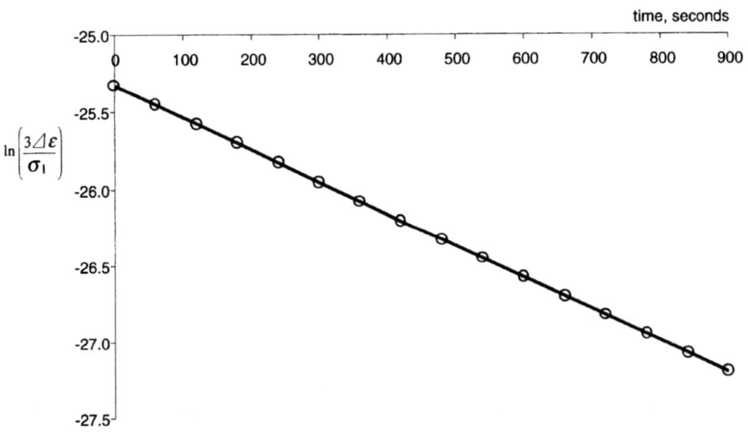

At time $t = 0$ we have

$$\Delta\varepsilon = \frac{\sigma_1}{3G_1}$$

from which we obtain $G_1 = \sigma_1/3\Delta\varepsilon = 55.0 \times 10^6/\left(3 \cdot 1.833 \times 10^{-4}\right) =$ 100 GPa. The best-fit straight line through the data has a gradient of -2.083×10^{-3}, from which we obtain
$\eta_1 = G_1/2.083 \times 10^{-3} = 100 \times 10^9/2.083 \times 10^{-3} = 48000$ MPa s.

For small strains the volumetric strain in the uniaxial test may be computed from

$$\varepsilon_V = \varepsilon_1 + 2\varepsilon_3,$$

and so the bulk modulus is given by

$$K = \frac{\text{mean stress}}{\text{volumetric strain}} = \frac{\frac{1}{3}(\sigma_1 + \sigma_2 + \sigma_3)}{\varepsilon_V} = \frac{\sigma_1}{3(\varepsilon_1 + 2\varepsilon_3)}.$$

Rearranging this gives

$$\varepsilon_1 = -2\varepsilon_3 + \frac{\sigma_1}{3K}$$

which shows how the bulk modulus may be found from the intercept of a plot of axial strain versus radial strain. This plot is shown below, and its intercept has been calculated as 733×10^{-6}, from which we obtain $K = \sigma_1 / (3 \cdot 733 \times 10^{-6}) = 55 \times 10^6 / (3 \cdot 733 \times 10^{-6}) = 25.0$ GPa. Notice that a large amount of extrapolation is required to find the bulk modulus, indicating that this is liable to be an inaccurate technique.

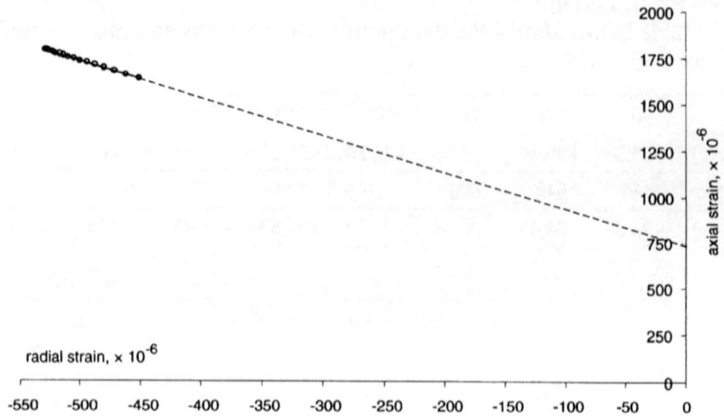

Finally, at $t = 0$ the constitutive equation for the material reduces to

$$\varepsilon_1(0) = \frac{\sigma_1}{3}\left\{\frac{2}{3K} + \frac{1}{G_2}\right\}$$

and rearranging this leads to

$$G_2 = \frac{1}{3\dfrac{\varepsilon_1(0)}{\sigma_1} - \frac{2}{3}\dfrac{1}{K}}$$

from which, by substituting values given above, we obtain $G_2 = 16.0$ GPa.

Q13.6 On a sketch of the complete stress–strain curve for rock in uniaxial compression, draw lines illustrating creep, relaxation and intermediate time-dependent straining along a line of slope −*K*

(a) on the ascending and descending sides of the curve, and

(b) then comment on the significance of your diagram for rock mass stability during time-dependent deformations for a single rock pillar of intact rock and for an abandoned chalk mine excavation (Smith and Rosenbaum, 1993[2]) as shown below.

---

[2] Smith G. J. and Rosenbaum M. S. (1993) Recent underground investigations of abandoned chalk mine workings beneath Norwich City, Norfolk. *Eng. Geol.*, **36**, 37–78.

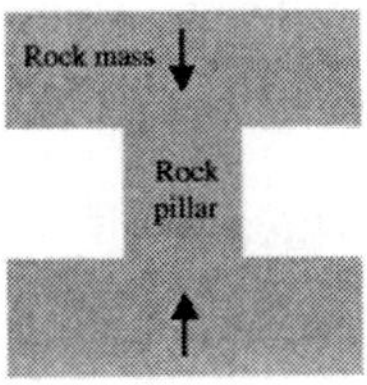

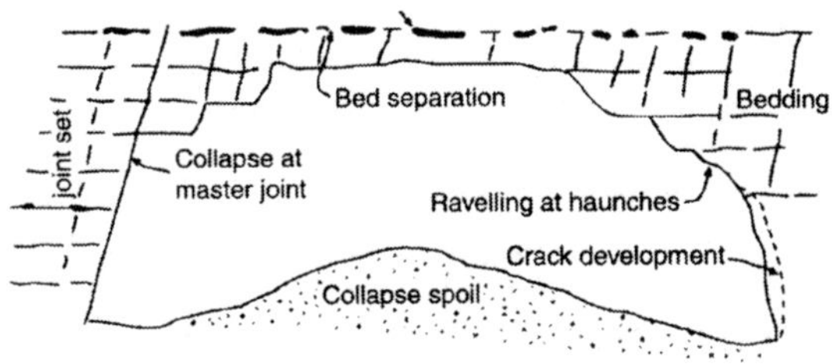

A13.6 (a) Creep is defined as continuing increase in strain at constant stress; relaxation is defined as stress reduction at constant strain. This results in the creep, relaxation and $-K$ lines shown on the diagram below.

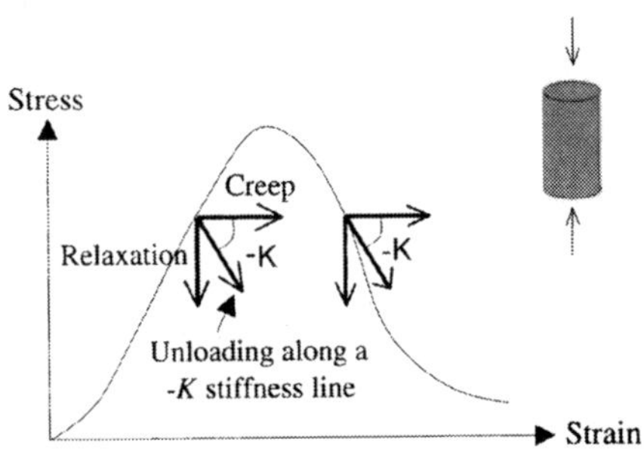

The significance of the diagram for the stability of the two rock masses referred to in the question is that the time-dependent effects of creep, relaxation and intermediate $-K$ creep–relaxation are all stable on the ascending side of the complete stress–strain curve. An exception is when the creep occurs near the top of the curve, so that the rock strains across the stress–strain curve, beneath the peak, to the descending portion whereupon it becomes unstable.

The $-K$ line is important because, in practical cases, the rock mass load is being applied by another part of the rock mass along this line, see sketch to the right. For dead weight loading, $-K$ will be zero, represented by a horizontal line on the stress–strain diagram; if the loading system is rigid, $-K$ tends to infinity, represented by a vertical line.

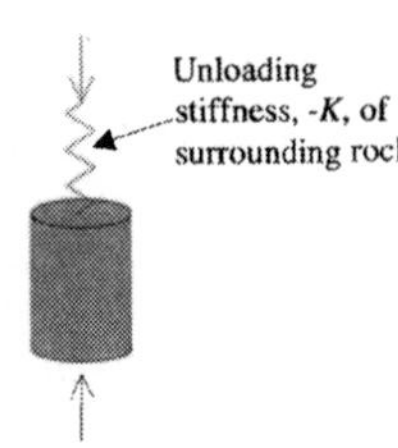

(b) The significance of the diagram for the stability of the two rock masses referred to in the question is that the time-dependent effects of creep and intermediate $-K$ creep–relaxation can lead to sudden collapse, depending on the value of $-K$ and the rock mass condition at any particular point on the complete stress–strain curve. The diagram of the rock pillar shown here was also shown in Q5.10, where the question was whether stress or strain was the cause of rock failure. We see from the $-K$ line in the first diagram above that, with a specified rock mass unloading stiffness, the cause of sudden failure will be a combination of stress and strain. In other words, and as indicated in the sketch on the next page, the failure of the pillar will be

- early and sudden if soft loading is applied, such as through soft rock strata (because the $-K$ line is above the descending portion of the complete stress–strain curve and hence applying more energy than the pillar can absorb as it fails), or
- later and more slowly for stiffer loading, such as through stiff strata, providing explosive phenomena such as rockbursts do not occur (because the $-K$ line will continue to be below the descending portion of the complete stress–strain curve for increasing strain values).

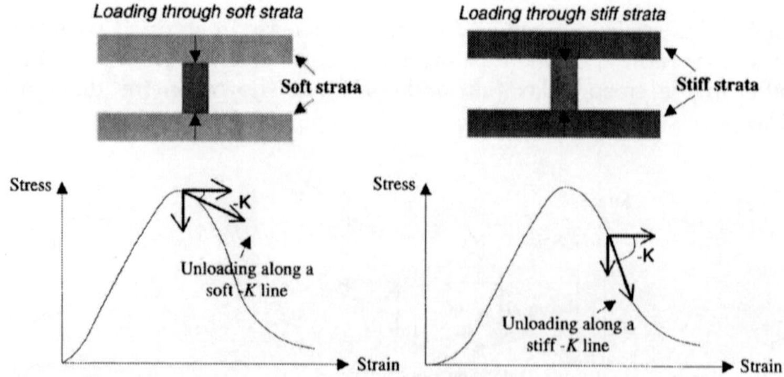

Moreover, such time-dependent effects *in situ* will be exacerbated by general degradation and weathering effects. The diagram to the right in the question illustrates progressive chalk mass deterioration in an underground excavation dating from Victorian times. Here there are several degradation mechanisms operating. The more weathering that occurs in the fractures, the softer and weaker the rock mass will become. Loose blocks fall out, and any bed separation causes the K value to reduce to close to zero. Smith and Rosenbaum also highlight the importance of groundwater flow on the deterioration mechanisms.

Another more general point is that the designers of Victorian rock engineering structures, e.g. for mining or for fresh and waste water systems, could not have anticipated the development of civilization and the current pounding of the ground surface by heavy traffic. So, in addition to all the factors mentioned, there are also the vibrations introduced by modern-day activities, whether as sudden damage due to high-intensity proximate rock blasting or the long-term cumulative effect of vibrations manifested as fatigue failure.

Q13.7 Laboratory fatigue testing results (Ray et al., 1999[3]) for intact Chunar sandstone from the lower Gondwana formation gave the following data.

[3] Ray S. K., Sarkar M. and Singh T. N. (1999) Effect of cyclic loading and strain rate on the mechanical behaviour of sandstone. *Int. J. Rock Mech. Min. Sci.*, **36**, 543–549.

Number of cycles from 0 to 50 MPa	Uniaxial compressive strength (MPa)
0	99.5
100	90.1
200	77.1
300	71.6
400	67.6
500	65.9
600	65.1

The rock samples were cycled for the number of times given in the left-hand column, and then tested for their uniaxial compressive strength, giving the values in the right-hand column.

Plot these results (number of cycles on x-axis, *UCS* on y-axis) and comment on the trend.

A13.7 As the number of cycles increases, so we see that the strength of the rock decreases. The effect of the cycling is therefore to damage the sandstone through fatigue.

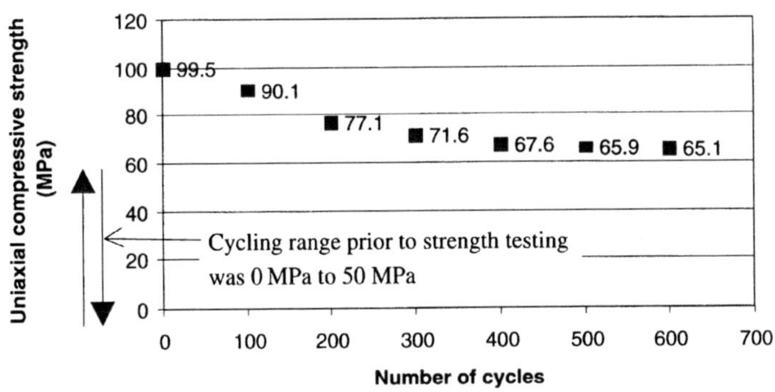

The initial cycles cause the most damage in terms of a percentage strength reduction, and the implication is that the rock is sensitive to such disturbances, despite the low level of stress cycling.

***Q13.8* How do fractures respond to stress waves? What do you think are the influences of a dynamic load on the fracture shear behaviour? What happens if the fractures pictured to the right are subjected to repeated shear movements?**

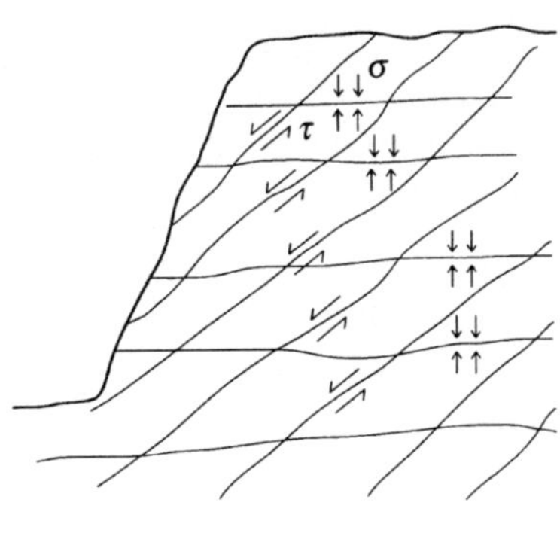

A13.8 Giani (1992) [4] has collated the influences of dynamic loading on fractures. Some of these are:

[4] Giani G. P. (1992) *Rock Slope Stability Analysis.* Balkema, Rotterdam, 361pp.

- shear resistance decreases with the number of forward and reverse fatigue loading cycles (in a directly analogous way to the results given in Q13.7);
- peak shear resistance increases for smooth fracture surfaces with increased loading application frequency;
- the contact area between opposing fracture surfaces increases with contact time;
- the friction angle of weak rocks increases for low normal loads up to a critical loading velocity factor; beyond this value, the friction angle remains constant up to a second critical loading velocity, beyond which shear resistance decreases.

The properties of the fractures in the rock slope will be modified by repeated application of dynamic loads and, in general, the modifications will weaken the rock mass and exacerbate slope instability. However, it is not possible to include all the detailed mechanisms directly into rock engineering design models, which is why we approximate such effects through 'effective rock properties'.

Q13.9 During the progression of a longwall mining face, bedded rock strata adjacent to the coal mine are being subjected to an extensile strain rate of 1×10^{-5} s^{-1} normal to the bedding planes. Assume that all the strain accumulated in the rock mass is concentrated in opening the bedding planes. How long will it be before the flow of water along the bedding planes is doubled?

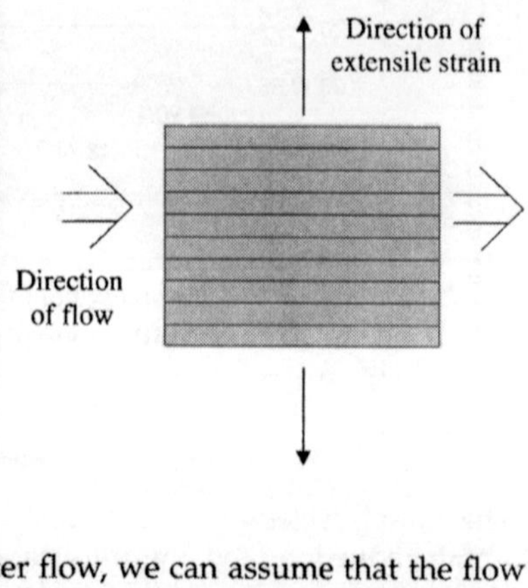

A13.9 Using the cubic law for water flow, we can assume that the flow of water is proportional to the cube of the bedding plane apertures. Then, for an initial aperture of e_i and a final aperture of ce_i, $(ce_i)^3 = 2(e_i)^3$, from which we find $c^3 = 2$ and hence $c = 1.26$. Thus, the aperture must increase in size from a value of 1 unit to a value of 1.26 units — in other words, the strain has to be 0.26 which will occur in
$0.26/1 \times 10^{-5}$ s $= 26000$ s $= 7.2$ h.

Q13.10 The diagram (Tan, 1993)[5] on the next page illustrates deformations in two mine rail tunnels (originally horseshoe-shaped) in China, which are located at 430 m depth in strongly fractured granite.

[5] from Tan T. J. (1993) The importance of creep and time-dependent dilatancy as revealed from case records in China, in *Comprehensive Rock Engineering* (J. A. Hudson, ed.). Vol. 3, Ch. 31, 709–744.

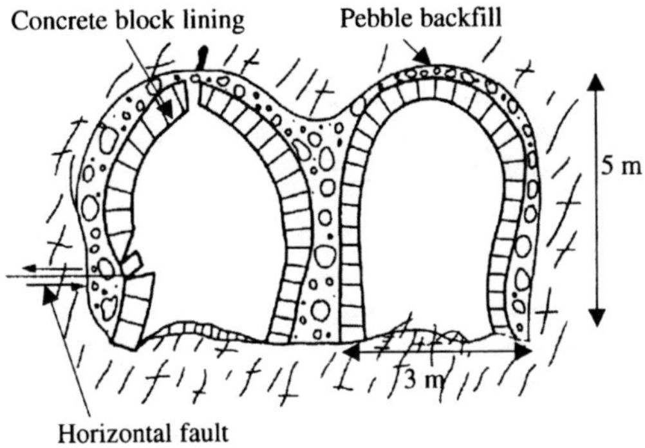

The tunnel axes are oriented perpendicular to the high intraplate horizontal tectonic stress, σ_H. The tunnel deformations have been described[5] as follows:

"Directly after excavation, the sidewalls started to bulge, horizontal cracks occurred which were growing in width and length; bottom heave was obvious. These processes increased in intensity with time. Void formation in the dilatancy process increasing with time leads to serious overall weakening of the rock structure; in addition it is aggravated by the suction and seepage of underground water. The horizontal displacements... were in the order of 50 cm within 90–150 days with an average rate of 5.4 mm/day. Usually the inwards motion of the corner areas near the bottom are largest, whereas the roof-top moved upwards so gradually the tunnel cross-section is squeezed into the 'peach' form... Bottom upheavals at some locations were about 40 cm within 131 days. Serious lateral motions of one sidewall have been observed."

Explain these observations in terms of the basic mechanisms involved.

A13.10 The observations are explained as follows.

"Directly after excavation, the sidewalls started to bulge, horizontal cracks occurred which were growing in width and length; bottom heave was obvious."

Firstly, consider the basic circumstances. The tunnels are located at 430 m depth so, knowing that 40 m depth $\approx$ 1 MPa, the vertical stress component will be about 430/40 $\approx$ 11 MPa. Also, we know that the tunnel axes are perpendicular to the maximal horizontal stress component, an orientation which causes the highest stress concentrations around the boundary of the excavations. Moreover, we can expect the maximal horizontal stress component to have a significantly greater value than the vertical stress component, thus causing the highest compressive stress concentrations to be located in the roof and floor.

Clearly, the ratio of rock stress to rock strength is high enough to cause significant displacements. The use of the words 'bulge' and 'bottom heave' to describe the sidewall and floor movements before the lining was installed implies that the rock was being squeezed *en masse* into the excavation. The horizontal cracks could have developed as a result of the granite structure, or because the horizontal stress was so high that a tensile stress developed at the axis level (although such tensile stresses in fractured rock are unlikely).

"These processes increased in intensity with time. Void formation in the dilatancy process increasing with time leads to serious overall weakening of the rock structure; in addition it is aggravated by the suction and seepage of underground water. The horizontal displacements... were in the order of 50 cm within 90–150 days with an average rate of 5.4 mm/day."

The fact that the processes increased in intensity with time implies that the rock mass in the direct vicinity of the tunnels was deteriorating and hence becoming even more susceptible to the action of the rock stresses. The rock mass — composed of heavily fractured granite and containing faults — is below the water table, which means that water will readily flow into the tunnels. The tunnels themselves have a concrete block lining with pebble backfill, which will be relatively permeable and so will not prevent this water flow. The movement of ground water through the rock mass towards the tunnels would therefore aggravate the time-dependent degradation.

"Usually the inwards motion of the corner areas near the bottom are largest, whereas the roof-top moved upwards so gradually the tunnel cross-section is squeezed into the 'peach' form... Bottom upheavals at some locations were about 40 cm within 131 days. Serious lateral motions of one sidewall have been observed."

The inward motion of the corner areas near the bottom was largest because the major principal stress is acting horizontally and the lining is not continuous across the floor of the tunnels. If the tunnels had been unlined, the roof would probably have been squeezed downwards, i.e. into the excavation, in the same way as the floor was being squeezed upwards. Thus, the lining must be having a significant effect in transferring the lateral force resulting from the horizontal stress into the roof and floor: the roof is pushed upwards, and the corner at the floor is pushed inwards. If the tunnels are not behaving symmetrically, as evidenced by the fact that serious lateral motions of one sidewall have been observed, it is likely that the fracturing in the granite is neither homogeneous nor isotropic, and that the rock mass is weaker on one side of the tunnels.

13.3 Additional points

All the subjects that have been covered in Chapters 4–13 — *in situ* stress, elasticity, the deformability, strength and failure of intact rock, fractures and rock masses, permeability, anisotropy and inhomogeneity, rock mass classification, *and* time dependency — can be incorporated into numerical programs, as demonstrated by the finite element stability analyses (Wittke, 1999 [6]) for the tunnels of the high-speed German railway line in which the elastoviscoplastic behaviour of both intact rock and fractures was incorporated. However, time dependency is a difficult subject and, although other programs can potentially incorporate time dependency, such as the 'tunnel support modeller' add-on to the FLAC program [7] which uses the Panet (1993) [8] convergence-confinement tunnel analysis, time-dependent behaviour is difficult to incorporate into rock mechanics modelling and is generally ignored. For example, the value of the design life, say 120 years, is not explicitly used in modelling calculations.

The reasons for the difficulty in incorporating time dependency into modelling for rock engineering design are that the mechanics involved in the rheology of rocks and rock masses is not fully understood, the range of strain rates in engineering applications is wide, ranging from blasting strain rates to creep strain rates, and it is difficult to establish the parameters associated with time-dependent behaviour.

[6] Wittke W. (1999) Stability analysis of tunnels of the new high speed Cologne–Rhine/Main Railway Line. *ISRM News Journal*, **5**, 3, 26–33.

[7] Distributed by ITASCA, software@itascag.com and www.itasca.com

[8] Panet M. (1993) Understanding Deformations in Tunnels, in *Comprehensive Rock Engineering* (J. A. Hudson, ed.). Vol. 1, Ch. 27, pp. 663–690.

14 Rock mechanics interactions and rock engineering systems

14.1 Interactions

When designing a structure to be built on or in a rock mass, it is necessary not only to consider individual factors such as the intact rock, fractures, rock stress, permeability, excavation and support, but also how these all interact together. As a means of linking the rock mechanics principles to the rock engineering applications, it is appropriate to consider how such interactions can be characterized. For rock mechanics modelling and rock engineering design for a specific project, we need to be able to identify the **relevant physical variables** and the **linking mechanisms**, and then consider their combined operation. Also, we need to ensure that all the relevant factors and their interactions will be taken into account.

For example, it may be determined that for a particular slope stability study the influence of changes in rock stress on fracture aperture and hence water flow does not need to be considered, and that the influence of water pressure and open fractures on the rock stress is not significant either. Thus, this aspect of hydro-mechanical coupling in the rock mass would not be considered in the design. The key point is that we should be aware that hydro-mechanical couplings and other interactions exist, and make an appropriate decision about whether to consider them or not.

A systematic method for thinking about all the interactions is to list them in a matrix. This is the basic device used by the **rock engineering systems** (RES) approach (Hudson, 1992[1]; Hudson and Jiao, 2000[2]). The principal factors considered relevant to the problem are listed along the leading diagonal of a square matrix (top left to bottom right) and the interactions between pairs of principal factors form the off-diagonal terms.

For example, in the case of just the two principal factors, Rock Stress (X) and Water Flow (Y), there are two interactions.

---

[1] Hudson J. A. (1992) *Rock Engineering Systems: Theory and Practice.* Ellis-Horwood, Chichester, 185pp.

[2] Hudson J. A. and Jiao Y. (2000) *Analysis of Rock Engineering Projects.* Imperial College Press, London (in preparation).

Interaction *XY*: the influence of rock stress on water flow, which is mainly by changing the apertures of water-bearing fractures.
Interaction *YX*: the influence of water flow on rock stress, which is mainly through the water pressure reducing the normal rock stress components in the vicinity of water-bearing fractures.

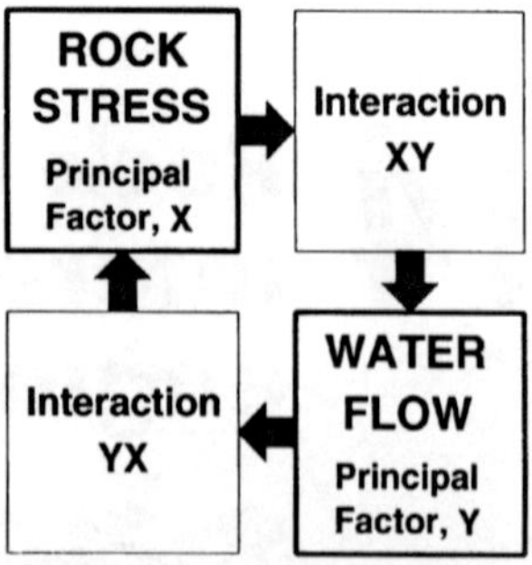

This 2 × 2 interaction matrix can be extended to any size matrix by adding further leading diagonal factors or variables and considering all the interactions. Because Interaction XY is not the same as Interaction YX, it is important to consider the cause and the effect in each case. For Interaction XY, principal factor X is influencing principal factor Y, whereas for Interaction YX, principal factor Y is influencing principal factor X. The way in which the rock stress affects the water flow is not the same as the way water flow affects rock stress. Thus, there will be pairs of complementary interactions, and a clockwise influencing convention has been used in the matrix presentation. A composite mechanism involving more than two principal factors is a series of paths through an interaction matrix with the appropriate numbers of principal factors included on the leading diagonal. Tracing these paths is a useful means of understanding the composite mechanisms, and we give an example in Q14.6.

Given that the interaction matrix (or some other presentational device) enables a systematic tabulation of the principal factors and their interactions, a **technical audit** of the content of an analysis and the analysis procedures can then be conducted to ensure that the analysis is adequate. Fig. 14.1 shows a three-level technical audit.

First-level audit

The first step is to ensure that the principal factors required to capture the essence of the problem are included in the modelling procedures. These are the leading diagonal terms of the interaction matrix. Then, the mechanisms linking the principal values are identified as the off-diagonal components. This provides the structure of the system. For example, if the principal factors are the physical variables involved, $x_1, x_2, \ldots, x_n$, then the interactions between pairs of variables can be established, as indicated in the top part of the diagram (Fig. 14.1).

Second-level audit

The second step is to consider how the system components are linked, either via the interaction matrix or as shown in the middle part of the diagram (Fig. 14.1).

Using an appropriate model, the system operation is modelled in this step so that an estimate can be made of the values of those parameters relating to the design criteria, e.g. the stress should not exceed 100 MPa, the displacement should not exceed 10 mm, the excavation overbreak should be less than, say, 5%, and so on.

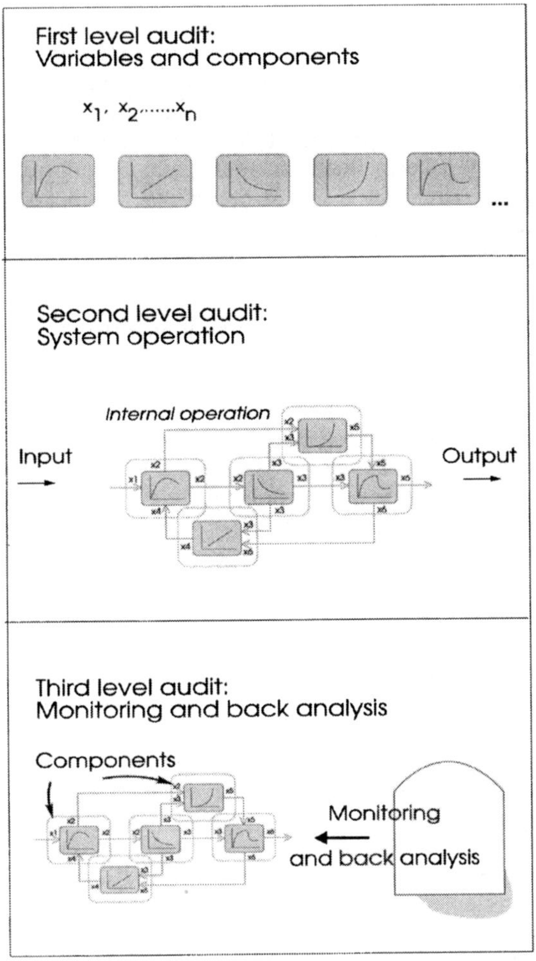

Figure 14.1 Stages in the technical auditing of rock mechanics modelling and rock engineering design.

Third-level audit

The third step is to monitor the actual operation of the system (the laboratory test, the *in situ* test, the construction process, the longer-term behaviour) which is the rock reality and compare the results with those indicated by the second step, as indicated in the lower part of the diagram. In this step, it is also advantageous to use back analysis procedures.

Currently, there is no formal requirement for technical auditing of either rock mechanics modelling or rock engineering design. Quality Assurance (QA) procedures are in use, but these refer only to the correct execution of the work, given specified site investigation, modelling and construction procedures. If critical components are missing from the modelling, and hence the rock engineering design, following a QA procedure will not prevent a disaster. Both the technical audit and the QA are required, and they must be performed in that order.

We anticipate greater emphasis on both the technical auditing and QA aspects in the years ahead. The questions and answers in this chapter are designed to highlight the process of thinking about the total problem, whether this is for rock mechanics modelling only or for modelling in order to develop a practical rock engineering design. It is of critical importance in engineering to have a **predictive capability**, and this can only be achieved if all the appropriate factors have been taken into account.

14.2 Questions and answers: rock mechanics interactions and rock engineering systems

Q14.1 **Explain with a one-sentence answer how, in a rock mass,**
(a) the rock stress can affect water flow in fractures,
(b) the water pressure in fractures can affect the rock stress,
(c) the fractures can affect water flow,
(d) water flow can affect the fractures,
(e) the fractures can affect rock stress, and
(f) the rock stress can affect the fractures.

A14.1

(a) Changes in rock stress open and close fractures, and the resulting changes in fracture aperture affect the water flow through the fractures.

(b) When water in fractures exerts a pressure on the fracture walls, the rock stress acting normal to the fractures is reduced by the value of the water pressure, which is the principle of effective stress.

(c) Water flows mainly through the fractures rather than the intact rock, because the hydraulic conductivity of a fracture is usually much greater than that of the intact rock.

(d) Continued water flow can either erode the fracture walls or cause other materials to block up the fractures, resulting in either increased or reduced water flow.

(e) Surfaces of open fractures are principal stress planes (because the air or water between cannot sustain a shear stress) and so the stress state in the rock adjacent to a fracture surface will be changed in both magnitude and orientation.

(f) Rock stresses can cause normal and shear fracture displacements and damage to fracture walls.

Q14.2 **Draw an interaction matrix having Fractures, Rock Stress, and Water Flow as the leading diagonal terms. Insert the letters representing the six mechanisms in Q14.1 in the appropriate off-diagonal boxes of the interaction matrix.**

A14.2 The required matrix is shown below. This is one example of a simple interaction matrix.

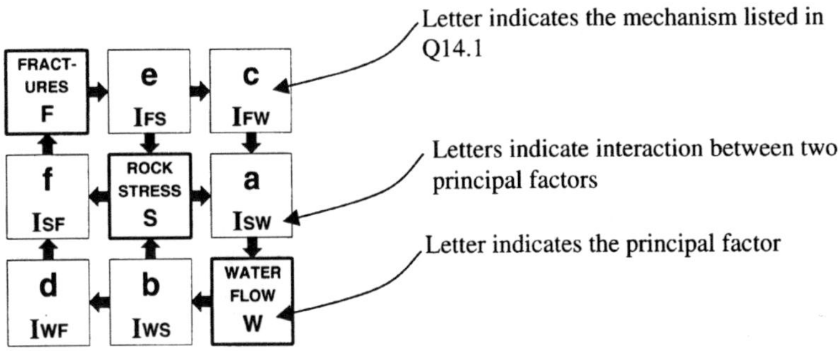

Note that the principal factors, Fractures, Rock Stress and Water Flow, could have different positions on the leading diagonal of the interaction matrix and the specific interactions would then be located in different off-diagonal positions. However, this would not alter the information contained in the interaction matrix.

Q14.3 Under what conditions is a matrix symmetrical? Why will almost all rock mechanics interaction matrices be asymmetrical?

A14.3 A matrix is symmetrical if the complementary off-diagonal terms are equal. For example, the matrix in A14.2 would be symmetrical if

Interaction FS were equal to Interaction SF, and
Interaction FW were equal to Interaction WF, and
Interaction SW were equal to Interaction WS.

Most rock mechanics interaction matrices will not be symmetrical because the influence of a principal factor X on a principal factor Y will not generally be the same as the complementary influence of principal factor Y on principal factor X. This is evidenced by the complementary interactions described in A14.1 and illustrated in the matrix in A14.2, which are clearly not the same.

Q14.4 In order to consider also the interaction of rock engineering activities with the rock mechanics factors, extend the 3 x 3 interaction matrix of Q14.2 to a 4 x 4 interaction matrix by adding the extra leading diagonal term CONSTRUCTION. Write a one-phrase explanation of the content of each of the six extra off-diagonal interaction boxes introduced by adding the leading diagonal term, CONSTRUCTION.

A14.4 The required 4 × 4 interaction matrix is shown below. The 'one-phrase explanation' in the boxes in the fourth column and fourth row are examples of potential interactions that might be involved; other examples are possible.

Note that the boxes in the fourth *column* represent influences that the rock mechanics factors have on construction, either in terms of the engineering design or the events that occur on site during rock

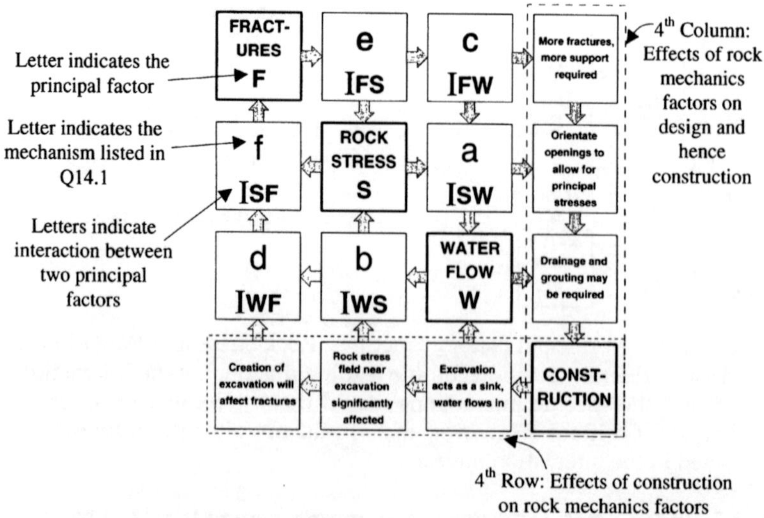

excavation. The boxes in the fourth *row* represent the complementary influences that the construction has on the rock mechanics factors.

Q14.5 The use of the interaction matrix is helpful for thinking about any problem involving manifold factors. The principal factors, located on the leading diagonal, are established first. Then, the interactions, i.e. the mechanisms linking the principal factors, are considered in the off-diagonal positions.

For example, when studying natural fractures in the rock mass, assume that we are interested in the four parameters[3] of fracture orientation, spacing, extent (or persistence) and roughness. Using these four parameters as the leading diagonal terms of a 4 x 4 interaction matrix, identify the content of the 12 off-diagonal boxes and hence show that the parameters are likely to be related.

A14.5 Example content for the off-diagonal boxes is shown in the matrix to the right. Note that the interactions listed are of several forms: direct cause and effect, commonly observed correlations, and effects caused by sampling. The existence of the interactions indicates that there will be relations between the orientations, spacings, extents and roughnesses of fractures in a given rock mass.

[3] The correct use of the terms 'parameter' and 'variable' is as follows. The term 'parameter' refers to the a, b and c in an expression such as $ax + by = c$, i.e. the coefficients making the relation specific. The term 'variable' refers to the unspecified, unknown x and y in the equation. However, these terms are used interchangeably in engineering literature. We have used the term 'parameters' here for the principal factors or leading terms of the interaction matrix following historical usage of the word, but the term 'variable' is more appropriate in some cases.

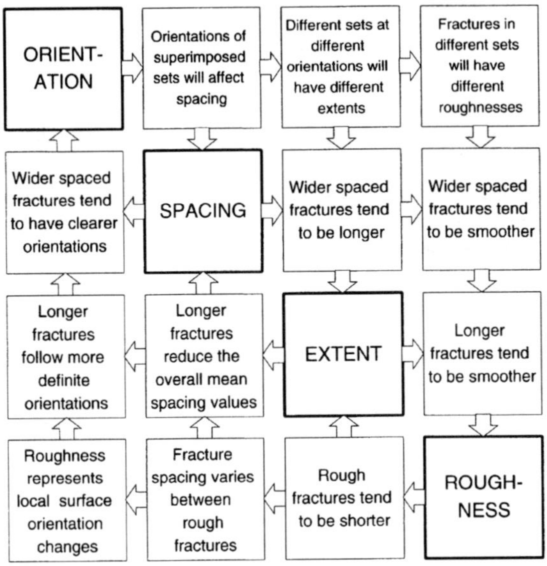

Q14.6 Each off-diagonal term in the interaction matrix represents a mechanism linking two principal factors. If a mechanism involves three or more principal factors, their combined interaction is represented by pathways through the interaction matrix. With reference to the interaction matrix required by Q14.4, show the example pathway representing how Construction blasting affects the Rock Stress around an excavation, which allows Fractures to open, leading to increased Water Flow which then affects Construction.

A14.6 The required pathway is shown in the interaction matrix to the left below, which is an abstracted representation of the matrix to the right below, i.e. the one developed for A14.4.

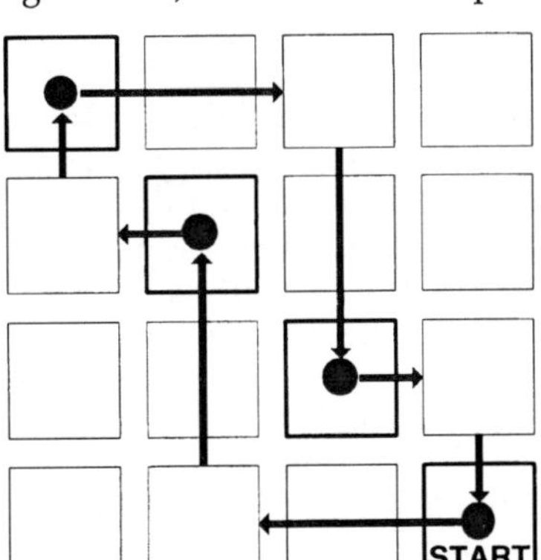

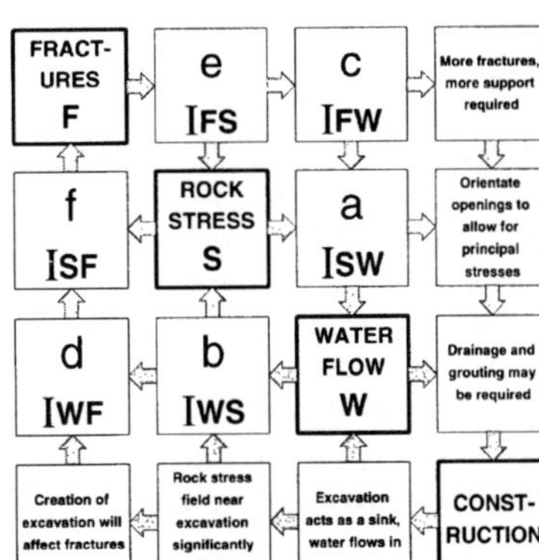

Many such mechanism pathways link the principal factors. Consideration of these pathways and how they might be inhibited by engineering actions can provide a systematic approach to hazard analysis and hence to the establishment of optimal construction sequences.

Q14.7 Assume that you have been asked to undertake a technical audit of a rock engineering design for a DIANE rock mass, where a CHILE numerical code has been used to determine the elastic rock stresses and displacements around an underground excavation. List briefly the likely sources of error in the determination of the rock stresses and displacements.

A14.7 We recall the meaning of the two acronyms, DIANE and CHILE:

DIANE: Discontinuous, Inhomogeneous, Anisotropic, and Not-Elastic
CHILE: Continuous, Homogeneous, Isotropic, and Linearly Elastic

The sources of error in modelling a DIANE rock with a CHILE numerical code will be the following.

(1) Specific influences of fractures are not taken into account: the rock has been assumed to be continuous when it is not.
(2) Variations in intact rock properties with location are not taken into account, e.g. in bedded strata.
(3) Variations of rock properties in different directions are not taken into account.
(4) Differences between the mechanical properties of the rock mass when it is being subjected to loading or unloading cycles are not taken into account, nor are time-dependent effects taken into account, such as strain rate, creep, relaxation and fatigue.

Thus, we can see that the rock stresses and displacements are not likely to be correct. There may well be other factors, such as the influence of water, which have not been taken into account and which will lead to further errors.

The engineering question is whether such CHILE solutions are of assistance for engineering design. Do they provide information and understanding compatible with the available resources and the engineering design objective? Are they a sufficiently good approximation for the problem in hand? Much of the seminal book by Jaeger and Cook (1979) [4] is based on the CHILE theory of elasticity and the associated applications have been successful, especially in those excavations around the world where at depth high rock stresses have closed the fractures and the rock mass is relatively homogeneous and isotropic. However, in near-surface excavations, where the rock stresses are lower, the fractures more frequent, and the strata more disturbed and weathered, there should be more concern about the validity of a CHILE model.

In short, algorithms in numerical codes should include the necessary interactions required to capture the essence of the problem as required by the engineering objective.

---

[4] Jaeger J. C. and Cook N. G. W. (1979) *Fundamentals of Rock Mechanics*. Chapman and Hall, London, 3rd edn., 593pp.

Q14.8 Assume that the interactions in the matrix required by Q14.4 (which has the principal factors of Fractures, Rock Stress, Water Flow, and Construction) have been considered for a specific rock mass and engineering objective. Each interaction has been assigned a number according to the following scheme:

0 – no interaction;
1 – weak interaction;
2 – medium interaction;
3 – strong interaction;
4 – critical interaction.

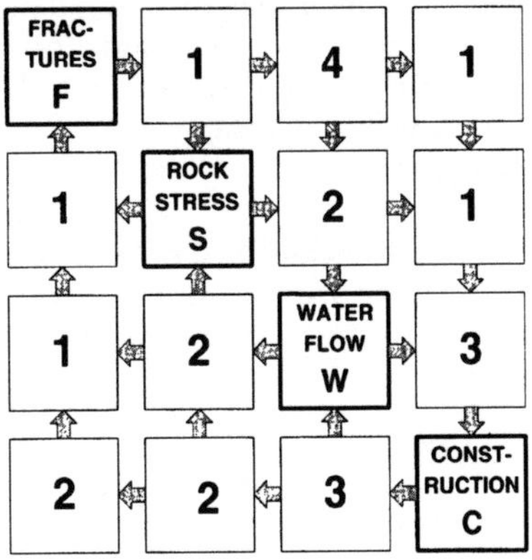

These numbers are shown in the matrix to the right.

For each principal factor, develop its 'Cause–Effect' (*C,E*) co-ordinates. These are the sums of the values in the row and column through each principal factor.

For example, the (*C,E*) co-ordinates for principal factor F are $C = 1 + 4 + 1 = 6$ and $E = 1 + 1 + 2 = 4$, i.e. (6,4). Hence establish the interaction intensity, $C+E$, and dominance, $C-E$, of each principal factor in the interactive system. Then plot the four principal factors using Cause and Effect axes.

A14.8 The required values are given in the following table.

Principal Factor	C	E	$C+E$ interactive intensity	$C-E$ dominance
Fractures, F	6	4	10	2
Rock Stress, S	4	5	9	–1
Water Flow, W	6	9	15	–3
Construction, C	7	5	12	2

The principal factors, F, S, W and C are plotted in Cause–Effect space in the diagram to the right.

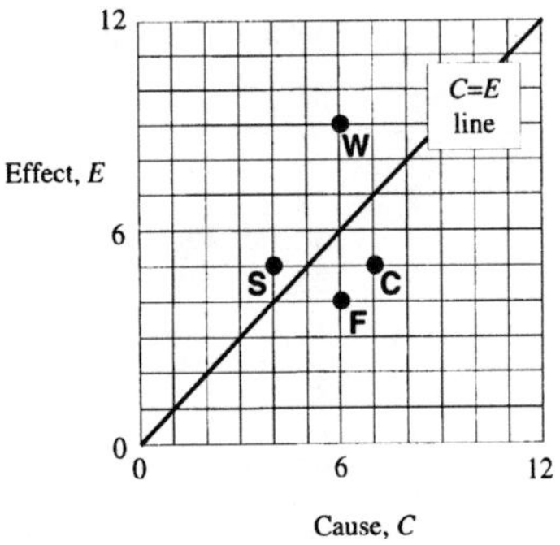

For this example, which is included to illustrate the Cause–Effect co-ordinates for principal factors in the interaction matrix, the maximum possible values of the co-ordinates are (12,12). Thus, the co-ordinate values of F(6,4), S(4,5), W(6,9) and C(7,5) indicate that the interaction matrix system structure has a medium interactive intensity. The principal factor W, Water Flow, has the greatest interactive intensity, represented by a $C + E$ value of 15.

In the plot above, more *interactive* factors plot further along the $C = E$

line, with W being the furthest having the highest $C + E$ value. The *dominant* factors, $C > E$, plot to the right of the $C = E$ line, as is the case for F and C.

The overall conclusion is that we have a moderately interactive system structure in which Water Flow has the strongest interaction. Fractures and Construction slightly dominate the system (being slightly to the right of the $C = E$ line) and Rock Stress and Water Flow are slightly dominated by the system (being slightly to the left of the $C = E$ line). These conclusions depend, of course, on the values assigned to the interactions.

Q14.9 In developing a rock mass classification system for estimating the relative instability of natural rock slopes in the Italian Alps (Mazzoccola and Hudson, 1996[5]), the following parameters were chosen.

P1. Geology	**P2. Folds**	**P3. Faults**	**P4. Rainfall**
P5. Freeze and thaw	**P6. Previous instability**	**P7. Intact rock strength**	**P8. Weathering**
P9. Number of sets	**P10. Orientation**	**P11. Aperture**	**P12. Persistence**
P13. Spacing	**P14. Mechanical properties**	**P15. Rock mass strength**	**P16. Hydraulic conditions**
P17. Slope orientation	**P18. Slope dimensions**	**P19. *In situ* stress**	**P20. Potential instability**

The 20th parameter, 'Potential instability', was added in a similar way to the Construction term in Q14.4 in order to consider the effects of the parameters on potential instability.

An interaction matrix was generated using these 20 parameters as the leading diagonal terms. The interactions in the off-diagonal terms were then assigned a value according to the same scheme as in Q14.8:

0 – no interaction; 1 – weak interaction; 2 – medium interaction; 3 – strong interaction; 4 – critical interaction.

This gave (C,*E*) co-ordinates and values of C + *E* and C – *E* as follows.

	P1	P2	P3	P4	P5	P6	P7	P8	P9	P10
C	24	34	52	17	21	22	15	9	24	28
E	5	4	3	4	17	42	8	27	13	12
C + *E*	29	38	55	21	38	64	23	36	37	40
C – *E*	19	30	49	13	4	–20	7	–18	11	16

	P11	P12	P13	P14	P15	P16	P17	P18	P19	P20
C	27	26	18	9	9	21	21	8	18	0
E	22	19	20	26	40	29	17	19	16	60
C + *E*	49	45	38	35	49	50	38	27	44	60
C – *E*	5	7	–2	–17	–31	–8	4	–11	2	–60

[5] Mazzoccola D. F. and Hudson J. A. (1996) A comprehensive method of rock mass characterization for indicating natural slope instability. *Q. J. Eng. Geol.*, **29**, 37–56.

Plot the 20 parameters using *C–E* axes, and hence determine for the natural slope system which three parameters are the most interactive, which three are the most dominant, and which three are the most subordinate.

A14.9 The parameters are plotted in $C–E$ space below. Note that, although the maximum possible (C, E) parameter co-ordinates in this example are $(19 \times 4, 19 \times 4) = (76, 76)$, it has not been necessary to extend the axes beyond (60,60).

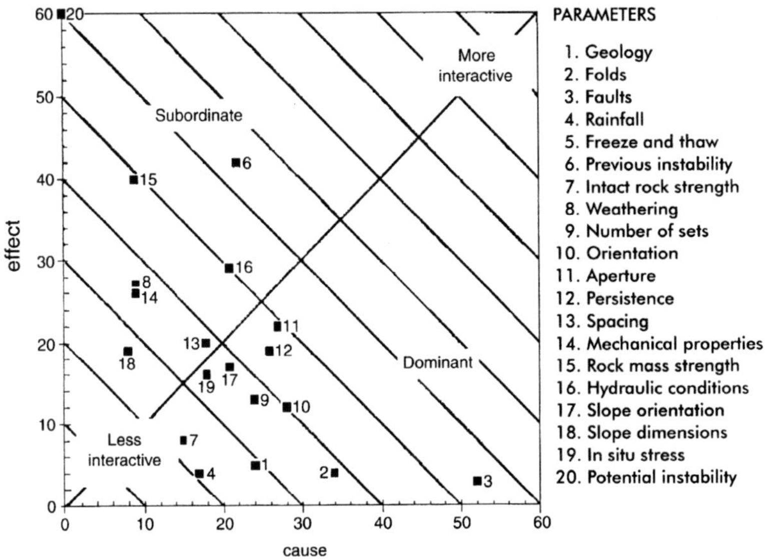

The most interactive parameters are those with the highest $C + E$ values. The three most interactive parameters are: P6, Previous Instability, with $C + E = 64$; P20, Potential Instability, with $C + E = 60$; and P3, Faults, with $C + E = 55$. Parameters with higher $C + E$ values plot further away from the origin along the direction of the main diagonal from (0,0) to (60,60).

Dominant parameters have positive $C - E$ values because their C value is greater than their E value, i.e. they have a greater effect on the system than the system has on them. The three most dominant parameters are: P3, Faults, with $C - E = 49$; P2, Folds, with $C - E = 30$; and P1, Geology, with $C - E = 19$. Dominant parameters, having a positive $C - E$ value, will plot to the right of the main diagonal.

Subordinate parameters have negative $C - E$ values because their E value is greater than their C value, i.e. they are affected more by the system than the system affects them. The three most subordinate parameters are: P20, Potential Instability, with $C - E = -60$ (this parameter was introduced originally as the purpose of the study); P15, Rock Mass Strength, with $C - E = -31$; and P6, Previous Instability, with $C - E = -20$. Subordinate parameters, having a negative $C - E$ value, will plot to the left of the main diagonal.

This type of study for a rock mechanics problem with many interacting parameters helps to clarify the circumstances, can give direct guidance for establishing a tailor-made rock mass classification, and will indicate the possibilities of applying a coupled mathematical model to represent the system (which for this example are remote because of the difficulty of representing all the interactions by explicit mathematical formulae). The analysis presented here was used to establish the parameter weighting factors in a rock mass classification scheme for assessing the relative instability of 20 natural slopes (Mazzoccola and Hudson, 1996).

Note that the approach considers the system structure, rather than the system operation, and therefore represents a simple first-level audit, as indicated in the top part of Fig. 14.1.

Q14.10 **One of the most important aspects of the systems approach is establishing when positive feedback in the system can occur, because positive feedback loops can cause instabilities.**

(a) Explain the meaning of the terms 'negative feedback' and 'positive feedback', and give examples of each in engineering rock mechanics.

(b) Also explain how engineering actions can mitigate or eliminate positive feedback (equivalent to altering the mechanisms in the off-diagonal boxes of the interaction matrix).

A14.10 (a) Consider the interaction matrix mechanism path shown to the right (from A14.6, but this could be any interaction matrix). A disturbance or perturbation at the Start position travels through the loop of mechanisms and arrives back at the Start position.

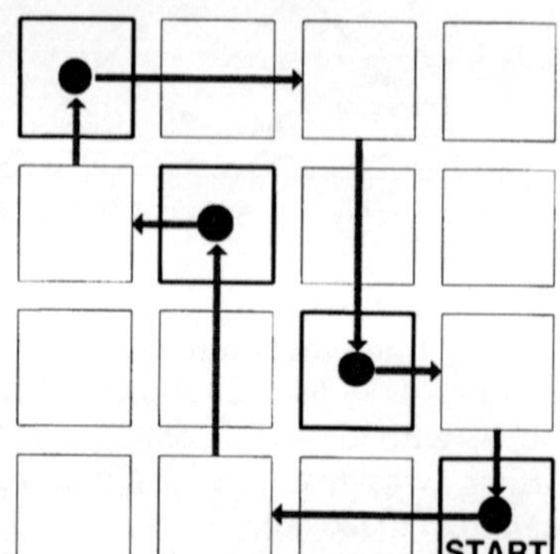

Negative feedback: the effects of the intervening mechanisms reduce the perturbation.

Positive feedback: the effects of the intervening mechanisms increase the perturbation.

In engineering rock mechanics, most mechanisms reduce such perturbations, and mechanism loops are generally associated with negative feedback. Generally, disturbances attenuate and die out. However, mechanism loops with positive feedback can occur. Because the disturbance is then enhanced through the operation of the feedback loop, instabilities can occur.

Examples of cases which can have either negative or positive feedbacks

Consider the three cases illustrated below: a rock block on a slope begins to slide, energy is released during rock failure around an excavation, and water is flowing through a rock fracture. There could be negative or positive feedback in each of these cases.

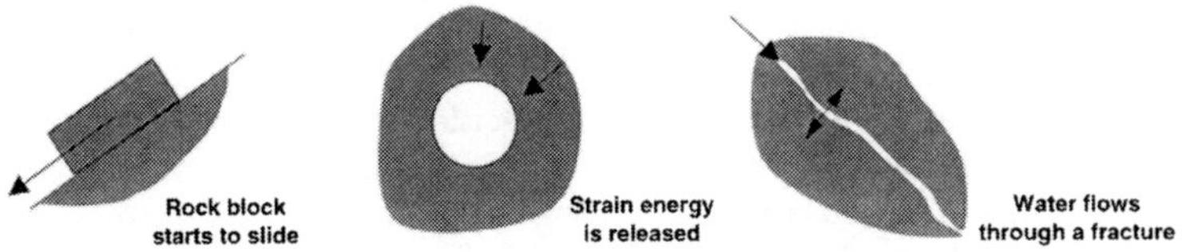

Rock block. If, as the block begins to slide, the irregularities of the rock block and rock slope surfaces tend to inhibit further movement, there is negative feedback. If, on the other hand, the block movement tends to remove such irregularities and reduce the interface friction, there is positive feedback and the block will accelerate.

Failure around an excavation. Assume that the stresses around an excavation are causing rock failure. If the rock failure causes a reduction in the stress concentrations and the energy released is absorbed in the rock failure process, then there is negative feedback and the process is self-stabilizing. If, however, the rock failure causes an increase in the stress concentration and not all the energy released is absorbed in the rock failure process, there is positive feedback which could lead to a rockburst.

Water flowing through a fracture. When water flows through a fracture over a period of time, it could deposit sediment in the fracture reducing the water flow. This is an example of negative feedback. Conversely, the action of the water could erode the fracture walls enabling more water to flow through the fracture. This is an example of positive feedback: an increase in water flow in the rock fracture will enlarge the fracture further, which in turn allows more water to flow, which could enlarge the aperture even further. . . .

Most rock mechanics mechanisms are of the attenuating or self-stabilizing type with negative feedback or a small positive feedback; if they were not, the actions of rock engineers would lead to mayhem because any type of engineering work would immediately run out of control. The art in rock engineering awareness is recognizing when positive feedbacks might occur and then avoiding them.

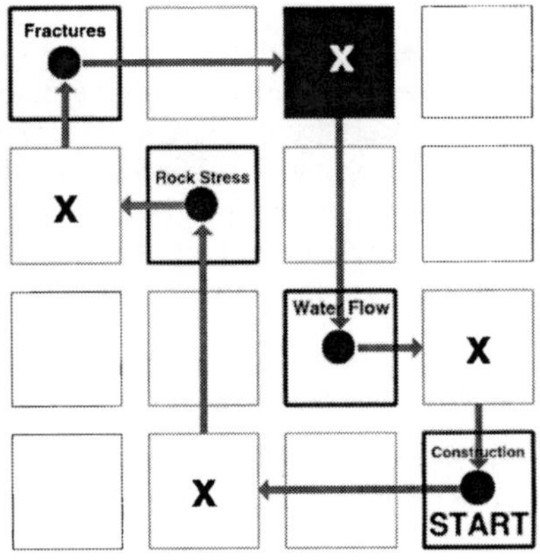

(b) The key engineering action is to identify the interactions associated with such positive feedbacks and to mitigate their effects. For example, if the mechanism path in the matrix to the right is associated with positive feedback, the mechanism path can be blocked by eliminating any of the mechanisms marked with an 'x'. In the case of the specific mechanism path in A14.5, the best option would probably be to grout the fractures (the black box). Other inhibiting actions, such as reinforcement or providing drains, will be appropriate to other circumstances depending on the system being studied.

This type of instability inhibition through identification and mitigation of the key mechanisms in potential positive feedback loops in the system structure will be effective in reducing engineering hazards. However, a complete study of all the positive feedback effects requires an understanding of the *operation* of the complete system. The off-diagonal terms in the interaction matrices presented so far represent the *structure* of the system, i.e. the interactions in isolation before the system is 'switched on'. For a linear system, it is possible to establish the coupled interactions (Jiao and Hudson, 1998 [6]), i.e. the interaction between two principal factors or variables when all interactions are operating and influencing one another. These coupled interactions are not the same as the isolated interactions.

14.3 Additional points

Once the variables and mechanisms required to capture the essence of a problem have been identified, modelling procedures, potential numerical codes available for use, and rock engineering designs can be technically audited for their content. Note that there can be no standardized modelling procedures, nor standardized site investigation and designs, because both will depend on the modelling and design objectives. Indeed the very essence of good modelling and design is to be flexible and adapt to the project circumstances. The result is that all projects are different and it is by no means clear in many cases that all the factors have been adequately taken into account in the design.

Thus, the implementation of technical auditing for all these purposes is to establish whether a model, particular numerical code, or design has incorporated the required variables and mechanisms needed to model the project circumstances. The principles of technical auditing are as follows (Hudson, 1999 [7]).

(1) Technical Auditing (TA) is the process of establishing whether rock engineering measurements, statements and conclusions can be considered to be valid.
(2) There is a requirement for a TA if there is a need to establish the reliability and credibility of information, or if there is a public interest dimension requiring public accountability.
(3) TA differs from Quality Assurance (QA) because QA *per se* is only concerned with following pre-determined procedures, regardless of whether the procedures are right or wrong. TA checks that the technical content is adequate for the purpose and hence is a pre-cursor to establishing the appropriate QA procedures.

---

[6] The methodology used for this technique is described in Jiao Y. and Hudson J. A. (1998) Identifying the critical mechanisms for rock engineering design. *Géotechnique*, **48**, 3, 319–335.

[7] Hudson J. A. (1999) Keynote lecture: technical auditing of rock mechanics modelling and rock engineering design, in *Proc. 37th US Rock Mechanics Symposium* held at Vail, CO, 1999 (B. Amadei, R. L. Kranz, G. A. Scott and P. H. Smeallie, Eds). Balkema, Rotterdam, pp. 3–12.

(4) The TA is made according to evidence, known criteria and the current scientific framework. The result is an opinion based on persuasive evidence.
(5) Materiality is the significance of any potentially relevant subject. Materiality determines the nature, quality and quantity of evidence required concerning any matter. Its expression will depend on the project being audited and the audit objective.
(6) The TA involves verification by evidence. It should have an independent status, be free from investigatory and reporting constraints, produce a benefit, and result in a report.
(7) The TA result will always be an opinion, thus the auditing must carry authority.

Understanding the rock mechanics and rock engineering system, with all its variables, interactions and operation, is crucial to establishing an adequate predictive modelling capability and hence establishing an adequate rock engineering design. We know that a simple CHILE model may not be enough. We know that we do not yet have fully coupled thermo-hydro-mechanico-chemical numerical codes. Thus, the ability to present all the potential variables and interactions and then winnow out those that we regard as relevant to the engineering design in hand is one of the ways forward and provides the necessary foundation for the technical auditing.

15 Excavation principles

15.1 Rock excavation

The objectives in **excavating rock** for civil, mining and petroleum projects are different, as indicated by the examples in Fig. 15.1.

In **civil engineering**, the objective is to support loads and to create surface or underground space for a specific engineering purpose, e.g. a foundation, a cutting for a road, a shaft for a hydroelectric scheme, a railway tunnel, a repository for disposing of radioactive waste. In **mining engineering**, the objective is to obtain the rock itself, either in bulk or for the minerals it contains. For example, in the case of a surface quarry or underground mine for limestone, most of the mined rock can be used, whereas in the case of an open-pit mine or underground mine for, say, copper or gold, the mined rock has to be crushed to release the mineral grains, a process known as comminution. In **petroleum engineering**, the objective is to transport the oil from the petroleum reservoir to the ground surface through boreholes, and there is a variety of associated rock mechanics problems such as enhancing flow from the reservoir into the boreholes and storing oil and gas in rock caverns.

Thus, for civil engineering, the rock only needs to be broken up sufficiently for the underground space to be created and for the rock to be removed. For mining engineering, the rock itself is required and the process of excavation is the first stage of comminution. For petroleum engineering, the main issues are drilling and maintaining open boreholes, and maximizing oil flow into the wellbores.

This leads to different design approaches for civil and mining engineering, and special problems for petroleum engineering in weak rocks and at significant depths. The geometry of a civil engineering excavation is usually clearly defined and subject to definite constraints. There has to be a clear design with a **project lifetime** of, say, 120 years. If possible, safety factors are established. All this significantly constrains the rock engineering design options. In mining, whilst the overall geometry of the mining will be determined by the orebody and mining leases, the

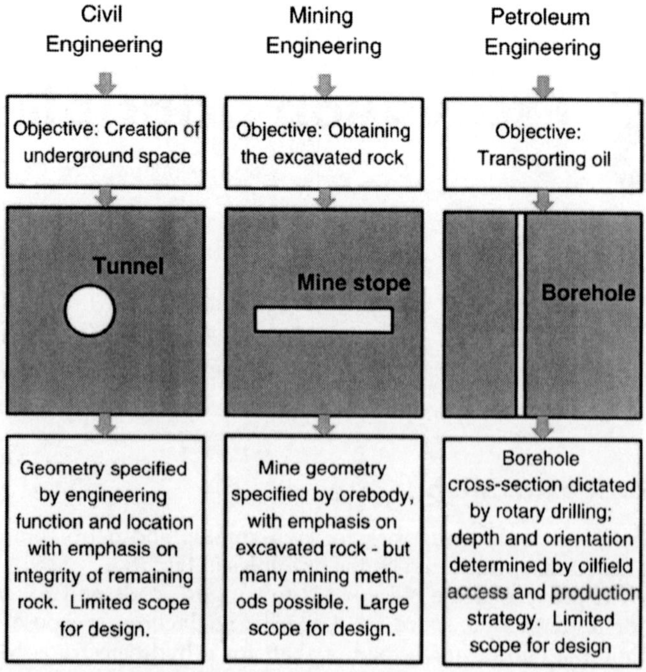

Figure 15.1 The differing objectives of civil, mining and petroleum engineering.

internal geometry can be specified by mining engineers. Also, the design life of different parts of a mine can vary from a few days (e.g. for a temporary access) to a hundred years (for the main access). This leads to many potential methods of excavating the rock in a mine, with different methods being suited to different circumstances. Many elegant mining methods have been developed, such as the 'longwall method', which consists essentially of a laterally moving tunnel. In a similar way, there can be considerable design flexibility in the quantity and orientation of wellbores used in petroleum engineering, but the remote access limits the design scope.

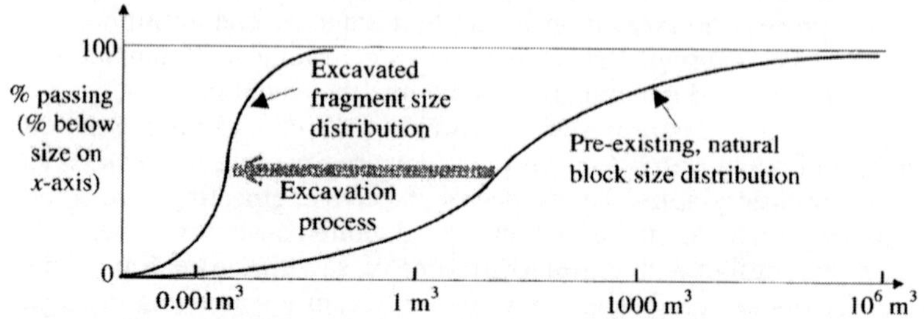

Figure 15.2 The process of rock excavation changes the pre-existing, natural, rock block size distribution to the excavated fragment size distribution.

The fundamental process of excavation, for civil or mining engineering, is to alter the **rock block size distribution** from the natural, pre-existing rock blocks to the fragments required for excavation, as indicated in Fig. 15.2. The energy required for this process is the energy required to create the new fracture surfaces, together with energy lost in, for example, stress waves and heat.

Rock can be excavated by **explosives** or by **machines**. With explosives (Fig. 15.3a) the excavation process is cyclical: the energy is put into the rock in seconds via detonation of the explosives, the gases have to be

Figure 15.3 (a) Blasted rock at quarry face. (b) Tunnel bored by machine.

cleared, the broken rock is removed, the next series of blastholes are drilled, and then charged with explosive, the explosives are detonated again, and so on. A typical cycle time is about 8 h.

Using a machine for excavation, say a bulldozer with a ripping attachment or a partial face or full-face tunnel boring machine (Fig. 15.3b), the energy is input to the rock at a much lower rate, but essentially continuously when the machine is cutting. Although neither the breakage of rock by explosives nor the breakage of rock by rock cutters is fully understood, we can apply basic mechanics and empirical rules remarkably effectively.

15.2 Questions and answers: excavation principles

***Q15.1* The complete stress–strain curve has been obtained for a cylindrical specimen of intact granite tested in uniaxial compression. The specimen is 100 mm long and 50 mm in diameter. Assume that, for the purposes of calculation, the curve can be approximated to the bilinear form below. The uniaxial compressive strength is reached at 0.1% strain and 200 MPa stress. When the curve reaches 0.2% strain, the rock microstructure has been destroyed and all that remains are small flakes of crushed mineral grains.**

(a) Calculate the energy under the equivalent complete force–displacement curve. This is the energy required to completely destroy the rock specimen. Express your answer in joules.

(b) For how long would a domestic 100-W light bulb have to be illuminated to use up the same amount of energy?

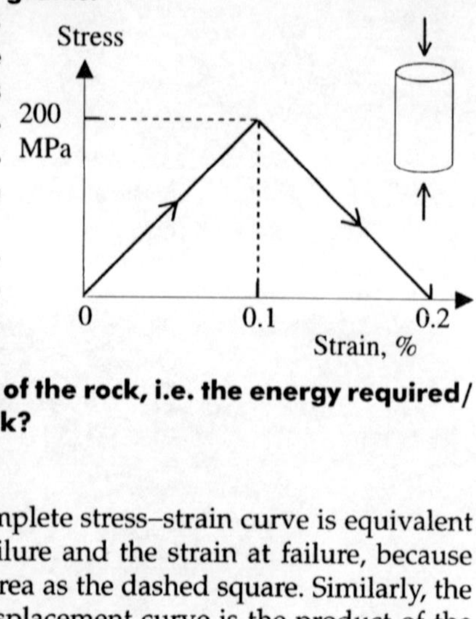

(c) What is the specific energy of the rock, i.e. the energy required/unit volume to pulverize the rock?

A15.1 (a) The area under the complete stress–strain curve is equivalent to the product of the stress at failure and the strain at failure, because the two triangles have the same area as the dashed square. Similarly, the area under the complete load–displacement curve is the product of the load at failure and the displacement at failure.

The load at failure = stress at failure × cross-sectional area = $200000000 \times \pi \times (50/2 \times 1000)^2 = 392699$ N. The displacement at failure = strain at failure × specimen length = $0.001 \times 0.1 = 0.0001$ m. Thus, the energy used $392699 \times 0.0001 = 3.93 \times 10^5 = 39$ N m. A joule is defined as the energy expended by 1 N moving through 1 m, or 1 J = 1 N m, so the energy required = 39 J.

(b) To establish for how long a domestic 100-W light bulb has to be illuminated in order to expend the same amount of energy, we need to express the energy in watt seconds. A joule is also the energy expended by 1 W for 1 s, or 1 J = 1 W s. Hence, 39 J is equivalent to 39 W s, and so the 100-W light bulb has to be illuminated for 39/100 s = 0.4 s.

Notice that if the compression test is conducted at a strain rate of 5×10^{-6} s^{-1}, the test will take $0.002/5 \times 10^{-6} = 400$ s, or just less than 7 minutes, to complete, representing a power requirement of 39/400 = 0.098 W, which is about the same as a small battery-powered radio.

(c) The specific energy is the energy required to pulverize the rock/unit volume = 39 J/(the volume of rock specimen). The volume is $\pi \times (0.05/2)^2 \times 0.1 = 2 \times 10^{-4}$ m^3. Therefore, the specific energy[1] is $39/(2 \times 10^{-4}) = 19.9 \times 10^4$ J/m^3 = 0.20 MJ/m^3.

Q15.2 During bench blasting in a quarry, it was found that 48.5 kg of explosive is required to break 125 m^3 of marble. Given that the explosive used was ammonium nitrate–fuel oil (ANFO) with an explosive energy of 3.92 MJ/kg, calculate the specific energy for these circumstances, state whether this is greater or less than the answer to Q15.1, and explain why there is a difference.

A15.2 The amount of energy used is $48.5 \times 3.92 = 190$ MJ. The amount of rock removed is 125 m^3. Therefore, the specific energy is 190/125 = 1.52 MJ/m^3, which is about 8 times greater than the 0.20 MJ/m^3 value obtained in A15.1.

The difference in the specific energies calculated in A15.1 and in this answer, results from the losses that are incurred in blasting. In the compression test, we anticipate that most of the energy used for the test will be absorbed in specimen degradation but, in any case, the energy actually absorbed by the specimen was used for the calculation, i.e. the energy represented by the area beneath the complete stress–strain curve. For the blasting calculation, the total energy was used, and much of this energy will be lost in the form of gas pressure and stress wave dissipation.

Q15.3 A 5-m-diameter tunnel is being excavated in limestone by a full-face tunnel boring machine (TBM), which operates by exerting a torque and a thrust.

(a) Calculate the specific energy required to break the rock in the circumstances given by the data in (i) and (ii) below.

(i) For the case where the TBM has four 500 V electrical motors providing the torque to drive the cutting head. Each motor uses 10 A when rotating freely, and 110 A when the head is cutting. Neglect the thrust for this case and take the cutting rate as 3 m/h.

(ii) For the same cutting information in part (i) but with an additional continuous thrust of 2.7 MN and an increased cutting rate of 3.6 m/h.

---

[1] Note that the specific energy computed here applies to this particular testing configuration and rock type. A change to the testing regime with the same rock type would result in a changed specific energy. Thus, specific energy is not an intrinsic material property.

(b) Does the work associated with the thrust contribute greatly to the specific energy required?

A15.3 (a) (i) As the cutting rate is 3 m/h, during a 1-m tunnel advance the TBM operates for 1/3 h or 1200 s. The amount of energy used over the 1-m advance is number of motors × volts × amps × time = $4 \times 500 \times 100 \times 1200$ W s = 2.4×10^8 W s. Since 1 J = 1 W s, we find that the amount of energy used over the 1-m advance is then 2.4×10^8 J = 240 MJ.

The volume of rock cut by the TBM during the 1-m advance is $1 \times \pi \times (5/2) = 19.63$ m^3. Thus, the specific energy = energy used/volume removed = 240/19.63 = 12.22 MJ/m^3.

(ii) The calculation of specific energy should also include the component supplied by the TBM thrust, which we now include. At the new cutting rate, the same calculation procedure as above gives the specific energy due to the cutting torque as 10.19 MJ/m^3. The thrust of 2.7 MN provides an additional 2.7 MN m of energy when applied over the 1-m advance. Since 1 N m = 1 J, this energy is 2.7 MJ. The extra specific energy related to the thrust is therefore 2.7/19.63 = 0.137 MJ/m^3, which gives a total specific energy when the thrust is included of $10.19 + 0.14 = 10.33$ MJ/m^3.

(b) From the figures in the answer above, the energy input via the thrust is only 0.14/10.33 = 0.01, or 1%, of the total energy required. It does not therefore contribute greatly to the specific energy required. It is interesting, however, that the provision of significant thrust has increased the cutting speed, which has had the effect of reducing the overall specific energy. In practice, we find that for a given TBM and rock condition there is an optimal balance between the torque and thrust which minimizes both the energy requirements and the machine vibrations, helping to prolong the machine life. At thrust force magnitudes greater or lower than the optimum, the cutters will, respectively, stall or not operate efficiently as they skid over the rock surface.

Q15.4 Comment on the magnitudes of the specific energy values obtained in Q15.1, Q15.2 and Q15.3 for the different circumstances of a laboratory compression test, blasting, and using a tunnel boring machine?

A15.4 *Laboratory compression test: 0.20 MJ/m^3.* This represents extreme microstructure degradation, but with few energy losses in the test. Note that a compressive strength of 200 MPa represents a strong rock.

Blasting: 1.52 MJ/m^3. The object of blasting is to change the pre-existing natural block size distribution to the required fragment distribution, but it is difficult to blast so that the breakage is uniform: thus, the specific energy is an average value. Also, there will be high energy losses in the components of the stress waves and gas pressure effects not contributing to rock breakage, although the blasting fracturing is far less dense than in the laboratory test which requires less energy. The rock type is a marble, and so it is probable that this will be a medium strength rock.

Tunnel boring machine: 10.33 MJ/m³. More fracturing is required to chip and remove the rock with a TBM than with blasting, and there is potential for large energy losses during rotation of the head through friction and vibrations. Limestone is usually a medium strength rock.

Overall comments. In the uniaxial compression test there are few energy losses. So, although a high microcrack density has been generated at the completion of the test, the specific energy is lowest because of the high energy transfer efficiency. In blasting and cutting by TBM, there are large energy losses: in both cases, the majority of the energy does not go into breaking the rock, i.e. in creating new fracture surfaces, but is lost in heat, sound, stress waves, kinetic energy, etc. This is why these specific energies are higher than for the laboratory test. The blasting is more efficient in terms of specific energy than the TBM because, *inter alia*, less fracturing of the rock takes place.

We would not have expected the specific energy for the three cases to be the same because different rock types and different systems of fracture development are involved. Thus, the term 'specific energy' should be suitably qualified to indicate the configuration and scale under which it is determined (which is similar to the tensile strength discussed in A6.9).

Q15.5 The objective of 'pre-splitting' when blasting is to create a continuous new fracture plane through the rock mass (which will become the final rock surface) before bulk blasting removes the rock up to the pre-split plane.

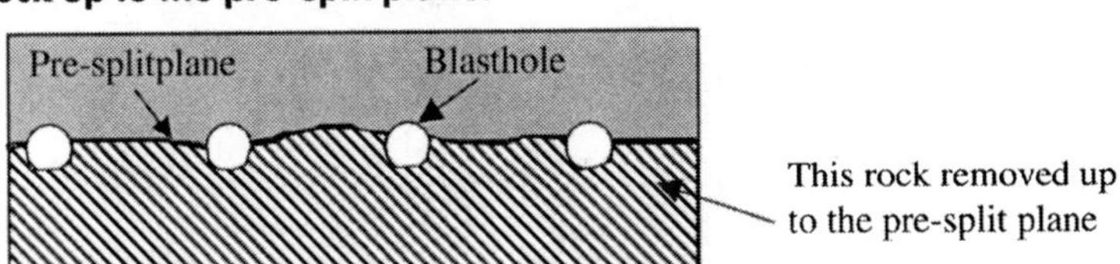

(a) List the most important factors that need to be controlled to ensure that the pre-split blasting techniques will indeed create a continuous pre-split fracture.

(b) The three photographs below show different rock slopes that have been pre-split. The pre-split plane has been created first; then bulk blasting has removed the rock up to the pre-split plane, as indicated in the sketch above. Thus, the photographs show one side of the pre-split 'plane' which is the final slope. Each of the photographs illustrates one of the following:

- **an acceptable pre-split rock face;**
- **a pre-split face where the fractures have adversely affected the pre-splitting; and**
- **a pre-split rock face where one or more of the necessary factors intimated in (a) above has not been properly controlled.**

Which photograph is which case?

Photo 1

Photo 2

Photo 3

A15.5 (a) The key requirements for ensuring a good pre-split face[2] (Langefors and Kihlström, 1963) are explained in ERM 1 as:

- the blastholes must be parallel and locally coplanar in order to form a plane;
- the blastholes should be closely spaced (distance between them should not be more than 10× blasthole diameter) in order to allow fractures driven by gas pressure to link the holes together;
- the blastholes should be lightly charged and the charges decoupled from the blasthole wall in order to avoid pulverizing the rock around the boreholes and to maximize the gas pressure effect; and
- the blastholes should be detonated simultaneously in order to maximize borehole interactions.

If these requirements are followed, the pre-split blasting will usually be successful: in fact, the technique is remarkably robust and a successful pre-split face can be achieved even under difficult conditions.

(b) *Photo 1* is an example where the guidelines have not been followed. The holes are clearly not parallel even over a short distance, and it is not certain that all the holes contained explosive. So, regardless of the fact that the presence of rock fractures can adversely affect the pre-split plane, insufficient care has been taken.

Photo 2 is an example of a rock face where the pre-existing fractures have adversely affected the pre-splitting. The large pre-existing fracture occurs at an orientation which has had an effect on the drilling of the blastholes, in that the holes have deviated from their intended direction to follow the fracture. Moreover, the pre-split plane has occurred along the fracture. In this case, the guidelines were probably followed, but the fractures in the rock mass have had a profound effect on the outcome.

Photo 3 illustrates an acceptable face. The boreholes are parallel and sufficiently closely spaced. The face has some irregularities, as there are dominant fracture sets at this site which have affected the pre-splitting. The purpose of the pre-splitting is to reduce damage behind the face compared to bulk blasting, and this will have been achieved, despite the rough final face.

Q15.6 In order to be able to estimate the blasting specific energy for different rock masses (and hence the different amounts of explosive required for blasting) a Blastability Index (*BI*) has been proposed (Jimeno et al., 1995[3]) using five rock properties. The *BI* is shown in

[2] The pre-splitting technique has been used for many years and is comprehensively described in the seminal rock blasting book, Langefors U. and Kihlström B. (1963) *The Modern Technique of Rock Blasting*. Wiley, New York, 405pp. The authors mention that "in conventional blasting the explosive is used in such a way as to destroy the quality of the remaining rock" and that with pre-splitting "the final contour appears almost as if it were cut out of the rock with a knife leaving the remaining rock practically undamaged". This final statement is, perhaps, an exaggeration, but it describes the technique perfectly.

[3] p. 168 in Jimeno C. L., Jimeno E. L. and Carcedo F. J. A. (1995) *Drilling and Blasting of Rocks*. Balkema, Rotterdam, 391pp. (The example in the question also comes from this reference.)

the table below, with the various ratings related to the specific rock mass properties

Rock Mass Description (*RMD*)	
Powdery or Friable	**10**
Blocky	**20**
Totally massive	**50**
Joint Plane Spacing (*JPS*)	
Close (<0.1 m)	**10**
Intermediate (0.1 to 1 m)	**20**
Wide (>0.1 m)	**50**
Joint Plane Orientation (*JPO*)	
Horizontal	**10**
Dip out of face	**20**
Strike normal to face	**30**
Dip into face	**40**
Specific Gravity Influence (*SGI*)	
$SGI = (25SG - 50)$ where *SG* is in tonnes/m^3	
Hardness (*H*)	
H = hardness on scale 1–10	

The value of the Blastability Index is given as

$$BI = 0.5(RMD + JPS + JPO + SGI + H)$$

and the amount of explosive required is 0.004 *BI* in kg ANFO/t, which is therefore a measure of specific energy.

(a) Calculate the specific energy that this index predicts for the following conditions: a highly laminated, soft ferruginous shale with horizontal to sub-horizontal bedding and *BI* rating values of *RMD* = 15, *JPS* = 10, *JPO* = 10, *SGI* = 10, *H* = 1. With reference to the values discussed in Q15.4, comment on the value obtained here.

(b) Comment on the advantages and disadvantages of this index.

A15.6 (a) The sum of the ratings is 46, so the Blastability Index $BI = 23$. The explosive requirement is then $0.004 \times 23 = 0.092$ kg ANFO/t. We know that 1 kg of ANFO releases 3.92 MJ, so the specific energy is $0.092 \times 3.92 = 0.361$ MJ/t.

However, we need this energy value in terms of cubic metres of rock, rather than the unit mass of rock. Because the SGI value is 10 and we know that SGI = (25SG–50), the SG is 2.4 t/m^3. Thus, the specific energy $= 0.361 \times 2.4 = 0.867$ MJ/m^3.

This value is much lower than the 12.2 MJ/m^3 for the blasting example presented in Q and A15.2 and discussed in A15.4, probably because here the intact rock (a highly laminated, soft ferruginous shale) is weak and the rock mass is well fractured.

(b) The *advantage* of a blastability index of this type is that it provides a rapid and cost-effective guide to the amount of explosive required for excavation in a given rock mass. Blasting is a complex mechanical process, and so some type of empirical classification approach is likely to be the most practical engineering method. The *disadvantage* of this specific index is that it is a rather coarse index. The Rock Mass Description and Joint Plane Spacing could have an increased number of categories: there is a ratings jump from 20 to 50 for both these properties, which is unnecessarily abrupt. The Hardness does not seem to have a high enough rating (assuming that this is the parameter used to represent the rock strength).

However, this *BI* may well be a useful tool even though many rock properties have been omitted in the index. Rock properties tend to be related and so the effects of omitted properties may be included to some degree, albeit indirectly. If the essence of the mechanical problem is captured (as with the rock mass classifications discussed in Chapter 12), the index will help with blasting assessments in rock engineering.

Q15.7 The rock engineering systems (RES) approach has been used to develop a comprehensive Blastability Designation (*BD*) index for rock masses (Latham and Lu Ping, 1999[4]), the purpose of which is to provide a quantitative assessment of the resistance of a rock mass to blasting. The index value is given as $BD = \sum_{j=1}^{n} W_j R_j$ where W_j is a parameter weighting value, R_j is the parameter rating, with the summation over the j parameters. Both the W_j and R_j values are evaluated for the rock mass under consideration. The greater the value of *BD*, the more difficult the rock is to blast. The twelve parameters used in the *BD* index are:

1. **strength (uniaxial compressive strength, *UCS*, MPa, also via Point Load Index);**
2. **resistance to fracturing (uniaxial tensile strength, σ_t, MPa);**
3. **sturdiness of the rock (density, ρ, t/m^3);**
4. **elasticity of rock (Young's modulus, *E*, GPa);**
5. **resistance of rock to dynamic loading (P-wave velocity, V_p, km/s);**
6. **hardness of rock (Schmidt rebound hardness value, *SHV*, rebound height scale);**
7. **deformability (Poisson's ratio, ν, dimensionless);**
8. **resistance of rock to breaking (fracture toughness, K_{1c}, MPa-m$^{1/2}$);**
9. ***in situ* block sizes (mean of block size distribution, mean);**
10. **fragility of rock mass (fractal dimension of rock block sizes, *D*);**
11. **integrity of rock mass (ratio of field: lab P-wave velocities, R_v, dimensionless);**
12. **fracture plane's strength (cohesion, *c*, MPa and friction angle, ϕ, degrees).**

Compile a table of these twelve parameters in three columns with headings Intact Rock Properties, Fracture Properties and Rock Mass Properties, and then comment on how well the rock mass has been characterized.

[4] Latham J.-P. and Lu Ping (1999) Development of an assessment system for the blastability of rock masses. *Int. J. Rock Mech. Min. Sci.*, **36**, 41–55.

A15.7 The table is as follows.

Intact Rock Properties	Fracture Properties	Rock Mass Properties
(1) Strength (*UCS*)	(12) Strength (c, ϕ)	(9) *In situ* block size ($\bar{x}$)
(2) Resistance to fracturing (σ_t)		(10) Fragility (D)
(3) Sturdiness (ρ)		(11) Integrity (R_v)
(4) Elasticity (E)		
(5) Resistance to dynamic loading (V_p)		
(6) Hardness (*SHV*)		
(7) Deformability (ν)		
(8) Resistance to breaking (K_{1c})		

In their article, Latham and Lu Ping (1999) point out that their work is related to "uncontrollable factors governed by *in situ* geological conditions and the term 'blastability' has been deliberately restricted to quantify this intrinsic resistance of the rock mass." In this scheme, the emphasis is on the intact rock properties, which we would expect because the purpose of blasting is to reduce the natural block size distribution to the required fragment size distribution.

Note that parameters 3, 4, 5 and 7 are intact rock properties not directly related to failure, but are included because they characterize the quality of the rock. Parameter 6 is more strongly correlated with the failure properties, and parameters 1, 2 and 8 are direct measurements of rock strength.

In terms of the pre-existing fracturing in the rock mass, parameters 9, 10 and 11 are indicators of the degree of fracturing present in the rock mass, but the only parameter explicitly representing the fracture failure properties is parameter 12, the Mohr–Coulomb values.

In their article, Latham and Lu mention that "one can include more, such as the discontinuity orientation related to the face to be blasted...water content in the rock mass, joint aperture and various other parameters". Let us consider then whether the following subjects — already discussed in this book — should be included in the *BD* index: the *in situ* rock stress, the nature of intact rock failure (as brittle or ductile), the number of joint sets, rock mass permeability, anisotropy and inhomogeneity, other rock mass classifications and specific energy. The rock stress is not usually a factor on exposed rock faces. The nature of rock failure is indirectly included because it is correlated with the other properties. Perhaps the number of joint sets could have been included, but the information is indirectly included in parameters 9–11. Permeability is relevant because it is related to the openness of the rock structure and the way in which the gas travels through the rock mass during blasting, but is difficult to include. Anisotropy and inhomogeneity are probably second-order factors here. The rock mass classification values, RMR and Q, could be included but the classifications were developed for tunnel support. And, finally, the specific energy is not an intrinsic property of the rock mass. Thus, it would not be easy to add any of these to the blastability index, except possibly the RMR and Q values, but much of the content of these classifications is already indirectly included.

The art in applying the engineering rock mechanics principles to rock engineering is to adopt a pragmatic approach by successfully capturing the essence of the problem without introducing unnecessary complications. The authors of the article have successfully achieved this.

Q15.8 In relation to rock tunnel boring machines (TBMs), two progress rates can be defined:

***Cutting rate*, C: the rate at which a tunnel is constructed *when the tunnelling machine is operating;* and**
***Advance rate*, A: the rate at which a tunnel is constructed.**

These two rates are not the same because the TBM is not continuously cutting; there is a *Utilization factor*, *U*, which is the amount of time (expressed as a proportion) that the machine is cutting, i.e. $A = U \times C$.

Over a short period, say 20 min, *A* can be equal to C. Over longer periods, of days, weeks or months, there can be a greater and greater discrepancy between C and *A*, as indicated in the diagram below (Barton, 1999[5]).

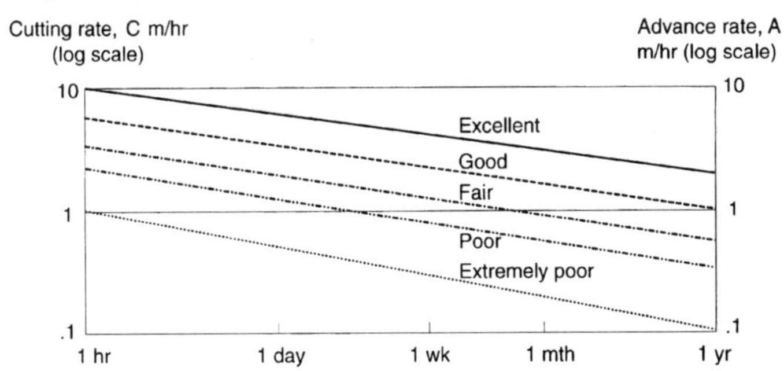

The terms 'Excellent' through to 'Extremely poor' refer to the rates, and Barton noted that the majority of TBM tunnels lie between the 'Good' and 'Poor' lines.Give at least five reasons why such lines occur, i.e. why the value of *U* decreases with time.

A15.8 Five reasons are as follows.

(1) The machine needs periods for scheduled maintenance, cutter replacements and repair. As time goes on, the cumulative effect will be to reduce the value of *U* because of additional damage events and failure of machine components.

(2) If the rock needs supporting, it may be necessary to stop the machine to allow roof and sidewall rock bolting and shotcreting to take place.

---

[5] This diagram was included in the General Report presented by Barton (1999) at the 9th ISRM Congress on Rock Mechanics held on 25–28 August 1999 in Paris, and is included in the article Barton N. (1999) TBM Performance Estimation in Rock Using Q_{TBM}. *Tunnels and Tunnelling*, September 1999, 30–34.

(3) Adverse ground conditions can cause large blocks to become unstable at the face, flooding and overbreak, all of which cause delays. The further one tunnels, the greater is the chance of encountering adverse ground conditions.
(4) If a hard rock machine encounters water-bearing fractures, these will also cause delays. Moreover, the further one tunnels, the greater is the chance of encountering a fracture discharging more water than any previously intersected.
(5) 'Unexpected events' such as power cuts, broken conveyors, derailments and strikes by the work force will also contribute to a lowering of the U value, and the chance of these occurring increases with tunnel length and time.

***Q15.9* When an underground excavation is made in a rock mass, there are three primary effects, all three of which result from Newton's 3rd law, i.e. for every action, there is an equal and opposite reaction. Thus, when new underground space is created, the resistance of the rock previously occupying the space becomes zero, forces cannot be transmitted from the rock into the space, and the water pressure is reduced to atmospheric pressure.**

Draw a sketch of these three effects and explain why they cause a so-called Excavation Disturbed Zone[6] (EDZ).

A15.9 The three primary effects of excavation are illustrated in the sketch below, taken from ERM 1. There are many potential complicating factors when an excavation is made but these complications can all be considered within the context of the three primary effects. For example, there can be 'coupled' effects when the alteration of the rock stresses opens fractures, which then conduct more water into the excavation. A major fracture in the excavation sidewall may exacerbate the effects and it may take some time for all the effects to be fully realized.

The three effects will always occur. Therefore, there will always be an EDZ around an excavation. The amount by which this zone of inevitable disturbance is further enlarged will depend on the excavation method. Thus, there is both an initial *inevitable* disturbance to the rock mass and an *additional* disturbance to the rock mass, the severity of the latter depending on the excavation method. It is incorrect to say that a TBM will not disturb the rock mass (because of the inevitable effects, which could have severe consequences), but a TBM will generally induce less *additional* disturbance than blasting.

Two of the effects depend on the size of the excavation. Other factors remaining the same, the inward rock movement does depend on the excavation size: if the size is doubled, the magnitude of the inward rock movement will also tend to be doubled. Also, larger excavations in the same host rock will have larger amounts of water inflow. However,

[6] The word 'Disturbed' is preferable to 'Damaged' for the term EDZ because the rock is indeed disturbed but not necessarily damaged. Whether there is damage or not depends on the engineering objective.

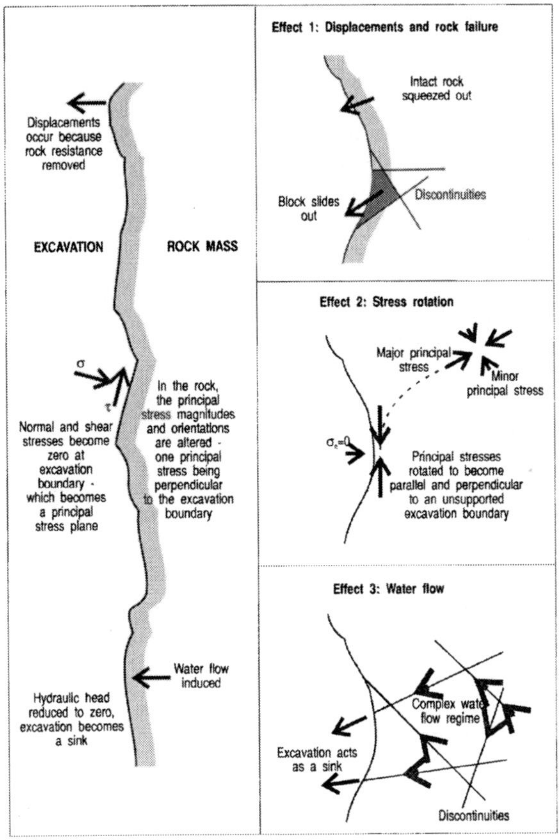

the stress concentrations around the excavation do not depend on the excavation size: for example, caverns (in a given stress field and elastic rock) with spans of 1 m, 10 m and 100 m (and having the same shape) will all have the same stresses induced around them.

Q15.10 (a) Can drill and blast excavation rates be significantly increased from those currently achieved?

(b) Can tunnel boring machine penetration rates be significantly increased?

(c) Are new methods of excavation likely to be introduced?

A15.10 (a) Drill and blast excavation rates (using customary equipment, methods and materials) are unlikely to be significantly increased, because after decades of use the drill and blast technique has already been optimized. Given the cyclical nature of the drill and blast process, significant improvements are unlikely.

(b) In principle, TBM *penetration* rates can be significantly increased because there is little difficulty in supplying large amounts of energy to these machines. The increased rates may occur, for example, by improved cutter technology or by combining the explosive and cutting/mucking processes in the machine's cutting head. The TBM *advance* rates will not increase so much because the whole tunnelling system (the thrust reaction grippers, the electro-hydraulic power system, installation of support elements, maintenance, etc.) has to keep up with the increased penetration rate, which involves many practical difficulties.

(c) New methods of excavation are not likely to be introduced in practice. There has been considerable research over the years into alternative methods of fragmenting and removing the rock and, whilst many of these methods (e.g. flame cutting, electrical and magnetic disintegration, nuclear techniques) have advantages, they also have significant practical disadvantages.

15.3 Additional points

The increasing demands on surface space, greater environmental awareness, and the technical and safety-related advantages of using underground excavations have led to considerable use of underground space for civil engineering purposes as evidenced by the list below.

Civil and military defence, nuclear shelters, aircraft hangers, emergency centres
Conventional and nuclear power stations
District heating
Drinking water transportation and storage
Dry docks
Hot water storage
Hydroelectric projects
Living quarters
Radioactive waste storage and disposal
Rail transport and stations
Refuse management and incineration
Research facilities, particle accelerators, wind tunnels
Road tunnels
Shopping malls
Sports, concert, theatre, religious and museum halls
Storage of food, drink, documents, aggregates, minerals
Storing petroleum products
Storm water drainage and storage
Waste water management and treatment plants

There will be increasing underground excavation in the years ahead to supply society with further civil and mineral resources. The maximal excavation size depends on the type of rock mass. *In a strong, unfractured granite,* large spans are possible, e.g. 100+ m spans, because the stress concentrations do not depend on excavation size. *In weaker, fractured rocks,* the spans are limited because of the need to support the rock, the support requirements increasing with excavation size. Most caverns to date have been excavated at depths shallower than 1000 m. The cavern span generally decreases with depth from 10–30 m spans in the 50–400 m depth range, reducing to 10–20 m spans in the 400–1000 m depth range. Deeper caverns are possible, with rock stress being the main mechanical constraint.

In this chapter, we have highlighted some of the key issues relating to the excavation of rock. The questions have tended to be descriptive and relatively simple in order to illustrate the principles. In engineering circumstances, rock excavation is a complex rock fragmentation process, highly dependent on the nature of the rock mass and the engineering techniques utilized. It is also critical to assess the excavatability of a rock mass and to tailor the excavation technique to the circumstances and the engineering objective. For example, if a TBM is purchased and subsequently found to be inadequate, a large amount of time and money may be lost. The complexity of the process is such that a great deal of work is still required in this field, involving a combination of theoretical, empirical and pragmatic approaches, all of which are enhanced by a thorough understanding of rock mechanics principles.

16 Rock reinforcement and rock support

16.1 The stabilization system

When an excavation is made in a rock mass, there may be a need to **stabilize** the excavation to reduce inward movement of the rock mass. The amount of such movement will depend on the quality and continuity of the host rock mass, and the significance of the movement will depend on the engineering objective. For a rail transport tunnel, there will be strict constraints on the allowable movement; for example, the design may require the rock displacements to be constrained to 10 mm or less over 120 years. In a mine stope, there may not be any constraints; and in the case of longwall coal mining the roof should collapse fully until the roof is in contact with the floor for the mining system to be effective.

The term **rock reinforcement** is used to indicate methods of enhancing the rock mass strength and hence improving the ability of the rock mass to contain the engineering excavation without deforming excessively. Rockbolts, wire mesh and shotcrete (mortar sprayed onto the excavation surface) are reinforcement methods because they strengthen the rock mass. The term **rock support** is used to indicate methods of applying supporting loads or displacement constraints as additional structural elements, so that the engineering excavation retains its integrity. Steel ribs, pre-cast concrete segmental lining and cast *in situ* concrete lining are support methods.

There can be some overlap between the two terms. In the case of a thick layer of shotcrete, for example, the shotcrete acts as a reinforcement in the sense that it inhibits rock block rotation and weathering, but it can also act as a structural element containing tangential stresses (Fig. 16.1, from Wilson, 1991 [1]). Alternatively, the use of pre-cast segments for support also has a reinforcing effect because the rotation of rock blocks and weathering is inhibited (Fig. 16.2, from Wilson, 1991) [1]. The term 'rock stabilization' refers to the use of either method, or the two in conjunc-

---

[1] Wilson D. (1991) *Breakthrough*. Random Century Group, London, 144pp.

Figure 16.1 Workers awaiting the breakthrough of the south seaward running tunnel TBM into the UK undersea shotcreted crossover cavern in the Channel Tunnel (from Wilson, 1991).

Figure 16.2 Use of precast lining segments in the UK north landward running tunnel of the Channel Tunnel showing spoil wagons running on temporary tracks (from Wilson, 1991).

tion, to ensure that the rock excavation retains its mechanical integrity.

Thus, the need to reinforce and support the rock mass surrounding an excavation and the method by which this can be achieved successfully depend on two key factors: the **purpose of the excavation**; and the concept of **allowing the rock mass to deform** in order to reduce the support loads. It is not necessary nor possible to stop all rock displacements; indeed, the rock may displace until it reaches a natural equilibrium position.

The questions in this chapter examine basic aspects of the rock stabilization concept. There are questions on rock reinforcement, rock support and an intermediate case. The questions in this chapter do not cover specific rock engineering design (these are in Chapters 18 and 20), and are intended to enhance understanding of the principles involved and hence which stabilization strategy is liable to be optimal for a given engineering objective and circumstances.

16.2 Questions and answers: rock reinforcement and rock support

Q16.1 When rockbolts are used for surface or underground excavations to reinforce a fractured rock mass, the rockbolts will be subjected to tension, shear and compressive forces. The diagram below (from Windsor, 1997[2]) indicates the types of forces on rockbolts a to f.

Recognizing that rockbolts are designed to operate in tension with little shear force applied, rate these rockbolt configurations in terms of their suitability on this criterion alone.

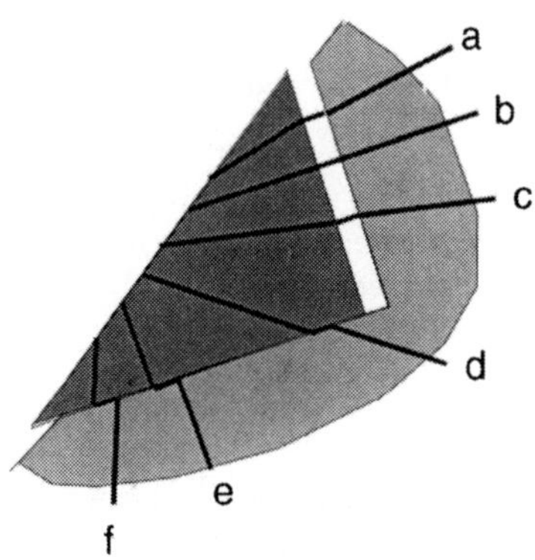

A16.1 The types of forces on the rockbolts are

a: tension and shear	b: tension only	c: tension and shear
d: shear and tension	e: shear only	f: shear and compression

Rating these configurations in order from best to worst according to the criterion 'tension with little shear force applied' gives: b, a and c, d, f, e.

Q16.2 A rock block of weight *W* is bolted to a plane inclined at angle α to the horizontal. Develop an expression for the bolt tension, *T*, in

[2] Windsor C. R. (1997) Rock Reinforcement Systems. *Int. J. Rock Mech Min. Sci.* **34**, 919–951.

terms of the angle, β, at which the bolt is installed relative to the plane, the angle of inclination of the plane and the Mohr–Coulomb friction properties, c and ϕ, of the block/plane interface. Hence, by differentiation, find the rockbolt angle at which the bolt tension is minimized.

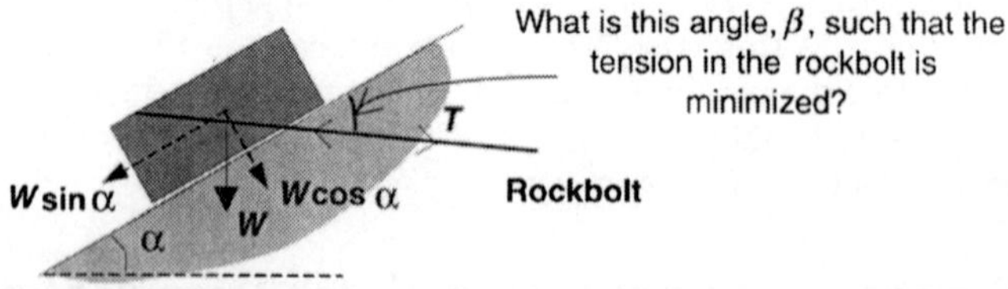

A16.2 The force, F_n, acting normal to the plane is the sum of the resolved block weight and rockbolt tension components, $F_n = W\cos\alpha + T\sin\beta$. The force, F_s, acting along the plane is similarly given by $F_s = W\sin\alpha - T\cos\beta$.

The limiting condition of the block/plane interface having reached peak strength gives

$$W\sin\alpha - T\cos\beta = cA + (W\cos\alpha + T\sin\beta)\tan\phi$$

where c is the cohesion[3] and A is the basal area of the block.

Rearranging gives the required answer of

$$T = \frac{W(\sin\alpha - \cos\alpha\tan\phi) - cA}{(\cos\beta + \sin\beta\tan\phi)}.$$

To find the minimal rockbolt tension, we set the partial derivative of T with respect to β to zero and solve the resulting equation for β. Recognizing that the expression for T will be minimized when the denominator is maximized, we have

$$\frac{\partial}{\partial\beta}(\cos\beta + \sin\beta\tan\phi) = 0, \text{ or } -\sin\beta + \cos\beta\tan\phi = 0.$$

Rearranging gives $\tan\phi = \sin\beta/\cos\beta = \tan\beta$, from which $\beta = \phi$.

Thus, the rockbolt tension is minimized when the rockbolt is installed at an angle to the slope which is the same as the angle of friction between the rock block and the slope. For a plane with a very low angle of friction, i.e., $\phi \approx 0°$, the rock block will hang from the rockbolt, which should be installed sub-parallel to the slope. For a very rough interlocking plane, i.e. $\phi \approx 90°$, the block is literally bolted down to the plane, with the rockbolts being perpendicular to the plane.

Q16.3 **The sketch below shows a rock block resting on a slope and subject to a stabilizing force from a rockbolt. The strength of the interface between the block and the slope is purely frictional, with an angle of friction ϕ.**

[3] Note that c is the cohesion per unit area, with units of stress.

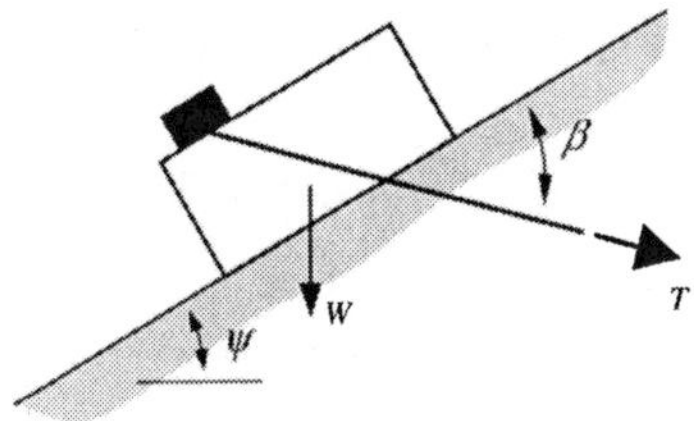

(a) The component of the bolt force acting parallel to the slope can be considered as either a positive restraining component or as a negative actuating component. Derive an expression for the factor of safety in both these cases.

(b) For the situation when $W = 1000$ kN, $\beta = 15°$, $\psi = 42°$ and $\phi = 36°$, what bolt tension is required to give a factor of safety of unity for each of the cases?

(c) Examine the behaviour of the two expressions for factor of safety as the bolt tension varies from 25 kN less than, to 25 kN more than, the force computed for each case above. Which of these two expressions should we use for design purposes, when we require a factor of safety in excess of 1?

(d) Rock bolts function by developing a tensile force within them, and are sometimes tensioned at the time of installation in order to generate this force. In the case of untensioned bolts, how is this force generated? Should we regard untensioned bolts as providing a positive restraining component or a negative actuating component?

(e) Imagine that we require a factor of safety of three for the rock block shown above. Should we use tensioned or untensioned bolts?

A16.3 (a) As the strength of the interface between the block and the slope is purely frictional, we can define the factor of safety

$$F = \frac{\text{forces resisting sliding}}{\text{forces causing sliding}} = \frac{N \tan \phi}{S}$$

where N and S are the shear and normal forces acting across the interface.

The free body diagram of the block, with the various forces acting on it resolved parallel and normal to the interface, is as shown to the right. From this we obtain

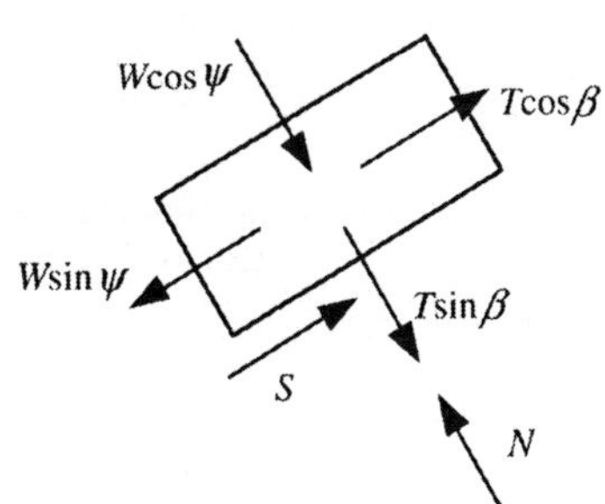

$$N = W \cos \psi + T \sin \beta$$

and

$$S = W \sin \psi - T \cos \beta.$$

If we consider that the component $T \cos \beta$ is a negative actuating force, these relations then lead to

$$F_a = \frac{(W \cos \psi + T \sin \beta) \tan \phi}{W \sin \psi - T \cos \beta}.$$

If we consider $T \cos \beta$ to be a positive restraining component, then we have

$$N = W \cos \psi + T \sin \beta, \text{ and } S = W \sin \psi$$

leading to

$$F_r = \frac{(W \cos \psi + T \sin \beta) \tan \phi + T \cos \beta}{W \sin \psi}$$

for the factor of safety.

(b) For the case of a factor of safety of 1, these expressions rearrange to

$$T_a = W \frac{\sin \psi - \cos \psi \tan \phi}{\cos \beta + \sin \beta \tan \phi} \text{ and } T_r = W \frac{\sin \psi - \cos \psi \tan \phi}{\cos \beta + \sin \beta \tan \phi}$$

which are identical. With $W = 1000$ kN, $\beta = 15°$, $\psi = 42°$ and $\phi = 36°$, the bolt tension required is 112.0 kN.

(c) Evaluating the two expressions for factor of safety over a range of values for the tension gives the results shown in the table below. These results are plotted in the following diagram.

T (kN):	87.0	97.0	107.0	117.0	127.0	137.0
F_a:	0.951	0.970	0.990	1.010	1.032	1.054
F_r:	0.957	0.974	0.991	1.009	1.026	1.043

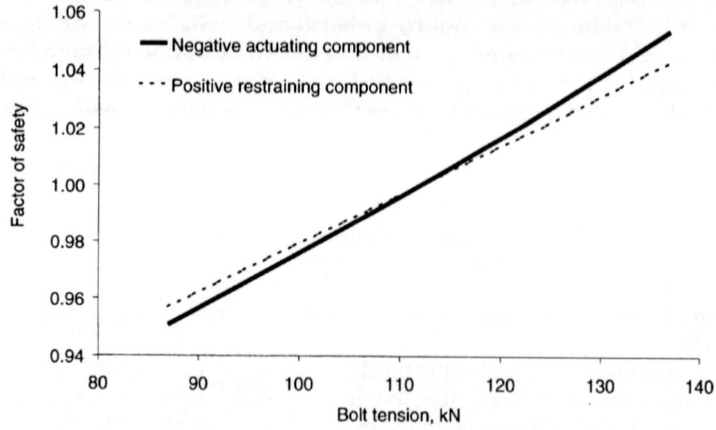

For those circumstances when we require a factor of safety greater than one, we should use the expression for a positive restraining component, as this gives the lower — and therefore most conservative — factor of safety.

(d) When two rock surfaces shear past one another, their geometric roughness means that dilation will tend to take place. If the pieces of rock on either side of the shearing surface are connected together with a rock bolt, then the bolt will tend to prevent the dilation. However, in doing this, a tensile force will be generated in the bolt. This indicates that untensioned bolts only act as stabilizing elements once some rock block shear displacement has occurred. No tensile force exists before this, and so untensioned bolts only provide restraint. As such, we consider untensioned bolts as providing a positive restraining component. Tensioned bolts, on the other hand, provide a component of force before any movement takes place, and so we regard these as providing a negative actuating component.

(e) On the basis of the definition of factor of safety we gave earlier, a factor of safety of 3 means that we are providing three times more strength to the interface than is required. This implies that we are preventing any movement occurring, and hence we must use tensioned bolts. The bolt force required to develop a factor of safety of 3, is then

$$T_a = W\frac{F_a \sin\psi - \cos\psi \tan\phi}{\sin\beta \tan\phi + F_a \cos\beta} = 476 \text{ kN}.$$

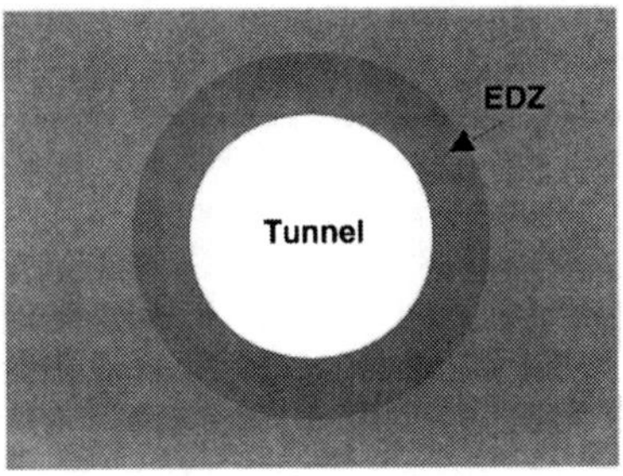

Q16.4 A circular tunnel is being excavated in a blocky rock mass by drilling and blasting. There is an Excavation Disturbed Zone (EDZ) around the excavated tunnel (defined on the basis of a blast-disturbed zone where there are loosened blocks which can fall into the tunnel under the action of gravity) which extends 0.75 m into the rock from the excavation surface.

What support pressure[4] is required at the crown to stabilize the loose blocks of the EDZ given that the unit weight of the rock, γ, is 25 kN/m^3?

A16.4 The support pressure, p, is calculated as W/A, where W is the weight of the loose blocks and A is the area of the tunnel being considered. Taking the EDZ volume, V, above a 1 m^2 area of tunnel roof, the weight of the EDZ is γV, so $W = 25 \times (0.75 \times 1 \times 1) = 18.75$ kN and $A = 1 \times 1 = 1$ m^2, giving $p = 18.75$ kN/m^2 = 18.75 kPa.

To provide some comparison with this value, a car tyre pressure is typically about 170 kPa, which is about nine times larger. Note that this calculation is only for the loosened blocks in the roof; blocks in other parts of the EDZ would be partially supported by the blocks beneath them, and so would require a lower support pressure.

Q16.5 (a) If the EDZ in Q16.4 is to be stabilized by the use of rockbolts inserted into the roof as a supporting method, and the

[4] Note that the words 'pressure' and 'stress' have different meanings. Pressure occurs in a fluid, is a scalar quantity, and acts equally in all directions. Stress occurs in a solid (because a solid can sustain a shear stress), is a tensor quantity and has different components in different directions. Here we use the word 'pressure' — in line with conventional usage for tunnel support — to mean the force/unit area acting upwards at the crown of the tunnel.

working capacity of each bolt, *T*, is 150 kN, what area of the roof will each bolt support?

(b) Would you anticipate any ancillary problems with this bolting pattern?

(c) In the circular tunnel, in which directions and at which locations would you install the bolts?

A16.5 (a) To find the area of roof each rock bolt will support, we equate the rockbolt load, 150 kN from A16.4, to the support pressure times the roof area required, A. Then $150 = 18.75A$, or $A = 150/18.75 = 8\ \mathrm{m}^2$ of roof per rockbolt. If a square rockbolt pattern is being used, there will be 2.8 m between rockbolts, as illustrated to the right (which is a sketch looking upwards at the tunnel roof).

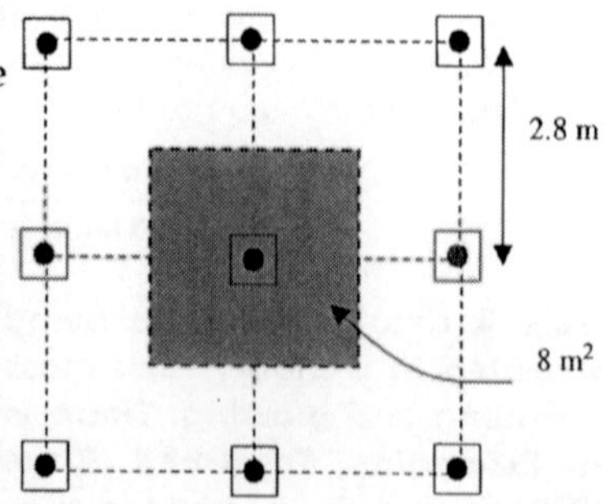

(b) The 2.8 m spacing between the bolts is sufficient to provide the necessary strategic rock support, but the rock is only supported at the heads of the rockbolts. This means that there are areas of exposed rock, >2 m across, from which rock blocks can fall. The detachment of the small blocks is avoided by using wire mesh that is attached to the bolt heads. The picture below illustrates such a square rockbolting array with wire mesh (although the bolt spacing shown here is less than 2.8 m).

Rockbolt heads

(c) The calculation refers only to the crown of the tunnel where the rock is free to fall. The support pressure can be reduced away from crown, because there is additional support from lower blocks, but bear in mind the potential sliding of the blocks as discussed in Q and A 16.2 and 16.3. As a general guide, radial bolting is appropriate for a circular tunnel. For the conditions assumed in this question, no rockbolts are needed in the floor, which is not as strange a comment as it may seem: sometimes rockbolts are installed in the floor to counteract the type of floor heave induced by a combination of high stress and weak rock, as illustrated to the left.

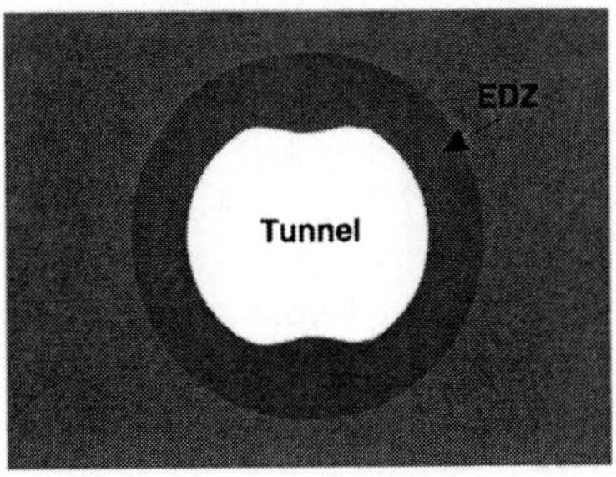

Q16.6 A long underground chamber is to be constructed in a rock mass with fractures dipping at 25° as shown. This problem is included in Goodman (1989)[5].

(a) Assuming the friction angle of the fractures, ϕ_j, is 50°, use the ϕ_j geometrical construction in two dimensions to locate zones of potential distress around the opening.

(b) Repeat this procedure with $\phi_j = 20°$ to investigate the effect of rock deterioration on interlayer slip around the excavation.

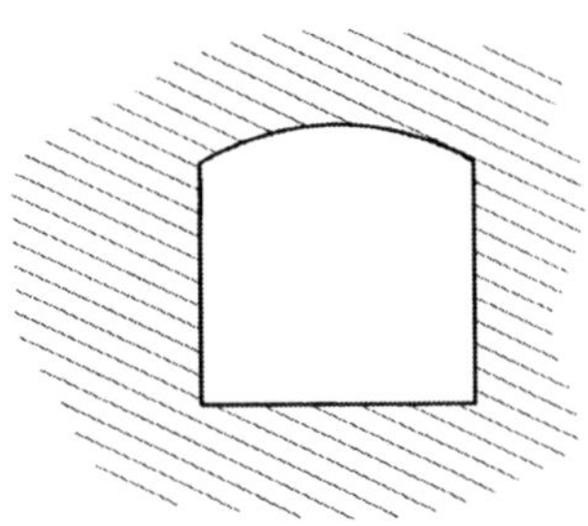

A16.6 The principle of the ϕ_j geometrical construction is indicated below.

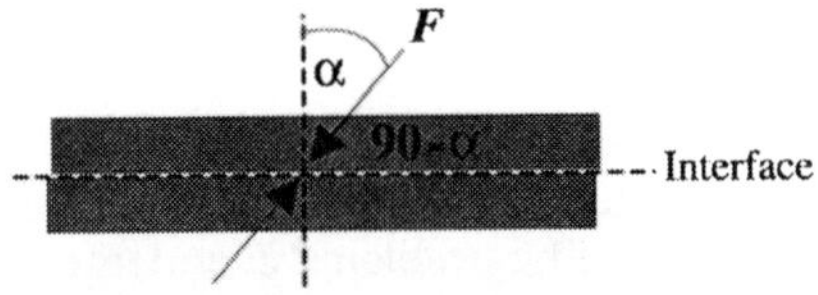

When a force is applied to an interface, slip can occur if the force, F, is applied at a sufficiently low angle, $90-\alpha$, to the interface, i.e. slip will occur for $F\sin\alpha > \tan\phi F\cos\alpha$, or $\tan\alpha > \tan\phi$, which means $\alpha > \phi$.

In the excavation peripheral rock, one principal stress is perpendicular to the excavation surface with a value of zero, and the other is parallel to the excavation surface.

This enables the construction below to be used in which the regions susceptible to interlayer slip can be readily identified — as those where the angle between the tangent to the excavation surface and the normal to the layers is greater than the angle of interlayer friction, ϕ_j.

[5] Goodman R. E. (1989) *Introduction to Rock Mechanics*. Wiley, New York, 2nd ed., 562pp.

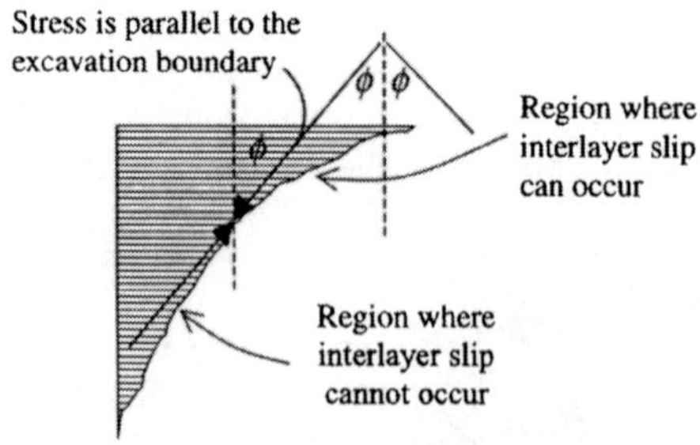

In the case for $\phi_j = 50°$, the left-hand diagram below shows that interlayer slip can occur over most of the roof and all the floor, but not at the walls. If the value of ϕ_j is reduced to 20°, however, interlayer slip can occur at any location around the tunnel roof, walls and floor. This is one example of the extra problems that can arise with, for example, weak rocks and changing water conditions.

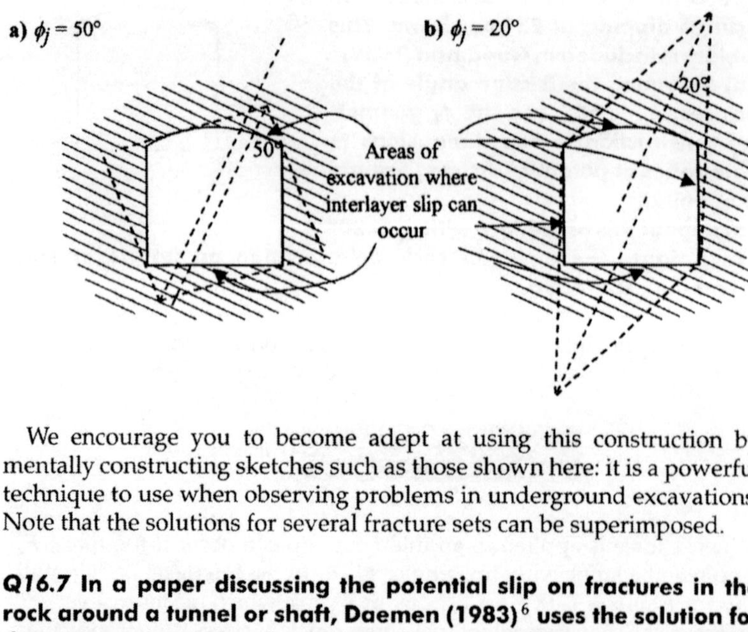

We encourage you to become adept at using this construction by mentally constructing sketches such as those shown here: it is a powerful technique to use when observing problems in underground excavations. Note that the solutions for several fracture sets can be superimposed.

Q16.7 In a paper discussing the potential slip on fractures in the rock around a tunnel or shaft, Daemen (1983)[6] uses the solution for the stresses around a circular opening in a CHILE rock to establish the normal and shear stresses on the fractures. Applying the Mohr–Coulomb criterion to these stresses enables the potential zones of distress around the excavation to be established.

The zones of distress shown in Daemen's diagram below are for an isotropic stress field and an angle of friction of $\phi = 20°$.

[6] Daemen J. J. K. (1983) Slip zones for discontinuities parallel to circular tunnels of shafts. *Int. J. Rock Mech. Min. Sci. Geomech. Abs.* **20**, 135–148.

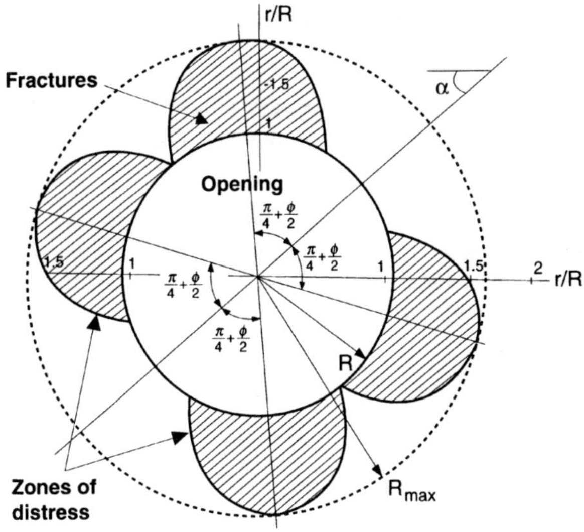

Show that the same result for the zone of distress at the excavation *boundary* can also be obtained using the ϕ_j theory.

A16.7 The construction shown below using the ϕ_j theory indicates zones of distress at the excavation boundary in the regions ABC and

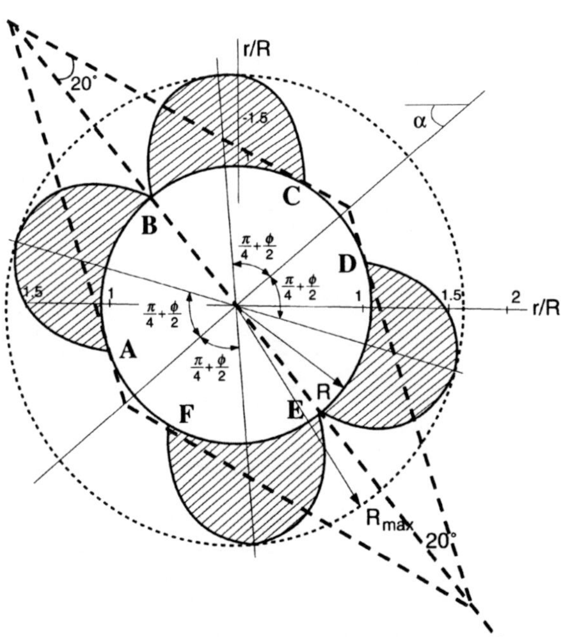

DEF. These zones coincide with those determined by Daemen, also shown in the diagram.

The ϕ_j method is quicker but only indicates the regions around the excavation *boundary* where there is the potential for interlayer slip. To establish the zones of distress within the rock mass, the rigorous approach is required.

Q16.8 Two types of 'ground response curve', A and B, are illustrated in the diagram below. The curves indicate the amount of support pressure required to maintain an excavated rock surface at a given value of displacement.

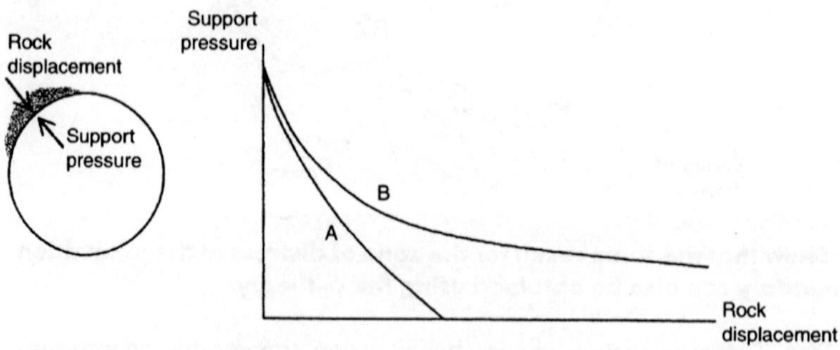

With reference to these curves, explain:

(a) why the *in situ* stress condition before excavation cannot be re-established by the support pressure applied after excavation;

(b) why it is inappropriate for an engineer to insist that the rock should be supported in such a way that no displacement whatsoever occurs.

A16.8 (a) The principal stresses of the *in situ* stress state before excavation will have certain magnitudes and directions. After excavation and on the excavation periphery, these magnitudes and directions will have changed, as indicated in A15.9 and in the sketch below.

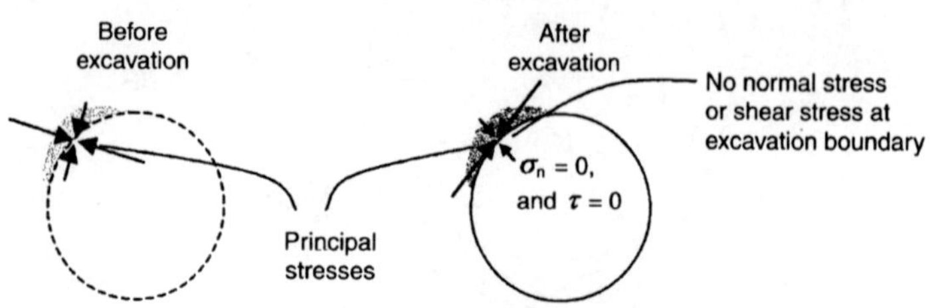

Recalling that stress is a tensor quantity with six independent components, in the general case it is not possible to re-establish the *in situ* stress by applying a normal stress, or pressure, to the excavation surface; it would also be necessary to apply shear stresses, which is not practical.

(b) For either ground response curve, A or B, although it is not possible to re-establish the *in situ* stress field by the application of support pressure, it is in theory possible to apply support pressure such that there is zero radial rock displacement at a given position, as indicated by the black circle on the support pressure axis in the sketch below. This could well involve considerable practical problems and would be expensive.

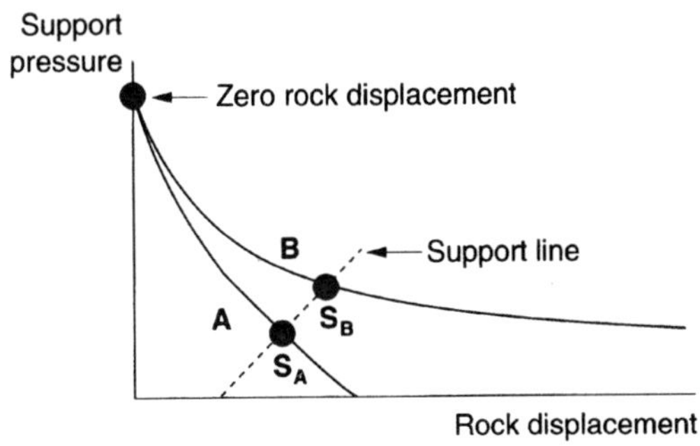

However, attempting to insist on zero rock displacement at the excavation periphery is unnecessary. In the case of curve A, the rock unloads to a stable position at zero support pressure, so no support is necessary. If support is installed after some displacement has occurred, as represented by the dashed line in the sketch, the rock displacement will be stopped at point S_A, where the support line meets the ground response curve, with less support pressure than that required for zero displacement.

In the case of curve B, assuming that the rock is not unloading to a stable position, the installation of support is necessary. If the same support represented by the dashed line in the sketch is used, the rock displacement will be stopped at point S_B, where the support line meets the ground response curve, again with less support pressure than that required for zero displacement.

Q16.9 A tunnel of radius 1.85 m is excavated in rock subjected to an initial hydrostatic stress field of 20 MPa and provided with a concrete lining of internal radius 1.70 m. Assuming elastic behaviour of the rock and lining, calculate the radial pressure and the radial displacement at the rock-lining interface if:

(a) the lining is installed at the face before any displacement of the rock has occurred; and

(b) the lining is installed following a radial convergence of 1 mm.

A16.9 (a) We solve this problem by plotting the ground characteristic (i.e. ground response curve) and the support characteristic (i.e. available support line) on the same axes, and identifying where they intersect. This is the operating or equilibrium point. Firstly, we are told that both the rock and the lining remain elastic. This means that both characteristics are straight lines.

To find the ground characteristic we need to identify the two end points of the line: one is the *in situ* condition of zero displacement at a radial stress of 20 MPa, the other is the elastic displacement induced when the radial stress is zero. We use the Kirsch solution to find this radial displacement and, for a hydrostatic stress state, it is given by

$$u_r = -\frac{pa}{2G}$$

where p is the value of the hydrostatic stress, a is the radius of the tunnel, and G is the shear modulus.

However, as we have not been given a value for the shear modulus, let us assume a value of 2 GPa. This gives a radial displacement of

$$u_r = -\frac{pa}{2G} = -\frac{20 \times 1.85}{2 \times 2000} = -0.0925 \text{ m.}$$

If we assume that the lining behaves as a thick-walled cylinder subject to radial loading, then the equation for the lining characteristic is

$$p_r = k\frac{u_r - u_o}{a}$$

where p_r is the radial support pressure, k is the lining stiffness, and u_o is the magnitude of the rock displacement when the lining is installed, and

$$k = \frac{E_c}{1 + \nu_c} \frac{a^2 - (a - t_c)^2}{(1 - 2\nu_c)\, a^2 + (a - t_c)^2} \tag{16.1}$$

where t is the thickness and the subscript c refers to the concrete lining.

To use Eq. (16.1) we assume values for E_c and ν_c as 30 GPa and 0.25, respectively. Using these figures, together with a = 1.85 m and $t_c = 1.85 - 1.70 = 0.15$ m, we find that k = 2.78 GPa. Thus, for a radial pressure of 20 MPa and u_o = 0 mm, the lining will deflect radially by

$$u_r = \frac{a}{k} p_r + u_o = \frac{1.85}{2.78 \times 10^3} 20 + 0 = 0.013 \text{ m.}$$

The operating point can now be found:

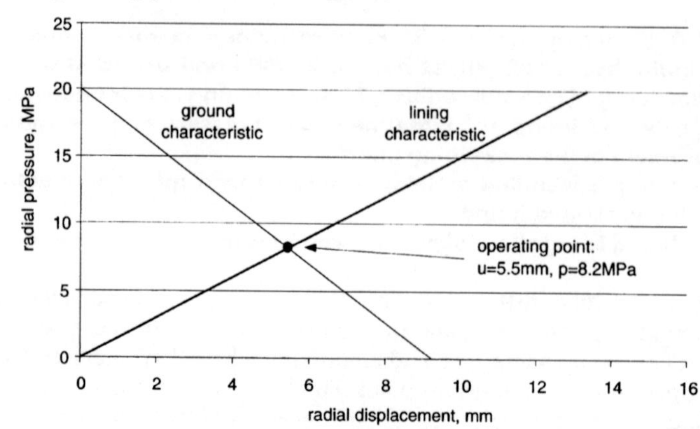

(b) If the lining is installed after a radial displacement of 1 mm has occurred at the tunnel boundary, then we have the following:

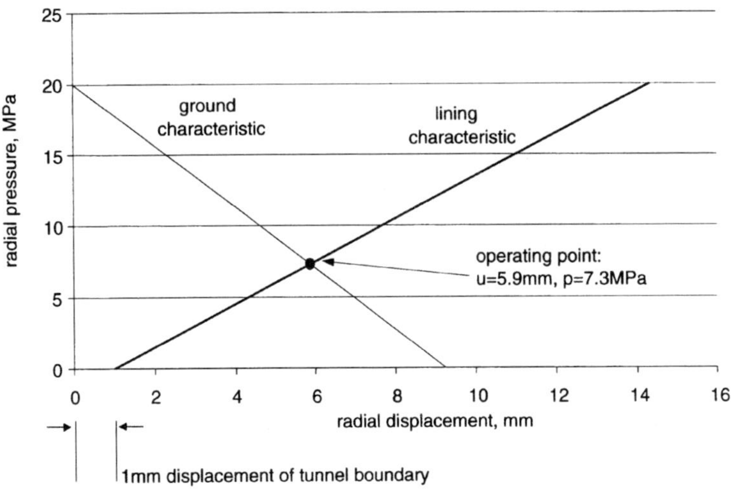

This shows how, by delaying the installation of the lining, we have reduced the pressure it is required to withstand, but at the expense of increasing the final radial displacement.

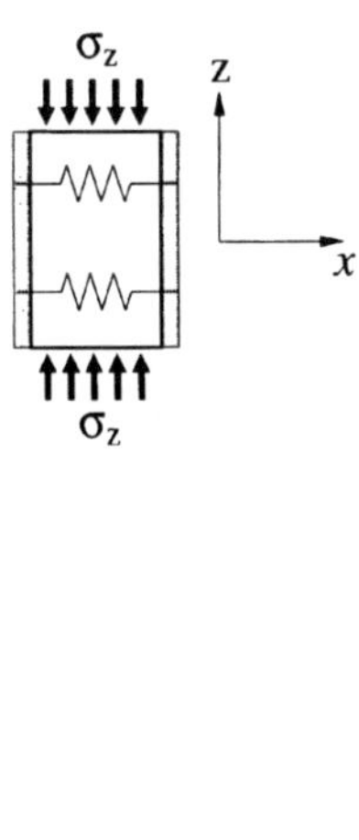

Q16.10 The sketch shows a pillar of rectangular cross section, which is very long in the *y*-direction (i.e. out of the plane of the paper) and is made of a linearly elastic isotropic rock. A support pressure is applied to the horizontal sides of the pillar through the action of springs that pass horizontally through the pillar.

(a) Develop an expression linking the vertical stress, σ_z, to the support pressure, *p*, in terms of the elastic constants of the pillar material and the spring stiffness. Use this expression to show that, for zero horizontal strain of the pillar, the ratio of vertical stress to support pressure is $(1 - \nu)/\nu$.

(b) If the strength of the pillar can be described by the Hoek–Brown criterion with $\sigma_c = 37$ MPa, $s = 1$ and $m = 15$, and the elastic modulus and Poisson's ratio of the rock are 75 GPa and 0.27, respectively, what spring stiffness is required if the pillar is to withstand a vertical stress of 55 MPa?

A16.10 (a) The generalized Hooke law for an isotropic material can be written directly as both

$$\varepsilon_x = \frac{1}{E}\left[\sigma_x - \nu\left(\sigma_y + \sigma_z\right)\right] \text{ and } \varepsilon_y = \frac{1}{E}\left[\sigma_y - \nu\left(\sigma_z + \sigma_x\right)\right].$$

Because the pillar is very long in the *y*-direction, we can say that it is in a state of plane strain and hence $\varepsilon_y = 0$. Substituting this and $\sigma_x = p$ in the second of these equations gives

$$\sigma_y = \nu(\sigma_z + p)$$

and then substituting this in the first equation leads to

$$\varepsilon_x = \frac{1}{E}\left[p\left(1 - \nu^2\right) - \sigma_z \nu\left(1 + \nu\right)\right].$$

If we use the usual geomechanics convention of compressive stresses and contractile strains as positive, then we can say that the relation between ε_x and p is

$$p = -k\varepsilon_x$$

where the spring stiffness, k, has units of stress. Note that the minus sign is required to generate a compressive support pressure from an extensile horizontal strain in the pillar.

Combining these last two equations gives

$$-\frac{1}{k}p = \frac{1}{E}\left[p\left(1-\nu^2\right) - \sigma_z\nu\left(1+\nu\right)\right]$$

which, upon rearranging, gives

$$\sigma_z = p\left[\frac{(1-\nu)}{\nu} + \frac{E}{\nu\left(1+\nu\right)}\frac{1}{k}\right].$$

The case of zero horizontal strain of the pillar is represented in this equation by an infinite spring stiffness. This condition reduces the above equation to

$$\sigma_z = p\left[\frac{(1-\nu)}{\nu}\right]$$

from which we see that the ratio of vertical stress to support pressure is $(1-\nu)/\nu$, as required.

(b) The Hoek–Brown criterion can be written as

$$\sigma_1 = \sigma_3 + \sqrt{m\sigma_c\sigma_3 + s\sigma_c^2}$$

which, upon substitution of the numerical values given here and assuming $\sigma_1 = \sigma_z$ and $\sigma_3 = p$, becomes

$$\sigma_z = 55 = p + \sqrt{555p + 1369}.$$

Rearranging this to give a quadratic in p leads to

$$p^2 - 665p + 1656 = 0$$

from which we find $p = 2.5$ MPa.

If we rearrange the equation for the pillar stiffness, we can obtain

$$k = \frac{E}{(1+\nu)}\frac{p}{\left[\nu\sigma_z - p\left(1-\nu\right)\right]}$$

and hence by substitution find $k = 11.3$ GPa.

To determine the area of steel bars required to provide this stiffness, we examine the force compatibility for a unit area of pillar face. The relation between strain and stress in the support is $\sigma = k\varepsilon$, and so the force required over a unit area of face of a pillar of unit width is $F = k\delta$, where δ is the displacement of the support. For a bar of unit length subjected to an axial force, the displacement is given by $\delta = F_b/A_bE_b$. Combining these equations gives

$$k = A_bE_b \text{ or } A_b = k/E_b.$$

From this we find that the area of steel bars required per square metre of pillar face is 11.3/210 = 0.054 m^2. In terms of bars whose diameter is 25 mm, this represents 110 bars, which could only be obtained by inserting the bars at a spacing of about 95 mm! However, although the quantity of bars is large, the stress induced in them is only p/A_b, or 46.3 MPa. As high tensile steel bars are capable of working at 300 MPa, this shows that this form of ***passive* reinforcement** is inefficient.

To improve the efficiency of the reinforcement, it would be appropriate to apply a tension to the bars during installation. This would generate the required support pressure directly, without requiring the pillar to deform elastically. This is known as ***active* reinforcement**.

16.3 Additional points

The book by Hoek et al. (1995)[7] explains many of the principles of rock reinforcement and support and associated subjects.

There is specific reference material available on rock reinforcement. For example, the Schlumberger Lecture Award paper by Windsor (1997)[8] summarizes the approach to rock reinforcement systems and highlights the components of the structural analysis, remembering that the main function of rock reinforcement is to enhance the strength of the discontinuous rock mass. Windsor provides a complete methodology for studying rock reinforcement, starting with the elements in the table below and including the development of safety factors.

Rock joint system components (in parallel)	*Reinforcement system components (in series)*
1. Areas of intact rock	1. Rock
2. Areas of asperities	2. Internal fixture
3. Areas of infill	3. Element
4. Areas of planar friction	4. External fixture

A publication by Villaescusa et al. (1999)[9] contains papers by Brown on the evolution of support and reinforcement philosophy, and by Hoek on support for very weak rock associated with faults and shear zones.

Our main emphasis in this book is on rock mechanics applied to civil and mining engineering, but the mechanical principles apply to all rock engineering. In petroleum engineering, the occurrence of unstable boreholes is a major problem — not only in inclined and sub-horizontal wells but also in vertical wells. The mechanical understanding of borehole in-

[7] Hoek E., Bawden W. F. and Kaiser P. K. (1995) *Support of Underground Excavations in Hard Rock*. Balkema, Rotterdam, 215pp.

[8] Windsor C. R. (1997) Rock Reinforcement Systems. *Int. J. Rock Mech. Min. Sci.* **34**, 6, 919–951.

[9] Villaescusa E., Windsor C. R. and Thompson A. G. (eds) (1999) *Rock Support and Reinforcement Practice in Mining*. Proceedings of the International Symposium on Ground Support, Kalgoorlie, Western Australia. Balkema, Rotterdam, 437pp.

Figure 16.3 (a) Isochromatics in photoelastic model indicating shear stresses around a square excavation, (b) using a slotting technique to move the shear stresses away from the excavation, thus reducing rock damage at the periphery of the excavation (from Roberts, 1977).

stability and a discussion of methods of reducing instability is included in Fjaer et al. (1992).[10]

The most elegant stabilization option is, if possible, to develop an excavation geometry such that no stabilization is necessary. In terms of rock fractures, an excavation can be oriented such that minimal rock block instability occurs, and, in terms of stress concentration (the amount by which an applied stress component is concentrated by the excavation), tunnels oriented parallel to the maximal principal stress experience the least stress concentrations on their boundaries. In an isotropic stress field, a circular tunnel cross-section reduces the maximal stress concentration to a factor of 2. In an anisotropic stress field, with the horizontal to vertical stress ratio being k, the maximal stress concentration is minimized by having an elliptical cross-section in which the ratio of width to height is also k.

Deviatoric stresses (i.e. the difference between the major and minor principal stresses) produce shear stresses which in turn cause failure. When the excavation is made, the stress normal to the excavation surface is reduced to zero, creating high shear stresses in the rock adjacent to the opening. Another stabilization option considered in the early days of rock mechanics was to move these shear stresses further back into the rock using a rock slotting technique (Fig. 16.3, from Roberts, 1977)[11]. Although it is an elegant solution, practical problems outweigh its usefulness, except in special circumstances where the slots can be easily cut.

---

[10] Fjaer E., Holt R. M., Horsrud P., Raaen A. M. and Risnes R. (1992) *Petroleum Related Rock Mechanics*. Elsevier, Oxford, 338pp.

[11] Roberts A. (1977) *Geotechnology*. Pergamon Press, Oxford, 347pp.

In Chapter 15, we noted that the possibility of introducing new methods of excavation is unlikely. Similarly, we do not anticipate that new methods of stabilization will be introduced. Failure of the rock mass occurs because of the presence of increased rock stresses, pre-existing fractures, water pressure and water flow, and rock deterioration over time. It is unlikely that stabilization improvements will be made because the factors listed above cannot be removed, and the existing techniques of rockbolts, shotcrete, lining and grouting offer direct solutions to the adverse effects.

17 Foundation and slope instability mechanisms

17.1 Near-surface instability

Rock foundations can be on a small scale, such as that of the concrete pier illustrated in the Frontispiece, or on a large scale, such as those of the majestic Hoover Dam, illustrated in Fig. 17.1, which is 221 m high and located on the Colorado River in the USA.

Rock slopes can be created for a variety of purposes. In Fig. 17.2a, we illustrate one region of a large open-pit copper mine. The slope geometry comprises a stepped series of benches, formed as the rock is excavated during the mining process. The height of each bench is governed by blasting and mining machinery parameters. In Fig. 17.2b, there is a highway cutting with rock slopes on either side, and a terminal slope where the road enters a tunnel. The methods of studying **slope instability** in these two cases may well be the same, but the consequences of slope failure in the two cases will be different: a single block rolling onto the road may well cause public alarm in the civil engineering case, but have little consequence in the mining engineering case.

Also, the methods of analysing slope instability will be a function of the rock type. For the granite mass illustrated in Fig. 17.3a, failure will be caused by the presence of the fractures: rock blocks will slide and topple, but a **new failure plane** is most unlikely to be generated through the strong intact granite. For the chalk illustrated in Fig. 17.3b, failure may similarly be caused by the blocky nature of the chalk, but it may also occur by a new failure plane being generated through the chalk mass because the chalk is relatively weak.

Foundations and slopes near the ground surface are often characterized by complex geological conditions, because of complex fracturing, a long history of groundwater movement, and other types of weathering. Moreover, compared to an underground excavation in a rock mass, the rock mass near to the surface is loaded by a 'soft system', often the dead weight of the rock mass itself, and there is greater kinematic potential for movement than there is around an underground excavation. All these

Figure 17.1 Hoover Dam on the Arizona–Nevada border, USA, (top) and the abutment rock (lower).

Figure 17.2 Slope stability in mining and civil engineering. (a) Slope stability during open-pit mining. (b) Slope stability in a road cutting.

factors lead to greater potential for rock instability in surface excavations compared to underground excavations.

The main factors involved in the stability of rock foundations and slopes are intact rock quality, fracture geometry, fracture mechanical properties, rock mass properties, rock stresses, hydraulic conditions, slope orientation, slope dimensions, proximate engineering disturbances, support and maintenance, and the effects of construction. In this chapter, we will consider the basic mechanisms that can lead to instabilities beneath foundations and in slopes, and we will concentrate on

Figure 17.3 The slope stability and the associated analysis is a function of the rock type. (a) Granite slope in a building stone quarry. (b) Chalk slope in a road cutting.

instabilities induced by the pre-existing fractures. Instabilities through the intact material are best tackled using soil mechanics principles, for which the reader is referred to Bromhead (1992) [1].

17.2 Question and answers: foundation and slope instability mechanisms

Q17.1 The diagram below (Serrano and Olalla, 1998 [2]) shows six basic potential mechanisms by which failure can occur beneath a foundation on a fractured rock mass containing one fracture set, indicated by the fine shading. Failure can occur through the intact rock, indicated by the letter R, and/or along a fracture plane, indicated by the letter P. Describe the six basic mechanisms illustrated.

A17.1 The mechanisms are as follows.

Case a: failure occurs through rock mass by creation of new failure surface in intact rock.
Case b: as (a), but with rock wedge also formed by failure along fractures.
Case c: as (a), but with rock wedge formed by fractures below loading area.

[1] Bromhead E. N. (1992) *The Stability of Slopes*. Blackie Academic, London, 2nd edn.
[2] The diagram is from Serrano A. and Olalla C. (1998) Ultimate bearing capacity of an anisotropic discontinuous rock mass, Part I. Basic modes of failure. *Int. J. Rock Mech. Min. Sci.*, **35**, 3, 301–324.

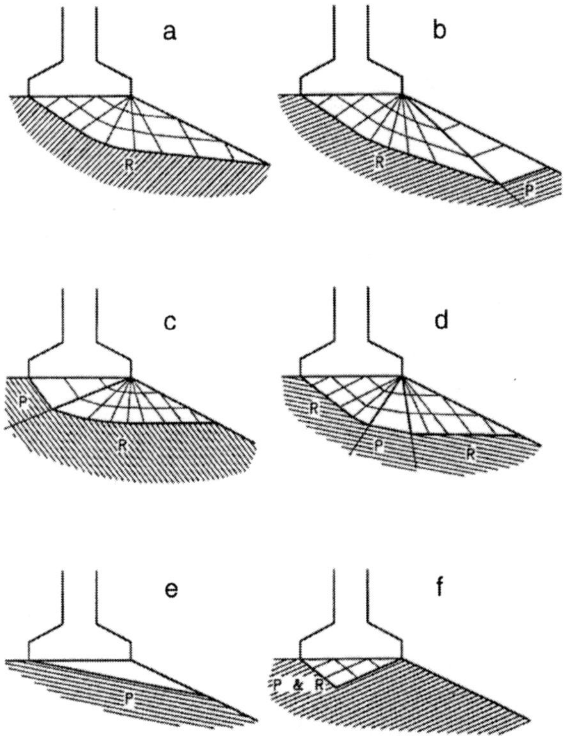

Case d: as (a) but with rock wedge between failure regions through intact rock.
Case e: plane sliding of single wedge formed by fracture.
Case f: failure in intact rock and along fracture.

These six mechanisms can be used to develop criteria for foundation failure, taking into account the possibilities of failure both through the intact rock and along the fracture planes.

This leads to more complex derivatives of the 'single plane of weakness' concept that was discussed in Chapter 7 for uniaxial loading of a single fracture in a rock specimen.

Q17.2 Consider extending the method of equilibrium analysis for plane failures (see Q16.3) to the case of retaining walls. The failing block, ABC in the sketch below, is restrained in three cases by the application of load to the vertical face, AB:

(a) by a horizontal force of magnitude *T* acting through the centroid of the block;

(b) by a horizontally acting uniform pressure distribution *p* from the top of the face to the point where the failure plane daylights; and

(c) by a pressure distribution varying linearly from zero at the top of the face to *q* at the point where the failure plane daylights.

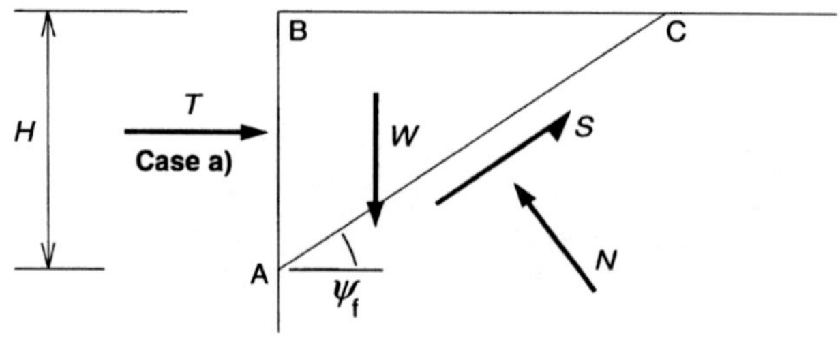

In each case, derive an expression for the factor of safety and, for the special case of $F = 1.0$, give an expression for T, p or q as appropriate.

A17.2 *Case (a) Horizontal force*

The forces T and W act through the centroid of the block, and the reaction forces S and N act at the midpoint of AC. However, to simplify the analysis we assume that they are all coincident with the centroid of the block, and hence we can ignore moment equilibrium.

For the geometry given, the volume of the block is

$$V = \tfrac{1}{2} L_{AB} L_{BC} = \tfrac{1}{2} H \frac{H}{\tan\psi_f} = \frac{H^2}{2\tan\psi_f}$$

and the weight of the block is

$$W = \gamma V = \frac{\gamma H^2}{2\tan\psi_f}.$$

Defining the factor of safety for this block as

$$\frac{\sum \text{forces resisting sliding}}{\sum \text{forces causing sliding}} = F,$$

and assuming that the strength of the plane AC is given by the Mohr–Coulomb criterion, by taking components of the various forces parallel and normal to AC we obtain

$$F = \frac{cL_{AC} + N\tan\phi}{W\sin\psi_f} = \frac{c\dfrac{H}{\sin\psi_f} + (W\cos\psi_f + T\sin\psi_f)\tan\phi}{W\sin\psi_f - T\cos\psi_f}.$$

There are many ways this relation can be rearranged, and one of them is

$$F = \frac{2cH + \left(\gamma H^2\cos^2\psi_f + 2T\sin^2\psi_f\right)\tan\phi}{\left(\gamma H^2 - 2T\right)\cos\psi_f\sin\psi_f}. \tag{17.1}$$

For the specific case of $c = 0$ and $T = 0$, this reduces to the relation for a block resting on a slope, i.e. $F = \tan\phi/\tan\psi_f$.

Finally, if we examine the case when $F = 1$, we can rearrange Eq. (17.1) to obtain

$$T = \frac{\gamma H^2(\sin 2\psi - (1+\cos 2\psi)\tan\phi) - 4cH}{2(\sin 2\psi + (1-\cos 2\psi)\tan\phi)} \tag{17.2}$$

and use this expression to investigate the requirements for the support force when $c = 0$. Rearranging gives

$$\frac{2T}{\gamma H^2} = \frac{(2\psi - (1 + \cos 2\psi)\tan\phi)}{(2\psi - (1 - \cos 2\psi)\tan\phi)}$$

and so we can plot contours of $2T/\gamma H^2$ for various angles of friction and dip of the failure plane. These are shown in the figure below.

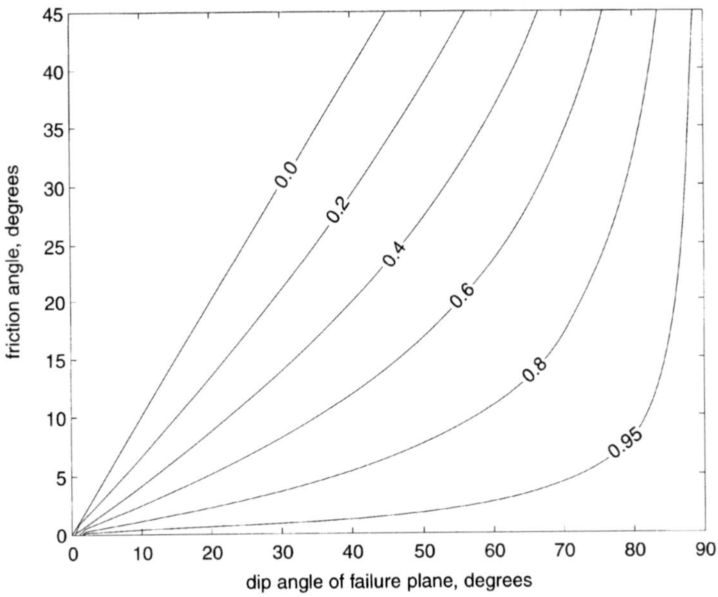

We can see from the plot that the required bolt force reduces to zero when the friction angle is greater than or equal to the dip angle of the failure plane, and it approaches the weight of the block for frictionless conditions.

Case (b) Uniform pressure

A uniform pressure is related to the horizontal force used in the above analysis through $T = pH$. Making this substitution in Eq. (17.1) leads to

$$F = \frac{2c + \left(\gamma H \cos^2\psi_f + 2p\sin^2\psi_f\right)\tan\phi}{(\gamma H - 2p)\cos\psi_f \sin\psi_f}.$$

Similarly, Eq. (17.2) becomes

$$p = \frac{\gamma H\,(\sin 2\psi - (1 + \cos 2\psi)\tan\phi) - 4c}{2\,(\sin 2\psi + (1 - \cos 2\psi)\tan\phi)}.$$

Case (c) Varying pressure distribution

A varying pressure distribution is related to the horizontal force T through $T = \frac{1}{2}qH$, and so we find that the factor of safety is given by

$$F = \frac{2c + \left(\gamma H \cos^2\psi_f + q\sin^2\psi_f\right)\tan\phi}{(\gamma H - q)\cos\psi_f \sin\psi_f}$$

and the pressure for $F = 1$ is given by

$$q = \frac{\gamma H\,(\sin 2\psi - (1 + \cos 2\psi)\tan\phi) - 4c}{(\sin 2\psi + (1 - \cos 2\psi)\tan\phi)}.$$

Q17.3 For the case of wedge instability in rock slopes, the factor of safety can be related to that of an equivalent plane instability (i.e. plane sliding in the same direction as that of the wedge) by

$$\boldsymbol{F_W = k_W \times F_P}$$

where the wedge factor, k_W, is computed from $k_W = \sin\beta/\sin\frac{1}{2}\xi$, and the angles β and ξ are defined as shown below.

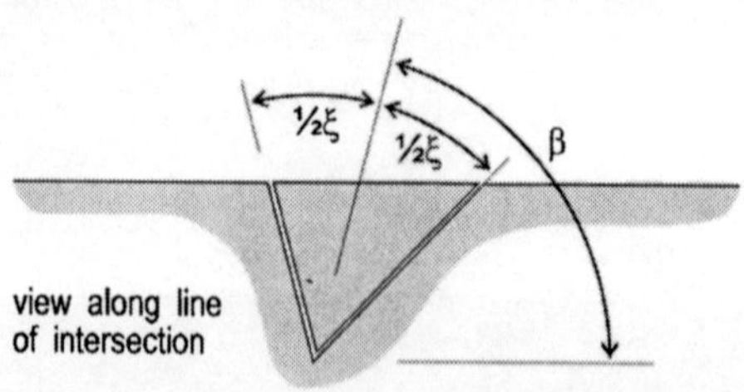

For the particular case of wedge instability in a slope of orientation 124/63 (dip direction/dip angle) with a horizontal top, intersected by two sets of fractures with orientations 182/52 and 046/69 and friction angle 29°, determine F_W.

A17.3 There are two main parts to the solution: to determine the factor of safety of the equivalent plane instability, and to determine the angles β and ξ.

Although the factor of safety for friction-only plane instability is a well-known result, it is easily confirmed by sketching a free body diagram of a block on a plane and noting the associated equilibrium conditions, as shown below.

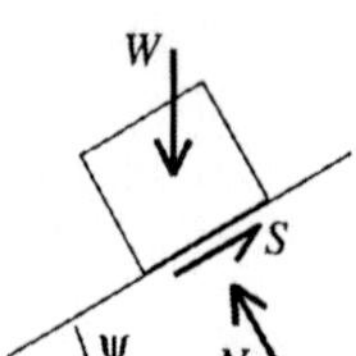

Resolving forces parallel and normal to the plane gives

$$S = W\sin\psi \text{ and } N = W\cos\psi.$$

For the equilibrium condition of $S = N\tan\phi$, we define the factor of safety as

$$F_P = \frac{N\tan\phi}{S}$$

from which we find

$$F_P = \frac{W\cos\psi\tan\phi}{W\sin\psi} = \frac{\tan\phi}{\tan\psi}.$$

This means that the equation for the factor of safety of the wedge instability can be written as

$$F_W = k_W \times F_P = \frac{\sin\beta}{\sin\frac{1}{2}\xi} \times \frac{\tan\phi}{\tan\psi}$$

from which we see that the three angles of β, ξ and ψ are required to describe the problem. These are established using the hemispherical projection, as shown below. In constructing this projection, it is important to realize that the angles β and ξ are measured in the plane that is perpendicular to the line of intersection that forms the wedge. The great circle representing this plane is shown dashed in the projection, and has as its normal the line of intersection of planes 1 and 2.

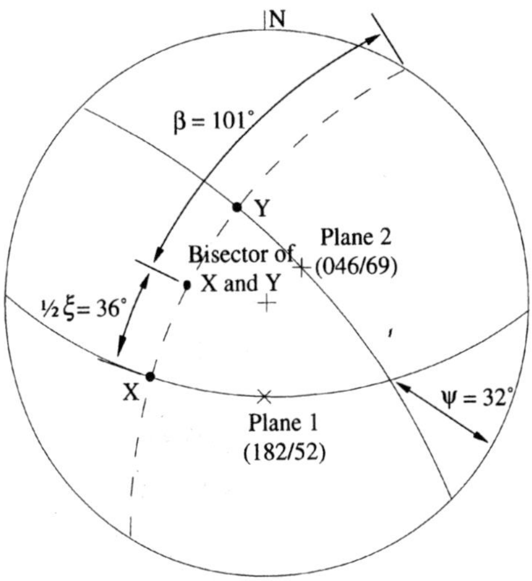

Substituting the various numerical values into the equation for the wedge factor of safety then gives

$$F_W = \frac{\sin\beta}{\sin\frac{1}{2}\xi} \times \frac{\tan\phi}{\tan\psi} = \frac{\sin 101}{\sin 36} \times \frac{\tan 29}{\tan 32} = \frac{0.982}{0.588} \times \frac{0.554}{0.625} = 1.48.$$

Notice that the magnitude of k_W is $0.982/0.588 = 1.67$, which shows that the confinement offered by the fractures forming the sides of the wedge contributes an additional 67% to the factor of safety, over and above that obtained from plane sliding. In this case, the plane case would be unstable.

The calculation presented here is approximate, because the various angles have been measured on the hemispherical projection. The equation for the factor of safety is, however, exact. To demonstrate this, if the various angles are computed using vector analysis, we find that $\beta = 100.9°$, $\xi = 72.0°$ and $\psi = 32.5°$, and from this we obtain $F_W = 1.454$. The answer computed using a wedge analysis algorithm, such as those presented by Hoek and Bray (1977), is 1.453.

Finally, we see that neither the orientation of the slope face nor the unit weight of the rock material enters into the calculation. This is because the stability of the wedge does not depend on its size, as it is a friction-only case. If either cohesion or water pressures are present, then the size of the wedge has to be taken into account. However, the slope face orientation should be such that a wedge is actually formed, and this geometric check should always be performed before computing the factor of safety.

Q17.4 Determine an upper bound for the collapse pressure, p, for the foundation shown below consisting of three rock wedges formed by the fracture sets in the rock mass.

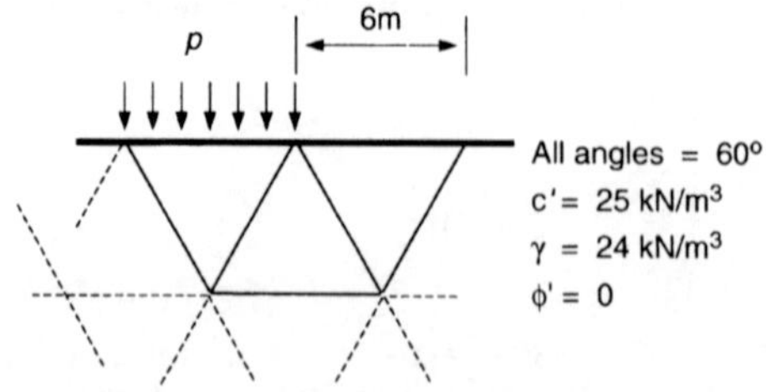

A17.4 We start by drawing free body diagrams of the blocks that comprise the foundation. It is important to remember that the forces drawn on a free body diagram are *those required to maintain the body in a state of equilibrium*. Also, when dealing with blocks that are in contact in the actual foundation, the forces on the common surface in the respective free body diagrams are in equilibrium, and so must be equal in magnitude and opposite in sense. When drawing the free body diagrams for a multi-block system such as this, it is usually easiest to start with the block furthest from the applied external load, as the direction of movement — and hence the sense of the inter-block forces — is generally evident. Using these principles, the free body diagrams for this particular foundation are then as shown below.

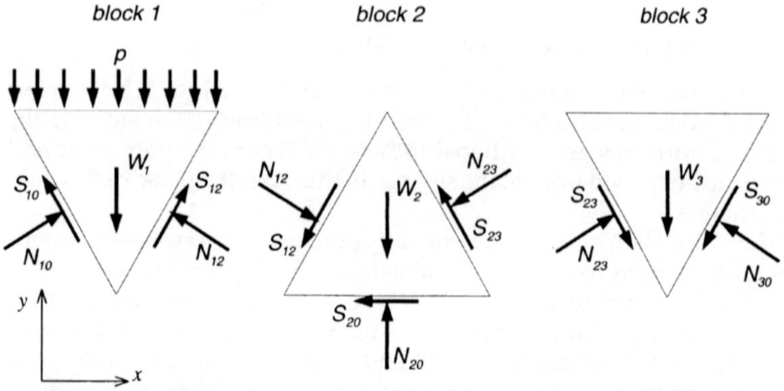

In these diagrams, each force that acts on the surface of a block is given two subscripts in order to identify fully the two blocks it acts between, with block 0 being the rock outside the foundation.

We assume each block to be in a state of limiting equilibrium, and that moment equilibrium can be ignored. In order to minimise the number of unknowns in the equilibrium equations, the solution begins with an analysis of block 3, moves on to block 2, and finishes at block 1.

Block 3

$\sum f_x = 0$, with forces directed to the right reckoned positive,

$$N_{23}\cos 30 + S_{23}\cos 60 - S_{30}\cos 60 - N_{30}\cos 30 = 0. \qquad (17.3)$$

Limiting equilibrium tells us that, for each face of the block, $S = cL + N\tan\phi$. For this particular foundation we know that $\phi = 0$, and so the limiting equilibrium condition is simply $S = cL$. Substituting this into Eq. (17.3) gives us

$$N_{23}\cos 30 + cL\cos 60 - cL\cos 60 - N_{30}\cos 30 = 0$$

and hence

$$N_{30} = N_{23}. \qquad (17.4)$$

$\sum f_y = 0$, with forces directed upwards reckoned positive,

$$N_{23}\sin 30 + N_{30}\sin 60 - S_{23}\sin 60 - S_{30}\sin 60 - W = 0.$$

Using the limiting condition of $S = cL$, this reduces to

$$N_{23}\sin 30 + N_{30}\sin 30 - cL\sin 60 - cL\sin 60 - W = 0$$

and substituting Eq. (17.4) into this gives

$$N_{23}\sin 30 + N_{23}\sin 30 - cL\sin 60 - cL\sin 60 - W = 0$$

which, on rearrangement and using the identities $\sin 30 = 1/2$ and $\sin 60 = \sqrt{3/2}$, leads to

$$N_{23} = W + cL\sqrt{3}. \qquad (17.5)$$

Block 2

$\sum f_x = 0$, with forces directed to the right being reckoned positive,

$$N_{12}\cos 30 - S_{23}\cos 60 - S_{12}\cos 60 - N_{23}\cos 30 - S_{20} = 0.$$

With the limiting condition of $S = cL$, this reduces to

$$N_{12}\cos 30 - cL\cos 60 - cL\cos 60 - N_{23}\cos 30 - cL = 0$$

and upon substitution of Eq. (17.5) and rearrangement we obtain

$$N_{12} = \frac{7}{\sqrt{3}}cL + W \qquad (17.6)$$

Block 1

$\sum f_x = 0$, with forces directed to the right reckoned positive,

$$N_{10}\cos 30 + S_{12}\cos 60 - S_{10}\cos 60 - N_{12}\cos 30 = 0.$$

With the limiting condition of $S = cL$ this reduces to

$$N_{10}\cos 30 + cL\cos 60 - cL\cos 60 - N_{12}\cos 30 = 0$$

and hence

$$N_{10} = N_{12}. \qquad (17.7)$$

$\sum f_y = 0$, with forces directed upwards reckoned positive,

$$N_{10}\sin 30 + S_{10}\sin 60 + N_{12}\sin 30 + S_{12}\sin 60 - W - pL = 0.$$

With the limiting condition of $S = cL$ this reduces to

$$N_{10} \sin 30 + cL \sin 60 + N_{12} \sin 30 + cL \sin 60 - W - pL = 0$$

and substituting Eq. (17.7) leads to

$$2N_{12} \sin 30 + 2cL \sin 60 - W - pL = 0.$$

Substituting Eq. (17.6) and rearranging results in

$$p = \frac{10}{\sqrt{3}}c = 5.774c. \tag{17.8}$$

For this foundation we have $c = 25\ \text{kN/m}^3$, and substituting this value into Eq. (17.8) gives a collapse pressure of

$$p = 144.3\ \text{kN/m}^2$$

and with $L = 6$ m we obtain a collapse load of

$$P = 866.0\ \text{kN/m}^2.$$

Notice that the collapse pressure (and hence collapse load) is independent of the weight of the blocks. This is because the friction angle is zero, but in general the collapse load increases as the unit weight of the rock increases.

If we perform the above calculations for a range of values for the friction angle and cohesion of the block boundaries, we can examine how the collapse pressure varies with these parameters. The graph below shows this, with the units of collapse pressure being MPa.

Notice that a logarithmic scale has been used for the curves of collapse pressure, indicating that its magnitude increases dramatically as both the cohesion and friction angle increase.

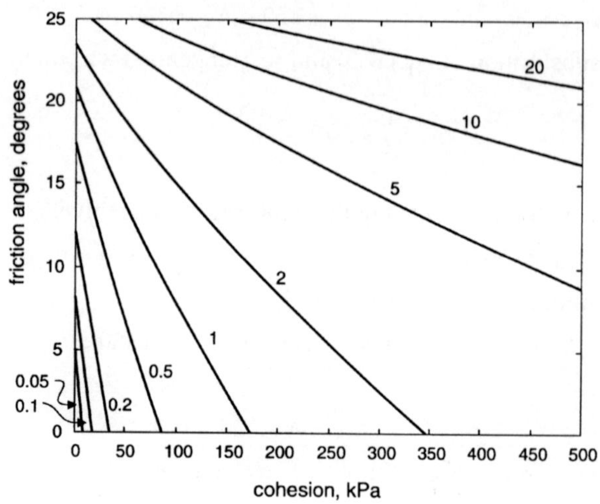

Q17.5 A cross-section through a rock slope is shown below. The rock is impermeable with a unit weight of 22 kN/m³ and the fractures AC and DE have an aperture of 1 mm. A drainage gallery at E reduces the water pressure at that point to the atmospheric pressure value. The drainage gallery, the fractures and the slope face all have the same strike.

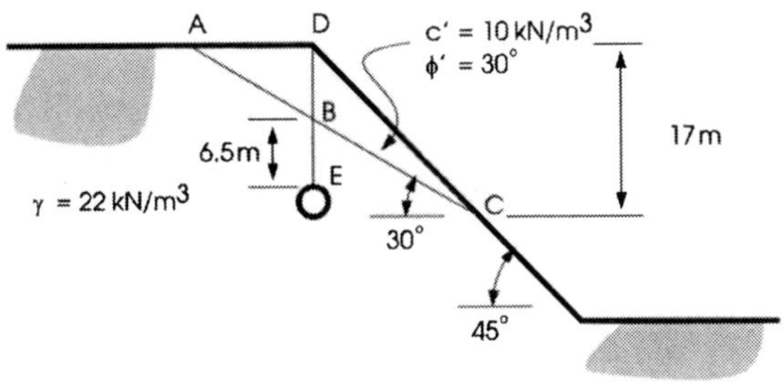

Assuming conditions of heavy rainfall, determine the water pressure at B, and hence the factor of safety against sliding of the block DBC. Take the kinematic viscosity of water to be 1.3 x 10^{-6} m²/s and assume block ABD is stable.

A17.5 We start by analysing the fracture network in order to determine the water pressure distribution along the fractures BD and BC. The first step is to determine the elevation of the nodes and the lengths of the channels from the network geometry. This is shown in the sketch below.

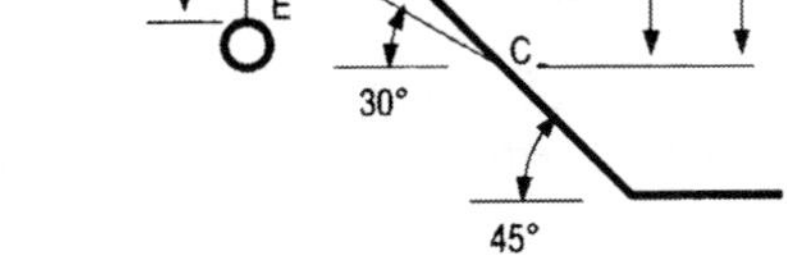

Node:	A	B	C	D	E
Elevation (m)	17.0	17.0 tan 30 = 9.815	0.0	17.0	9.815 − 6.5 = 3.315

Channel:	AB	BC	BD	BE
Length (m)	17.0/sin 30 − 17.0/cos 30 = 14.37	17.0/cos 30 = 19.63	17.0 − 17.0 tan 30 = 7.185	6.5

We then draw up a table of heads:

Node:	A	B	C	D	E
Pressure head, p/γ_w	0	p_B/γ_w	0	0	0
Elevation head, z	17.0	9.815	0	17.0	3.315
Total head, H	17.0	H_B	0	17.0	3.315

and a table of channel conductance:

Channel:	AB	BC	BD	BE
L (m)	14.37	19.63	7.185	6.5
c (ms^{-1}, $\times 10^{-6}$)	43.76	32.03	87.52	96.75

The fundamental equation for computing the head at a node is

$$H_i = \frac{\sum c_{ij} H_j}{\sum c_{ij}}$$

and for node B in the network this is

$$H_B = \frac{c_{BA} H_A + c_{BC} H_C + c_{BD} H_D + c_{BE} H_E}{c_{BA} + c_{BC} + c_{BD} + c_{BE}},$$

which is evaluated as

$$H_B = \frac{(43.76 \times 17.0) + (32.03 \times 0.0) + (87.52 \times 17.0) + (96.75 \times 3.315)}{43.76 + 32.03 + 87.52 + 96.75}$$

$$= 9.815 \text{ m}.$$

The pressure at node B is then

$$p_B = \gamma_w (H_B - z_B) = \gamma_w (9.815 - 9.815) = 0,$$

and so we can see that the drain works perfectly: the pressure is reduced to zero at node B, and as a result the water pressure is zero everywhere along BD and BC.

Examining the stability of block DBC, as block ABD is stable it will play no part in the analysis. If we define a factor of safety as

$$F = \frac{\sum \text{resisting forces}}{\sum \text{driving forces}},$$

then we have for block DBC

$$F = \frac{l_{BC} \cdot c' + (W_{DBC} \cos 30) \tan \phi'}{W_{DBC} \sin 30}$$

where W_{DBC} is the weight of block DBC. To compute this we determine the area of DBC as follows:

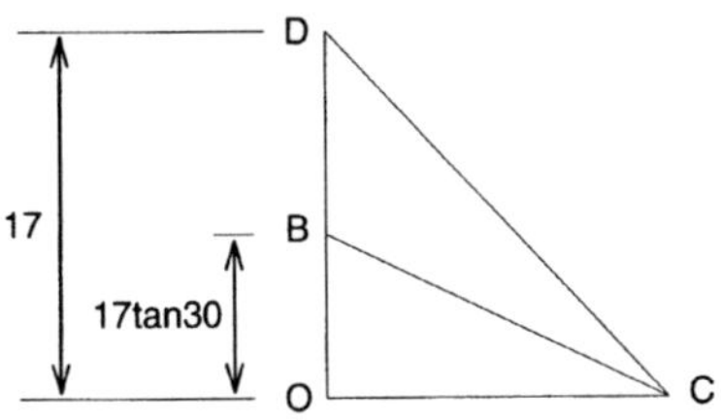

$$A_{DBC} = A_{DOC} - A_{BOC}$$
$$= \tfrac{1}{2} \times 17 \times 17 - \tfrac{1}{2} \times 17 \times 17 \tan 30$$
$$= 61.07\ \text{m}^2$$

and thus $W_{DBC} = 22.0 \times 61.07 = 1343.6$ kN.

The factor of safety is then

$$F = \frac{\dfrac{17.0}{\cos 30} \times 10 + (1343.6 \cos 30) \tan 30}{1343.6 \sin 30} = 1.29.$$

Q17.6 Assume the same circumstances in Q17.5 and consider the case when the block DBC has moved downhill a limited amount, resulting in a considerable increase in the aperture of DB. Calculate the factor of safety against sliding of the block DBC.

A17.6 In the circumstance that the aperture of DB is much larger than 1 mm, far more water will enter at D than can be drained away through BE, and as a result the pressure at B will increase towards a hydrostatic value. If we assume that there will also be a linear pressure variation along BC, from hydrostatic at B to zero at C, then we can compute the new factor of safety as

$$F = \frac{l_{BC} \cdot c' + (W_{DBC} \cos 30 - U - V \sin 30) \tan \phi'}{V \cos 30 + W_{DBC} \sin 30},$$

where U and V are the forces due to the water pressure along BC and BD, respectively. These forces are determined thus:

$$p_B = \gamma_w \times l_{BD} = 9.81 \times 7.185 = 70.49\ \text{kN/m}^2;$$
$$V = \tfrac{1}{2}\gamma_w l_{BD}^2 = \tfrac{1}{2} p_B l_{BD} = 253.2\ \text{kN};$$
$$U = \tfrac{1}{2} p_B l_{BC} = 691.8\ \text{kN}.$$

Hence, the factor of safety is computed as

$$F = \frac{(17/\cos 30) \times 10 + (1343.6 \cos 30 - 691.8 - 253.2 \sin 30) \tan 30}{253.2 \cos 30 + 1343.6 \sin 30} = 0.44.$$

Clearly, increasing the aperture of DB has a significant effect on the stability of the block, as demonstrated by performing the computation for a range of aperture values and plotting the results:

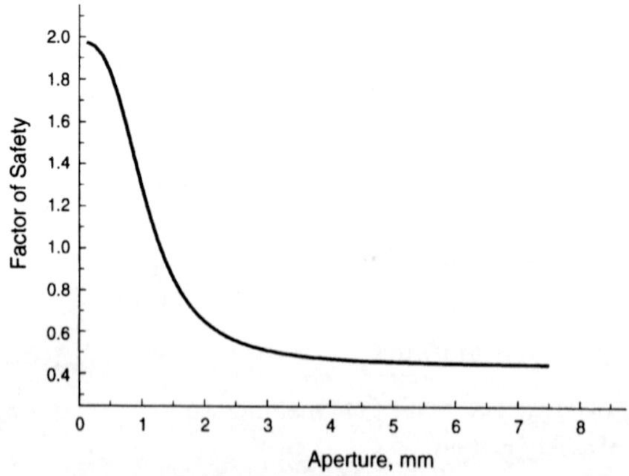

The engineering significance of this result is that, unless we have total confidence in our assessment of the aperture of BD, we should assume that it is much greater than 1 mm, and accept that the slope as currently configured has a factor of safety of about 0.44.

Q17.7 A rectangular foundation base, 6 m x 10 m in plan, carries a vertical load of 180 MN which it transmits to the rock as a uniform bearing pressure. The rock can be taken to be CHILE, with a modulus of elasticity $E = 800$ MPa and Poisson's ratio $\nu = 0.2$.

Ignoring the weight of the rock, evaluate the vertical settlement at the centre, C, of the base and the vertical stress at a point 5 m below C.

A17.7 The loaded area we are analysing is shown to the right.

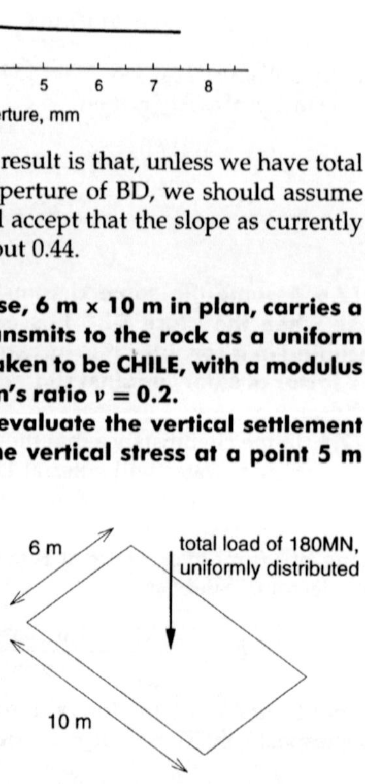

The solution to problems involving loaded areas on CHILE foundations can be found from application of the integral form of the Boussinesq equations. However, these equations are defined over one quadrant, with the point of interest being at some depth on the z-axis. As a result, we can only apply the equations over one quarter of the loaded area, as follows:

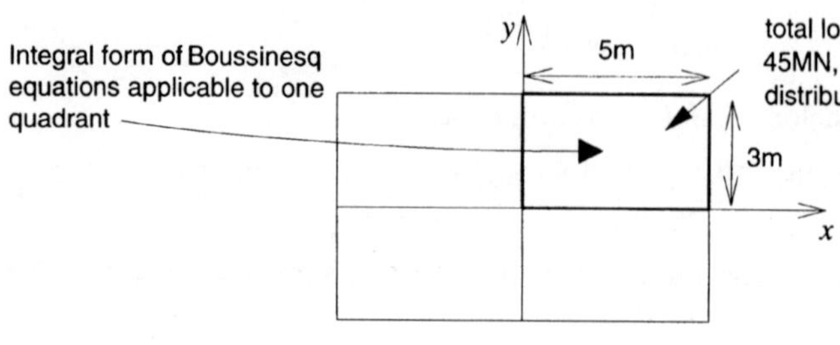

To obtain the stresses and displacements due to the complete loaded area, we multiply the results for a single quadrant by 4. This is possible because each quadrant of the loaded area has no structural influence on any of the others.

For the single quadrant shown above, the displacement at a point on the z-axis is given by

$$u_z = \frac{p}{4\pi G}\left[I_u(5,3) + I_u(0,0) - I_u(5,0) - I_u(0,3)\right] \tag{17.9}$$

where

$$I_u(x,y) = 2(1-\nu)\left[x\ln(R+y) + y\ln(R+x)\right] - (1-2\nu)z\tan^{-1}\left(\frac{xy}{Rz}\right)$$

and

$$R = \sqrt{x^2+y^2+z^2} \tag{17.10}$$

For a position on the ground surface, we have $z = 0$, and for this particular foundation we have $E = 800$ MPa and $\nu = 0.2$ (giving $G = 333.3$ MPa), with the applied pressure p being given by

$$p = \frac{P}{A} = \frac{180}{10\times 6} = 3 \text{ MPa}.$$

Substituting these values into Eq. (17.9) we obtain

$$\begin{aligned} u_z &= \frac{p}{4\pi G}[28.86 + 0.00 - 12.88 + 5.27] \\ &= \frac{3}{4\pi \times 333.33} \times 10.71 = 7.672 \times 10^{-3} \text{ m} \end{aligned}$$

with the total displacement then being $4u_z = 4 \times 7.672 \times 10^{-3} = 0.031$ m.

The stress at a point on the z-axis for the single quadrant shown above is given by

$$u_z = \frac{p}{4\pi}[I_\sigma(5,3) + I_\sigma(0,0) - I_\sigma(5,0) - I_\sigma(0,3)] \tag{17.11}$$

where

$$I_\sigma(x,y) = \tan^{-1}\left(\frac{xy}{Rz}\right) + \frac{xyz}{R}\left(\frac{1}{x^2+z^2} + \frac{1}{y^2+z^2}\right)$$

and R is defined in Eq. (17.10).

For a depth of 5 m we have $z = -5$, and substituting this, together with the appropriate values of x, y, and p into Eq. (17.11), we obtain

$$\begin{aligned} \sigma_z &= \frac{p}{2\pi}[0.85 + 0.00 - 0.00 + 0.00] \\ &= \frac{3}{2\pi} \times 0.85 = 0.41 \text{ MPa} \end{aligned}$$

with the total stress then being $4\sigma_z = 4 \times 0.41 = 1.63$ MPa.

We can examine the distribution of both displacement and stress across the extent of the loaded area by substituting the appropriate values of x and y into Eq. (17.9) or Eq. (17.11), respectively. The asymmetry of the problem under these circumstances means that we need to compute the influence of each of the four rectangular areas explicitly, rather than computing a single quadrant and multiplying the result by 4. Plots of these distributions are shown below.

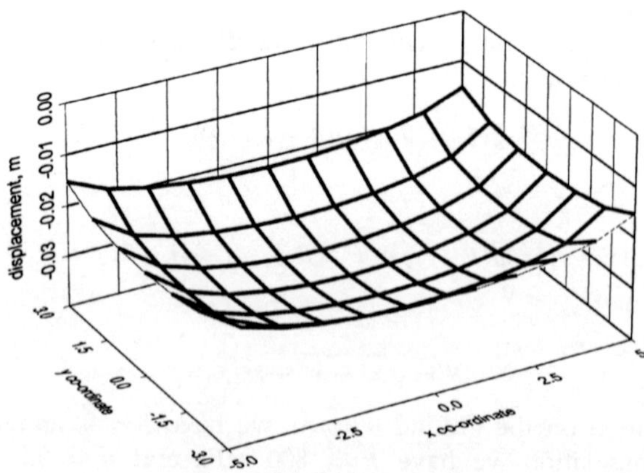

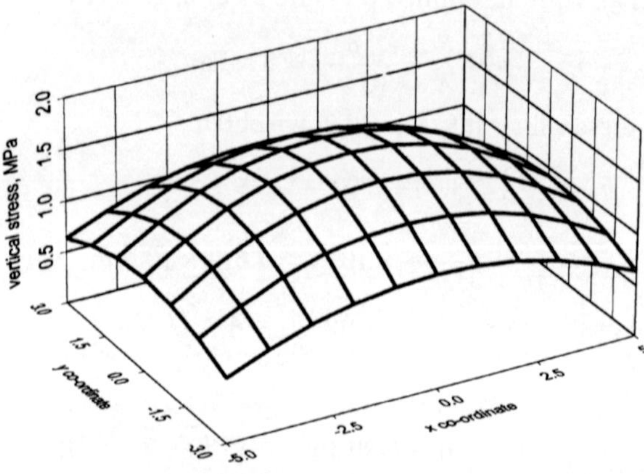

Q17.8 The expression for the radial stress induced beneath a line load of magnitude *P* acting normal to the surface of a semi-infinite half space is

$$\sigma_r = \frac{2P\sin\theta}{\pi r}.$$

(a) Show on a sketch the geometry of this problem, indicating clearly the angle θ and the orientation of the stress components σ_θ and σ_r. What are the magnitudes of σ_θ and $\tau_{r\theta}$ in this problem?

(b) Sketch the locus of constant σ_r for varying θ, and show how the expression above leads to the concept of a 'bulb of pressure'. For the case when the applied line load is not normal to the surface, show on a sketch how the bulb of pressure changes its nature, indicating clearly any zones where tensile radial stresses may be induced.

(c) This analysis is for CHILE materials. By making use of the ϕ_j theory, show qualitatively how the bulb of pressure may be expected to change when the ground beneath the load consists of a layered discontinuous rock mass dipping at some arbitrary angle.

(d) On the basis of these results, what are the ramifications this analysis has for site investigation in layered rocks?

A17.8 (a) This is a problem in which the stress components are defined in polar co-ordinates, which are expressed in terms of a radius r and an angle θ. If we examine the expression for radial stress,

$$\sigma_r = \frac{2P\sin\theta}{\pi r},$$

we see that it is a maximum when θ is 90°. This maximum must also occur when the direction of action of the radial stress is parallel to the applied load, and so we conclude that the geometry of the problem is as shown below.

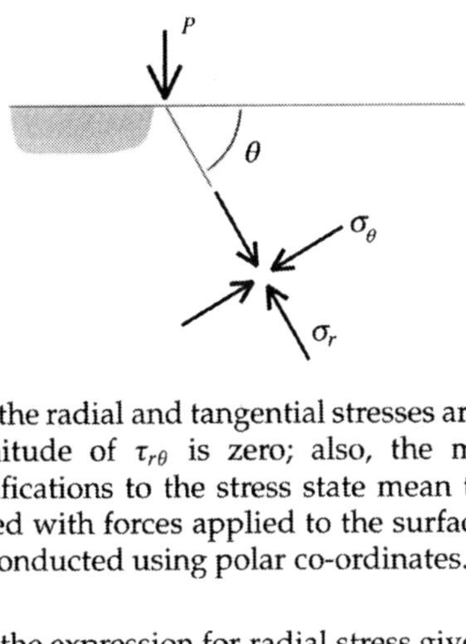

In this problem, the radial and tangential stresses are principal stresses, and so the magnitude of $\tau_{r\theta}$ is zero; also, the magnitude of σ_θ is zero. These simplifications to the stress state mean that the solution to problems associated with forces applied to the surface of a semi-infinite body are usually conducted using polar co-ordinates.

(b) Rearranging the expression for radial stress gives

$$r = \frac{2P}{\pi\sigma_r}\sin\theta,$$

and so, knowing that P and σ_r are constant, we can find radius values for given angles. If we assume that $2P/\pi\sigma_r = 1$, the following results are obtained.

θ	0	15	30	45	60	75	90
r	0.000	0.259	0.500	0.707	0.866	0.966	1.000

These are plotted in the diagram below, and show how a cylindrical contour of constant radial stress develops (remembering that a line load is being applied). It is the cylindrical form of this contour that gives rise to the term 'bulb of pressure'.

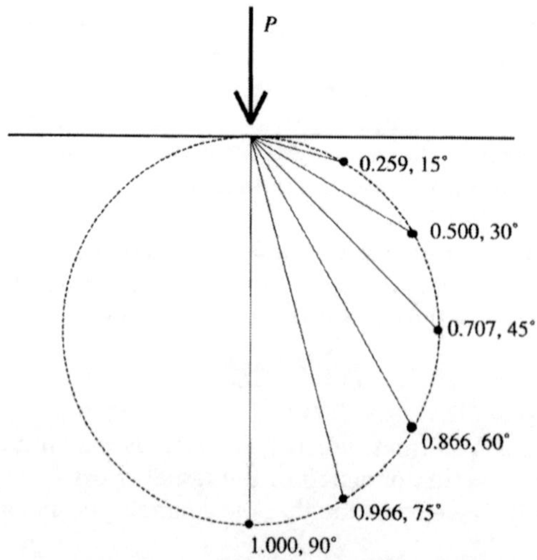

The maximal radius for any given contour is always co-linear with the applied force. This means that for the case where the applied forces are not normal to the surface, the contours rotate as shown below.

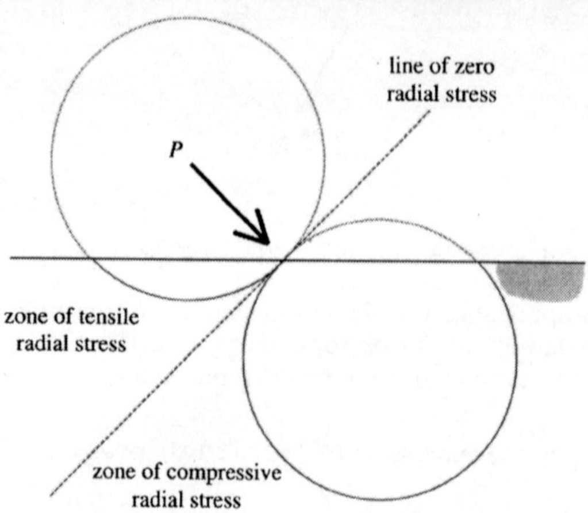

(c) The ϕ_j theory utilizes the fact that slip occurs on a fracture if the applied stress subtends an angle greater than the friction angle to the fracture normal (see A16.6). If we superimpose fracture orientations on a diagram of radial stress contours, then we can identify where the radial stress is acting at such an angle that slip is likely to occur. Once slip does occur, then the stress field is deformed so that it extends deeper into the rock mass. This is shown diagrammatically below following Goodman (1989)[3].

[3] Goodman R. E. (1989) *Introduction to Rock Mechanics*. Wiley, Chichester, 2nd edn., 562pp.

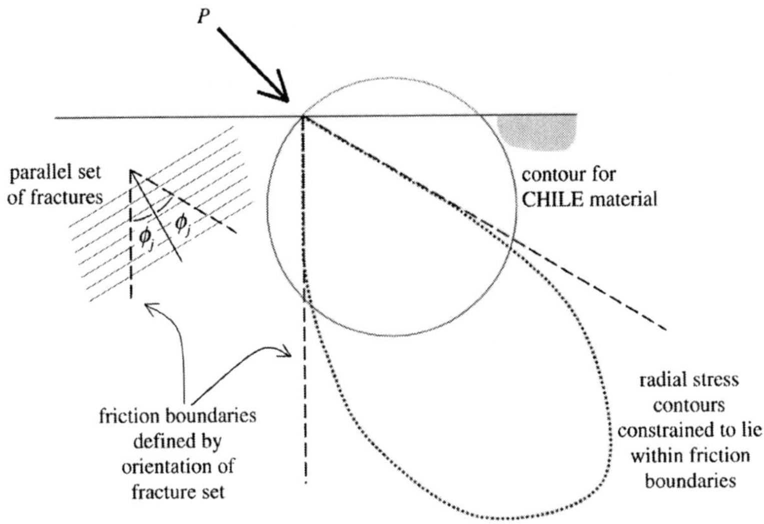

(d) The diagram above illustrates that whenever engineering is taking place in DIANE rocks, we should expect the geometry of the induced stress fields to be different from those predicted using CHILE models. In the case of site investigation in layered rocks, the depth to which a given stress is induced can be much greater than that predicted using a CHILE model, indicating that the rock mass should be investigated over a wider and deeper zone than a CHILE model would suggest.

Q17.9 The system of rock blocks shown in the sketch below is to be used in the verification procedure of a computer code for analysing progressive failure of rock slopes, and for this a manually derived solution is required.

The system is in limiting equilibrium with block A tending to topple

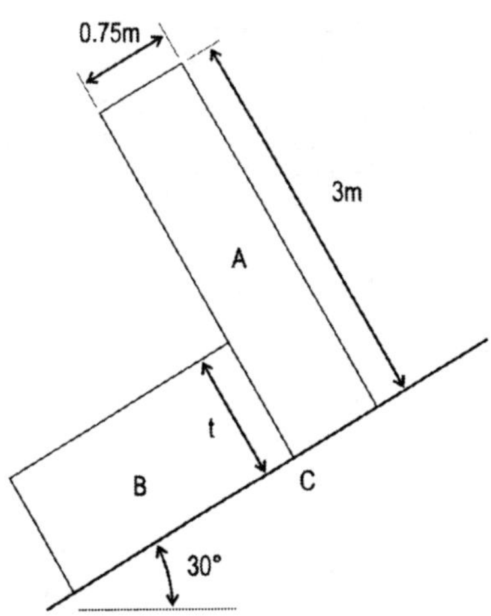

about the corner C, while block B is on the point of sliding downhill. The shear resistance on all surfaces is purely frictional with $\phi = 35°$.

Given that B is twice as heavy as A, determine the thickness *t* of block B. Also show that there is no tendency for block A to slip at the corner C.

A17.9 In order to solve problems of this type we start by drawing a free body diagram, marking on all of the forces acting on the various elements. In this case we know that, if the force due to the weight of block A is W, then the force due to the weight of block B is $2W$. Both of these forces act through the centroid of the respective blocks. There are also shear and normal forces acting on all of the interfaces. This information leads to the diagram shown below.

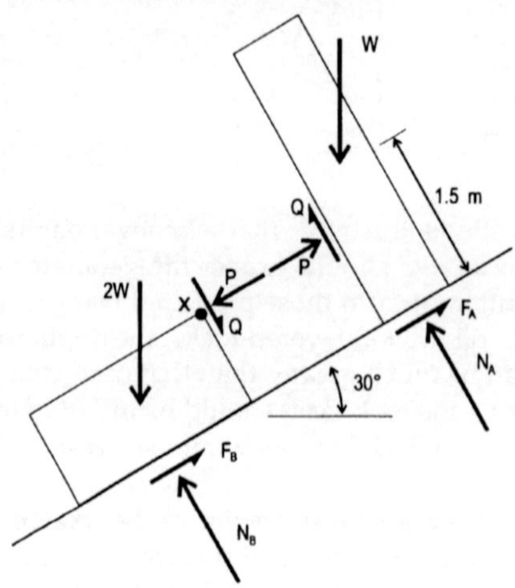

As the system is in a state of limiting equilibrium with block B on the point of sliding, we can write $F_B = N_B \tan 35$. Furthermore, if block B moves down the slope but block A does not slide, then sliding must occur at point X and block A must rotate about point C. This information gives us both a relation between P and Q and the position at which they act: $Q = P \tan 35$, acting at a distance t above the plane.

Because block A is toppling, we can write down the equation of moment equilibrium for this block by taking moments (anticlockwise positive) about point C:

$$W \sin 30 \times 1.5 - W \cos 30 \times \frac{0.75}{2} - Pt = 0.$$

Rearranging this gives

$$P = \frac{1}{t} W \,(1.5 \sin 30 - 0.375 \cos 30)\,. \qquad (17.12)$$

We know that block B is sliding and hence $F_B = N_B \tan 35$. To determine the forces in this relation, we write down the force equilibrium equations (parallel and normal to the plane) for block B:

$$F_B - 2W \sin 30 - P = 0 \text{ and } N_B - W \cos 30 - Q = 0$$

from which we obtain

$$(P + 2W \sin 30) = (Q + 2W \cos 30) \tan 35.$$

P and Q are related through $Q = P \tan 35$ because sliding is taking place at point X. This simplifies the above equation to

$$(P + 2W \sin 30) = (P \tan 35 + 2W \cos 30) \tan 35$$

from which we find

$$P = 2W \frac{\cos 30 \tan 35 - \sin 30}{1 - \tan^2 35}. \tag{17.13}$$

This can now be equated to Eq. (17.12) to give

$$2W \frac{\cos 30 \tan 35 - \sin 30}{1 - \tan^2 35} = \frac{1}{t} W \,(1.5 \sin 30 - 0.375 \cos 30)\,.$$

Rearranging this gives the value of t:

$$t = \tfrac{1}{2} \frac{(1.5 \sin 30 - 0.375 \cos 30)\left(1 - \tan^2 35\right)}{\cos 30 \tan 35 - \sin 30} = 1.019 \text{ m}.$$

Finally, in order to confirm that block A is not sliding, we must examine the relation between the forces acting at point C. The force equilibrium equations (parallel and normal to the plane) for block A are

$$F_A + P - W \sin 30 = 0 \text{ and } N_A + Q - W \cos 30 = 0$$

and hence

$$F_A = W \sin 30 - P \text{ and } N_A = W \cos 30 - Q.$$

These give us a relation for the angle of friction required at A to prevent slip:

$$\tan \phi = \frac{F_A}{N_A} = \frac{W \sin 30 - P}{W \cos 30 - Q}.$$

As before, because sliding is occurring at point X, this can be simplified to

$$\tan \phi = \frac{W \sin 30 - P}{W \cos 30 - P \tan 35}.$$

Substituting the value of P found in Eq. (17.13) and cancelling throughout to remove W gives the required friction angle as

$$\phi = \tan^{-1}(0.144) = 8.2^\circ$$

As this value is much lower than the actual friction angle of 35°, we can see that sliding will not occur.

***Q17.10* Solve Q17.5 numerically using the sector method, and comment on any discrepancies between the numerical solution and the analytical solution.**

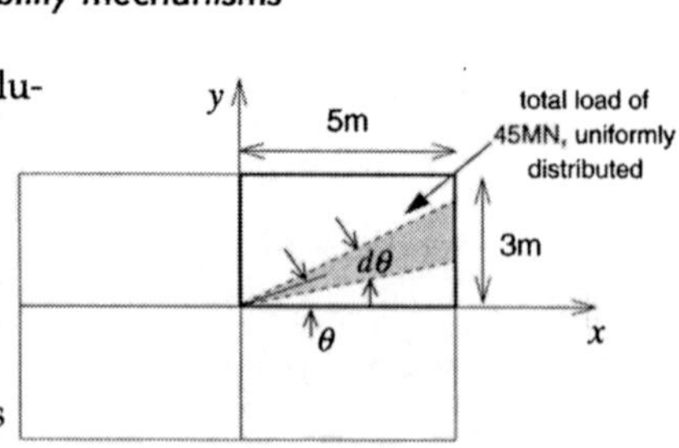

A17.10 As with the analytical solution, we are able to make use of the symmetry of the problem, and so will only analyse one quarter of the area, with the final stresses and displacements being found by multiplying the computed results by 4. A typical sector within the quadrant under analysis is illustrated above.

For a sector such as that shown above, the equations for displacement at the surface and vertical stress at a depth z are

$$u_z = p\frac{1-\nu}{2\pi G}\sum_{i=1}^{n} r_i \mathrm{d}\theta \tag{17.14}$$

and

$$\sigma_z = \frac{p}{2\pi}\sum_{i=1}^{n}\left[1-\frac{z^3}{\left(z^2+r_i^2\right)^{3/2}}\right]\mathrm{d}\theta \tag{17.15}$$

where r is the sector radius and $\mathrm{d}\theta$ is the sector included angle.

To use Eqs. (17.14) and (17.15), we decide on the number of sectors n, compute the resulting value of $\mathrm{d}\theta$, and then compute the radius of each sector before substituting into the equations. Computation of each radius can be awkward, and may involve considerable geometry. In the case shown here, we will also need to decide which of the two edges of the loaded area is appropriate for a given sector. However, as the method is most likely to be implemented on a computer, it is more convenient to calculate the radius for each sector relative to the lines representing both edges, and then take the minimum of the two:

$$r = \min\left(\text{radius relative to side, radius relative to top}\right). \tag{17.16}$$

If we examine the geometry of the area, we find that for all sectors i, Eq. (17.16) can be written as

$$r_i = \min\left(\frac{a}{\cos\theta_i}, \frac{b}{\sin\theta_i}\right)$$

where a and b are the half width and half height of the area (i.e. 5 m and 3 m for this particular example), respectively.

For the case of 5 sectors in the quarter area, the computation is as follows:

i	θ_i (°)	r_i (m)	$r_i\mathrm{d}\theta$ (m)	$1-\frac{z^3}{\left(z^2+r_i^2\right)^{3/2}}$
1	9	5.062	1.590	0.205
2	27	5.612	1.763	0.222
3	45	4.243	1.333	0.175
4	63	3.367	1.058	0.135
5	81	3.037	0.954	0.118
Total			6.698	0.855

and from these we find that

$$u_z = 4 \times \left[p \frac{1-\nu}{2\pi G} \times 6.698 \right] = 0.0307 \text{ m}$$

and

$$\sigma_z = 4 \times \left[\frac{p}{2\pi} \times 0.855 \right] = 1.632 \text{ MPa.}$$

The analytical results for these values were 0.0307 m and 1.633 MPa; clearly, in this case the sector method converges to the analytical results with only a small number of sectors. This convergence is demonstrated in the following graph, and shows how, once the number of sectors exceeds about 8, the errors are very small indeed.

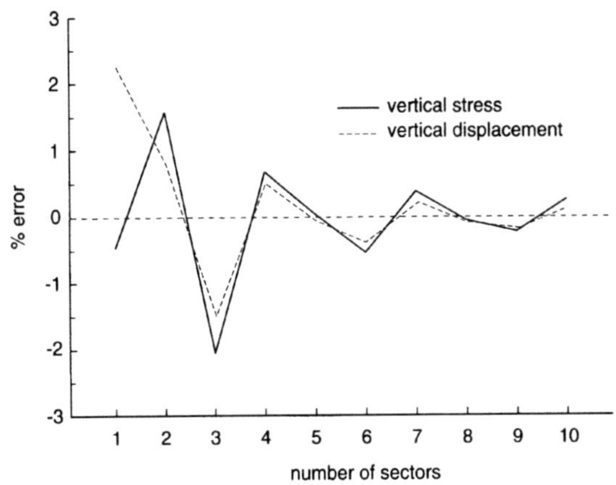

17.3 Additional points

An International Workshop on Rock Foundations was held in association with the ISRM 8th Congress in Tokyo in 1995, and the 70 papers are published in Yoshinaka and Kikuchi (1995)[4]. In their preface, the editors note that Terzaghi's theory for the bearing capacity of soil was systemized in the early 1940s, but related work on rock foundations did not develop then because rock was assumed to be a sound material compared to soil. However, the Malpasset Dam failure in 1959, which was caused by the failure of foundation rock, stimulated the necessary work.

Yoshinaka and Kikuchi also note that our modern infrastructure includes "high dams constructed on complex formations, highway bridges built on slopes consisting of weathered and/or fissured rocks, high-rise pylons for electric transmission lines, strait crossing long-span bridges, and nuclear power stations" and all of these could be subjected to earthquake or other types of loading. Moreover, the processes of foundation

[4] Yoshinaka R. and Kikuchi K. (eds) (1995) *Rock Foundation*. Proceedings of the International Workshop on Rock Foundation, Tokyo, Balkema, Rotterdam, 457pp.

and superstructure construction are intimately connected and it is not easy to change the completed work nor implement remedial procedures. This makes the final design of foundations more important than other rock engineering structures, such as tunnels and underground caverns, where a 'design as you go', or observational, approach is often adopted.

The seminal book[5] on rock slope engineering is Hoek and Bray (1977). The subjects covered in that book are economic and planning considerations, basic mechanics of slope failure, graphical presentation of geological data, geological data collection, shear strength of rock, groundwater, plane failure, wedge failure, circular failure, toppling failure, blasting and extra topics.

[5] Hoek E. and Bray J. W. (1977) *Rock Slope Engineering*. Institution of Mining and Metallurgy, London, 402pp.

18 Design of surface excavations

18.1 The project objective

We have emphasized that the rock engineer must establish the **engineering objective**, consider the relevant rock mechanics mechanisms, obtain the necessary site investigation information, conduct appropriate modelling in order to develop a predictive capability, and then decide on the design. In the following photographs, we highlight the point that the rock engineer can be faced with **different objectives** and hence different tasks.

In the previous chapter, in Fig. 17.1 we illustrated the rock on the downstream side of the right abutment of the finished Hoover Dam. In Fig. 18.1, the overall process of dam construction is shown as the Alto Lindoso Dam was being constructed. The rock engineering problems

Figure 18.1 Construction of the Alto Lindoso Dam on the Portugal–Spain border.

Figure 18.2 Rock slope on the A82 road by Loch Lomond in Scotland.

here are foundation loading at the base, and abutment loading at the sides. The other main issue is the secondary permeability of the rock mass, given the final water level in the completed dam: we have to be able to assess the potential leakage paths under and around the dam.

Fig. 18.2 shows a finished rock slope adjacent to a road in Scotland, which was designed after a concerted study of the rock mass fractures. Here the considerations involved the long-term stability and any required maintenance of the rock mass, both in terms of its overall stability and the possibility of avoiding rockfalls on to the road.

Sometimes it is not possible to engineer the slopes to avoid slope instability because of the limitations of the terrain. Fig. 18.3 shows a portion of Highway 22, the main highway between India and Tibet. Here, the road winds its way up the valleys, hugging the valley sides, and so there is almost no scope for any design variations. It is also not practical in terms of manpower resources and materials to provide support measures or rockfall protection along the whole highway.

In mining, there is much greater scope for design variations, especially in large quarries and open-pit mining of large orebodies, because the whole mine geometry can be arranged to optimize the extraction conditions. Also, mining slopes only have to be stable in so far as any instabilities do not inhibit the mining operations — which means that some forms of minor instabilities are acceptable. An example of the large extent of slopes present in some mining operations is illustrated in Fig. 18.4, which shows the Kalgoorlie 'Superpit' in Western Australia.

Although mining operations provide good opportunities for obtaining information about the rock mechanics properties, they are often located in faulted or contorted geology — almost by definition in the case of orebodies because the desired metal has been concentrated by geological processes. In the case of the Kalgoorlie area, faults are present with

Figure 18.3 Unstable rock masses along Highway 22 from India to Tibet.

the maximum principal stress being horizontal and parallel to the main fault. In Fig. 18.5, one of the china clay quarries in Cornwall, UK, is shown. The action of carbonic acid has caused the feldspathic minerals in the original granitic rock mass to alter to the clay kaolin, which is being mined. However, the decomposition process was not uniform, leading to some areas of residual granite (possessing residual jointing) and some areas of clay. Needless to say, the prediction of slope stability in such an inhomogeneous rock/soil mass is much more difficult than in the hard rock mass illustrated in Fig. 18.2.

The engineering objective could also be to preserve old rock structures such as the castle in Spain illustrated in Fig. 18.6. The edge of a cliff is a good military location for a castle wall, but not such a good location from the rock engineering point of view. The rock face below the castle was shotcreted in recent times, and the shotcrete has since become separated from the rock and is now in a parlous state. Or, the rock engineer might be asked to evaluate the stability of natural slopes such as that illustrated in Fig. 18.7, which was the theme of the rock mass classification question in Q14.9.

Figure 18.4 Rock slopes at the Kalgoorlie gold mining 'super pit' in Western Australia.

Figure 18.5 China clay mining in Cornwall, UK.

18.2 Questions and answers: design of surface excavations

Given the range of possibilities described in the previous section, we cannot cover all the techniques used in the design of surface excavations, but the principles highlighted by the questions and answers in the previous chapters, plus the techniques illustrated in this section, will provide you with the necessary background.

Figure 18.6 Cliff-top castle wall in Spain.

Figure 18.7 Mountain slope in the Italian Alps.

Q18.1 A foundation is proposed on a rock mass consisting of a series of horizontal strata. The results of a site investigation and associated laboratory testing programme are shown in the table below.

Depth (m)	**Description**	***RQD* (%)**	**Intact rock elastic modulus (GPa)**	**Fracture deformation modulus (GPa/m)**
0	**Moderately weathered sandstone**	**61**	**4.0**	**18.0**
3	**Slightly weathered sandstone**	**98**	**5.0**	**20.0**
8	**Slightly weathered limestone**	**80**	**39**	**115**
10	**Fresh, massive limestone**	**100**	**40**	**120**

End of hole at 12 m

(a) Calculate the effective elastic modulus for the uppermost 10 m of the rock mass.

(b) A plate-loading test using a 0.5 m diameter rigid plate at the ground surface has indicated an elastic modulus of 1.0 GPa. By considering the moduli of the various strata, explain why this might be. What value of modulus should be used in the design?

(*Aide memoire*: For a rock mass consisting of *i* strata, each of which contains a single set of planar, parallel fractures, oriented parallel to the strata boundaries, the effective elastic modulus normal to the fractures, E_m, may be calculated using the expression

$$E_m = \frac{\sum t_i}{\sum \left(\frac{1}{E_i} + \frac{\lambda_i}{E_{di}}\right) t_i}$$

where, for each of the *i* strata, t_i = thickness, E_i = Young's modulus of the intact rock, λ_i = fracture frequency, E_{di} = modulus of deformation of the fractures.)

A18.1 (a) If we assume a negative exponential distribution for discontinuity spacings, then the relation between *RQD* and fracture frequency is

$$RQD = 100\,(\lambda t + 1)\exp(-\lambda t)$$

For the moderately weathered sandstone, the *RQD* of which is 61% (using the standard threshold value of $t = 0.1$ m), the corresponding fracture frequency is 13.48 m^{-1}. For the slightly weathered sandstone, for which $RQD = 98\%$, the corresponding frequency is 2.15 m^{-1}. The slightly weathered limestone has an *RQD* of 80% and hence a frequency of 8.24 m^{-1}. The frequency of the fresh, massive limestone is not required, as we only need to compute the modulus for the uppermost 10 m.

From these values we find that the modulus for the rock mass is

$$E_{m} = \frac{\sum t_i}{\sum \left(\frac{1}{E_i} + \frac{\lambda_i}{E_{di}} \right) t_i}$$

$$= \frac{3+5+2}{\left(\frac{1}{4.0}+\frac{13.48}{18.0}\right)\cdot 3+\left(\frac{1}{5.0}+\frac{2.15}{20.0}\right)\cdot 5+\left(\frac{1}{39}+\frac{8.24}{115}\right)\cdot 2} = 2.11 \text{ GPa.}$$

(b) The calculated value is greater than the value measured in the plate-loading test of 1.0 GPa.

If we consider greater thicknesses of the rock mass in turn, then we find the following:

$$E_{m}\,(0\text{ m}-3\text{ m}) = \frac{3}{\left(\frac{1}{4.0}+\frac{13.48}{18.0}\right)\cdot 3} = 1.00 \text{ GPa}$$

$$E_{m}\,(0\text{ m}-8\text{ m}) = \frac{3+5}{\left(\frac{1}{4.0}+\frac{13.48}{18.0}\right)\cdot 3+\left(\frac{1}{5.0}+\frac{2.15}{20.0}\right)\cdot 5} = 1.76 \text{ GPa.}$$

This shows that the plate-loading test has determined the modulus of only the uppermost 3 m of the rock mass. This is a feature of small diameter plate loading tests, as a consideration of the vertical stress induced on the plate axis shows.

For a loaded area (rather than a rigid plate, for which there is no unique analytical solution) the magnitude of this stress is given by

$$\sigma_v = p\left[1 - \frac{z^3}{\left(a^2+z^2\right)^{3/2}}\right]$$

where z is the depth below the surface and a is the radius of the loaded area. A plot of σ_v/p against z/a shows how the induced stress rapidly reduces with depth, with the 5% influence depth being $5.4a$. For the case of a 0.5 m diameter plate, this represents a depth of 1.35 m.

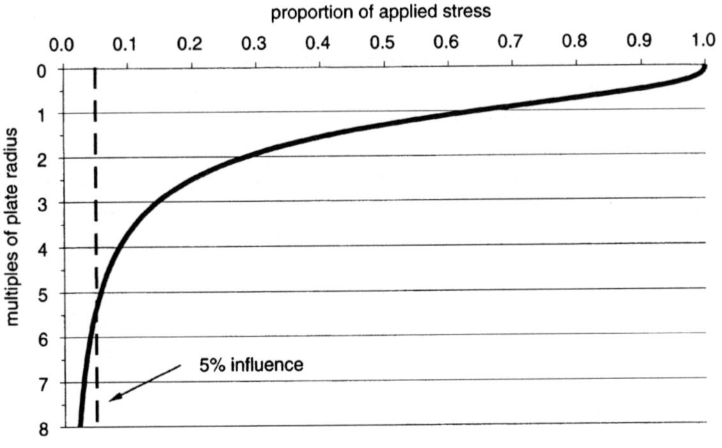

In order to determine the modulus at greater depths, it is usually necessary to perform plate-loading tests at the bottom of large diameter boreholes. Although large diameter plate-loading tests would, in theory, provide the correct values, the enormous loads required for such tests mean that the engineering is prohibitively costly. This also shows that, for a small foundation, the plate loading test will give an appropriate modulus, although for a larger foundation the composite modulus may be more appropriate.

Q18.2 In a rock mass containing a single set of vertically dipping fractures, consider how the fractures might adversely affect the profile of a vertical slope excavated using pre-split blasting.

A18.2 The diagram below, from Jimeno et al. (1995)[1], shows qualitatively how the profile of a pre-split plane is affected by the relative orientations of the plane and the fracture set.

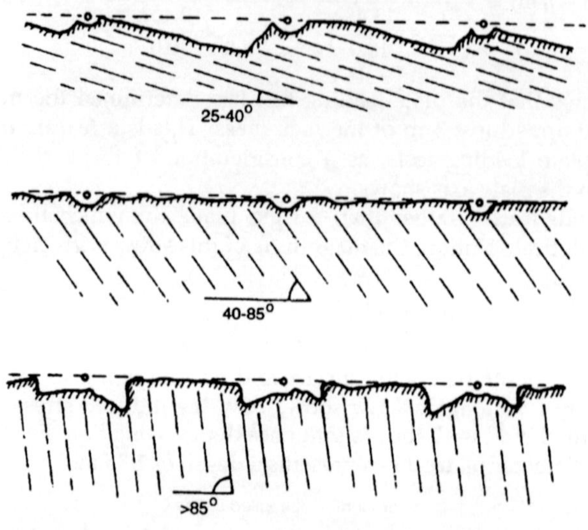

Two mechanisms are acting during pre-split blasting: stress waves and high gas pressures. Each of these contributes to the generation of the pre-split plane, although the effect of the former is attenuated through the use of decoupled charges.

Although the gas pressure effect will allow the formation of the pre-split plane, the reflection and refraction of the stress waves will locally deviate the plane according to the rock structure — causing the irregularities illustrated in the diagram.

Q18.3 A quarry is to be opened in a rock mass which contains four fracture sets with dip directions and dip angles as follows:

[1] Jimeno C. L., Jimeno E. L. and Carcedo F. J. A. (1995) *Drilling and Blasting of Rocks*, Balkema, Rotterdam, 391pp.

Set	**1**	**2**	**3**	**4**
Dip direction/Dip angle	**292/64**	**151/37**	**052/76**	**020/16**

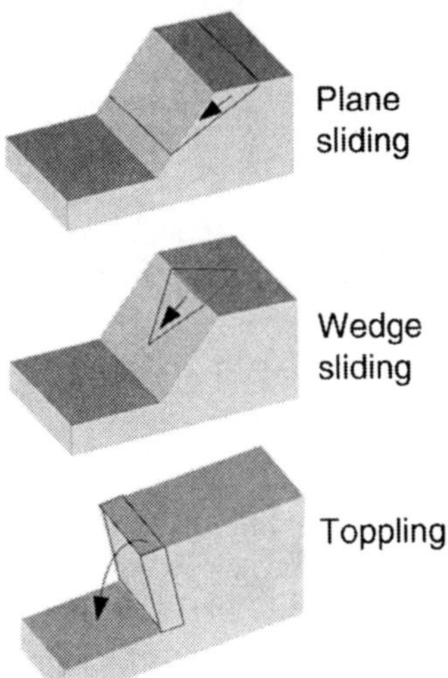

The rock mass can be considered dry, and the angle of friction for all fractures is 30°. Consider the primary potential modes of instability (plane, wedge and direct toppling) at 15° intervals of dip direction (i.e. 0°, 15°, 30°, ..., 345°, 360°) and use kinematic feasibility techniques to prepare a table showing the steepest safe slope and the respective critical failure mode at each azimuth.

A18.3 The fracture data are first plotted on the hemispherical projection. We need to plot both fracture normals (for the analysis of plane sliding and direct toppling) and great circles (in order to locate the intersections, which are required for both the analysis of wedge sliding and direct toppling). This plot is shown below.

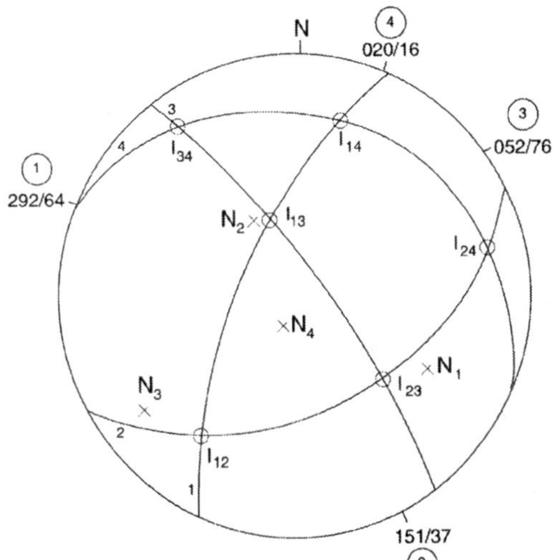

Three kinematic feasibility overlays are required: one for each of the instability modes of plane sliding, wedge sliding and direct toppling. For a given slope dip direction, the plane sliding and wedge sliding overlays lie on opposite sides of the hemispherical projection, and this means that they can both be drawn on the same sheet of tracing paper and used simultaneously. This combined overlay is shown below, see ERM 1.

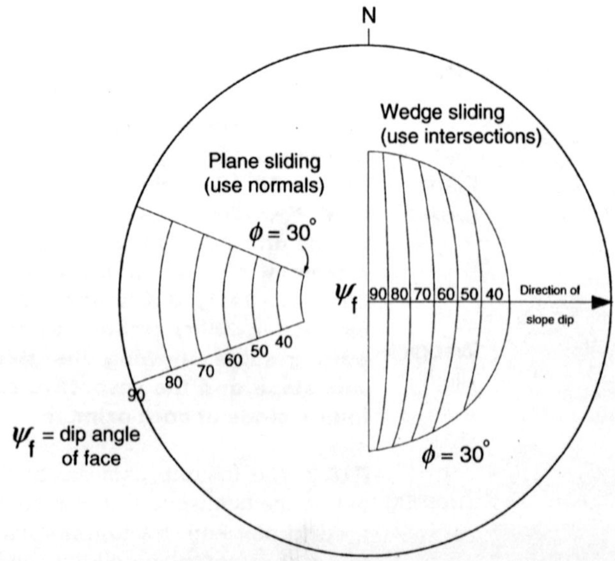

Note that, for each of the instability modes, the lower limit for slope angle is given by the friction angle, which in this case is 30°. The upper limit for slope angle will be selected during the analysis in order to avoid instability.

The toppling instability overlay is shown below, with the friction angle forming one of the limits to the overlay.

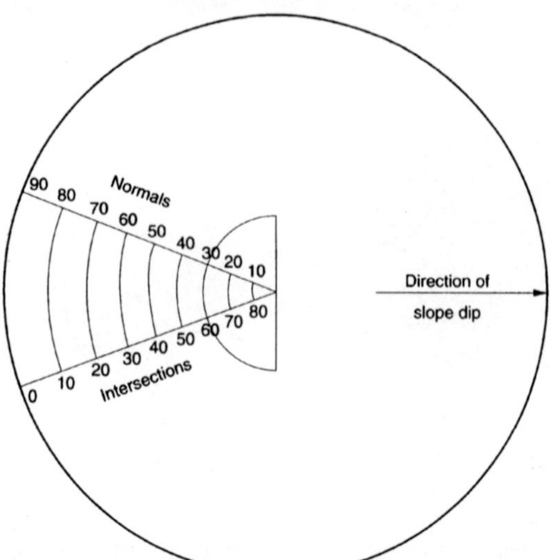

To identify the maximum permissible slope angle to avoid instability at each azimuth, we use each of the overlays in turn (see ERM 1). For each of the azimuth values under consideration, we identify the maximum slope angle that will avoid instability for each of the three instability mechanisms. These are then tabulated, as shown below.

Slope dip direction	Maximum slope angle		Critical angle	Critical mode
	plane	wedge		
000	90	55	55	wedge
015	90	57	57	wedge
030	90	63	63	wedge
045	75	70	70	wedge
060	75	65	65	wedge
075	90	52	52	wedge
090	90	45	45	wedge
105	90	38	38	wedge
120	90	35	35	wedge
135	35	35	35	plane
150	35	38	35	plane
165	35	40	35	plane
180	90	47	47	wedge
195	90	57	57	wedge
210	90	75	75	wedge
225	90	90	90	plane
240	90	90	90	plane
255	90	87	87	wedge
270	90	75	75	wedge
285	65	68	65	plane
300	65	60	60	wedge
315	90	57	57	wedge
330	90	53	53	wedge
345	90	52	52	wedge

All angles in degrees.

Toppling instability has not been included in this table because there is little, if any, incidence of it in this rock mass. This may be readily seen, because the two principal requirements for direct toppling — i.e. two fracture sets which possess a near-vertical line of intersection, and a horizontal fracture set — are not present. The only likelihood of toppling instability is with those blocks formed by sets 1, 2 and 3. For a friction angle of 30° these blocks should suffer sliding instability on set 2 but, if they are prevented from doing so by some form of 'keying', and at the same time are sufficiently high in the direction of the intersection of sets 1 and 3 to form unstable blocks, then toppling may occur.

***Q18.4* Prepare a similar table as in Q18.3 but for the case when the friction angle is not known.**

A18.4 In the case when the friction angle is unknown, we must select a slope angle such that the kinematic feasibility of all instability modes is prevented. This is the same as determining the maximum permissible slope angle to prevent instability when the friction angle is zero. Again, three overlays are required: one for each of the instability modes of plane sliding, wedge sliding and direct toppling. The maximum slope angles are given in the following table.

Slope dip direction	Maximum slope angle		Critical angle	Critical mode
	plane	wedge		
000	15	10	10	wedge
015	15	13	13	wedge
030	15	13	13	wedge
045	75	12	12	wedge
060	75	10	10	wedge
075	90	8	8	wedge
090	90	10	10	wedge
105	90	11	11	wedge
120	90	12	12	wedge
135	38	20	20	wedge
150	38	35	35	wedge
165	38	28	28	wedge
180	90	24	24	wedge
195	90	22	22	wedge
210	90	20	20	wedge
225	90	20	20	wedge
240	90	22	22	wedge
255	90	23	23	wedge
270	90	15	15	wedge
285	65	12	12	wedge
300	65	10	10	wedge
315	90	9	9	wedge
330	90	8	8	wedge
345	90	9	9	wedge

All angles in degrees.

Again, toppling instability is found to be unimportant. However, what is clear now is just how troublesome wedge instability has become. We find that, over the entire range of azimuths, wedge instability is the critical mode, and that avoiding it drives the slope angle down to very low values.

Q18.5 **Use the results from Q18.3 to draw a plan of the slope crest around the quarry excavation assuming that the floor of the quarry is circular. Determine the best orientation for a *radial* access road to the quarry floor (assume that the road can be constructed in any direction and that the optimal orientation is for a road with the steepest possible side slopes — so that excavation associated with the road is minimized).**

Repeat part (b) for an unknown friction angle, as was the case in Q18.4.

A18.5 If we assume that the walls of the quarry are of uniform height H, then the horizontal distance from the crest of the slope to the toe is given by $H/\tan\psi$, where ψ is the slope angle. In addition, the volume of

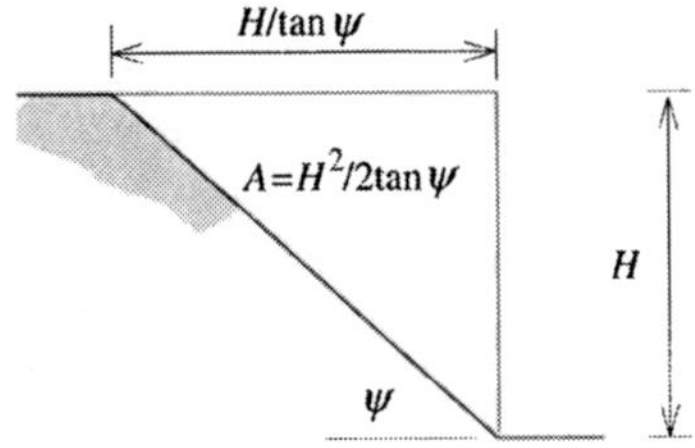

excavation required for this slope is given by $H^2/2\tan\psi$ per unit length of slope crest.

Thus, to sketch the quarry we can draw a circle of some convenient diameter to represent the quarry floor, and then mark a radial distance outwards from this at each of the azimuths in the stability tables, such that the distance is proportional to $1/\tan\psi$. Connecting the points thus marked will give an approximation to the crest of the slope. In order to determine the optimal trend for the access road, we simply use the equation above to compute the area of excavation required for the two side slopes, and choose the direction for which this is a minimum. The result of these calculations is given in the table below, together with a sketch of the slope crest around a circular quarry.

Slope dip direction	000	015	030	045	060	075	090	105	120	135	150	165
Slope angle, ψ	55	57	63	70	65	52	45	38	35	35	35	35
$1/\tan(\psi)$	0.70	0.65	0.51	0.36	0.47	0.78	1.00	1.28	1.43	1.43	1.43	1.43
$1/2\tan(\psi)$	0.35	0.32	0.25	0.18	0.23	0.39	0.50	0.64	0.71	0.71	0.71	0.71
Slope dip direction	180	195	210	225	240	255	270	285	300	315	330	345
Slope angle	47	57	75	90	90	87	75	65	60	57	53	52
$1/\tan(\psi)$	0.93	0.65	0.27	0.00	0.00	0.05	0.27	0.47	0.58	0.65	0.75	0.78
$1/2\tan(\psi)$	0.47	0.32	0.13	0.00	0.00	0.03	0.13	0.23	0.29	0.32	0.38	0.39
Opposing slope dip directions	000, 180	015, 195	030, 210	045, 225	060, 240	075, 255	090, 270	105, 285	120, 300	135, 315	150, 330	165, 345
Combined excavation volume	0.82	0.65	0.39	0.18	0.23	0.42	0.63	0.87	1.00	1.04	1.09	1.10

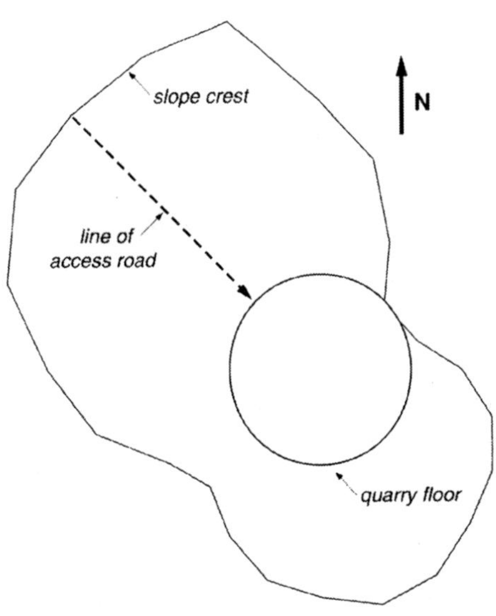

The minimal volume of side slope excavation is required by the two facing slopes whose dip directions are 045° and 225°, which means that the optimal trend for the road is either 135° or 315°. If we examine the quarry faces at these azimuths we see that they are 35° and 57°, respectively, and on this basis it makes sense to use the slope whose dip direction is 135° for the access road.

We can undertake the same process for the case when the friction angle is unknown, and the results of this analysis are given in the table below. In this case, the minimal volume of side slope excavation is required by the two facing slopes whose dip directions are 015° and 195°, which means that the optimal trend for the road is either 105° or 285°. The quarry faces at these azimuths have slope angles of 11° and 12°, respectively, and so either orientation may be used for the road.

Slope dip direction	0	015	30	45	60	75	90	105	120	135	150	165
Slope angle, ψ	10	13	13	12	10	8	10	11	12	20	35	28
$1/\tan(\psi)$	5.67	4.33	4.33	4.70	5.67	7.12	5.67	5.14	4.70	2.75	1.43	1.88
$1/2\tan(\psi)$	2.84	2.17	2.17	2.35	2.84	3.56	2.84	2.57	2.35	1.37	0.71	0.94
Slope dip direction	180	195	210	225	240	255	270	285	300	315	330	345
Slope angle	24	22	20	20	22	23	15	12	10	9	8	9
$1/\tan(\psi)$	2.25	2.48	2.75	2.75	2.48	2.36	3.73	4.70	5.67	6.31	7.12	6.31
$1/2\tan(\psi)$	1.12	1.24	1.37	1.37	1.24	1.18	1.87	2.35	2.84	3.16	3.56	3.16
Opposing slope dip directions	000, 180	015, 195	030, 210	045, 225	060, 240	075, 255	090, 270	105, 285	120, 300	135, 315	150, 330	165, 345
Combined excavation volume	3.96	3.40	3.54	3.73	4.07	4.74	4.70	4.92	5.19	4.53	4.27	4.10

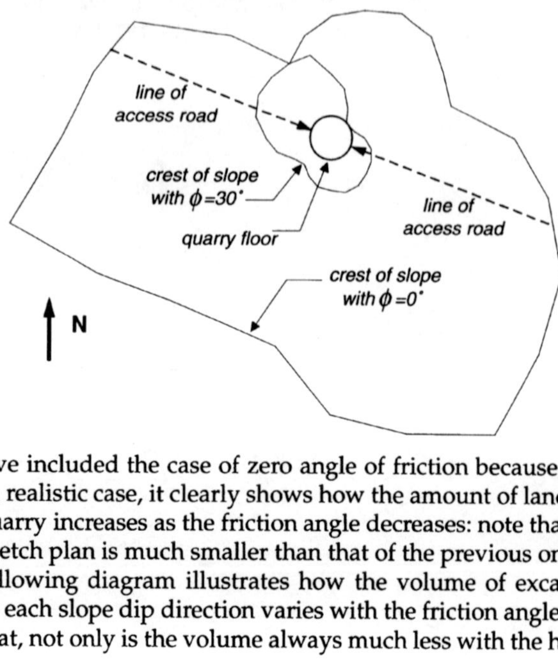

We have included the case of zero angle of friction because, although it is not a realistic case, it clearly shows how the amount of land required by the quarry increases as the friction angle decreases: note that the scale of this sketch plan is much smaller than that of the previous one.

The following diagram illustrates how the volume of excavation required at each slope dip direction varies with the friction angle. This plot shows that, not only is the volume always much less with the higher friction angle, but the volume requirements are also much more anisotropic. This gives greater scope for the designers of the excavation to minimize the volume of additional (probably wasted) excavation, by preferentially selecting those dip directions (i.e. 045° and 225°) for the excavation faces.

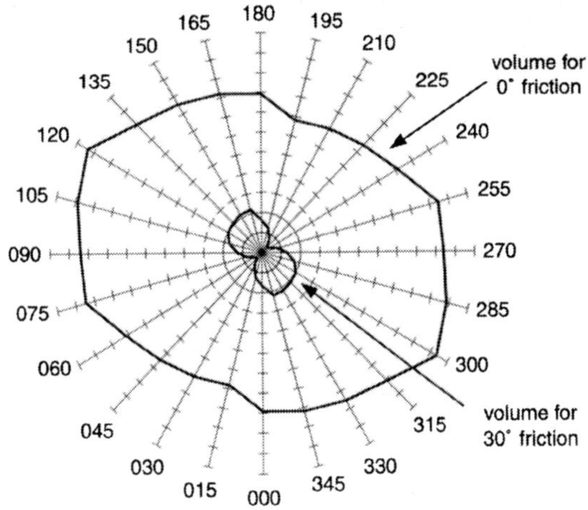

Q18.6 Add toppling due to interlayer slip to the analysis of Q18.3.

A18.6 We use the hemispherical projection plot of fracture normals for flexural toppling, with a simplified overlay as we are only analysing the stability of the rock mass when the friction angle is 30°. This overlay is shown below, with the radial scale calibrated directly in the required slope dip angle.

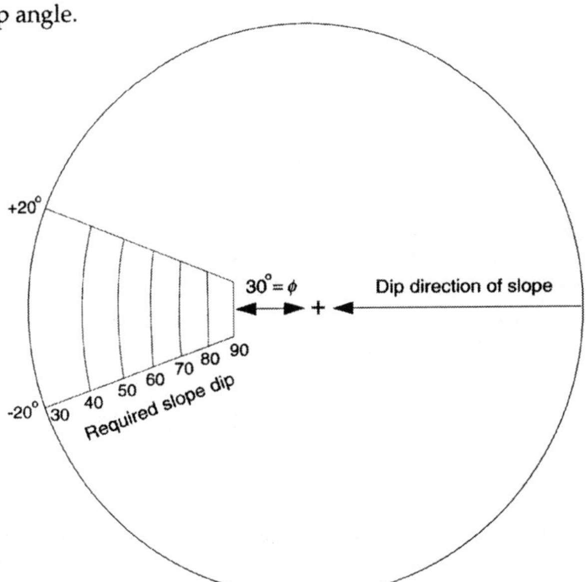

For each of the azimuth values under consideration, we identify the maximum slope angle that will avoid instability for this particular mechanism, and then tabulate the results, as shown below.

Slope dip direction	Maximum slope angle flexural toppling	Critical angle all modes	critical mode
000	90	55	wedge
015	90	57	wedge
030	90	63	wedge
045	90	70	wedge
060	90	65	wedge
075	90	52	wedge
090	90	45	wedge
105	90	38	wedge
120	55	35	wedge
135	90	35	plane
150	90	35	plane
165	90	35	plane
180	90	47	wedge
195	90	57	wedge
210	90	75	wedge
225	40	40	flexural toppling
240	40	40	flexural toppling
255	90	87	wedge
270	90	75	wedge
285	90	65	plane
300	90	60	wedge
315	80	57	wedge
330	80	53	wedge
345	80	52	wedge

All angles in degrees.

We see that flexural toppling does indeed become a dominant potential failure mode at two of the azimuth orientations, namely 225° and 240°.

***Q18.7* A trial design for a proposed 100 m deep roadstone quarry can be approximated in plan by a circle. The quarry has been designed with a number of 10 m high benches, such that the overall slope angle is 45° and the individual bench angles are 60°.**

Preliminary mapping of the site has established that the rock mass contains four dominant sets of fractures, having the following mean orientations:

Feature	Cleavage	Joint set A	Joint set B	Bedding
Dip	36°	75°	45°	28°
Dip direction	015°	330°	080°	260°

(a) Using kinematic feasibility techniques, determine whether or not conditions exist that could lead to instability in the proposed slopes.

(b) Indicate, in general terms, the design changes required to overcome any predicted instability problems.

A18.7 A schematic section through the quarry face reveals that the profile of the face is as below.

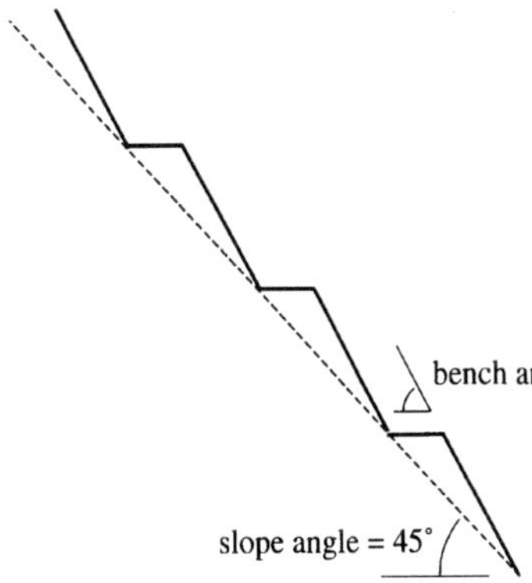

This indicates that, in general, sliding instability involving the entire face will occur on features (either fracture planes for plane sliding or lines of intersection of two planes for wedge sliding) that dip at an angle that exceeds the angle of friction but is less than the face angle of 45°. Instability involving individual benches will occur for features that dip at angles greater than 45° but less than 60°.

If we assume that the fractures in the rock mass are cohesionless and possess a friction angle of 35°, then we can construct the following two kinematic overlays for plane sliding and wedge sliding instability.

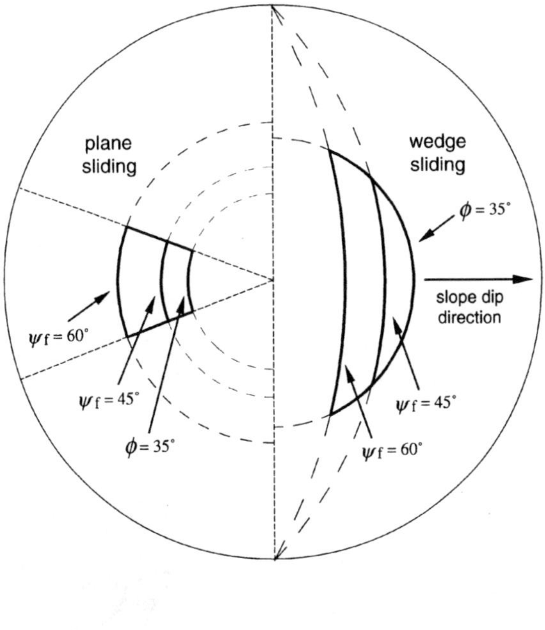

These overlays are used in conjunction with a hemispherical projection plot of both the great circles (for wedge sliding) and normals (for plane sliding) of the fracture data, which is shown below.

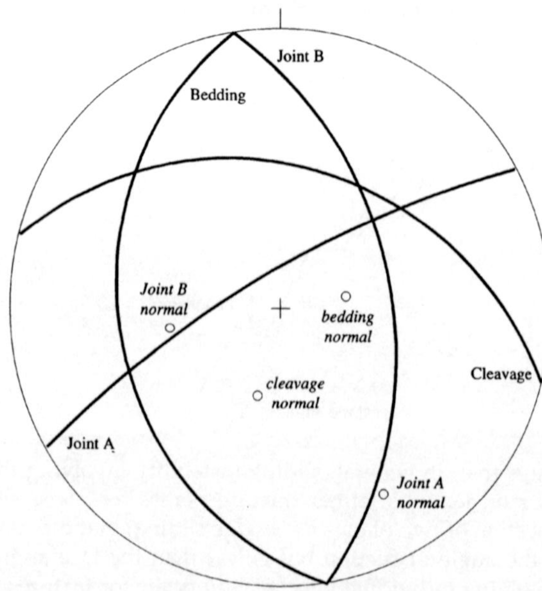

The result of this kinematic assessment is as shown in the table below.

Slope dip direction	Plane sliding	Wedge sliding
000	normal to cleavage, entire face	Joints A and B, individual benches
030	normal to cleavage, entire face	Joints A and B, entire face
060	normal to Joints B, entire face	Joints A and B, entire face
090	normal to Joints B, entire face	Joints A and B, individual benches
120	×	×
150	×	×
180	×	×
210	×	×
240	×	×
270	×	×
300	×	×
330	×	×

Instability due to toppling has not been included in this table as the conditions for its development (i.e. normal to one fracture set and the intersection of two other fracture sets both sub-vertical) are absent from the rock mass.

The sketch below summarizes the instability conditions, with the overlay being used to determine more exactly the angular boundaries between the various instability conditions. It is clear from this sketch that almost the entire southwest quadrant of the quarry will suffer from sliding instability that will effect the entire face. There are regions to the extremes of this range where sliding on individual benches may occur. This will be controlled by the local geometry of the fractures, and so will need to be identified *in situ*. Usually these small instabilities can either be reinforced, or the unstable blocks removed. However, the principal problem is how to overcome the instability of the entire face.

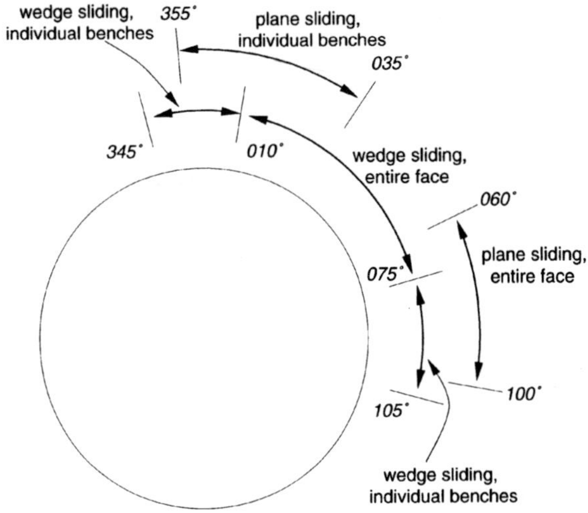

The engineer has essentially three options to reduce the effect of this sliding instability: flatten the slopes, reinforce the slopes, or change the shape of the quarry to reduce the amount of face that is subject to instability.

The first two of these options will usually be unacceptable for an asset such as a quarry. Flattening the slopes will reduce the amount of material that can be extracted from the quarry, and the installation of reinforcement over a large part of the quarry faces will be too expensive to be economic. The third option — changing the shape of the quarry — may be possible, and leads to some interesting design issues.

The design problem is to reduce the length of quarry face that is subject to sliding instability. One way of doing this would be to use a rectangular quarry, with its major axis aligned along, say, 055°–235°, as shown in the sketch below. If the rectangle has an aspect ratio of 1:2, then with such an arrangement the length of unstable face is reduced to only 1/6, or 17%, of the entire face length. This compares to 25% for the case of the circular quarry.

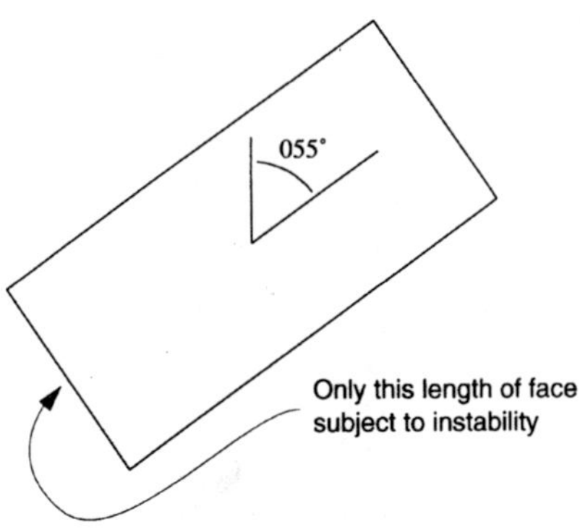

It may now be economic to reduce the overall slope angle for the southwest face of the quarry, and avoid all instability.

Q18.8 A rock slope with a face angle of ψ_f is cut into a rock mass containing a single set of fractures dipping into the slope at an angle β. The strength of the fractures is purely frictional. Assume that $\tan\phi$, ψ_f and β can all be considered as normally distributed variables with the following parameters:

Variable	$\tan\phi$	ψ_f	β
Mean	0.55	50°	60°
Standard deviation	0.15	5°	10°

Investigate the variability of the factor of safety due to interlayer slip for this slope, using either standard normal random values obtained from statistical tables, or values produced by your computer/calculator. Perform as many trials as you have the patience for (but at least 35).

A18.8 A straightforward way of solving this is to use a computer spreadsheet. We start by entering standard random variables into three columns (one column for each of the random parameters) and then using these to compute random values of the parameters using the relation *random variable = (standard random variable × standard deviation) + mean.* Each triplet of the random values is then used to compute a factor of safety. The table will look like the one below.

			$\tan(\phi)$	ψ_f	β	
Mean:			0.55	50°	60°	
Standard deviation:			0.15	5°	10°	
Standard normal randoms			Random values			Factor of safety
−0.258	−0.083	−0.103	0.51	48.7	57.4	1.77
−0.514	2.382	2.006	0.47	47.4	54.9	2.17
1.282	−2.004	−0.529	0.74	56.4	72.8	0.91
0.394	1.054	−1.315	0.61	52.0	63.9	1.25
−0.919	−0.360	0.067	0.41	45.4	50.8	3.78
0.547	0.097	−1.085	0.63	52.7	65.5	1.18
−1.381	0.972	−0.283	0.34	43.1	46.2	−27.47
0.926	−0.768	−2.011	0.69	54.6	69.3	1.03
0.283	−1.761	0.903	0.59	51.4	62.8	1.32
0.350	−1.133	1.050	0.60	51.8	63.5	1.28
−0.424	0.731	0.541	0.49	47.9	55.8	2.00
−1.854	−0.516	−2.065	0.27	40.7	41.5	−1.98
0.507	0.784	−0.854	0.63	52.5	65.1	1.20
−0.153	−0.319	1.328	0.53	49.2	58.5	1.65
−0.795	−0.055	−2.569	0.43	46.0	52.1	3.04
−0.611	1.601	0.448	0.46	46.9	53.9	2.39
−1.495	−0.842	0.546	0.33	42.5	45.1	−7.69
2.059	−0.805	−0.276	0.86	60.3	80.6	0.70
0.911	−0.371	−1.502	0.69	54.6	69.1	1.03
0.322	−0.018	−0.133	0.60	51.6	63.2	1.29
0.997	0.147	0.340	0.40	45.0	50.0	4.54
1.431	0.167	−1.072	0.76	57.2	74.3	0.87
−0.172	1.903	0.366	0.52	49.1	58.3	1.67

Standard normal randoms			Random values			Factor of safety
0.353	1.084	−0.232	0.60	51.8	63.5	1.28
2.271	0.021	−0.003	0.89	61.4	82.7	0.65
0.310	−0.448	0.643	0.60	51.6	63.1	1.30
−0.361	0.741	0.414	0.50	48.2	56.4	1.91
−1.140	0.665	−0.193	0.38	44.3	48.6	7.48
−0.746	0.269	0.834	0.44	46.3	52.5	2.83
0.321	−1.190	0.429	0.60	51.6	63.2	1.29
−0.962	−0.337	−0.306	0.41	45.2	50.4	4.16
0.975	−1.552	1.076	0.70	54.9	69.8	1.01
1.332	−2.155	0.310	0.75	56.7	73.3	0.89
1.306	−1.412	−1.343	0.75	56.5	73.1	0.90
−1.607	−2.746	0.186	0.31	42.0	43.9	−4.30
2.140	−1.312	1.109	0.87	60.7	81.4	0.68
		Means:	0.57	50.5	61.0	

Notice that some of the factors of safety are negative. This is due to the combination of random values used, and such a factor of safety is clearly wrong. In rigorous Monte Carlo simulation they have to be accounted for, but here we will simply ignore them.

The factor of safety computed using the means of the random values (shown at the foot of the table) is 1.43, whereas if the actual mean values quoted at the head of the table are used the factor of safety is 1.51. This discrepancy is simply due to the limited number of trials performed. However, if the mean of all of the positive factors of safety is taken, the result is 1.86, which is a very different value. This indicates that the mean factor of safety taken over all of the trials is not the same as the factor of safety computed using the mean values of the random variables, and shows why Monte Carlo simulation should be performed.

The factors of safety are plotted as a histogram below.

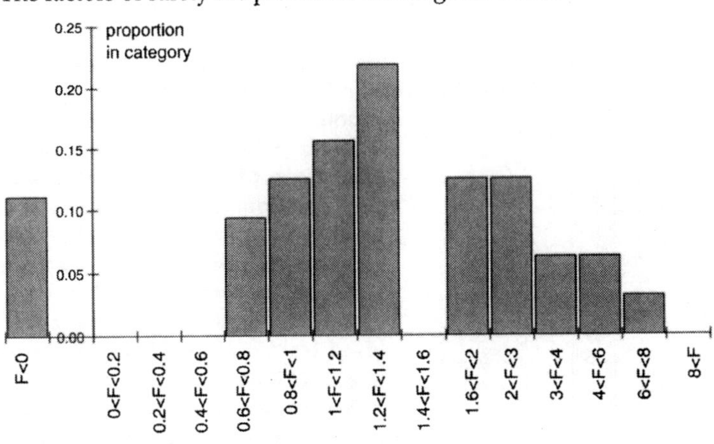

Even if we ignore the number of trials that result in a negative factor of safety, there is a substantial proportion with a factor of safety less than one. Indeed, for these results we have $N(0 < F) = 32$, $N(0 < F < 1.0) = 7$, and $N(0 < F < 1.5) = 19$, giving $\Pr(0 < F < 1.0) = 7/32 = 0.22$ and $\Pr(0 < F < 1.5) = 19/32 = 0.59$. Using the probabilistic criteria given by Priest and Brown (1983) [2], the slope can be regarded as unstable, despite the mean factor of safety being 1.86.

Finally, in order to investigate the form of the distribution of factors of safety more precisely, we must perform many more trials. For 10000 trials, the histogram looks like that shown below, and gives a mean factor of safety of 2.19, with probabilistic assessments of $\Pr(0 < F < 1.0) = 0.29$ and $\Pr(0 < F < 1.5) = 0.54$. The distribution can be seen to be lognormal in form, the mean factor of safety has increased, but the slope remains unstable when judged against the probabilistic criteria.

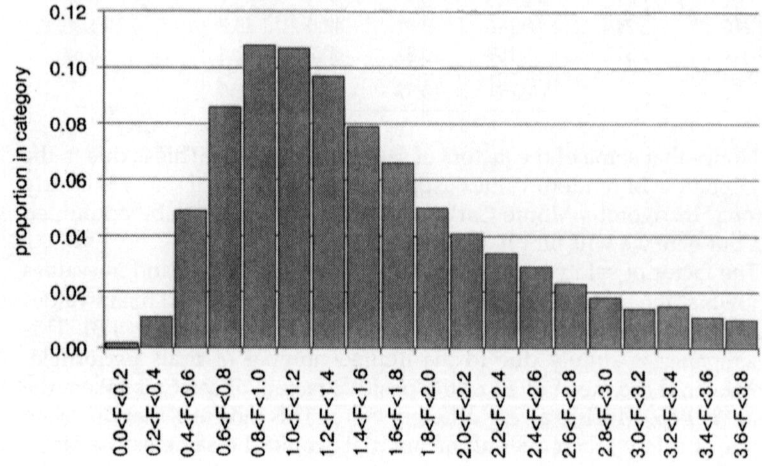

Q18.9 A proposed dam will exert a vertical force of 9400 kN/m and a horizontal force of 4400 kN/m on its horizontal foundation, as shown in the figure below. At a depth of 6 m below the founding surface there is a horizontal bedding plane, the strength of which is purely frictional with $\phi = 32°$. The unit weight of the rock above this plane is 22 kN/m^3.

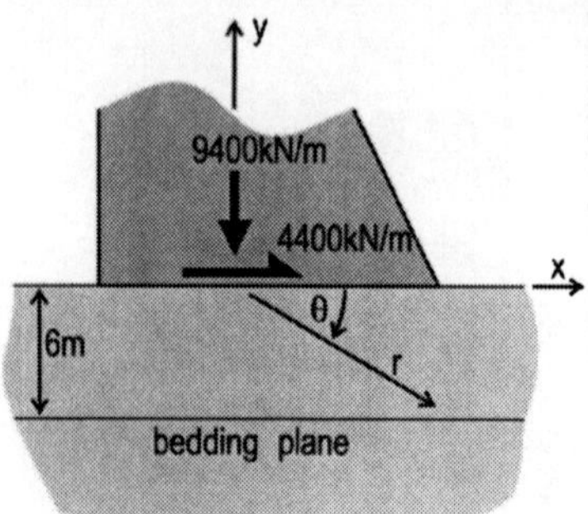

For a polar co-ordinate system, the origin of which coincides with the rectangular co-ordinate system as shown in the diagram, the components of radial stress due to a vertical line load *P* and a horizontal line load *Q* may be calculated from $\sigma_{r(P)} = (2P\sin\theta)/\pi r$ and $\sigma_{r(Q)} = (2Q\cos\theta)/\pi r$, respectively.

At what value of x-ordinate is the radial stress on the bedding plane

[2] Priest S. D. and Brown E. T. (1983) Probabilistic stability analysis of variable rock slopes. *Trans. Inst. Min. Metall.* **92**, A1–12.

zero, and will the stress state on the bedding plane to the left of this point be sustainable? Over what range of x-ordinate values to the right of this point will slip occur on the bedding plane? What conclusions can be drawn about the overall stability of the proposed dam?

A18.9 The forces P and Q are applied to the ground through the material of the dam, and hence exert normal and shear stresses on the foundation, rather than line loads. However, in order to make a preliminary analysis we will assume that these forces are, in fact, applied to the foundation as line loads.

Although we could find the position of zero radial stress algebraically, it is both more convenient and instructive to sketch the general form of the shape of the contours of radial stress (the so-called 'bulbs of pressure') beneath an inclined line load, and use this to identify an approximate location for it. This analysis will be approximate because it does not take into account the direct stress in the radial direction due to the weight of the rock overlying the bedding plane. The radial stress contours take the following form:

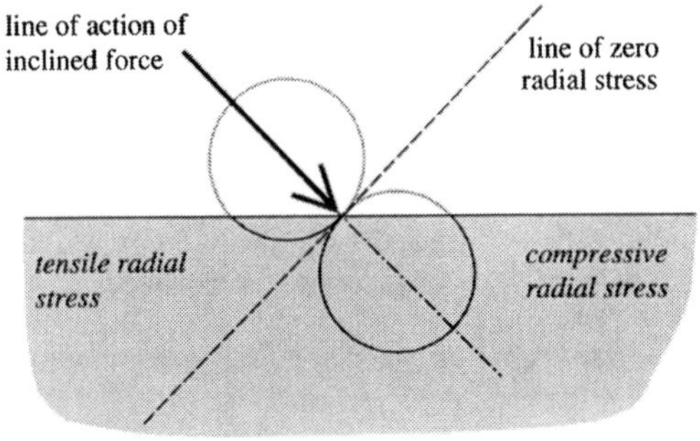

This shows that the position of zero radial stress is given simply by the orientation of the line load. Sketching the geometry of the problem allows us to evaluate this:

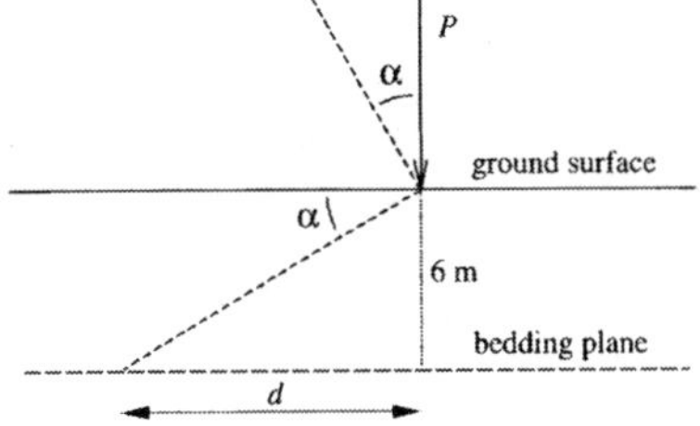

Thus we find that

$$\tan\alpha = \frac{Q}{P} = \frac{4400}{9400} = \frac{6}{d}$$

from which we obtain

$$d = 6 \times \frac{9400}{4400} = 12.8 \text{ m}$$

and hence the x-ordinate of the approximate position of zero radial stress is $x = -12.8$ m.

The sketch of the contours of radial stress shows that, in general, the stress state to the left of this point will not be sustainable due to the presence of the tensile radial stresses. However, we would need to undertake a detailed analysis that took into account the vertical stress due to the weight of the rock overlying the bedding plane and any water pressures induced due to the operation of the dam to make certain of this.

In order to determine the range of x-ordinate values to the right of this position over which the bedding plane is stable, we need to compute the stress state induced on the plane and compare this with its strength parameters. Again, although we could determine a solution for the position of the specific point at which the plane is in limiting equilibrium, it is more useful and instructive to determine the conditions at a number of points and hence plot the variation with distance.

This first step in this analysis is to sketch the geometry of a general point on the bedding plane, in order to determine the angles required for computation of the radial stresses (from the equations given) and the stress transformations:

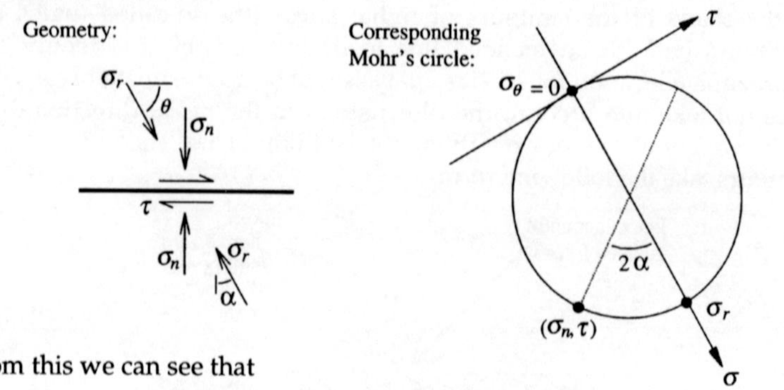

From this we can see that

$$\sigma_n = \frac{\sigma_r}{2}(1 + \cos 2\alpha) \text{ and } \tau = \frac{\sigma_r}{2}\sin 2\alpha.$$

A table of results can now be constructed, as shown below.

x, m	−15	−10	−5	0	5	10	15	20	25	30
r, m $\left(=\sqrt{6^2+x^2}\right)$	16.16	11.66	7.81	6.00	7.81	11.66	16.16	20.88	25.71	30.59
α, deg $(=\tan^{-1}(6/x))$	−68.2	−59.0	−39.8	0.0	39.8	59.0	68.2	73.3	76.5	78.7
θ, deg $(=90^\circ-\alpha)$	158.2	149.0	129.8	90.0	50.2	31.0	21.8	16.7	13.5	11.3
$\sigma_{r(P)}$, kN/m^2	137.6	264.0	588.6	997.4	588.6	264.0	137.6	82.4	54.3	38.4
$\sigma_{r(Q)}$, kN/m^2	−161.0	−206.0	−229.6	0.0	229.6	206.0	161.0	128.5	105.9	89.8
$\sigma_{r(\text{combined})}$, kN/m^2	−23.4	58.0	359.0	997.4	818.2	470.0	298.6	210.8	160.3	128.1
σ_v, kN/m^2	−3.2	15.4	211.9	997.4	482.9	124.4	41.2	17.4	8.7	4.9
τ, kN/m^2	8.1	−25.6	−176.6	0.0	402.4	207.3	102.9	58.0	36.4	24.6
σ_n, kN/m^2 $(=\sigma_v+6\gamma)$	128.8	147.4	343.9	1129.4	614.9	256.4	173.2	149.4	140.7	136.9
ϕ_{required}, deg $(=\tan^{-1}(\tau/\sigma_n))$	3.6	−9.9	−27.2	0.0	33.2	39.0	30.7	21.2	14.5	10.2

The results for these 10 points are plotted on the following graph, together with the curve that joins them. This curve represents the variation in required friction angle, and from this we can see that there is a region from about $x = 4.5$ m to $x = 14.0$ m where slip will occur (the actual values, found algebraically, are $x = 4.69$ m and $x = 14.38$ m). Notice that no slip occurs for negative values of x, although the change in sign of the friction angle indicates that the sense of the shear stress changes at $x = 0$.

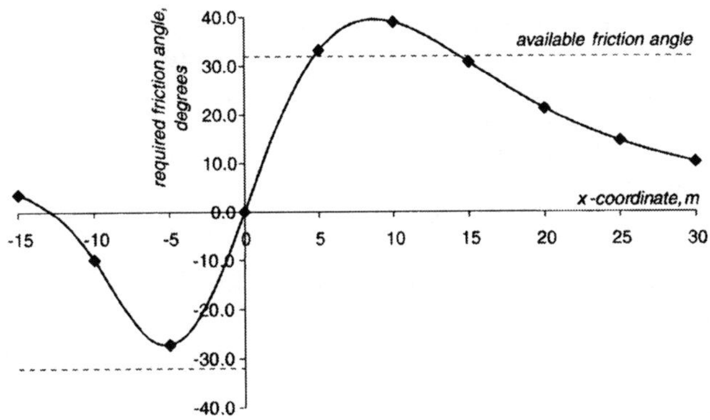

Overall, we can conclude that the stability of the dam will be marginal. Firstly, the negative radial stresses to the left of the dam will not be sustainable (as the tensile strength of bedding planes can be taken to be zero). Secondly, slip will take place in the region to the right of the dam where the frictional strength of the bedding plane is exceeded. Although a more sophisticated analysis may show that the dam is, theoretically, stable, we must never forget that the consequences of instability in a structure such as a dam will usually be catastrophic.

Q18.10 Using the principle of virtual work, determine the value of *P* for limiting equilibrium for the foundation shown below.

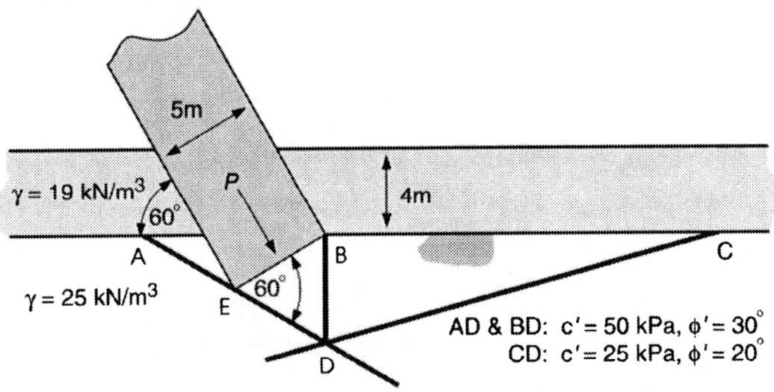

A18.10 We start by drawing a free body diagram of the entire foundation, and mark vectors representing the virtual displacements taking place on the various block boundaries.

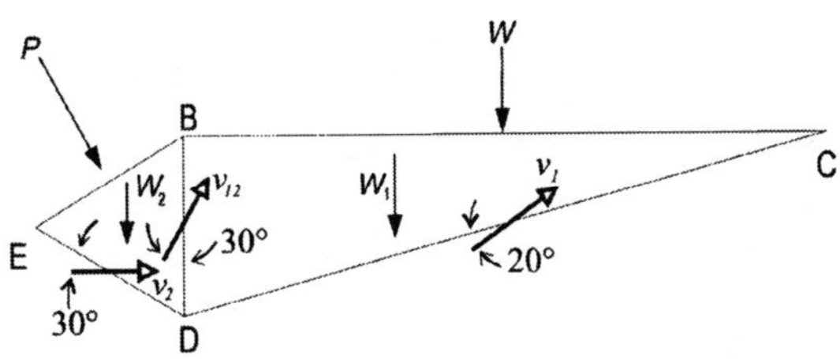

In this diagram the force W represents the weight of the material which overlies the foundation, and W_1 and W_2 represent the weight of the two blocks 1 and 2, respectively.

Having drawn the free body diagram, it is necessary to compute the relations between the various virtual displacements. This is most easily done by accurately drawing the vectors on a separate diagram.

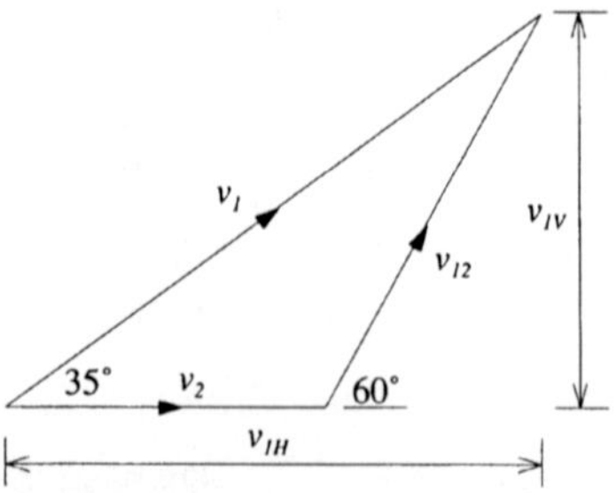

One of these vectors must be given an arbitrary value in order to scale the remainder. Here, we set $\nu_1 = 1.0$, and so we find

$$\frac{\nu_1}{\sin 120} = \frac{\nu_2}{\sin 25} \Rightarrow \nu_2 = 0.4880,$$

$$\frac{\nu_{12}}{\sin 35} = \frac{\nu_1}{\sin 120} \Rightarrow \nu_{12} = 0.6623,$$

$$\nu_{1V} = \nu_1 \sin 35 \Rightarrow \nu_{1V} = 0.5736,$$

and

$$\nu_{1H} = \nu_1 \cos 35 \Rightarrow \nu_{1H} = 0.8192.$$

We can now write down an expression for the external virtual work:

$$P \sin 30 \cdot \nu_2 + P \cos 30 \cdot \nu_{2V} + W_2 \cdot \nu_{2V} - (W + W_1) \cdot \nu_{1V} = EVW$$

which, because $\nu_{2V} = 0.0$, simplifies to

$$P \sin 30 \cdot \nu_2 - (W + W_1) \cdot \nu_{1V} = EVW.$$

The internal virtual work is given by

$$50 \times 5 \cos 30 \cdot \nu_2 + 50 \times 5 \cos 30 \cdot \nu_{12} + \frac{5}{\sin 15} \times 25 \cos 20 \cdot \nu_1 = IVW.$$

For equilibrium we have $EVW = IVW$, and so

$$250 \cdot 30 \cdot (\nu_2 + \nu_{12}) + 125 \frac{\cos 20}{\sin 15} = P \sin 30 \cdot \nu_2 - (W_1 + W) \cdot \nu_{1V},$$

from which we have

$$P = \frac{250 \cdot 30 \cdot (\nu_2 + \nu_{12}) + 125 \frac{\cos 20}{\sin 15} + (W_1 + W) \cdot \nu_{1V}}{\sin 30 \cdot \nu_2}. \quad (18.1)$$

We now need to compute W_1 and W, the weight of block 1 and the weight of the overburden:

$$W_1 = \frac{1}{2} \times 5 \times \frac{5}{\tan 15} \times 25 = 1166.3 \text{ kN/m},$$

and

$$W = \frac{5}{\tan 15} \times 4 \times 19 = 1418.2 \text{ kN/m}.$$

Thus, substituting into Eq. (18.1), we have

$$P = \frac{250 \cos 30 \times (0.4880 + 0.6623) + 453.8 + 2584.4 \times 0.5736}{\sin 30 \times 0.4880}$$

$$= 8956 \text{ kN/m}.$$

18.3 Additional points

Foundations and rock slopes are generally at the Earth's surface and hence will be excavated in rock masses that have been subjected to some form of **weathering**. The range of this weathering varies from that in cold regions and the effects of permafrost to that in hot regions, where high temperatures and rainfall can cause mineralogical alteration and associated mechanical degradation to significant depths. The weathering may be expected to extend to a depth of up to 35 m — which is of the same order as the dimension of an engineered slope. Moreover, the weathering is likely to be non-uniform, and so the slope will have different properties at different locations, making the analysis and design more complex. An example of this is the heterogeneous rock and soil strata forming the slopes of the Panama Canal, for which precise geotechnical predictions of the performance of the slopes are difficult.

The consequence of the near-surface conditions is that rock property characterization for slope design is always critically important, as are sensitivity analyses of the influences of the various parameters on stability. Furthermore, although we have given examples of quarries in which slopes occur at all azimuths, many rock engineering structures may contain slopes at fixed locations and orientations. Such a circumstance may occur with the cuttings for a transport route through mountainous terrain. Under these conditions, the engineer has little scope for adopting a design that minimizes the risk of instability, and must instead adopt measures that protect the transport corridor and its users. This will probably involve the installation of fences and rock fall traps. Development of techniques for designing such protection from rock fall instability is currently a major activity.

The principal difficulty in the design and analysis of rock slopes and foundations is the DIANE nature of near-surface rock masses. As engineers are called upon to make ever more ambitious designs, so the need to incorporate the DIANE features into a design becomes paramount. As a result, more sophisticated techniques for characterization of the rock mass and analysis of the engineered structures are required. For example, all of the questions and answers we have presented use sets of parallel fractures, so that the analyses can be simplified for convenience. More advanced techniques using actual field fracture data are currently being developed, and these will assist in improved design.

19 Underground excavation instability mechanisms

19.1 Underground instability

In Chapters 17 and 18, we presented information relating to the potential for **rock instability** at or close to the ground surface. When an excavation in rock is made entirely underground, there are somewhat different rock mechanics circumstances primarily related to the nature of the rock mass and stiffness of the loading system — as indicated in Fig. 19.1.

At depth, the weathering and rock fracturing will be less intense, so the rock mass will be more coherent, stiffer and stronger. However, the rock stress and water pressure increase with increasing depth, so there will be a greater potential for stress-induced damage and water pressure effects such as water pressure-induced block movements. At the ground

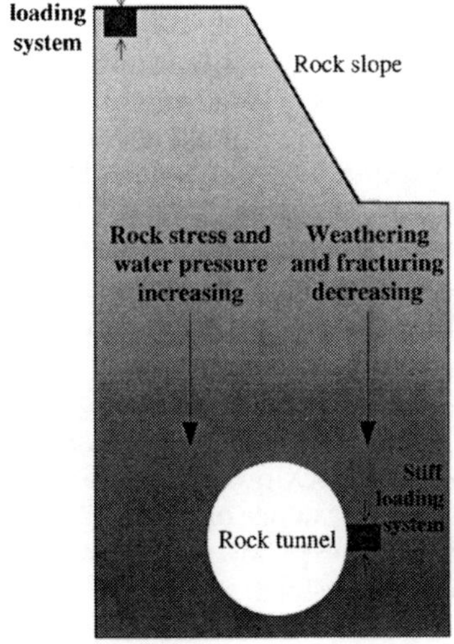

Figure 19.1 Rock engineering at the surface and at depth.

Figure 19.2 Measuring fracture surface roughness.

surface, e.g. the element at the top left of Fig. 19.1, the loading system is soft: the load will not decrease significantly with displacements of the rock mass, because of the 'dead weight' nature of the loading. Conversely, at depth, for example an element of rock around the tunnel being loaded by the surrounding rock, the loading system can be much stiffer.

It is more difficult to obtain information about the rock mass properties underground than at the surface. As we noted in Q2.8, the amount of rock core obtained during a site investigation programme may be only of the order of 0.0005% of the rock mass being considered. Also, in addition to this small sampling volume for underground design (compared to the information available from surface rock outcrops for surface design), some of the rock properties are much more difficult to measure, for example the persistence of fractures. These properties, such as the fracture surface roughness shown being measured in Fig. 19.2, are often limited to the core dimensions.

The rock mass will be disturbed by the rock excavation process, but the elegant pre-splitting technique used to protect the remaining rock

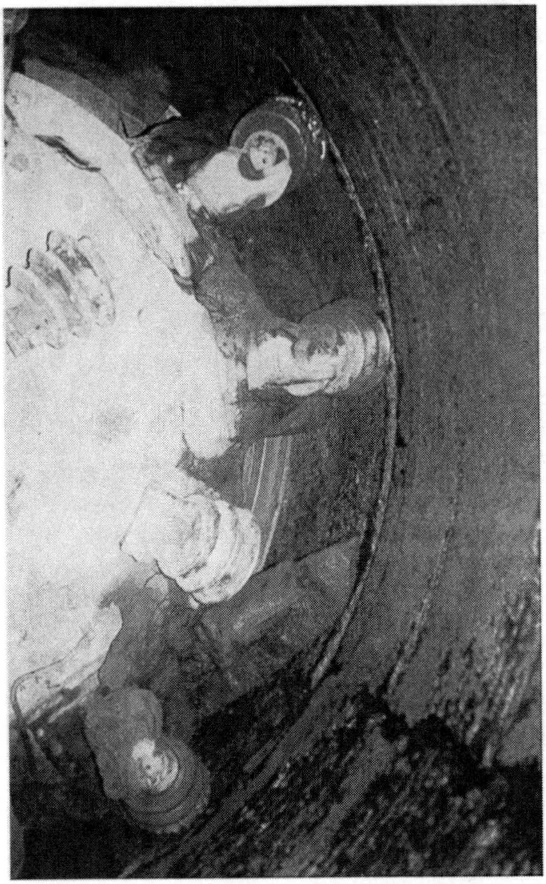

Figure 19.3 Tunnel boring machine.

in surface blasting cannot be used so effectively underground because of the influence of the rock stress. Although the use of tunnel boring machines, as shown in Fig. 19.3, causes less additional disturbance compared to blasting, they are susceptible to significant changes in rock stiffness and strength. If the rock becomes too soft, there is insufficient resistance for the gripper pads to hold to provide the forward thrust; if the rock becomes too strong, the machine will not be able to cut the rock adequately — and both these factors could occur at the same time.

Thus we see that, for rock engineering underground, more information may well be required about the rock conditions than for engineering at the surface, but generally less is available.

Similarly, it is difficult to predict the support that will be required in a rock tunnel without detailed information on the rock conditions. The tunnel may need no support in some regions and a stiff, strong support in other areas, as illustrated in Fig. 19.4. A 'design as you go' approach

Figure 19.4 Pilot tunnel with steel arch support.

Figure 19.5 Inflow of toxic water.

can be used for tunnel support but not for TBM excavation, because the TBM cannot easily be changed.

There may be other factors involved depending on the purpose of the tunnel. The performance requirements for a conveyor tunnel, road tunnel, rail tunnel, gas storage tunnel, pressurized water tunnel, radioactive waste repository tunnel, explosive storage tunnel, defence headquarters tunnel and public access museum tunnel will all be different. There may also be environmental problems, such as the inflow of contaminated water, as illustrated in Fig. 19.5.

Thus, we need to understand firstly that creating the excavation causes the three primary effects of the rock moving in, the alteration of the rock stress field, and increased flow of water, and secondly we should design with a knowledge of the instability mechanisms that can occur, given the project objective and hence the excavation requirements. The two main factors causing instability are the **movement of rock blocks** and the disturbance caused by the **induced stresses.** It is essential therefore to understand the method for assessing the kinematic feasibility of rock blocks, and how stress concentrations around underground excavations can cause rock failure.

19.2 Questions and answers: underground excavation instability mechanisms

Questions 19.1 to 19.6 are a set of questions all concerning rock block instability in the same underground rock cavern, and so we recommend that they are approached in the correct order.

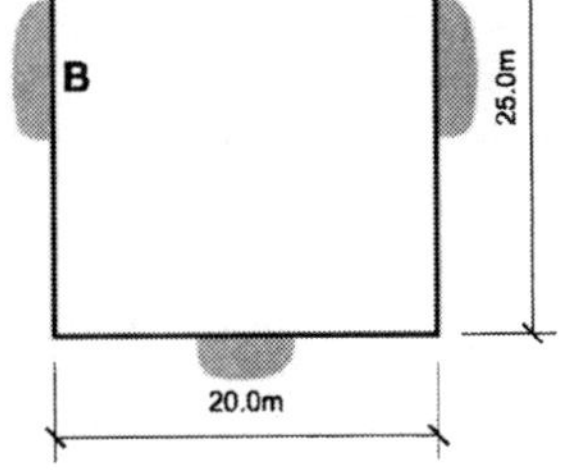

***Q19.1* The diagram on the right shows the cross-section of an underground machine hall. The hall is to be excavated in a sequence of metamorphosed argillaceous rocks, of unit weight 22 kN/m³, which contains five fracture sets with the following orientations:**

Set	Dip direction	Dip angle
1	058°	54°
2	195°	70°
3	127°	81°
4	160°	32°
5	335°	64°

Each set has a friction angle of approximately 30° and zero cohesion. It is proposed that the axis of the machine hall will be oriented on a heading 030° and will be horizontal.

Determine the kinematic feasibility of all the tetrahedral blocks formed by the intersection of the fracture sets and the *roof*.

A19.1 To solve this question we use hemispherical projection techniques (see Chapter 7 for an illustration of the techniques and note that a hemispherical projection sheet is included in Appendix B).

(1) Plot the fracture data on the hemispherical projection. Mark a tick on the periphery at the appropriate azimuth for each of the five fracture sets, and label them with the fracture set number and the set orientation.

(2) Draw the great circle for each fracture set in turn, labelling each great circle with the set number.

(3) Draw a concentric circle representing the friction angle. In this case, this is a circle 30° in from the periphery. The completed projection is as follows:

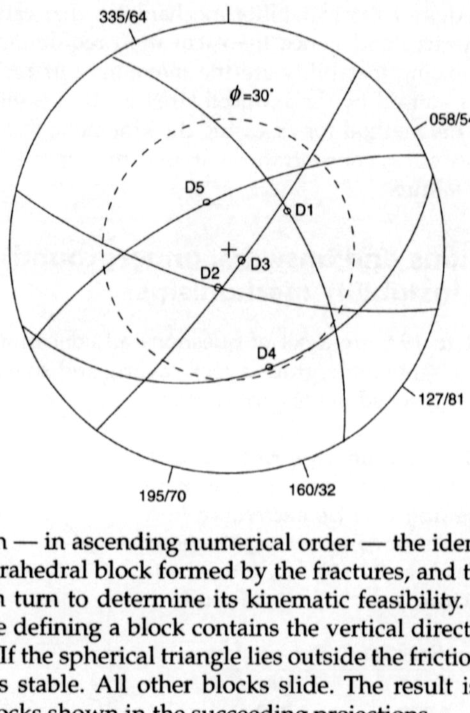

(4) Write down — in ascending numerical order — the identifying code for each tetrahedral block formed by the fractures, and then examine each one in turn to determine its kinematic feasibility. If the spherical triangle defining a block contains the vertical direction, then the block falls. If the spherical triangle lies outside the friction circle, then the block is stable. All other blocks slide. The result is as follows, with the blocks shown in the succeeding projections.

Block	Instability	Block	Instability	Block	Instability
123	sliding	234	sliding	345	falling
124	sliding	235	falling		
125	sliding	245	sliding		
134	sliding				
135	sliding				
145	falling				

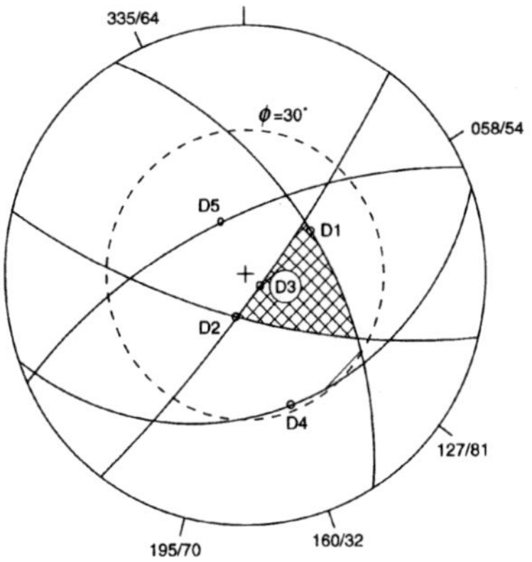

Block 123

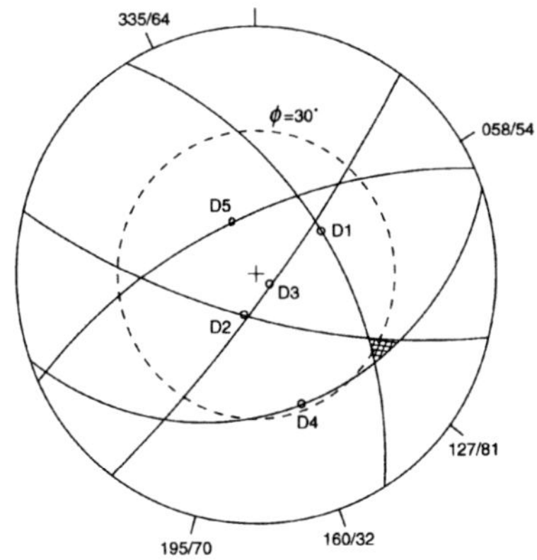

Block 124

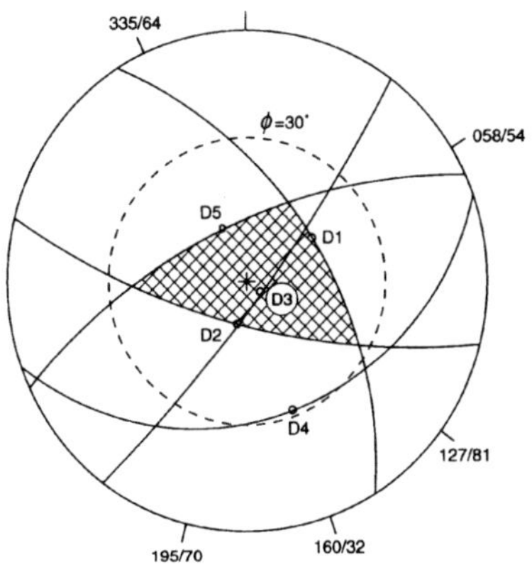

Block 125

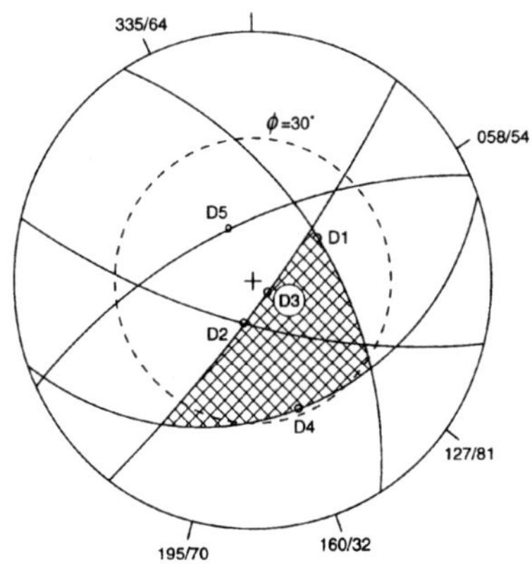

Block 134

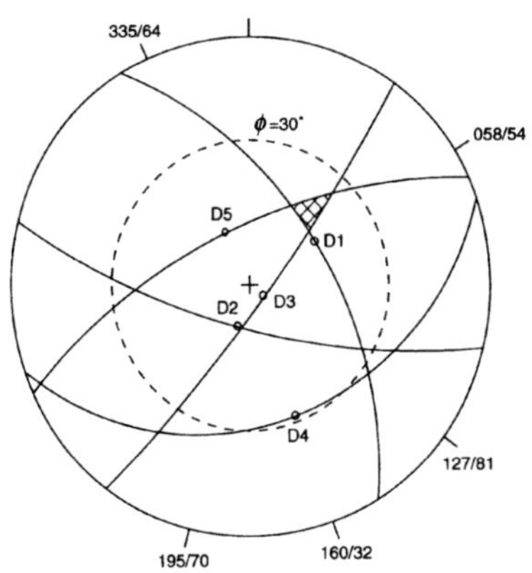

Block 135

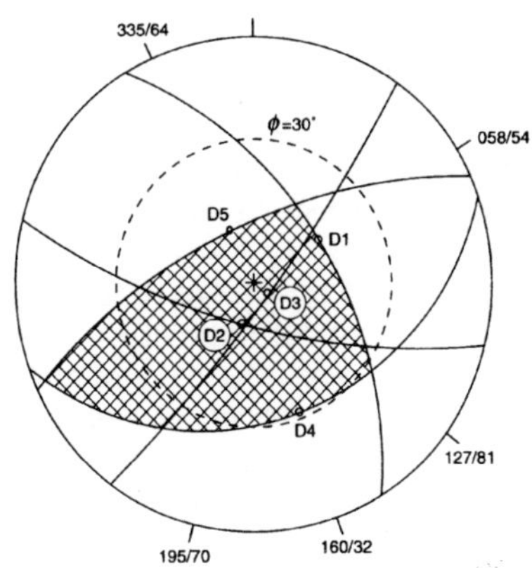

Block 145

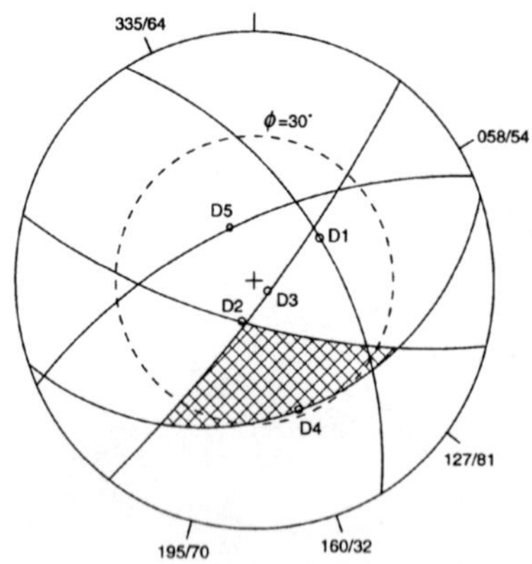

Block 234

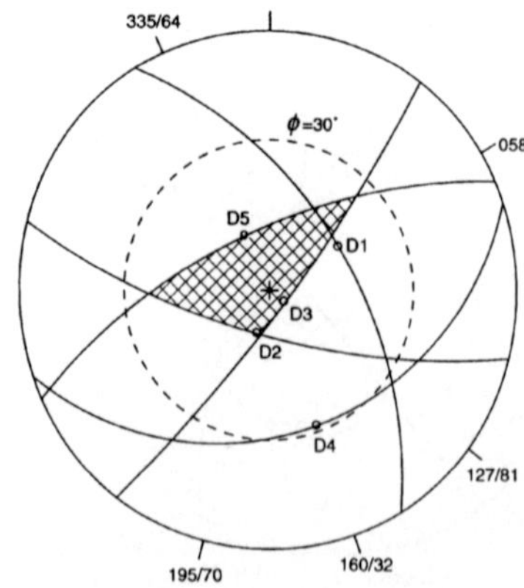

Block 235

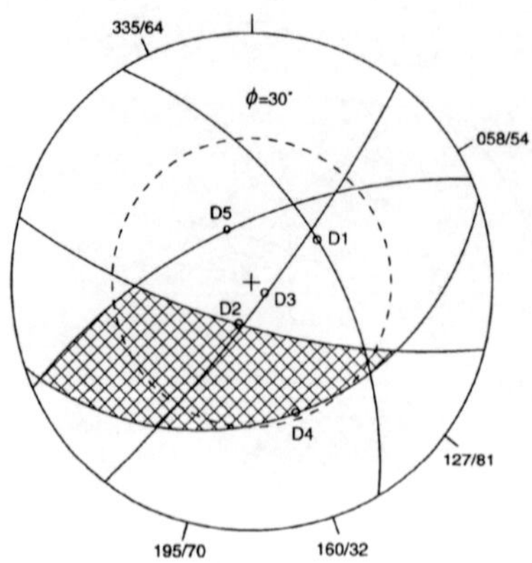

Block 245

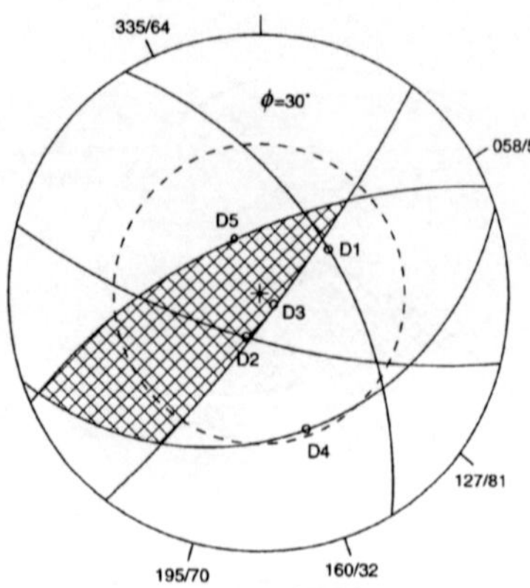

Block 345

Q19.2 For the unstable roof blocks established in the answer to Q19.1, determine the direction of movement, and plane(s) of sliding where relevant.

A19.2 For each block which has been identified as being unstable due to sliding, we need to identify the sliding direction. This will be the most steeply dipping of:

(i) one of the three lines of intersection of two planes which make up the faces of the block; or
(ii) the line of maximum dip of one of the planes which forms the block, *providing the line of maximum dip lies on the periphery of the spherical triangle that defines the block.*
(1) Identify and label each of the intersections shown on the projection, and determine the orientation of each one. The intersections are as follows:

Intersection	Orientation	Intersection	Orientation	Intersection	Orientation
12	119/34	23	191/70	34	214/20
13	049/54	24	114/24	35	044/37
14	126/27	25	268/39		
15	029/50			45	246/02

(2) For each block already identified as being unstable due to sliding, write down the candidate sliding directions and their orientations. The most steeply dipping of these is the sliding direction.
The candidate sliding directions, together with the critical direction, are given in the table below. Note that only block 134 has three candidate planes for plane sliding, and that blocks 124 and 135 have none: this reflects the fact that the line of maximum dip of a plane can only be the candidate for a sliding direction if it is located on the spherical triangle representing the block.

Block	Planes			Intersections			Critical
123	d_1 185/54		d_3 150/81	I_{12} 119/34	I_{23} 191/70	I_{31} 049/54	d_3 150/81
124				I_{12} 119/34	I_{24} 114/24	I_{41} 126/27	I_{12} 119/34
134	d_1 185/54	d_3 150/81	d_4 160/32	I_{13} 049/54	I_{34} 214/20	I_{41} 126/27	d_3 150/81
135				I_{13} 049/54	I_{35} 044/37	I_{51} 029/50	I_{13} 049/54
234			d_4 160/32	I_{23} 191/70	I_{34} 214/20	I_{42} 114/24	I_{23} 191/70
245	d_2 195/70	d_4 160/32		I_{24} 114/24	I_{45} 246/02	I_{52} 268/39	d_2 195/70

Q19.3 For the unstable roof blocks established in the answer to Q19.1, determine the volume of the largest block that could fall from the roof and hence calculate the necessary support pressure to ensure stability.

A19.3 Four blocks — 125, 145, 235 and 345 — have been identified as falling from the roof. We need to draw each of these to scale in order to determine their volumes, which requires the plan traces of the lines of intersection that make up their roof edges. In addition, we need the orientations of these lines of intersection in order to compute the block volumes. For the blocks listed above, we need intersections 12, 14, 15, 23, 25, 34, 35, and 45.

(1) For all of the required intersections, identify and label each one on the projection and determine its orientation. Then draw a radial dotted line from each one to the periphery. The intersections are as follows:

I_{12}	I_{14}	I_{15}	I_{23}	I_{25}	I_{34}	I_{35}	I_{45}
119/34	126/27	029/50	191/70	268/39	214/20	044/37	246/02

(2) For each plane that forms a block face, draw its strike on the projection, using a dashed line.
(3) For each block in turn, draw a pair of lines representing the boundary edges of the roof. Within these draw the outline of the face triangle, using the dashed strike lines plotted on the projection to determine their trends. Also draw the intersections representing the roof edges,

using the dotted intersection lines on the projection to determine their trends. The three roof lines should meet at a point: if they do not, your drawing is not accurate enough.

(4) Measure the length of the three face lines and the three roof edges. Compute the included angle at each face apex using the dip directions of the fractures that make up the roof planes, and write down the dip angle of each roof edge. Using these results, compute the area of the face triangle, the height of the block and hence its weight and the average support pressure.

(5) The results of the calculations for these blocks are shown in the series of diagrams below. From these we see that block 235 is the largest, at 2616 kN, and that the support pressure required to ensure stability of the various falling blocks is 53.3 kN/m^2.

It is worthwhile comparing the weight of each block with the size of the block's spherical triangle on the projection: this clearly shows that there is no relation between the two, and hence block weight cannot be estimated by studying the projection alone. Note also that, for blocks that deviate widely from a regular tetrahedron, the three individual values of face area and height are liable to differ markedly. This is an inherent problem with the technique, caused by the difficulty of drawing the blocks and scaling the measurements sufficiently accurately. For critical work, a computational method must be used.

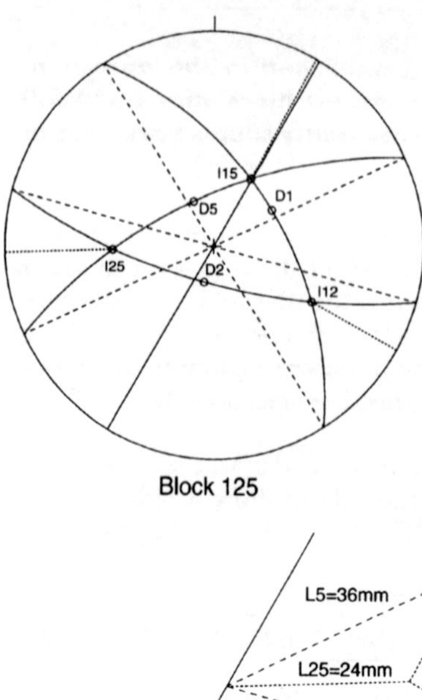

Block 125

Angles taken from the hemispherical projection:

$\theta_{12} = |\alpha_1 - \alpha_2| = |058 - 195| = 137$

$\theta_{25} = |\alpha_2 - \alpha_5| = |195 - 335| = 140$

$\theta_{5-} = |\alpha_5 - \alpha_1| = |335 - 058| = 277$

$\beta_{12} = 34$

$\beta_{25} = 39$

$\beta_{51} = 50$

Scale drawing of block (original scale 5 mm: 1 m):

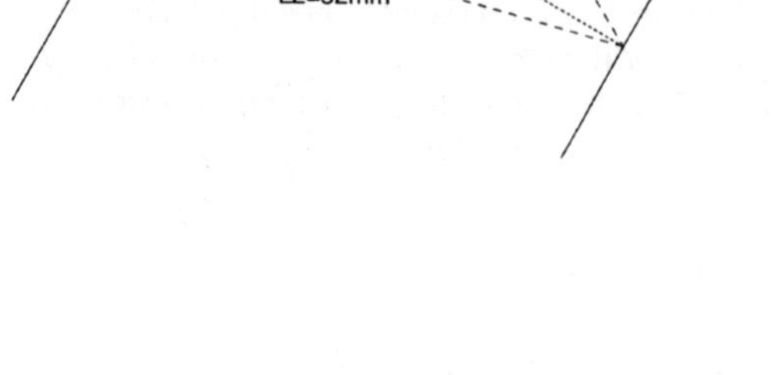

$$A_f = \tfrac{1}{3}(\tfrac{1}{2}l_1l_2|\sin\theta_{12}| + \tfrac{1}{2}l_2l_5|\sin\theta_{25}| + \tfrac{1}{2}l_5l_1|\sin\theta_{51}|)$$
$$= \tfrac{1}{3}(\tfrac{1}{2}\cdot 34\cdot 52\cdot|\sin 137| + \tfrac{1}{2}\cdot 52\cdot 36\cdot|\sin 140| + \tfrac{1}{2}\cdot 36\cdot 34\cdot|\sin 97|)$$
$$= \tfrac{1}{3}(602.9 + 601.6 + 607.4) = 604.0\ \text{mm}^2 \equiv 24.16\ \text{m}^2$$

$$h = \tfrac{1}{3}(l_{12}\tan\beta_{12} + l_{25}\tan\beta_{25} + l_{51}\tan\beta_{51})$$
$$= \tfrac{1}{3}(29\cdot\tan 34 + 24\cdot\tan 39 + 16\cdot\tan 50)$$
$$= \tfrac{1}{3}(19.6 + 19.4 + 19.1) = 19.37\ \text{mm} \equiv 3.9\ \text{m}$$

$$W = \tfrac{1}{3}\gamma A_f h = \tfrac{1}{3}\times 22\times 24.16\times 3.9 = 691.0\ \text{kN, and}$$
$$p = \tfrac{1}{3}\gamma h = \tfrac{1}{3}\times 22\times 3.9 = 28.6\ \text{kN/m}^2.$$

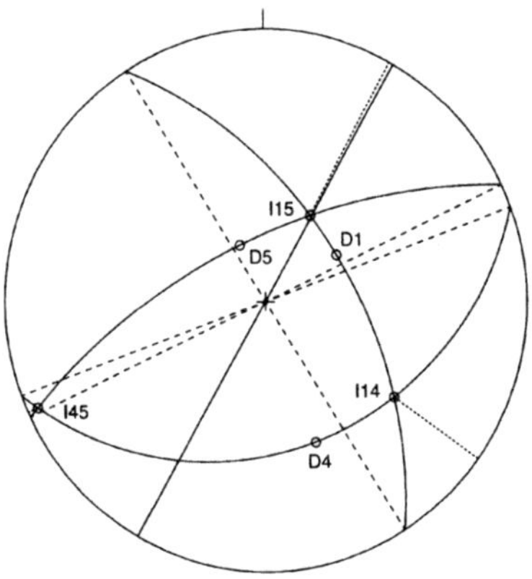

Block 145

Angles taken from the hemispherical projection:

$\theta_{14} = |\alpha_1 - \alpha_4| = |058 - 160| = 102$

$\theta_{45} = |\alpha_4 - \alpha_5| = |160 - 335| = 175$

$\theta_{51} = |\alpha_5 - \alpha_1| = |335 - 058| = 277$

$\beta_{14} = 27$

$\beta_{45} = 2$

$\beta_{51} = 50$

Scale drawing of block (original scale 5 mm: 1 m):

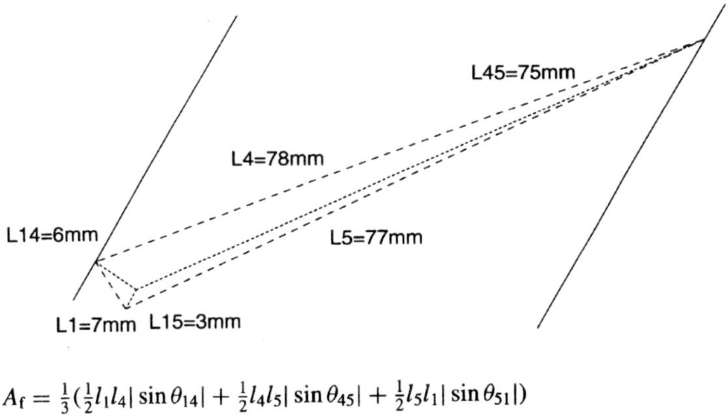

$$A_f = \tfrac{1}{3}(\tfrac{1}{2}l_1l_4|\sin\theta_{14}| + \tfrac{1}{2}l_4l_5|\sin\theta_{45}| + \tfrac{1}{2}l_5l_1|\sin\theta_{51}|)$$
$$= \tfrac{1}{3}(\tfrac{1}{2}\cdot 7\cdot 78\cdot|\sin 102| + \tfrac{1}{2}\cdot 78\cdot 77\cdot|\sin 175| + \tfrac{1}{2}\cdot 77\cdot 7\cdot|\sin 227|)$$
$$= \tfrac{1}{3}(267.0 + 261.7 + 267.5) = 265.4\ \text{mm}^2 \equiv 10.62\ \text{m}^2$$

$$\begin{aligned} h &= \tfrac{1}{3}(l_{14}\tan\beta_{14} + l_{45}\tan\beta_{45} + l_{51}\tan\beta_{51}) \\ &= \tfrac{1}{3}(6\cdot\tan 27 + 75\cdot\tan 2 + 3\cdot\tan 50) \\ &= \tfrac{1}{3}(3.0 + 2.6 + 3.6) = 3.1\ \text{mm} \equiv 0.61\ \text{m} \end{aligned}$$

$$W = \tfrac{1}{3}\gamma A_j h = \tfrac{1}{3}\times 22\times 10.62\times 0.61 = 47.5\ \text{kN, and}$$
$$p = \tfrac{1}{3}\gamma h = \tfrac{1}{3}\times 22\times 0.61 = 4.5\ \text{kN/m}^2.$$

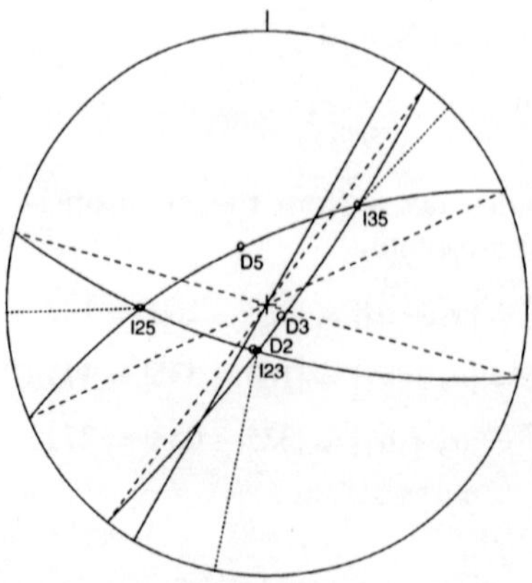

Block 235

Angles taken from the hemispherical projection:

$\theta_{23} = |\alpha_2 - \alpha_3| = |195 - 127| = 68$

$\theta_{35} = |\alpha_3 - \alpha_5| = |127 - 335| = 208$

$\theta_{52} = |\alpha_5 - \alpha_2| = |335 - 195| = 140$

$\beta_{23} = 70$

$\beta_{35} = 37$

$\beta_{52} = 39$

Scale drawing of block (original scale 5 mm: 1 m):

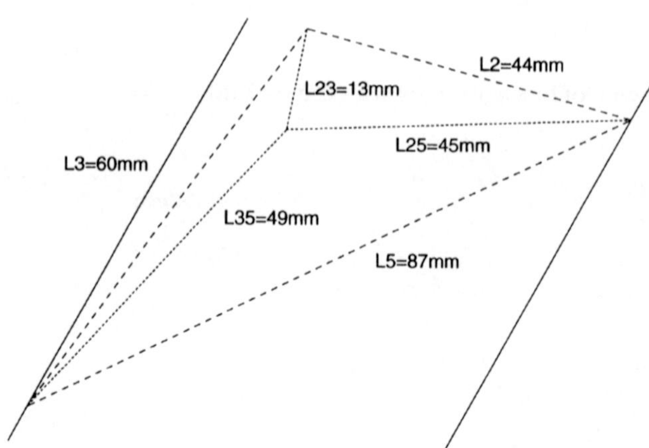

$$\begin{aligned} A_f &= \tfrac{1}{3}(\tfrac{1}{2}l_2 l_3|\sin\theta_{23}| + \tfrac{1}{2}l_3 l_5|\sin\theta_{35}| + \tfrac{1}{2}l_5 l_2|\sin\theta_{52}|) \\ &= \tfrac{1}{3}(\tfrac{1}{2}\cdot 44\cdot 60\cdot|\sin 68| + \tfrac{1}{2}\cdot 60\cdot 87\cdot|\sin 208| + \tfrac{1}{2}\cdot 87\cdot 44\cdot|\sin 140|) \\ &= \tfrac{1}{3}(1223.9 + 1225.3 + 1230.3) = 1226.5\ \text{mm}^2 \equiv 49.06\ \text{m}^2 \end{aligned}$$

$$
\begin{aligned}
h &= \tfrac{1}{3}(l_{23}\tan\beta_{23} + l_{23}\tan\beta_{35} + l_{52}\tan\beta_{52})\\
&= \tfrac{1}{3}(13\cdot\tan 70 + 49\cdot\tan 37 + 45\cdot\tan 39)\\
&= \tfrac{1}{3}(35.7 + 36.9 + 36.4) = 36.3\ \text{mm} \equiv 7.27\ \text{m}
\end{aligned}
$$

$$
\begin{aligned}
W &= \tfrac{1}{3}\gamma A_j h = \tfrac{1}{3}\times 22\times 49.06\times 7.27 = 2615.6\ \text{kN, and}\\
p &= \tfrac{1}{3}\gamma h = \tfrac{1}{3}\times 22\times 7.27 = 53.3\ \text{kN/m}^2.
\end{aligned}
$$

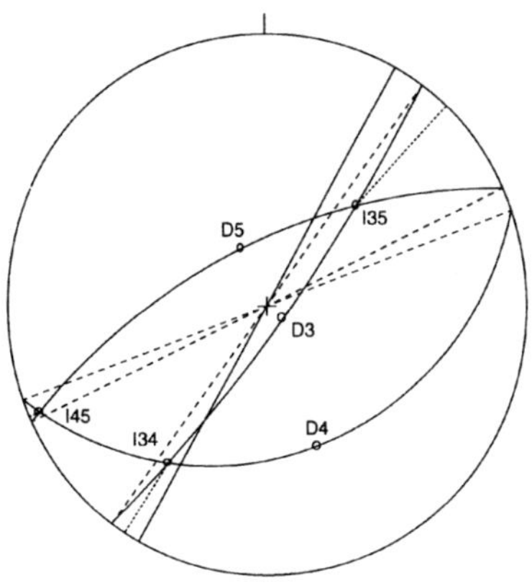

Block 345

Angles taken from the hemispherical projection:

$$
\begin{aligned}
\theta_{34} &= |\alpha_3 - \alpha_4| = |127 - 160| = 33\\
\theta_{45} &= |\alpha_4 - \alpha_5| = |160 - 335| = 175\\
\theta_{53} &= |\alpha_5 - \alpha_3| = |335 - 127| = 208\\
\beta_{34} &= 20\\
\beta_{45} &= 2\\
\beta_{53} &= 37
\end{aligned}
$$

Scale drawing of block (original scale 5 mm: 1 m):

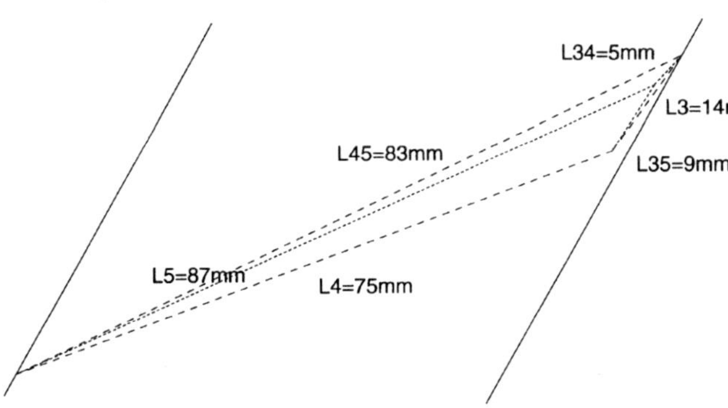

$$
\begin{aligned}
A_f &= \tfrac{1}{3}(\tfrac{1}{2}l_3 l_4|\sin\theta_{34}| + \tfrac{1}{2}l_4 l_5|\sin\theta_{45}| + \tfrac{1}{2}l_5 l_3|\sin\theta_{53}|)\\
&= \tfrac{1}{3}(\tfrac{1}{2}\cdot 14\cdot 75\cdot|\sin 33| + \tfrac{1}{2}\cdot 75\cdot 87\cdot|\sin 175| + \tfrac{1}{2}\cdot 87\cdot 14\cdot|\sin 208|)\\
&= \tfrac{1}{3}(286.0 + 284.3 + 285.9) = 285.4\ \text{mm}^2 \equiv 11.42\ \text{m}^2
\end{aligned}
$$

$$
\begin{aligned}
h &= \tfrac{1}{3}(l_{34}\tan\beta_{34} + l_{34}\tan\beta_{45} + l_{53}\tan\beta_{53})\\
&= \tfrac{1}{3}(5\cdot\tan 20 + 83\cdot\tan 2 + 9\cdot\tan 37)\\
&= \tfrac{1}{3}(1.82 + 2.90 + 6.78) = 3.83\ \text{mm} \equiv 0.77\ \text{m}
\end{aligned}
$$

$$W = \tfrac{1}{3}\gamma A_j h = \tfrac{1}{3} \times 22 \times 11.42 \times 0.77 = 64.5 \text{ kN, and}$$

$$p = \tfrac{1}{3}\gamma h = \tfrac{1}{3} \times 22 \times 0.77 = 5.6 \text{ kN/m}^2.$$

Q19.4 Repeat Qs 19.1 and 19.2 for the haunch A (as illustrated in Q19.1).

A19.4 We use the same method but with an inclined projection (see Chapter 7 for examples of the use of an inclined projection) giving the following results.

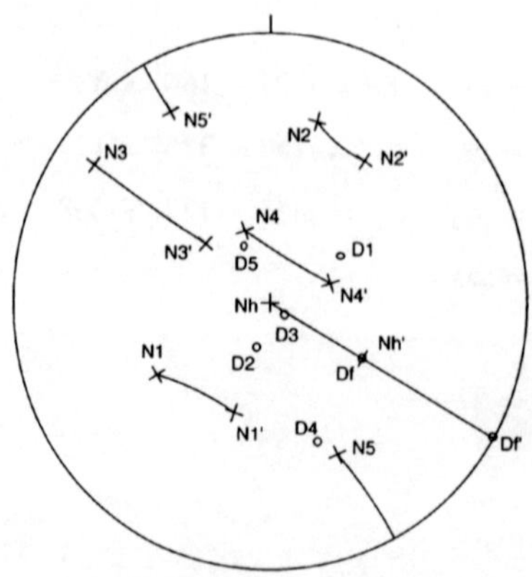

Inclination of principal elements

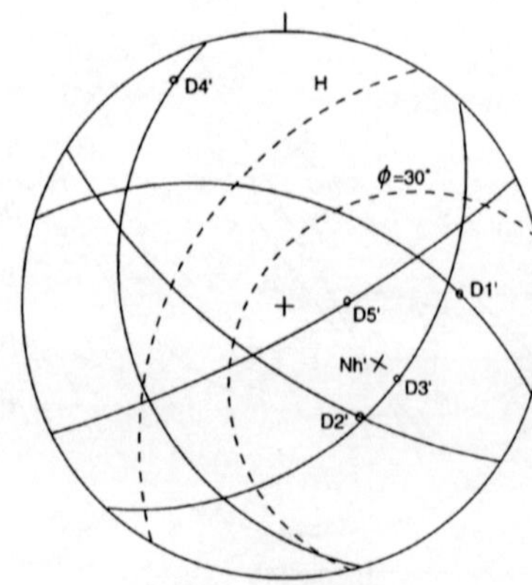

Inclined projection

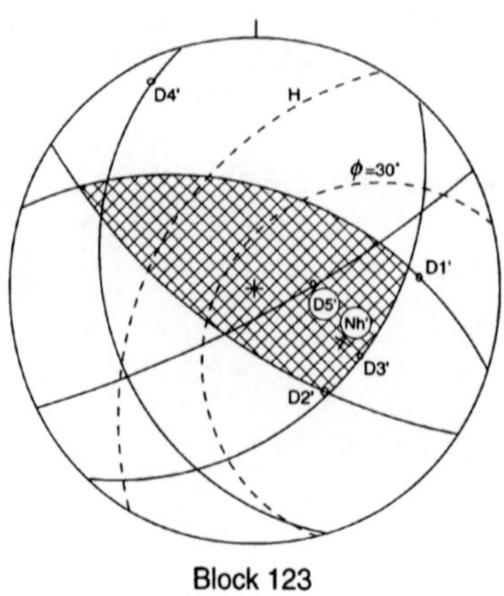

Block 123

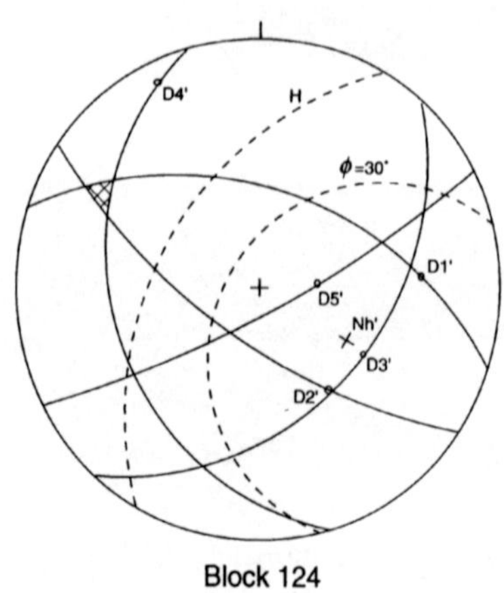

Block 124

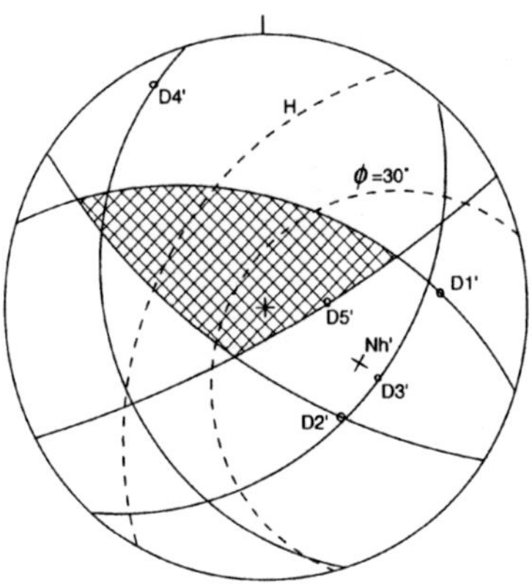

Block 125

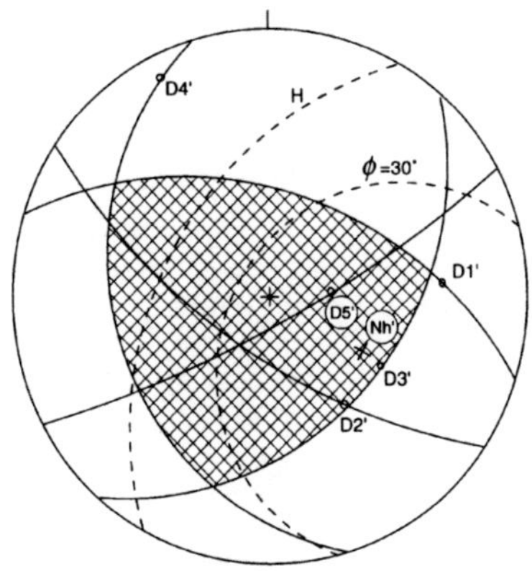

Block 134

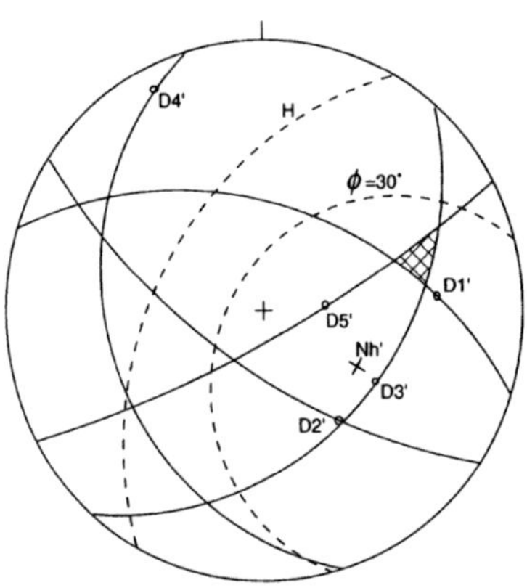

Block 135

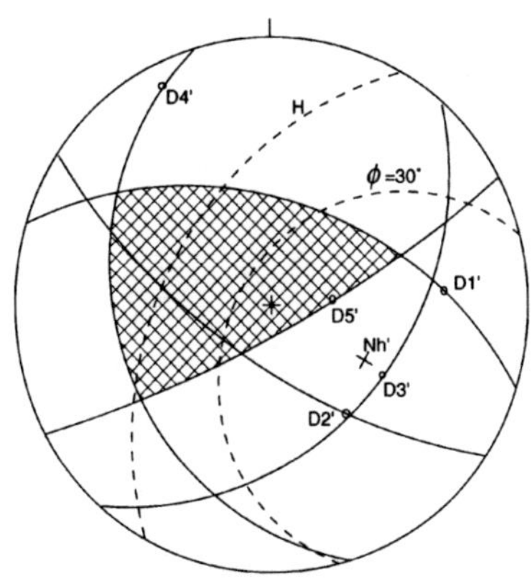

Block 145

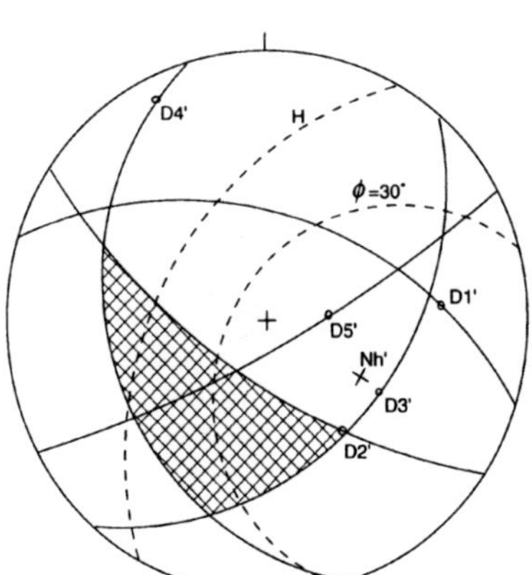

Block 234

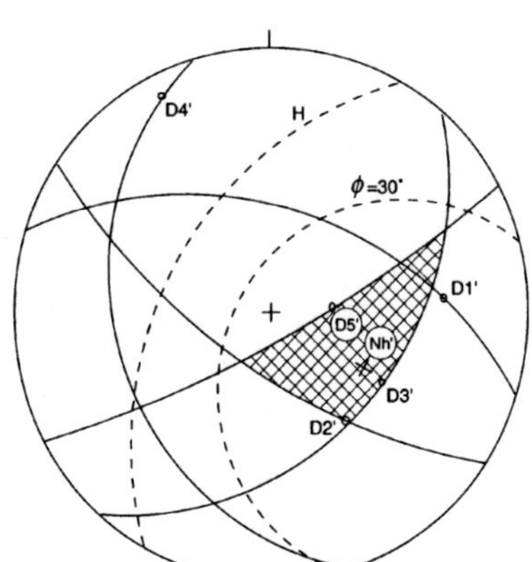

Block 235

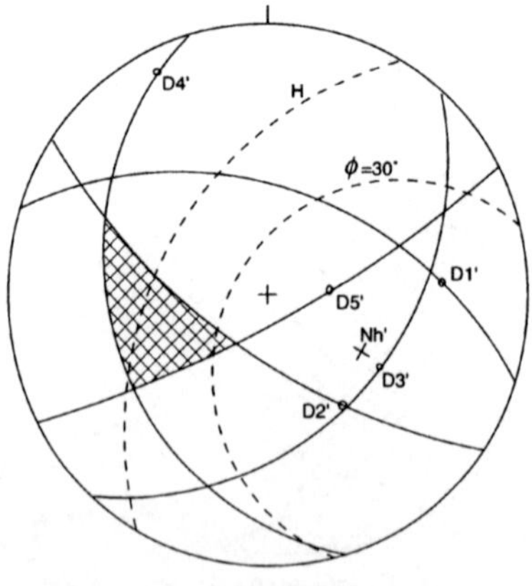

Block 245

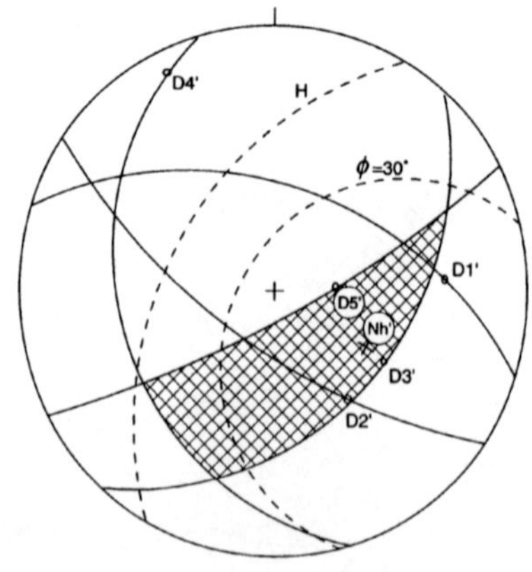

Block 345

Block	Stability condition	Direction
123	falls	
124	stable (block is directed upwards)	
125	slides along line of maximum dip of set 5	355/64
134	falls	
135	slides along intersection of sets 1 and 3	049/54
145	slides along line of maximum dip of set 5	355/64
234	slides along line of maximum dip of set 2	195/70
235	falls	
245	slides along intersection of sets 2 and 5	268/39
345	falls	

Q19.5 Repeat Qs 19.1 and 19.2 for side wall B (as illustrated in Q19.1).

A19.5 Using the same methodology but with an inclined projection, the following results are obtained.

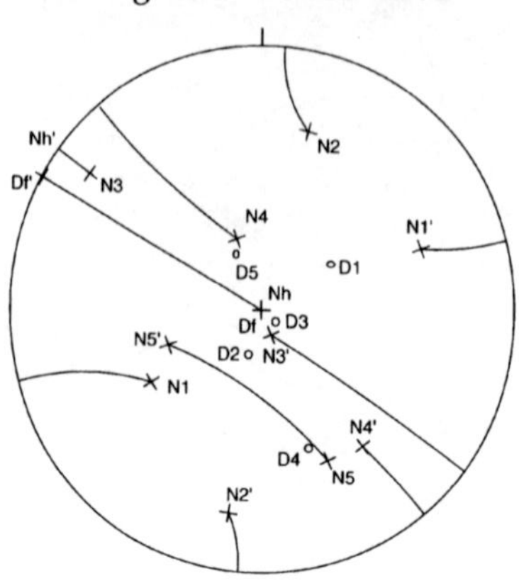

Inclination of principal elements

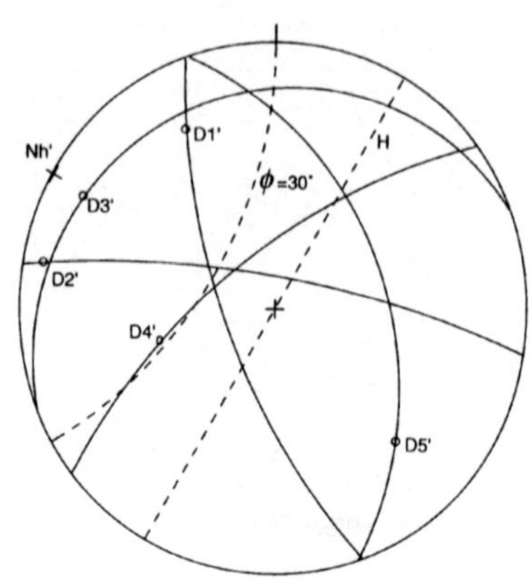

Inclined projection

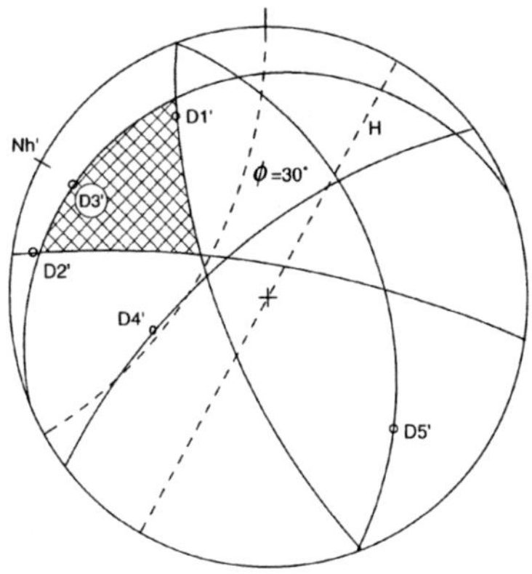

Block 123

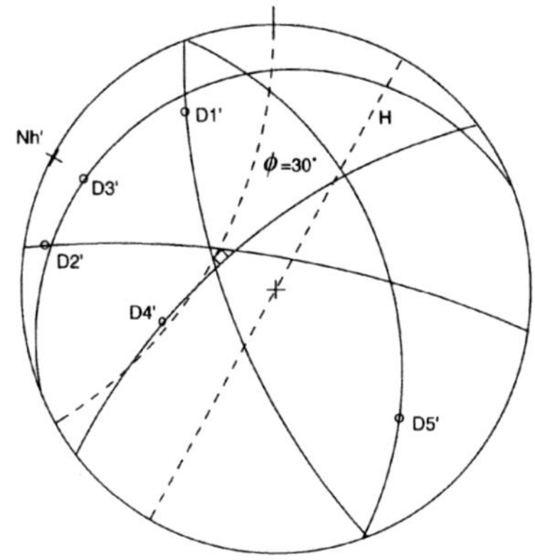

Block 124

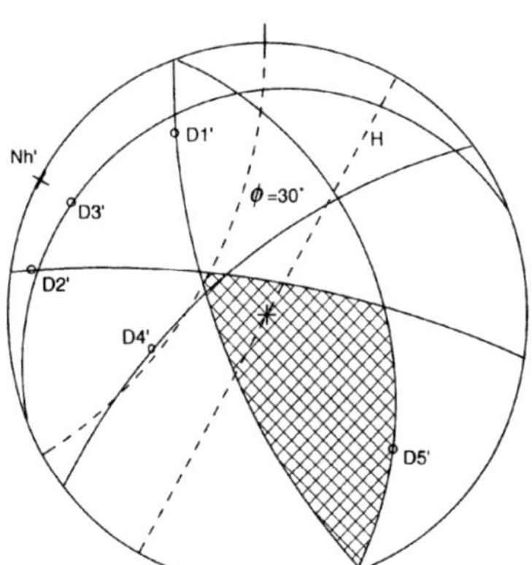

Block 125

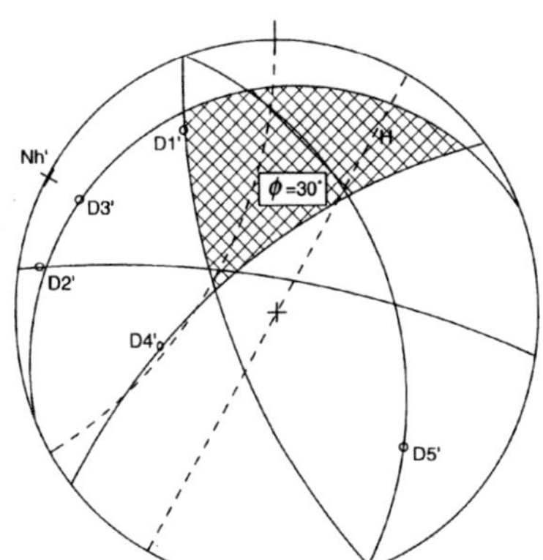

Block 134

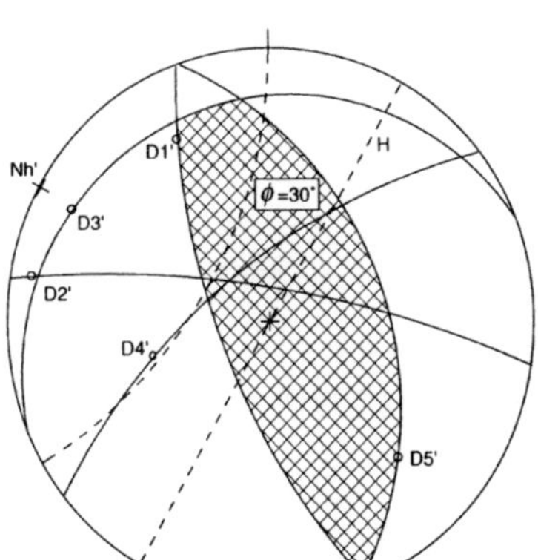

Block 135

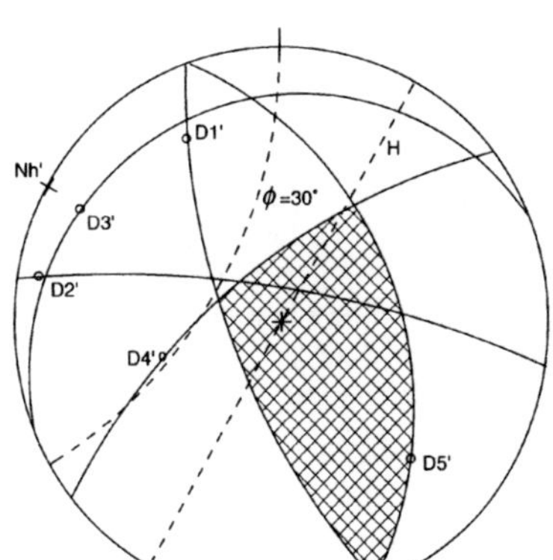

Block 145

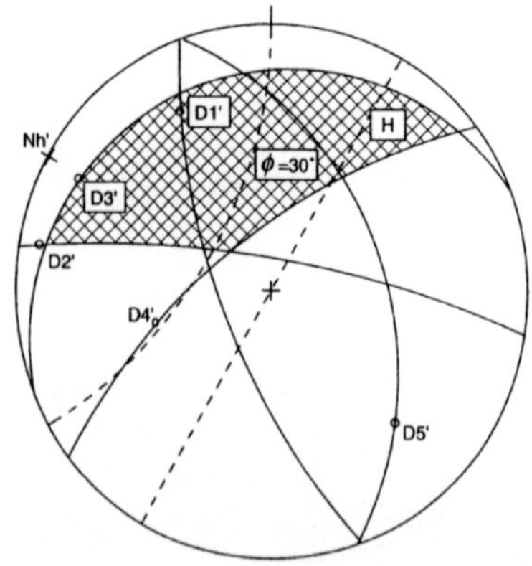

Block 234

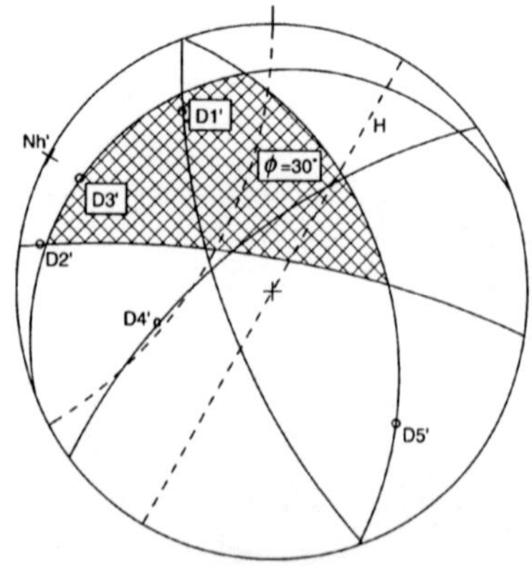

Block 235

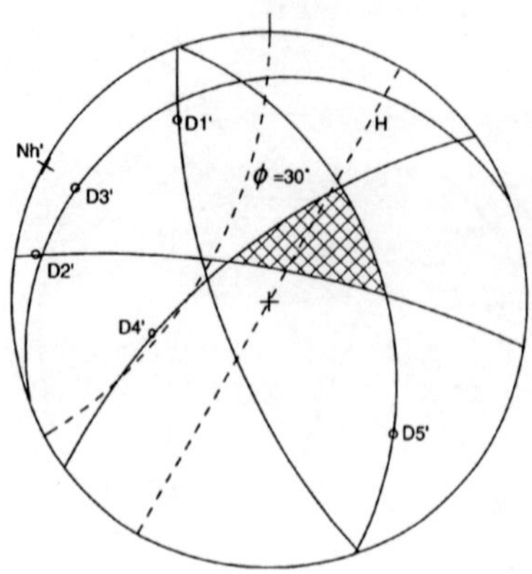

Block 245

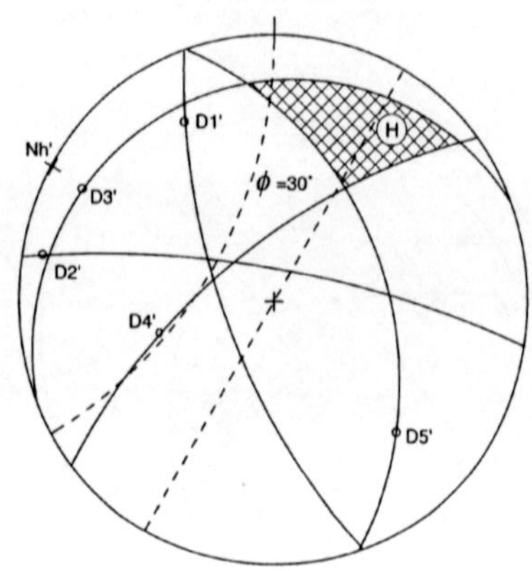

Block 345

Block	Stability condition	Direction
123	slides along line of maximum dip of set 3	127/81
124	slides along intersection of sets 1 and 2	119/34
125	slides along intersection of sets 1 and 2	119/34
134	slides along line of maximum dip of set 1	058/54
135	slides along line of maximum dip of set 1	058/54
145	stable (no part of the block dips in excess of the friction angle)	
234	slides along line of maximum dip of set 3	127/81
235	slides along line of maximum dip of set 3	127/81
245	stable (no part of the block dips in excess of the friction angle)	
345	slides along intersection of sets 3 and 5	044/37

Q19.6 For both the haunch A and the side wall B studied in Qs 19.4 and 19.5, and with the information given in Q19.1, calculate the factor of safety of the block formed by fracture sets 1, 2 and 5.

A19.6 In the south east haunch of the excavation, block 125 suffers from sliding along the line of maximum dip of fractures from set 5 (i.e. in the direction 335/64). We are told that the strength of the fractures is purely frictional, with $\phi = 30°$. If we furthermore assume that there is zero water pressure, then the stability problem reduces to that of a block sliding on a plane, and is independent of the size and shape of the tetrahedral wedge. The only controlling factors in this type of problem are the friction angle and the dip angle of the fracture, and the factor of safety is therefore given by

$$F_{\mathrm{p}} = \frac{\tan\phi}{\tan\psi} = \frac{\tan 30}{\tan 64} = \frac{0.577}{2.050} = 0.28.$$

In the north west wall of the excavation, block 125 exhibits sliding along the line of intersection of fracture sets 1 and 2. As with the case above, the wedge stability problem reduces to that presented earlier for rock slopes in which we used the concept of the wedge factor to determine the factor of safety.

As with rock slopes, the orientations of the excavation surface and the third plane making up the tetrahedral wedge (fracture set 4 in this case) do not enter into the calculation, other than to ensure the kinematic feasibility of the wedge (and this, of course, was checked in A19.5). The geometry of the wedge and the angles required for the wedge factor are shown in the lower hemispherical projection below. The factor of safety is then given by

$$F_{\mathrm{w}} = k_{\mathrm{w}} \times F_{\mathrm{p}} = \frac{\sin\beta}{\sin\frac{1}{2}\xi} \times \frac{\tan\phi}{\tan\psi} = \frac{\sin 80}{\sin 35} \times \frac{\tan 30}{\tan 34} = \frac{0.985}{0.574} \times \frac{0.577}{0.675} = 1.47.$$

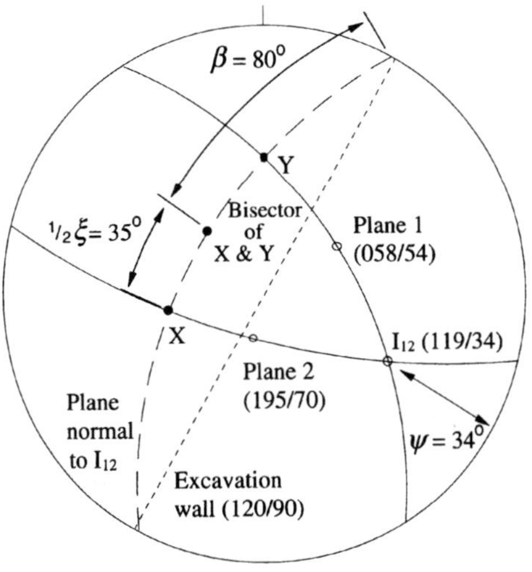

If the factor of safety is computed analytically, its value is found to be 1.50. The discrepancy between this and the value given above is due to the limited accuracy of the hemispherical projection. However, as the error is only some 2%, we can see that hemispherical projection techniques are a practical means of making preliminary assessments of wedge sliding stability.

Q19.7 At a depth of 450 m, a 3-m diameter circular tunnel is driven in rock having a unit weight of 26 kN/m³ and uniaxial compressive and tensile strengths of 60.0 MPa and 3.0 MPa, respectively. Will the strength of the rock on the tunnel boundary be reached if
(a) $k = 0.3$, and
(b) $k = 2.5$?

A second tunnel, of 6 m diameter, is subsequently driven parallel to and at the same centre line level as the first, such that the centre line spacing of the two tunnels is 10 m. Comment on the stability of the tunnels for the field stresses given by (a) and (b) above.

A19.7 This is a problem to be solved using the Kirsch solution for circular openings in a state of plane strain. We are asked to examine the stability of the rock on the boundary of the tunnel. As the tunnel has neither a support pressure nor an internal pressure applied to it, the rock on the boundary is subjected to a uniaxial state of stress, with the local $\sigma_3 = \sigma_r = 0$ and local $\sigma_1 = \sigma_\theta$.

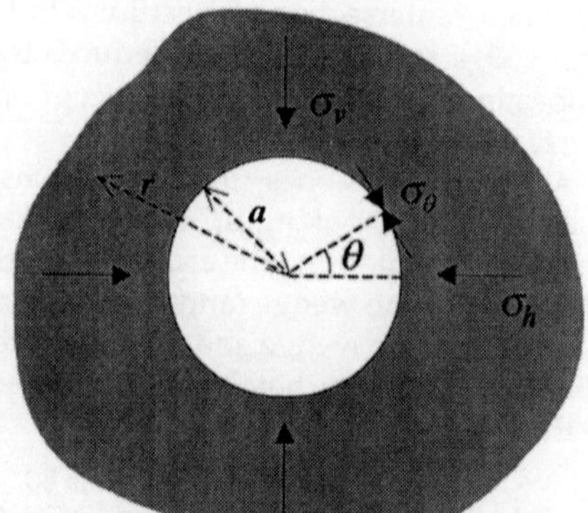

The Kirsch solution for the circumferential stress is

$$\sigma_\theta = \tfrac{1}{2}\sigma_v\left[(1+k)\left(1+\frac{a^2}{r^2}\right) + (1-k)\left(1+3\frac{a^4}{r^4}\right)\cos 2\theta\right]$$

and for a location on the tunnel boundary, where $a = r$, this reduces to

$$\sigma_\theta = \sigma_v\left[(1+k) + 2(1-k)\cos 2\theta\right].$$

We assume that the vertical stress is caused by the weight of the overburden, in which case we have

$$\sigma_v = \gamma \cdot z = 0.026 \cdot 450 = 11.70 \text{ MPa}.$$

The extreme values of induced stress occur at positions aligned with the principal *in situ* stresses, and so in order to compute the stress induced in the crown and invert of the tunnel we use $\theta = 90^\circ$, and for the sidewalls we use $\theta = 0^\circ$. For the case of $k = 0.3$, the stresses are then found to be -1.17 MPa (i.e. tensile) and 31.59 MPa, respectively. For the case of $k = 2.5$, the induced stresses are 76.05 MPa and 5.85 MPa. Thus, we see that the compressive strength of the rock, 60 MPa, is reached at the crown and invert of the tunnel for the case of $k = 2.5$.

After the second tunnel has been driven, we can find an approximate solution to the problem through a multiple application of the Kirsch solution. We start by determining whether the tunnels are inside each other's zone of influence. Given the approximate nature of the solution, a rigorous computation of this is not justified. Instead, we take the distance of the 5% zone of influence for the hydrostatic case and examine the tunnel spacings on this basis.

For the first tunnel we have $a = 1.5$ m, from which we find that $r(5\%) = a\sqrt{20} = 1.5\sqrt{20} = 6.7$ m. For the second tunnel, for which $a = 3.0$ m, we find $r(5\%) = 13.4$ m. The centre-to-centre spacing of the tunnels is 10 m. Thus, on the basis of the 5% zone of influence, we see that the first tunnel is affected by the second, but the second is not affected by the first. This means that the stresses induced on the boundary of the second tunnel are as computed above, but for the first tunnel we need to use an approximation. We do this by computing the stress state induced by the second tunnel at the position of the centre of the first tunnel, and then using this computed stress state to determine the stresses induced on the boundary of the first tunnel.

As the centre-to-centre distance of the tunnels is 10 m, and as the two tunnels are at the same elevation, we use $r = 10$ m and $\theta = 0^\circ$ in the Kirsch equations. As the second tunnel has a radius $a = 3$ m, for the case of $k = 0.3$ we find $\sigma_\theta = 12.48$ MPa and $\sigma_r = 4.20$ MPa. The shear stress is zero. This means that, for the first tunnel, we have $\sigma'_v = 12.48$ MPa and $k' = 4.20/12.48 = 0.336$. The stress induced in the crown and invert is then found to be

$$\begin{aligned}\sigma_\theta &= \sigma'_v\left[(1+k') + 2(1-k')\cos 2\theta\right] \\ &= 12.48\,[(1+0.336) + 2(1-0.336)\cos 180] = 0.12 \text{ MPa}\end{aligned}$$

and the stress induced in the sidewalls is

$$\begin{aligned}\sigma_\theta &= \sigma'_v\left[(1+k') + 2(1-k')\cos 2\theta\right] \\ &= 12.48\,[(1+0.336) + 2(1-0.336)\cos 0] = 33.25 \text{ MPa.}\end{aligned}$$

For the case of $k = 2.5$ the stresses induced at the centre of the first tunnel by the second tunnel are $\sigma_\theta = 13.33$ MPa and $\sigma_r = 24.46$ MPa, which means that, for the first tunnel, we have $\sigma'_v = 13.33$ MPa and $k' = 24.46/13.33 = 1.84$. The stress induced in the crown and invert is then found to be

$$\begin{aligned}\sigma_\theta &= \sigma'_v\left[(1+k') + 2(1-k')\cos 2\theta\right] \\ &= 13.33\,[(1+1.84) + 2(1-1.84)\cos 180] = 60.05 \text{ MPa}\end{aligned}$$

and the stress induced in the sidewalls is

$$\begin{aligned}\sigma_\theta &= \sigma'_v\left[(1+k') + 2(1-k')\cos 2\theta\right] \\ &= 13.33\,[(1+1.84) + 2(1-1.84)\cos 0] = 15.53 \text{ MPa.}\end{aligned}$$

As a check of the accuracy of the approximate method, we can analyse the problem using a computational method. For the case of a CHILE material under plane strain, a simple 2D boundary element method is ideal for computing the induced stresses. Here, we have used the program BOUND, as presented by Beer and Watson (1992) [1].

[1] Beer G. and Watson J. O. (1992) *Introduction to Finite Element and Boundary Element Methods for Engineers*. Wiley, Chichester. Note that the program BOUND is a simple educational tool and ideal for investigating problems such as the one in Q19.7.

The discretization of the problem is shown below. We have taken advantage of vertical symmetry, and used four boundary elements on the half-boundary of each tunnel.

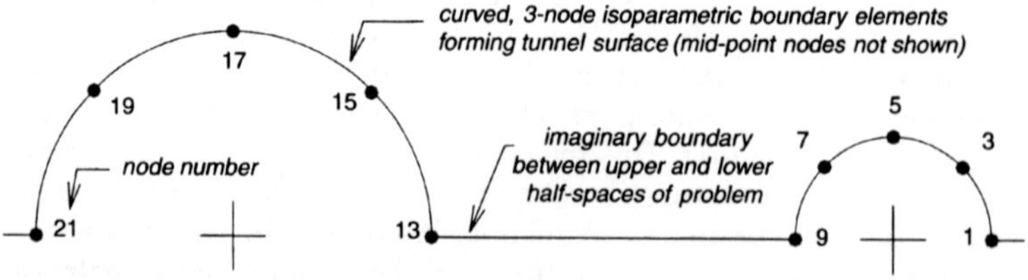

We initially examine the stresses induced around each of the tunnels individually, and the stresses induced when they both exist. The results are given below.

3.0 m diameter tunnel				6.0 m diameter tunnel				θ	Kirsch solution	
$k = 0.3$		$k = 2.5$		$k = 0.3$		$k = 2.5$			$k = 0.3$	$k = 2.5$
node	circum-ferential stress (MPa)	node	circum-ferential stress (MPa)	node	circum-ferential stress (MPa)	node	circum-ferential stress (MPa)		circum-ferential stress (MPa)	
1	31.50	1	5.90	13	31.50	13	5.90	0°	31.59	5.85
5	−1.14	5	76.00	17	−1.14	17	76.00	90°	−1.17	76.05
9	31.50	9	5.90	21	31.50	21	5.90	180°	31.59	5.85

These results show how the induced stresses are independent of the size of the tunnel, and how the induced stresses are symmetric for each tunnel. They are also essentially identical with the results obtained using the Kirsch solution.

We now analyse the problem when both tunnels are in place. The results are shown in the table below and in the two graphs following.

Boundary element 3.0 m and 6.0 m diameter tunnel				Kirsch-based approximation 3.0 m and 6.0 m diameter tunnel			
$k = 0.3$		$k = 2.5$		$k = 0.3$		$k = 2.5$	
node	circum-ferential stress (MPa)	node	circum-ferential stress (MPa)	node	circum-ferential stress (MPa)	node	circum-ferential stress (MPa)
1	33.10	1	12.20	1	33.25	1	15.53
5	−0.26	5	62.50	5	0.12	5	60.05
9	33.70	9	19.70	9	33.25	9	15.53
13	32.10	13	10.90	13	31.59	13	5.85
17	−1.12	17	73.90	17	−1.17	17	76.05
21	31.90	21	6.94	21	31.59	21	5.85

The results here are revealing and instructive. Two points need to be emphasized.

(a) In a general sense, the approximation gives reasonable results (it does, for example, correctly predict zones of tension and compression).
(b) The approximation predicts the stresses around both tunnels equally well.

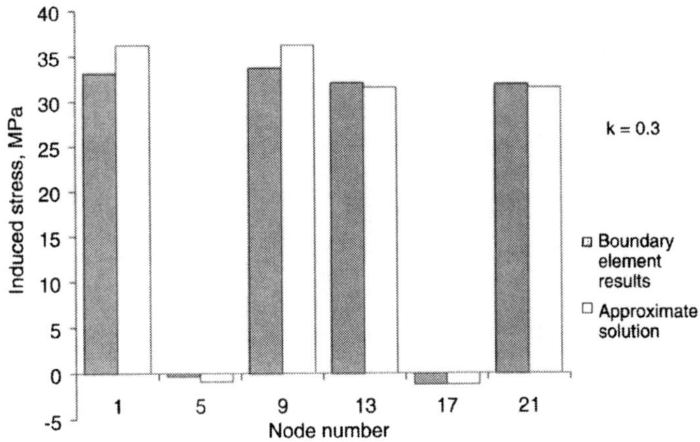

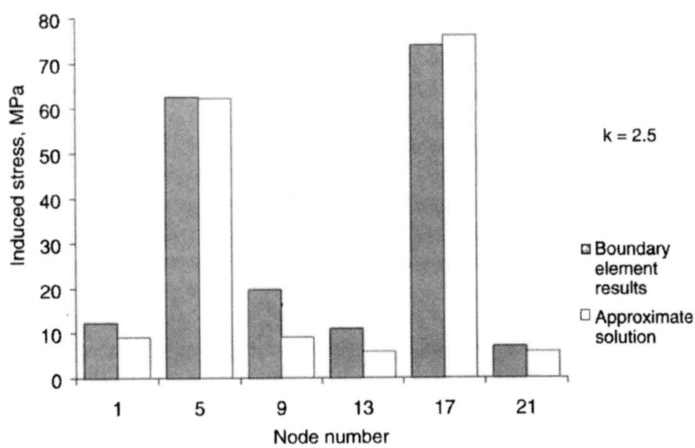

Q19.8 The diagram shows the relative positions of two parallel horizontal tunnels, each 3 m in diameter. Prior to excavation, the principal stresses in the area were $p_x = p_y = p_z = 11$ MPa.

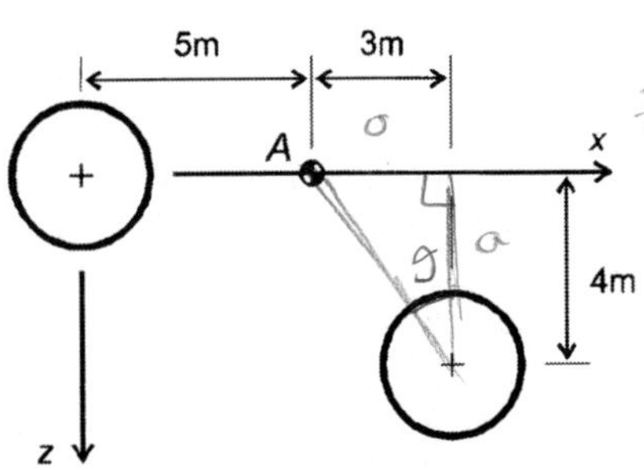

(a) Determine the principal stresses and their directions at point A after excavation has been completed.

(b) A horizontal fault coincident with the x-axis passes through A. If the shear strength of the fault is purely frictional with $\phi = 20°$, will slip on the fault occur at A?

A19.8 Here we will apply the Kirsch solution to each excavation and superpose the results in order to estimate the state of stress at point A.

For the left-hand tunnel and point A, the parameters required for the Kirsch solution are $k = 1.0$, $\sigma_v = 11.0$ MPa, $a = 1.5$ m and $r = 5$ m. The induced stresses at point A are then

$$\sigma_\theta = \frac{1}{2}\sigma_v\left[(1+k)\left(1+\frac{a^2}{r^2}\right)+(1-k)\left(1+3\frac{a^4}{r^4}\right)\cos 2\theta\right]$$
$$= \sigma_v\left(1+\frac{a^2}{r^2}\right) = 11.99 \text{ MPa}$$

and

$$\sigma_r = \frac{1}{2}\sigma_v\left[(1+k)\left(1-\frac{a^2}{r^2}\right)-(1-k)\left(1-4\frac{a^2}{r^2}+3\frac{a^4}{r^4}\right)\cos 2\theta\right]$$
$$= \sigma_v\left(1-\frac{a^2}{r^2}\right) = 10.01 \text{ MPa.}$$

For the right-hand tunnel and point A, the parameters required for the Kirsch solution are identical to those used above. Furthermore, as the *in situ* stress state is isotropic, then the stresses induced at point A due to the right-hand tunnel are the same as those due to the left-hand tunnel, i.e. $\sigma_\theta = 11.99$ MPa and $\sigma_r = 10.01$ MPa.

In order to superpose the above results, we must first ensure that the stress states are referred to the same co-ordinate system. Later, we are asked to consider the stability of a horizontal fault through point A, and on this basis it is appropriate to use the system of axes shown. As a result, we must transform the stresses computed for the right-hand tunnel. Here for simplicity we will use stress transformation equations based on Mohr's circle.

At point A and for the right-hand tunnel, the angle to be turned through to rotate from the direction of the major principal stress (i.e. σ_θ) to the direction that is normal to the fault (i.e. vertical) is

$$\alpha = \arctan\left(\tfrac{4}{3}\right) = 53.1°.$$

The transformed stresses are then

$$\sigma_z = \frac{\sigma_\theta+\sigma_r}{2}+\frac{\sigma_\theta-\sigma_r}{2}\cos 2\alpha$$
$$= \frac{11.99+10.01}{2}+\frac{11.99-10.01}{2}\cos(106.2) = 10.72 \text{ MPa}$$

$$\sigma_x = \frac{\sigma_\theta+\sigma_r}{2}+\frac{\sigma_\theta-\sigma_r}{2}\cos 2(\alpha+90)$$
$$= \frac{11.99+10.01}{2}+\frac{11.99-10.01}{2}\cos(286.2) = 11.28 \text{ MPa}$$

$$\tau_{xz} = -\frac{\sigma_\theta-\sigma_r}{2}\sin(2\alpha) = -\frac{11.99-10.01}{2}\sin(106.2) = -0.95 \text{ MPa}$$

When we superpose stress states computed using the Kirsch solution, we must remember to subtract the additional field stresses that are included in each solution. The final stress state is then given by

$$\begin{bmatrix} \sigma_x & \tau_{xz} \\ \tau_{xz} & \sigma_z \end{bmatrix}_{\text{point A}} = \begin{bmatrix} \sigma_r & 0 \\ 0 & \sigma_\theta \end{bmatrix}_{\text{left tunnel}} + \begin{bmatrix} \sigma_x & \tau_{xz} \\ \tau_{xz} & \sigma_z \end{bmatrix}_{\text{right tunnel}} - \begin{bmatrix} k\sigma_v & 0 \\ 0 & \sigma_v \end{bmatrix}_{\text{field stresses}}$$

which evaluates to

$$\begin{bmatrix} \sigma_x & \tau_{xz} \\ \tau_{xz} & \sigma_z \end{bmatrix}_{\text{A}} = \begin{bmatrix} 10.01 & 0 \\ 0 & 11.99 \end{bmatrix} + \begin{bmatrix} 11.28 & -0.95 \\ -0.95 & 10.72 \end{bmatrix} - \begin{bmatrix} 11.0 & 0 \\ 0 & 11.0 \end{bmatrix}$$

$$= \begin{bmatrix} 10.29 & -0.95 \\ -0.95 & 11.71 \end{bmatrix} \text{MPa.}$$

and from this the friction angle required to maintain stability is computed as

$$\phi = \arctan\left(\frac{\tau_{xz}}{\sigma_z}\right) = \arctan\left(\frac{-0.95}{11.71}\right) = -4.6^\circ$$

where the negative sign is an indicator of the sense of shearing. As the friction angle for the fault is 20°, it is therefore stable.

We can perform the calculation sequence given above for a number of different points (say, by using a computer spreadsheet) in order to examine the stress distribution along the fault. The results of this are shown in the figure below, together with stress distributions computed using a boundary element program. Notice that there is very good agreement between the two methods.

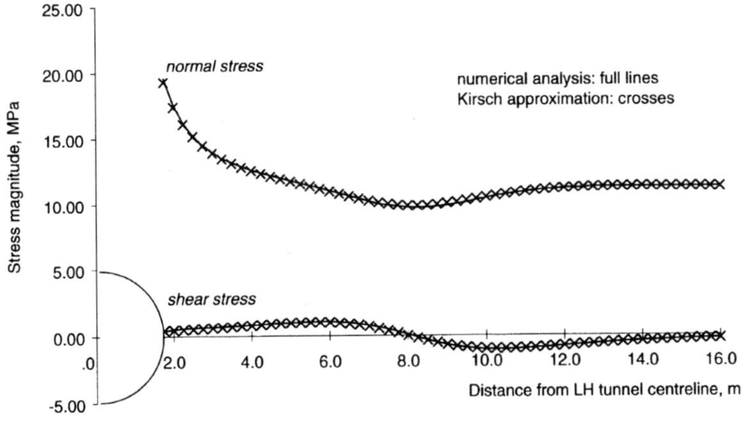

The figure below shows the variation of required friction angle, computed using both the Kirsch approximation and a boundary element program. The errors involved are relatively small — no more than half of one degree — and show how the Kirsch approximation can be used for initial assessments of stability.

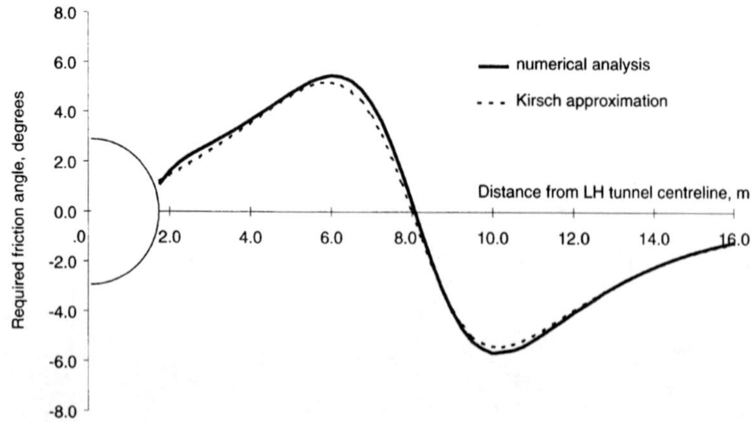

Q19.9 An ovaloid excavation at a depth of 750 m has in vertical section its major axis horizontal, and the ratio of its width to height is 4. The radius of curvature of its ends is equal to half its height. Assume that the *in situ* stress state can be calculated on the basis of complete lateral restraint in a CHILE medium (ERM 1, Section 4.6.2) with $\gamma = 28.0\ \mathrm{kN/m^3}$ and $\nu = 0.3$.

An elastic boundary element analysis for $k = 0$ shows that the stress in the centre of the roof is −20.5 MPa, and in the side wall is 104 MPa. An analysis with $k = 1$ gives corresponding stresses of 4.59 MPa and 84.2 MPa. What stresses would the boundary element analysis give for the *in situ* stress state?

Using the equations for stresses in terms of radius of curvature:

(a) compare the stress in the centre of the roof of the excavation with that for an elliptical excavation with the same width/height ratio; and

(b) compare the stress in the side wall with that for a boundary of equal radius of curvature.

A19.9 As the numerical analysis used an elastic model, we can determine the stresses induced for any value of k, the ratio of horizontal to vertical stress magnitudes, *pro rata*. For a case of complete lateral restraint, the ratio of horizontal to vertical stress magnitude is given by[2]

$$k = \frac{\sigma_H}{\sigma_V} = \frac{\nu}{1-\nu}$$

which means that in this case we have $k = 0.3/(1 - 0.3) = 0.429$. As we already have results for $k = 0$ and $k = 1$, the stress magnitudes for the *in*

[2] This simple formula is a traditional one in rock mechanics but note that, in a specific field case, the ratio between the horizontal and vertical stress components cannot be estimated using this formula, because of the influence of other factors, such as erosion and tectonic stresses as discussed in Chapter 4.

situ case can be computed using

$$\frac{\sigma_{k=0.429} - \sigma_{k=0}}{0.429 - 0} = \frac{\sigma_{k=1} - \sigma_{k=0}}{1 - 0},$$

and so for the roof we find

$$\sigma_{\text{roof}} = 0.429\,[4.59 - (-20.5)] + (-20.5) = -9.74 \text{ MPa}$$

and for the side wall we find

$$\sigma_{\text{side wall}} = 0.429\,[84.2 - 104] + 104 = 95.5 \text{ MPa}.$$

In addition, the magnitude of the vertical stress is $\sigma_V = \gamma z = 0.028 \times 750 = 21$ MPa.

For an elliptical opening oriented with its major axis horizontal, the circumferential stress induced in the boundary at the top and bottom of the opening is

$$\sigma_{\text{top}} = \sigma_v \left(k - 1 + k\sqrt{\frac{2H}{\rho_{\text{top}}}} \right) \text{ where } \rho_{\text{top}} = \frac{W^2}{2H}.$$

Combining these equations and substituting $q = W/H$ leads to

$$\sigma_{\text{top}} = \sigma_v \left[\frac{2k}{q} + k - 1 \right]$$

and from this, knowing that $q = 4$ for this particular opening, we find that

$$\sigma_{\text{top}} = \sigma_v \left(\frac{2k}{q} + k - 1 \right) = 21 \left(\frac{2 \cdot 0.429}{4} + 0.429 - 1 \right) = -7.5 \text{ MPa}.$$

The stress induced at the ends of this elliptical opening is given by

$$\sigma_{\text{end}} = \sigma_v \left(1 - k + \sqrt{\frac{2W}{\rho_{\text{end}}}} \right)$$

where $\rho_{\text{end}} = H/2$ because the opening is an ovaloid. These relations lead to

$$\sigma_{\text{end}} = \sigma_v \left(1 - k + 2\sqrt{q} \right)$$

from which we obtain

$$\sigma_{\text{end}} = 21 \left(1 - 0.429 + 2\sqrt{4} \right) = 96.0 \text{ MPa}.$$

These approximate solutions agree reasonably well with the results found from the rigorous numerical analysis. The first graph below gives a visual assessment of how the various computed stresses compare, and shows that for the side walls the agreement is very good, although for the roof and floor the agreement is not so good. The second graph below shows how the stresses compare for values of $k = 0$ and $k = 1$ for a range of values of q. With the exception of the case for stress induced in the roof when $k = 1$, the agreement is very good. Taken as a whole, these results show that the approximate solution can give a good insight into the behaviour of the opening.

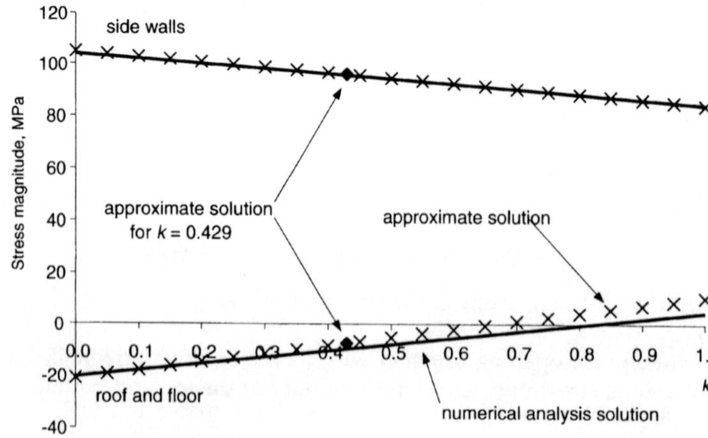

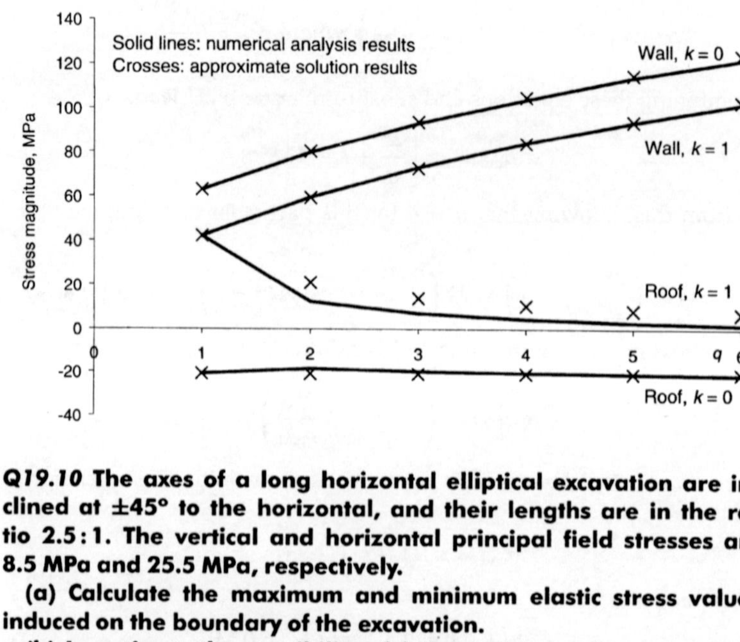

Q19.10 The axes of a long horizontal elliptical excavation are inclined at ±45° to the horizontal, and their lengths are in the ratio 2.5 : 1. The vertical and horizontal principal field stresses are 8.5 MPa and 25.5 MPa, respectively.

(a) Calculate the maximum and minimum elastic stress values induced on the boundary of the excavation.

(b) Investigate the possibility of failure of the rock on the excavation boundary, on the basis that the strength of the rock in compression is given by c = 30 MPa and $\phi = 40°$, and in tension is zero.

A19.10 The equation for the circumferential stress induced in the material at the boundary of an elliptical excavation is

$$\sigma_\theta = \frac{p}{2q}\left\{\begin{array}{l}(1+k)\left[(1+q^2)+(1-q^2)\cos 2(\chi-\beta)\right] \\ -(1-k)\left[(1+q)^2\cos 2\chi+(1-q^2)\cos 2\beta\right]\end{array}\right\}$$

where p is the vertical stress magnitude, k is the ratio of horizontal to vertical stresses, q is the ratio of the two axes of the ellipse, β is the inclination of the long axis to the horizontal, and χ is a parameter for locating position on the boundary. These parameters are shown in the sketch below, together with another location parameter θ.

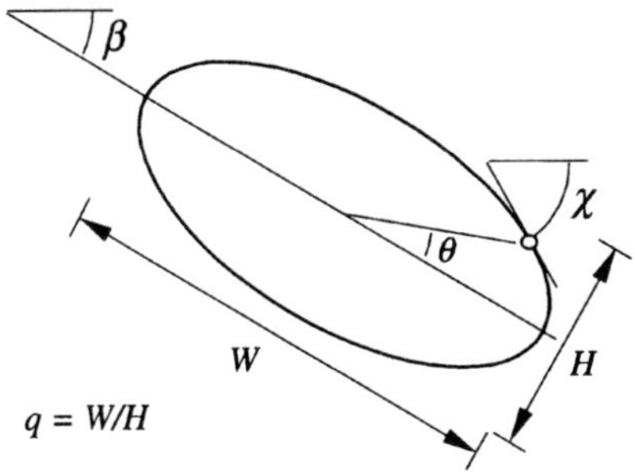

For the case when $\beta = 45^\circ$, then $\cos 2(\chi - \beta) = \cos(2\chi - 90) = \sin 2\chi$ and $\cos 2\beta = \cos 90 = 0$. Substituting these values into the equation above, reduces it to

$$\sigma_\theta = \frac{p}{2q}\left\{(1+k)\left[(1+q^2)+(1-q^2)\sin 2\chi\right]-(1-k)\left[(1+q)^2\cos 2\chi\right]\right\}.$$

In order to determine the locations on the boundary where the circumferential stress is a maximum or a minimum, we can differentiate this expression with respect to χ, set the result to zero and solve for χ. Thus, we have

$$\frac{\mathrm{d}\sigma_\theta}{\mathrm{d}\chi} = \frac{p}{2q}\left\{2(1+k)(1-q^2)\cos 2\chi + 2(1-k)(1+q)^2\sin 2\chi\right\}.$$

and, by equating this to zero and rearranging, we find

$$\tan 2\chi = -\frac{(1+k)(1-q)}{(1-k)(1+q)}.$$

For this particular excavation, we have $q = 2.5$ and $k = 25.5/8.5 = 3$, which leads to

$$\chi = \tfrac{1}{2}\arctan\left[-\frac{(1+k)(1-q)}{(1-k)(1+q)}\right]$$
$$= \tfrac{1}{2}\arctan\left[-\frac{-6}{-7}\right] = \tfrac{1}{2}\arctan[-0.857] = -20.3^\circ.$$

The form of the tangent function means that both $\chi = -20.3^\circ$ and $\chi = 90^\circ + (-20.3^\circ) = 69.7^\circ$ satisfy this equation. If we substitute these into the equation for σ_θ in order to find the magnitude of the circumferential stress at these orientations, we obtain

$$\sigma_\theta = \frac{8.5}{2\cdot 2.5}\left\{(1+3)\left[(1+2.5^2)+(1-2.5^2)\sin(-40.6^\circ)\right]\right.$$
$$\left. -(1-3)\left[(1+2.5)^2\cos(-40.6)\right]\right\} = 104.2 \text{ MPa}$$

and

$$\sigma_\theta = \frac{8.5}{2\cdot 2.5}\left\{(1+3)\left[(1+2.5^2)+(1-2.5^2)\sin(139.4^\circ)\right]\right.$$
$$\left. -(1-3)\left[(1+2.5)^2\cos(139.4^\circ)\right]\right\} = -5.6 \text{ MPa}.$$

Thus, we see that the first of these angles represents the location where the rock is subjected to the greatest compressive stress, and the second is the location where it is subjected to the greatest tensile stress.

In order to determine the compressive strength of the material, we use the relation

$$\sigma_c = \frac{2c}{\tan\left(45 - \frac{\phi}{2}\right)}$$

which was derived in A6.3 to find

$$\sigma_c = \frac{2 \cdot 30}{\tan\left(45 - \dfrac{40}{2}\right)} = \frac{60}{\tan 25} = 128.7 \text{ MPa}.$$

Thus, as the compressive strength of the rock is 128.7 MPa and its tensile strength is zero, we can see that the excavation boundary is stable at the location of maximum compressive stress, but unstable at the location of greatest tensile stress.

It is important to realize that the parameter χ is the angle the tangent to the boundary of the opening makes to the horizontal. A more useful parameter for locating position is the angle a point subtends to the major axis, and this is shown by θ in the sketch above. From the geometry of an ellipse, we find that these angles are related through the relation

$$\tan\theta = -\frac{1}{q\tan(\chi - \beta)}$$

although, in order to take account of the correct quadrant of the trigonometrical functions, the angle θ is better evaluated using the atan2 function[3]

$$\theta = \text{atan2}\left[-q\sin(\chi - \beta), \cos(\chi - \beta)\right].$$

Using this relation, we find that the locations of the points of maximum and minimum boundary stress relative to the major axis of the ellipse are then

$$\theta_{max} = \text{atan2}\left[-2.5\sin(-20.3^\circ - 45^\circ), \cos(-20.3^\circ - 45^\circ)\right]$$
$$= \text{atan2}(2.271,\ 0.418) = 10.4^\circ$$

and

$$\theta_{max} = \text{atan2}\left[-2.5\sin(69.7^\circ - 45^\circ), \cos(69.7^\circ - 45^\circ)\right]$$
$$= \text{atan2}(-1.045,\ 0909) = 139.0^\circ$$

The following diagram shows the complete boundary stress distribution, together with the locations of the maxima and minima. The lines drawn normal to the boundary represent the stress magnitude at that position, with lines inside the excavation profile indicating tensile stress.

A useful technique for qualitatively assessing stress distributions is the so-called 'streamline analogy', whereby we picture the opening as an obstruction standing in flowing water. For the case under consideration

[3] Programming languages such as Fortran include the atan2 function in order to resolve the issue of correct quadrant. The same effect can be obtained on a calculator using 'rectangular to polar' co-ordinate conversions.

here, the water is flowing horizontally across the page as this is the direction of the major principal stress. We can see that the boundary stress is greatest in those regions where the flowing water would first come into contact with the obstruction, and that the regions of tensile stress coincide with where we would expect eddies, or turbulence, to occur behind the obstruction.

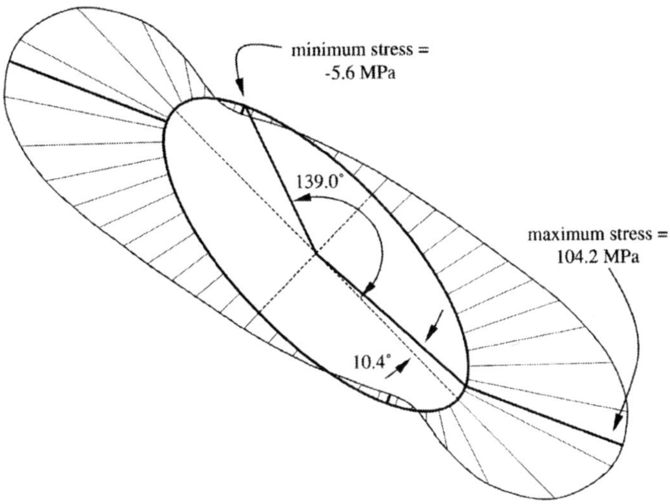

19.3 Additional points

We have emphasized the basic principles and techniques involved in considering underground excavation instability mechanisms. There is no substitute for this basic understanding of the 3-D geometry of rock

Figure 19.6 Visualizing a rock mass dissected by four fracture sets.

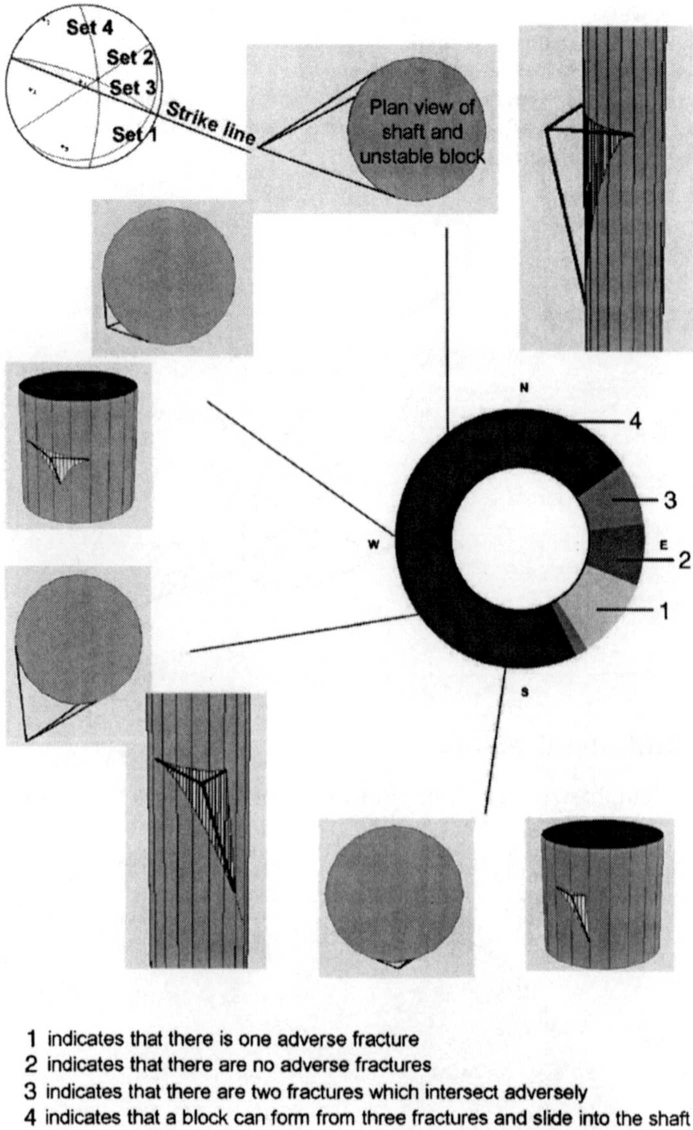

Figure 19.7 Example composite diagram showing the type of presentation that can be prepared to assist a rock engineer in deciding on the necessity and type of rock support.

blocks and the nature of stress distributions, but **computer programs** are commercially available now which can support the analysis with graphics and do most, if not all, of the tedious computation. For example, the rock mass visualization shown in Fig. 19.6 was obtained using 3DEC [4].

---

[4] 3DEC and other programs are available from Itasca at www.itascacg.com

Computer programs are available for many rock mechanics purposes including the analysis of rock mass geometry, the blocks which will be intersected by an excavation and which blocks will be unstable, the stresses and strains induced by excavation, and other subjects such as coupled thermo–hydro–mechanical predictions. The diagram in Fig. 19.7 of the predicted unstable blocks in shaft excavation was compiled from outputs of UNWEDGE[5].

The seminal book on the design of underground excavations is Hoek and Brown (1980), which includes chapters on planning considerations, classification of rock masses, geological data collection, graphical presentation of geological data, stresses around underground excavations, strength of rock and rock masses, underground excavation failure mechanisms, underground excavation support design, rockbolts and shotcrete and mesh, blasting in underground excavations, and instrumentation.

[5] UNWEDGE and other programs are available from www.rocscience.com

20 Design of underground excavations

20.1 The project objective

The purpose of the engineered structure will influence its design. For example, the stabilization measures required for an excavation such as that illustrated in Fig. 20.1 will depend on whether it is to be a permanent structure for a civil engineering project or a temporary structure for a mine. In civil, mining and petroleum engineering, there will be different constraints on the tolerable disturbance caused by excavation, on the rock displacements, on the type of instability that can be allowed to occur, on the type of support that could be installed, etc.

We have to consider what types of instability might be expected. Will it be instability of rock blocks defined by pre-existing fractures,

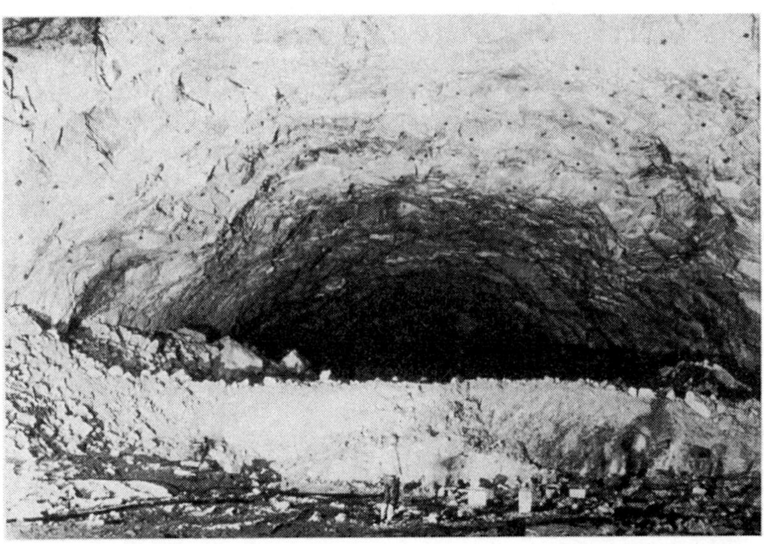

Figure 20.1 Construction of an underground excavation for a hydro-electric scheme.

Figure 20.2 Interaction of high stresses and rock mass fractures has combined to cause failure of the rock mass and the installed support in an underground gold mine.

stress-induced failure of the intact rock, or a combination of the two? The consequences of interaction between high stresses and rock mass structure are illustrated in Fig. 20.2, where we can see that the individual rock blocks have been displaced by a considerable amount and the installed support has been significantly distorted.

As with the design of surface structures on rock masses, it is necessary to establish the project objective before deciding on the type of analysis that will be conducted. The **project objective** determines the performance requirements of the structure, the modelling that should be used to provide the necessary predictive capability for rock engineering, and hence the rock mechanics information required. Although knowledge of, for example, the structural geology of the host rock and its basic mechanical properties will always be useful, there is no such thing as a standard modelling procedure and hence no standard site investigation procedure for rock engineering.

We encourage you to apply the basic principles of rock mechanics and rock engineering to all potentially complex analyses, as most of the trends in rock behaviour can be predicted from basic principles. The type of approach illustrated by the list of questions following in Section 20.2 is a good way to begin. It is only when the overall rock behaviour is not clear, or when large numbers of numerical values are required, that more sophisticated analyses should be implemented. Accordingly, we concentrate in this chapter on the application of the basic principles and techniques to a variety of rock engineering circumstances.

20.2 Question and answers: design of underground excavations

Q20.1 A vertical shaft is to be sunk through a rock mass in which the magnitudes of the two horizontal principal stresses have a ratio of 3:1. A clear space of 5 m x 5 m must be maintained within the shaft for equipment installation. The three excavation cross-sections shown in the sketch are under consideration for the shaft design.

Discuss the relative design merits of each cross-section in terms of rock mechanics principles, and hence provide recommendations for the optimal shape (and in the case of profile (b) and (c), optimal orientation) for the excavation.

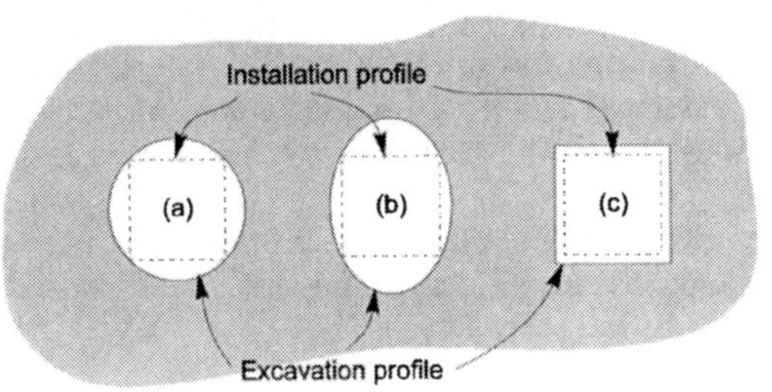

A20.1 **(a) Circular shaft**

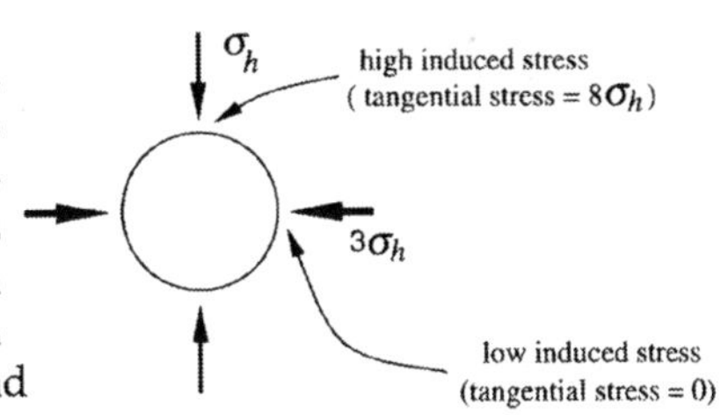

A CHILE analysis of the stresses around the circular shaft using the Kirsch solution shows that there will be a variation in the stress concentrations, from a value of 8 to a value of 0. Such large variations in stresses may lead to rock failure, and it is difficult to design the shaft with a single support or reinforcement solution.

(b) Elliptical shaft

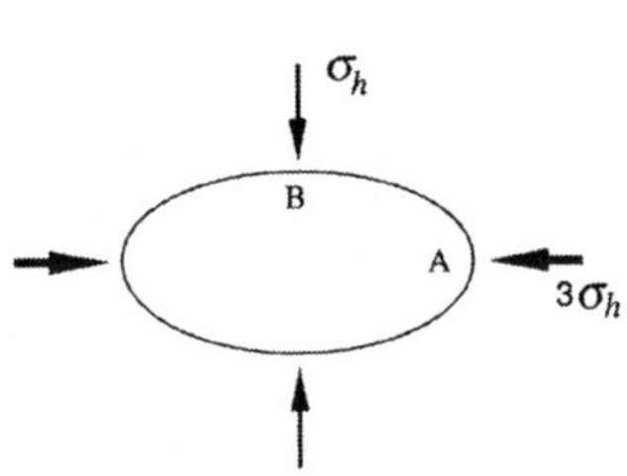

When the shaft is elliptical, if the ratio of the major axis to the minor axis is made equal to the stress ratio, and the excavation is aligned with its long axis parallel to the major principal stress, then the tangential stresses at points A and B are given by $\sigma_A = \sigma_B = (1 + k)\sigma_h$. Thus, if in this case the axial ratio is made equal to 3, then the stresses at points A and B are both equal to $4\sigma_h$. In fact, this tangential stress magnitude is constant around the periphery of the opening.

Providing that this stress is sufficiently less than the strength of the rock, the elliptical geometry is preferred because the uniform, induced stress field provides a zone of uniform tangential compression around the opening. This stabilizes the rock, in contrast to the stresses induced around the circular shaft.

(c) Square shaft

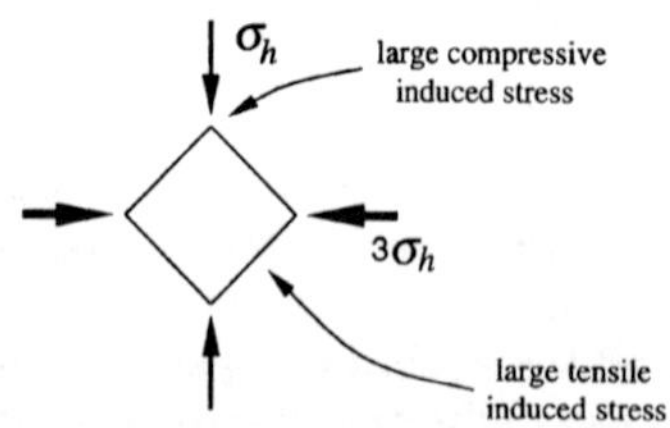

A closed form solution for the stresses induced around a square opening is not available, because of the discontinuous geometry at the corners. However, if we use the streamline analogy, together with knowledge of the relation between induced stresses and radius of curvature of the boundary of the opening, we can make an estimate of the stability of the excavation.

The small radius of curvature at the corners indicates that the tangential stress induced at this location will be large and compressive. Along each wall of the opening, the radius of curvature is effectively infinite, and so we can expect low, and probably tensile, stresses to be induced at these locations. This will be the case regardless of the orientation of the square within the stress field, or the exact values of the stresses.

Thus, the design strategy is to use an elliptical shaft oriented with its major axis parallel to the major principal stress, and the axial ratio of the opening equal to the stress ratio, i.e. 3.

Q20.2 You have been asked for an initial decision on the orientation of an underground structure. The structure comprises ten parallel tunnels, each 5 km long, to be built in a rock mass containing definite fracture sets. You have also been asked about the best shape for the cross-section of the tunnels.

The dip direction and dip of the fracture sets are 270/45 and 90/45; the direction of the maximum principal stress is horizontal, west–east, and its magnitude is twice the value of both the vertical and the other horizontal stress component. The ratio of the maximum principal stress component to the intact rock uniaxial compressive strength is 0.5.

What would be your initial design based on this information alone?

A20.2 The initial design should be based on avoiding instability of rock blocks and instability of the intact rock due to rock stress failure.

Given that there are only two main fracture sets, there may not be any potential for block failure (the minimum number of faces a rock block can have is four — a tetrahedral block — which requires the excavation surface and three fractures). However, it is good rock engineering policy to tunnel perpendicular to fractures where possible, because this reduces the problems of block instability, which would indicate a tunnel orientation of west–east.

The ratio of rock stress to rock strength is high: for example, a stress concentration value of 2 will raise the induced stress to the value of the rock strength. To reduce stress concentration problems, the tunnel should be parallel to the maximum principal stress, i.e. in a west–east direction. Since this is the same direction as indicated by avoidance of fracture problems, the choice of a west–east orientation should definitely be made based on this information alone.

The stress components perpendicular to the tunnel axis would then be equal. This means that a circular tunnel cross-section would be preferable because it will minimize stress concentrations.

Such design thinking, based on the rock mechanics principles alone, can often be all that is required to ensure the success of a rock engineering project. Wang and Fan (2000)[1] give an example of the strategic optimization of new coal mine development in soft rock in China based on such principles. The optimization procedure has enabled coal mines to be operated much more effectively than before because the roadway maintenance has been significantly reduced.

This example illustrates the significance of the concept of stress as a tensor. Although clear from the principles discussed, we may not be intuitively expecting that horizontal tunnelling in different directions through a homogeneous rock mass will produce different effects: in one direction the tunnel is stable and in another direction it is unstable. The fundamental reason for this phenomenon is that rock can sustain a shear stress.

Q20.3 A site is under investigation for use in a geothermal energy project. It is planned to drill a number of boreholes to a depth of about 2000 m and use these for injecting water at high pressure into the naturally fractured rock mass. The water will take heat from the surrounding rock, and will then be extracted from production boreholes and used for heating. A key question is the stability of the fractured rock mass during this process.

Extensive site investigations have revealed that the rock mass contains three pervasive sets of fractures, oriented at 000/00, 165/90 and 245/90, all of which have a friction angle of 44°. Also, a stress determination programme has established that the three principal stresses have orientations 000/90, 040/00 and 130/00. Their magnitudes (in MPa) are given by $26z$, $6 + 12z$ and $15 + 28z$ respectively, where z is the depth below ground surface. Although the intact rock is essentially impermeable, the rock mass is so highly fractured that it is naturally saturated with groundwater and hence effective stress conditions exist within the fractures.

The water injection holes have to be located relative to the production hole such that the pressurized water will flow along the natural fractures between them. For the target depth, examine the effective normal stress induced on a fracture from each of the three sets, and identify which set will be jacked open by the injected water. Hence, propose a layout for the injection boreholes. Bear in mind that the boreholes can be deviated during drilling such that they run horizontally, so that a vertical flow regime can be established if necessary.

[1] Wang Tongliang and Fan Qiuyan (2000) Optimization of Soft Rock Engineering with Particular Reference to Coal Mining. *Int. J. Rock Mech. Min. Sci.* 37.

Compute the water pressure required to open the critical fracture set (1) 250 m above, (2) 250 m below, and (3) at the target horizon.

Using these results, predict the flow regime and overall stability of the rock mass when water is injected at the target horizon.

Does this affect the layout of the boreholes and operation of the system, and if so, how?

A20.3 The relations given for the *in situ* stress magnitudes are

$$\sigma_z = 26z, \ \sigma_h = 6 + 12z \text{ and } \sigma_H = 15 + 28z.$$

The unit weight of rock is usually about 25 kN/m^3, and so it appears from these equations that the depth, z, has units of kilometres, with the constant of proportionality having units of MPa/km. The constant of proportionality in the relation for σ_z is comparable in magnitude to the unit weight of rock, and so shows that these are total stress relations. To convert them into effective stress relations, we need to subtract the unit weight of water, giving

$$\sigma'_z = (26 - 10)z = 16z, \quad \sigma'_h = 6 + (12 - 10)z = 6 + 2z, \text{ and}$$
$$\sigma'_H = 15 + (28 - 10)z = 15 + 18z.$$

At the target horizon, which is situated at a depth of 2 km, the *in situ* effective stresses are then $\sigma'_z = 32$ MPa, $\sigma'_h = 10$ MPa and $\sigma'_H = 51$ MPa.

In order to determine which of the three fracture sets is most susceptible to being jacked open by the pressurized water, we need to determine the effective normal stress acting on each of the sets. For the horizontal set, this is the vertical stress, giving $\sigma'_{n1} = 32$ MPa. For the two vertical sets, we need to apply a stress transformation in the horizontal plane to determine the normal stresses. The geometry of the problem in the horizontal plane is as shown below.

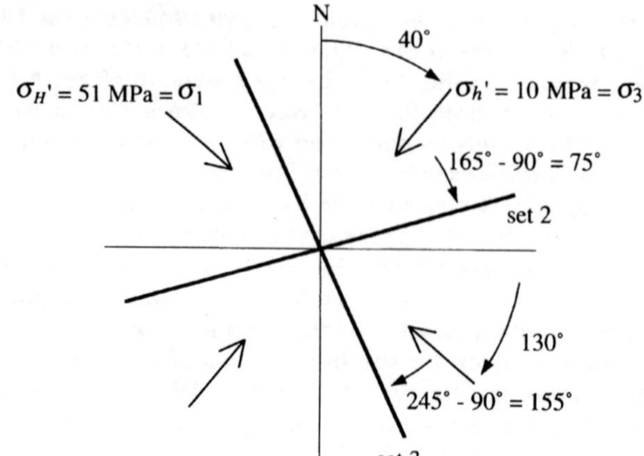

This diagram shows that the angle between the major principal stress component in the plane, σ'_H, and the normal to set 2 is $130^\circ - (75^\circ + 90^\circ) = -35^\circ$, and the angle to the normal to set 3 is 130° – (155° – 90°) = 65°, reckoning anticlockwise as positive. The Mohr circle diagram associated with this geometry is shown below.

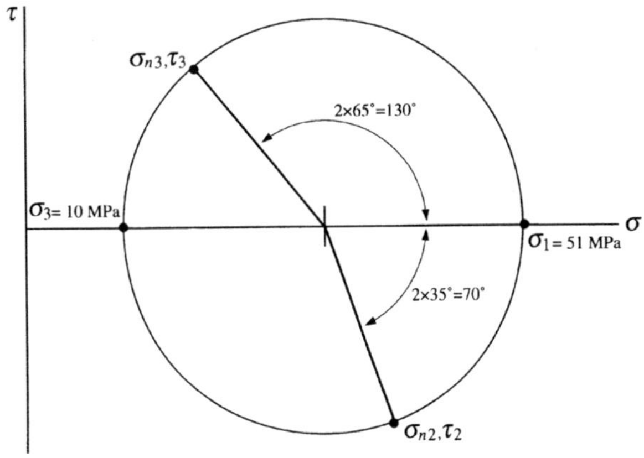

From this diagram we can see that set 3 has the lowest normal stress acting on it, and will therefore be the fracture set that is first jacked open by the pressurized water. By inspection of the diagram we see that the normal stress is

$$\sigma_{n3} = \frac{\sigma'_H + \sigma'_h}{2} + \frac{\sigma'_H - \sigma'_h}{2} \cos 130 = 30.5 + 20.5 \cos 130 = 17.3 \text{ MPa}.$$

As a consequence, the boreholes will need to be drilled such that they intersect these fractures as near to perpendicular as possible; the required borehole orientation is then either 065/00 or 245/00.

We can now examine the behaviour of these fractures at the three required depths of 1.75 km, 2.25 km and 2.00 km. We start by computing the normal and shear stress on the fractures, and then determine the fluid pressure required to induce shearing on the fracture. These calculations are shown in the table below, and in these the angle θ is 130°. The limiting normal stress for a given induced shear stress is calculated from the geometry of the strength criterion for the fractures, i.e. $\tau = \sigma_n \tan\phi = \sigma_n \tan 44$.

Depth					Induced normal stress,	Induced shear stress,	Limiting normal stress,	Critical fluid pressure,
	σ'_H	σ'_h	$c = \frac{\sigma'_H + \sigma'_h}{2}$	$r = \frac{\sigma'_H - \sigma'_h}{2}$	$c + r\cos\theta$	$\tau = r\sin\theta$	$\sigma_{lim} = \frac{\tau}{\tan 44}$	$\sigma - \sigma_{lim}$
(km)	(MPa)	(MPa)	(MPa)	(MPa)	(MPa)	(MPa)	(MPa)	(MPa)
1.75	46.5	9.5	28.0	18.5	16.11	14.17	14.68	1.43
2.00	51.0	10.0	30.5	20.5	17.32	15.70	16.26	1.06
2.25	55.5	10.5	33.0	22.5	18.54	17.24	17.85	0.69

From this table, we see that the critical fluid pressure decreases with depth. This means that, once the water pressure at a given horizon approaches the maximum allowable, it will induce shearing immediately below the horizon and hence the injected water will flow downwards. As a result, if the water is to be retrieved, the production boreholes will need to be located below the injection boreholes.

Q20.4 A circular tunnel of radius 4 m is to be driven in a weak rock mass subjected to an *in situ* hydrostatic stress field of 9 MPa. The triaxial compressive strengths of the rock mass in its initial and fractured states are given by $\sigma_1 = 8 + 4\sigma_3$ (MPa) and $\sigma_1 = 3\sigma_3$, respectively.

A preliminary analysis for design is to determine the relative instabilities of the roof, sidewalls and floor.

Plot the ground response curves for the roof, sidewalls and floor, taking $G = 2.1$ GPa, $f = 1.4$ and $\gamma = 25$ kN/m³, and comment on the appropriate support pressure.

A20.4 The ground characteristic for this tunnel has two parts: a linear part, extending from the *in situ* conditions to the point at which fracturing starts at the tunnel boundary; and a nonlinear part thereafter. We start by finding the support pressure below which a fracture zone develops. This is given by

$$p_1 = \frac{2p - C_o}{1 + b}$$

and for this case we have $p = 9$ MPa, $C_o = 8$ MPa and $b = 4$ (because the strength of the unfractured rock is $\sigma_1 = 8 + 4\sigma_3 = C_o + b\sigma_3$). The critical pressure is then

$$p_1 = \frac{2p - C_o}{1 + b} = \frac{2 \times 9 - 8}{1 + 4} = \frac{10}{5} = 2 \text{ MPa.}$$

The radial deformation at the boundary of the fractured zone is given by

$$u_{re} = -\frac{p - p_1}{2G} r_e$$

where r_e is the radius of the fractured zone. For the case of the critical pressure, which is when the fracture zone just starts to develop, we have $r_e = a = 4$ m and $p_1 = 2$ MPa. With $G = 2.1$ GPa we find that

$$u_{re} = -\frac{9 - 2}{2 \times 2.1 \times 10^3} \times 4 = \frac{28}{4.2 \times 10^3} = 0.0067 \text{ m.}$$

The linear part of the ground characteristic therefore extends from (0, 9) to (0.0067, 2) — in (displacement, pressure) co-ordinates.

For the nonlinear part of the ground characteristic, we choose a series of support pressure values, and then compute r_e, u_r, p_{roof} and p_{floor} for each value. The appropriate equations are

$$r_e = a \left[\frac{2p - C_o}{(1 + b) p_i} \right]^{1/(d-1)} \qquad u = -\frac{a(p - p_1)}{G(1 + f)} \left[\frac{f - 1}{2} + \left(\frac{r_e}{r} \right)^{1+f} \right]$$

$$p_{roof} = p_i + (r_e - r)\gamma_{rock} \qquad p_{floor} = p_i - (r_e - r)\gamma_{rock}.$$

We are told that $f = 1.4$, $d = 3$ (because $\sigma_1 = 3\sigma_3 = d\sigma_3$) and $\gamma_{rock} = 25\ \text{kN/m}^2$. Sample calculations are shown in the table, with the three ground characteristics shown in the diagram below.

p_i (MPa)	r_e (m)	u (m)	p_{roof}	p_{floor}
9.00		0.0000		
2.00	4.000	0.0067	2.00	2.00
1.80	4.216	0.0074	1.81	1.79
1.60	4.472	0.0084	1.61	1.59
1.40	4.781	0.0096	1.42	1.38
1.20	5.164	0.0114	1.23	1.17
1.00	5.657	0.0139	1.04	0.96
0.80	6.325	0.0178	0.86	0.74
0.60	7.303	0.0247	0.68	0.52
0.40	8.944	0.0394	0.52	0.28
0.20	12.649	0.0892	0.42	−0.02

The diagram shows how the floor stabilizes without a support pressure after a radial displacement of about 85 mm. However, the roof and sidewalls never stabilize and will always require supporting. It is also interesting to see from the table of results how the thickness of the zone of fractured rock increases with reducing support pressure. If an engineering requirement is to minimize the size of the fractured zone, then the support pressure needs to be close to 2 MPa.

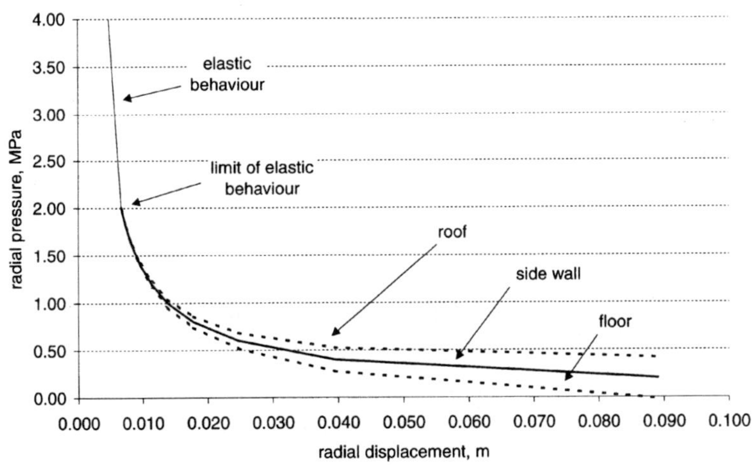

Q20.5 For the design of part of a large underground civil defence facility in a rock mass, there are two competing excavation geometries, as shown in the sketches of the vertical cross-sections given below. Both geometries consist of excavated rooms separated by rock pillars.

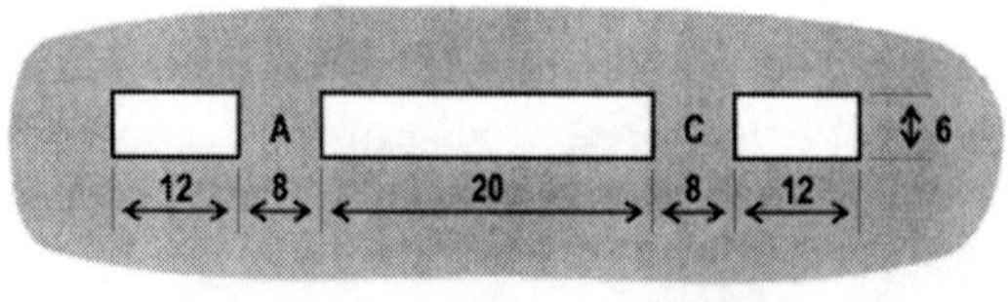

$$\delta_A = (56.20 - 3.304p_A - 0.533p_C) \times 10^{-3}$$
$$\delta_C = (56.20 - 0.533p_A - 3.304p_C) \times 10^{-3}$$

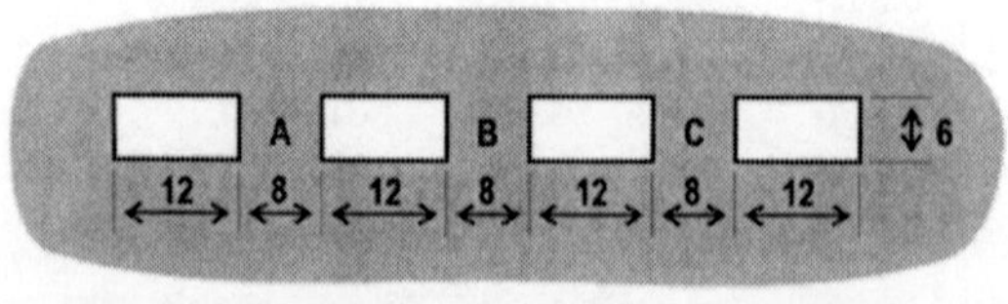

$$\delta_A = (56.20 - 3.304p_A - 1.121p_B - 0.533p_C) \times 10^{-3}$$
$$\delta_B = (66.52 - 1.121p_A - 3.602p_B - 1.121p_C) \times 10^{-3}$$
$$\delta_C = (56.20 - 0.533p_A - 1.121p_B - 3.304p_C) \times 10^{-3}.$$

Analysis of these geometries has been undertaken using a CHILE boundary element program, in order to determine the relation between displacement and support pressure at the various pillar locations. These relations are given below the sketches with δ representing displacement and p representing support pressure.

The support pressure is to be supplied by the rock pillars. Perform a rock-support interaction analysis for each of the two geometries in order to determine which is the preferred design. The stress–strain characteristics of the two pillars are as given below:

σ (MPa)	5.0	8.0	10.0	11.1	11.4	10.9	10.0	8.7	3.0	0.5
$\varepsilon \times 10^{-3}$	0.5	1.0	1.5	2.0	2.5	3.0	3.5	4.0	7.0	12.0

A20.5 Two-pillar system

For the two-pillar layout, we have the following two ground characteristics (i.e. one for each pillar):

$$\delta_A = (56.20 - 3.304p_A - 0.533p_C) \times 10^{-3}$$
$$\delta_C = (56.20 - 0.533p_A - 3.304p_C) \times 10^{-3}.$$

However, as the layout of the rooms is symmetrical, we can put $p_A = p_C = p$ and $\delta_A = \delta_C = \delta$ to obtain one equation

$$\delta = (56.20 - 3.837p) \times 10^{-3}. \tag{20.1}$$

This equation for the displacement of the boundary now has to be converted into an equation for strain in the pillar. This strain has two components: an initial strain due to the effect of the *in situ* stress before the excavation was made; and a subsequent strain induced by displacement of the boundary. It is therefore given by

$$\varepsilon = \varepsilon_i + \varepsilon_d = \varepsilon_i + \frac{\delta}{h} \tag{20.2}$$

where h, the height of the pillar, is 6 m in this case.

To determine the initial strain, we use the general form of Hooke's law, i.e.

$$\varepsilon_i = \frac{1}{E}\left[\sigma_i - \nu(\sigma_j + \sigma_k)\right]. \tag{20.3}$$

Now, as these pillars can be considered to be very long out of the plane of the cross-section, we can assume that they are in a state of plane strain. Taking the y-direction to be out of the plane of the cross-section, this allows us to write

$$0 = \frac{1}{E}\left[\sigma_y - \nu(\sigma_z + \sigma_x)\right]$$

and hence discover that

$$\sigma_y = \nu(\sigma_z + \sigma_x)$$

which, upon substitution into Eq. (20.3) leads to

$$\varepsilon_z = \frac{1+\nu}{E}\left[(1-\nu)\sigma_z - \nu\sigma_x\right].$$

Substituting $\nu = 0.2$ and $E = 9.12$ GPa, together with the initial stress state of $\sigma_x = 1$ MPa and $\sigma_z = 5$ MPa, gives the initial strain in the vertical direction as 0.5×10^{-3}. Eq. (20.2) then becomes

$$\varepsilon = 0.5 \times 10^{-3} + \frac{\delta}{6}. \tag{20.4}$$

Substituting Eq. (20.1)) into Eq. (20.4) leads to

$$p = -\frac{6000}{3.837}\varepsilon + \frac{59.20}{3.837} = -1564\varepsilon + 15.43$$

which is the ground characteristic in terms of strain. This can be plotted along with the pillar characteristic to find the operating point, as shown in the diagram below. The operating point is then found to be at a pillar pressure of 11.19 MPa but, much more importantly, is beyond the peak strength of the pillar. As the figure shows, the system is only marginally stable.

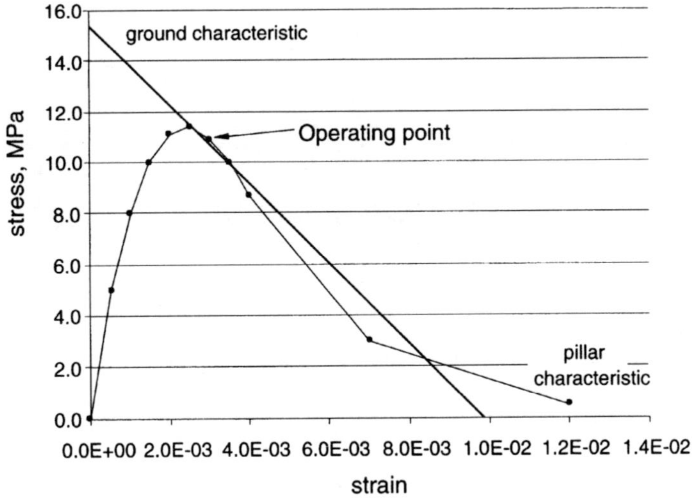

Three-pillar system

For the three-pillar layout, we have the following three ground characteristics (i.e. one for each pillar)

$$\delta_A = (56.20 - 3.304p_A - 1.121p_B - 0.533p_C) \times 10^{-3}$$

$$\delta_B = (66.52 - 1.121p_A - 3.602p_B - 1.121p_C) \times 10^{-3}$$

$$\delta_C = (56.20 - 0.533p_A - 1.121p_B - 3.304p_C) \times 10^{-3}.$$

As the layout is symmetric, we can put $p_C = p_A$ to give

$$\delta_A = (56.20 - 3.837p_A - 1.121p_B) \times 10^{-3}$$

$$\delta_B = (66.52 - 2.242p_A - 3.602p_B) \times 10^{-3}$$

and from these we find (by using Eq. (20.4))

$$\varepsilon_A = (9.867 - 0.6395p_A - 0.1868p_B) \times 10^{-3}$$

$$\varepsilon_B = (11.587 - 0.3737p_A - 0.6003p_B) \times 10^{-3}$$

Rearranging these equations into the usual format of $y = mx + c$ gives

$$p_A = -\frac{1000}{0.6395}\varepsilon_A + \frac{9.867 - 0.1868p_B}{0.6395} \quad (20.5)$$

$$p_B = -\frac{1000}{0.6003}\varepsilon_B + \frac{11.587 - 0.3737p_A}{0.6003} \quad (20.6)$$

In this case we have four unknowns (ε_A, ε_B, p_A, p_B) and four equations (the two ground characteristics above and a pillar characteristic for each of the two pillars). If we had linear equations for the pillar characteristics, we would be able to solve the system of equations directly. However, as the pillar characteristics are nonlinear this means that an iterative

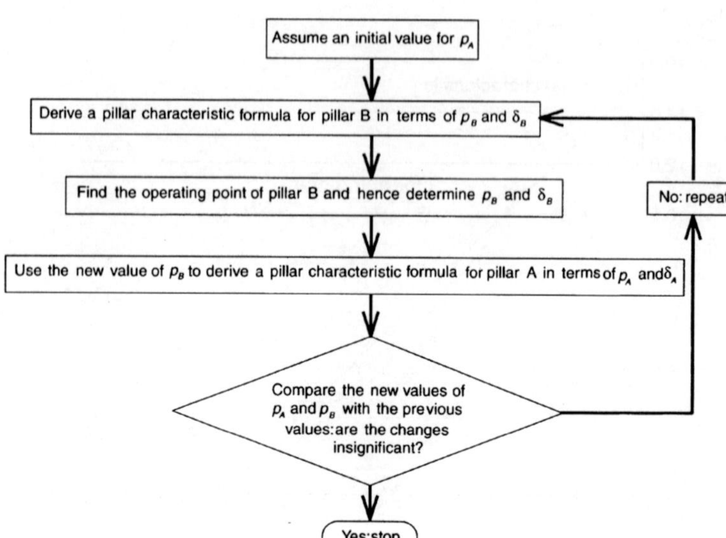

approach is required to solve the system. The iteration follows the steps as shown in the diagram above.

Although it is easiest to determine the operating point of each pillar graphically and perform the iteration manually, it is reasonably straightforward to use a spreadsheet program to perform the iteration automatically. In this case, calculations are required in order to find an operating point in terms of the intersection of a ground characteristic and the appropriate segment of the pillar characteristic.

An example of the solution, starting with an assumed pressure of 25 MPa at pillar A, is given in the table below. This particular starting pressure has been selected simply for demonstration: a better initial estimate could have been made on the basis of tributary area theory.

Iteration	Pillar A				Pillar B			
	m (MPa)	c (MPa)	pressure (MPa)	strain	m (MPa)	c (MPa)	pressure (MPa)	strain
1			25.00					
2					−1666	3.739	3.178	3.368×10^{-4}
3	−1564	14.50	11.18	2.126×10^{-3}				
4					−1666	12.345	9.891	1.473×10^{-3}
5	−1564	12.54	10.11	1.552×10^{-3}				
6					−1666	13.01	10.29	1.631×10^{-3}
7	−1564	12.42	10.05	1.521×10^{-3}				
8					−1666	13.05	10.31	1.642×10^{-3}
9	−1564	12.42	10.04	1.519×10^{-3}				
10					−1666	13.05	10.31	1.643×10^{-3}
11	−1564	12.42	10.04	1.519×10^{-3}				
12					−1666	13.05	10.31	1.643×10^{-3}

The operating conditions are therefore pillar pressures of 10.0 MPa and 10.3 MPa at A and B, respectively. The figures below show how the system converges.

Ground and pillar characteristics at iterations 2 and 3:

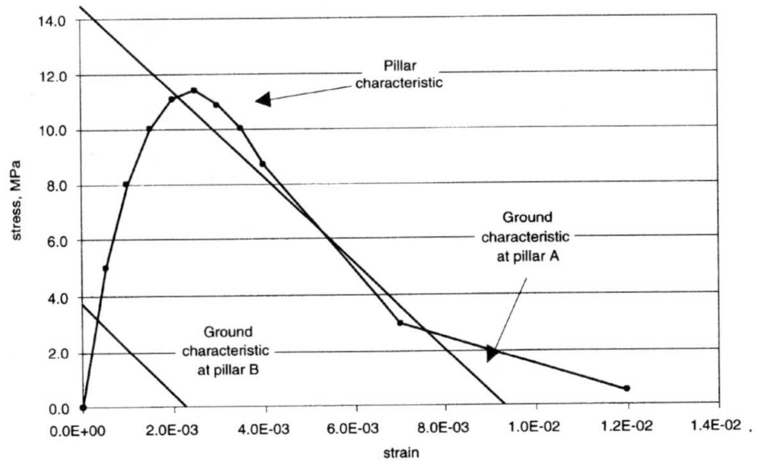

Ground and pillar characteristics at iterations 11 and 12:

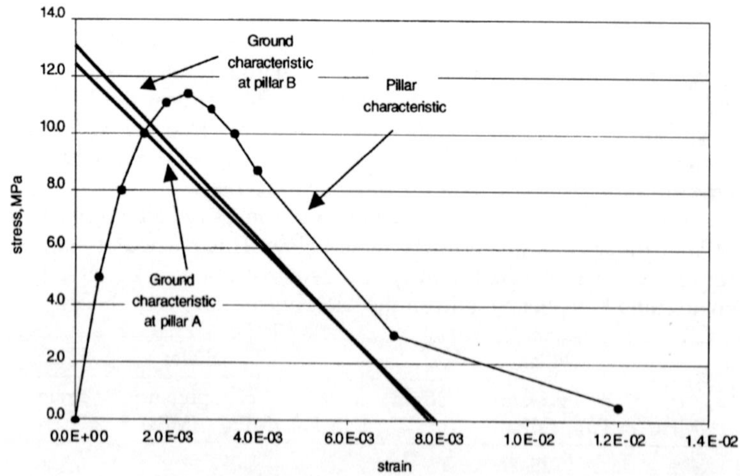

From the graph of the final state, we can see along which segment of the pillar characteristic the operating point lies. This allows us to demonstrate how the problem could be solved were each pillar characteristic a linear equation.

For both pillars the equation of the appropriate segment of the pillar characteristic is

$$p_A = 2200\varepsilon_A + 6.700 \tag{20.7}$$

$$p_B = 2200\varepsilon_B + 6.700 \tag{20.8}$$

and so Eq. (20.5) through to Eq. (20.8) can be written as the following system of linear equations:

$$\begin{aligned}
1 \cdot p_A + 0.2921 p_B + 1563\varepsilon_A + 0 \cdot \varepsilon_B &= 15.43 \\
0.6225 p_A + 1 \cdot p_B + 0 \cdot \varepsilon_A + 1666\varepsilon_B &= 19.30 \\
1 \cdot p_A + 0 \cdot p_B - 2200\varepsilon_A + 0 \cdot \varepsilon_B &= 6.70 \\
0 \cdot p_A + 1 \cdot p_B + 0 \cdot \varepsilon_A - 2200\varepsilon_B &= 6.70
\end{aligned}$$

In matrix form these are

$$\begin{bmatrix} 1 & 0.2921 & 1563 & 0 \\ 0.6225 & 1 & 0 & 1666 \\ 1 & 0 & -2200 & 0 \\ 0 & 1 & 0 & -2200 \end{bmatrix} \times \begin{bmatrix} p_A \\ p_B \\ \varepsilon_A \\ \varepsilon_B \end{bmatrix} = \begin{bmatrix} 15.43 \\ 19.30 \\ 6.70 \\ 6.70 \end{bmatrix},$$

and the solution of this matrix equation is

$$\begin{bmatrix} p_A \\ p_B \\ \varepsilon_A \\ \varepsilon_B \end{bmatrix} = \begin{bmatrix} 10.04 \\ 10.31 \\ 1.519 \times 10^{-3} \\ 1.643 \times 10^{-3} \end{bmatrix}.$$

These values can be seen to be identical to those obtained by iteration.

Clearly, the second of the two designs is preferable on the basis of pillar stability. Adoption of this geometry also reduces the roofspan and hence will mitigate instability in the excavation roof.

Q20.6 A new mining operation is planned which can extract from one of two coal seams. The first seam is at a depth of 250 m and has a thickness of 3 m, and the second seam is at a depth of 500 m with a thickness of 1 m. A preliminary design for the longwall mining panels calls for the upper seam to be mined with a panel width of 200 m, and the lower seam to have a panel width of 400 m (see the sketch below for typical geometry, in which the mining direction is into the page).

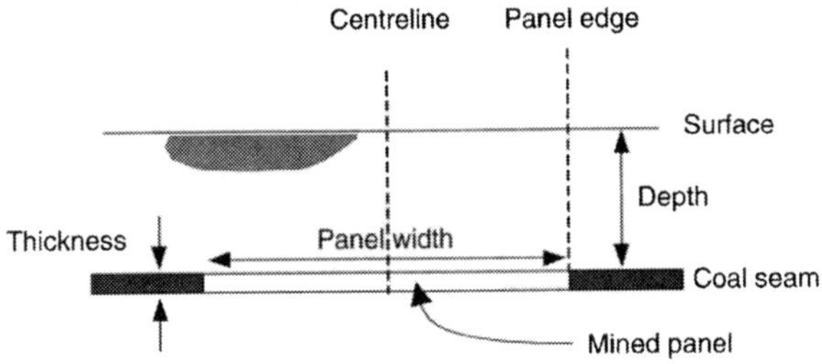

It has been suggested that the *Subsidence Engineers' Handbook* (SEH) (National Coal Board, Mining Department, 1975)[2] can be used to provide estimates of the surface subsidence profile and horizontal strain distribution transverse to the direction of mining.

Assess the suitability of the extraction geometries for each seam (based on estimates of surface subsidence and horizontal strain), using the following data taken from the SEH.

The SEH gives values of the ratio of maximal subsidence to extraction thickness of about 0.75 and 0.72 for the upper and lower seams, respectively. The proportion of subsidence that occurs at a given position along the transverse profile for both seams is given by the table below, taken from the SEH.

Subsidence proportion	**0.95**	**0.90**	**0.80**	**0.7**	**0.6**	**0.5**	**0.4**	**0.3**	**0.2**	**0.1**	**0.0**
Ratio of distance from panel centre to depth	**0.08**	**0.1**	**0.17**	**0.2**	**0.25**	**0.29**	**0.32**	**0.38**	**0.42**	**0.62**	**1.11**

Similarly, the strain proportion that occurs at a given position along the transverse profile for both seams is given by the table below, also taken from the SEH.

[2] National Coal Board, Mining Department (1975) *Subsidence Engineer's Handbook*. National Coal Board, Mining Department. London, 2nd ed. (revised).

Strain proportion	**−0.70**	**−0.80**	**−1.00**	**−0.80**	**−0.40**	**0.00**	**0.40**	**0.80**	**1.00**	**0.80**	**0.60**	**0.40**	**0.20**	**0.00**
Ratio of distance from panel centre to depth	**0.00**	**0.02**	**0.11**	**0.19**	**0.24**	**0.29**	**0.31**	**0.38**	**0.42**	**0.49**	**0.52**	**0.59**	**0.67**	**1.11**

In addition, the multipliers for the maximum contractile and extensile strains for these panel width to panel depth ratios are both equal to about 0.7.

A20.6 **Width of panel = 200 m, depth of panel = 250 m, seam thickness = 3 m.**

As the ratio of maximal subsidence to extraction thickness is about 0.75, for an extraction thickness of 3 m the maximal subsidence — directly above the centre of the panel — will be about 2.25 m. This figure is now used to scale the dimensionless profile data taken from the SEH, in order to produce the subsidence profile.

The subsidence profile data are converted into actual subsidence values by multiplying the values in the first row by 2.25 m (the magnitude of the maximum subsidence) and actual distances by multiplying the values in the second row by 250 m (the depth to the panel). This results in the following:

Subsidence (m)	2.138	2.025	1.800	1.575	1.350	1.125	0.900	0.675	0.450	0.225	0.000
Distance (m)	20.0	25.0	42.5	50.0	62.5	72.5	80.0	95.0	105.0	155.0	277.5

These results can then be plotted, as shown in the following diagram. Notice from this diagram that, above the panel edge, the subsidence is greater than 0.5 m, and that the overall width of the subsidence trough is more than 500 m.

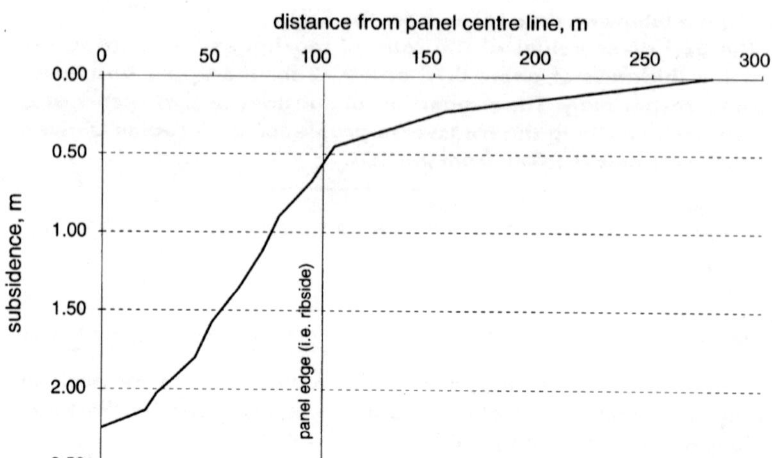

The transverse strain profile is produced in the same way as the subsidence profile. The results are given in the table and diagram below.

Notice that the strain profile is drawn using the sign convention given in the SEH, which is *tension positive*.

Strain	-5.0×10^{-3}	-5.8×10^{-3}	-7.2×10^{-3}	-5.8×10^{-3}	-2.9×10^{-3}	$0.0 \times 10^{+0}$	2.9×10^{-3}	5.8×10^{-3}	7.2×10^{-3}	5.8×10^{-3}	4.3×10^{-3}	2.9×10^{-3}	1.4×10^{-3}	$0.0 \times 10^{+0}$
Distance (m)	0.0	5.0	27.5	47.5	60.0	72.5	77.5	95.0	105.0	122.5	130.0	147.5	167.5	277.5

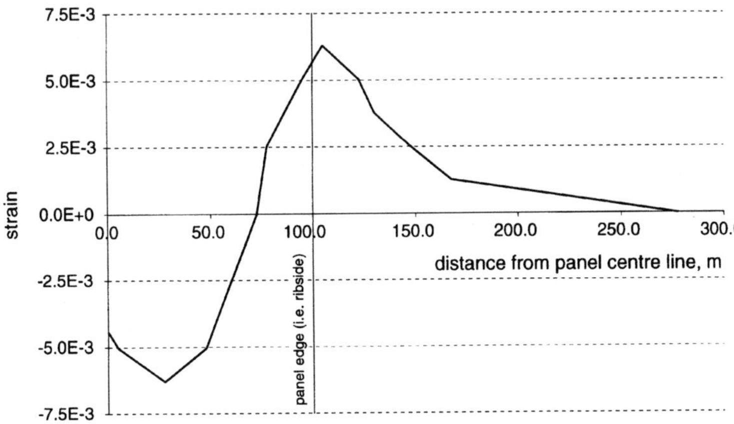

Width of panel = 400 m, depth of panel = 500 m, seam thickness = 1 m
As the ratio of maximum subsidence to extraction thickness is about 0.75, for an extraction thickness of 1 m the maximum subsidence will now be about 0.72 m. Using a multiplier for distances of 500 m, the subsidence profile is given by the values in the table below, and is shown in the diagram following. Notice from the diagram that, above the panel edge, the subsidence is greater than 0.5 m, and that the overall width of the subsidence trough is now more than 1000 m.

Subsidence (m)	2.138	2.025	1.800	1.575	1.350	1.125	0.900	0.675	0.450	0.225	0.000
Distance (m)	20.0	25.0	42.5	50.0	62.5	72.5	80.0	95.0	105.0	155.0	277.5

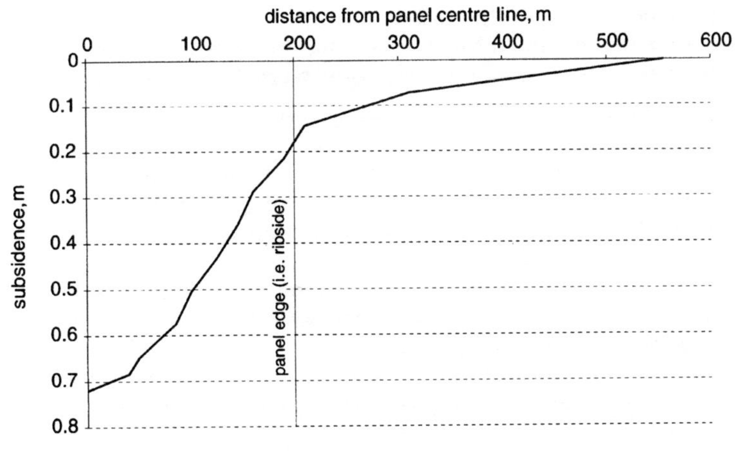

With the depth to the panel being 500 m, and the maximal subsidence being 0.72 m, the strain profile is given in the table and the diagram below. Again, the strain profile is drawn using the sign convention of *tension positive*.

Strain	-7.1×10^{-4}	-8.1×10^{-4}	-1.0×10^{-3}	-8.1×10^{-4}	-4.0×10^{-4}	$0.0 \times 10^{+0}$	4.0×10^{-4}	8.1×10^{-4}	1.0×10^{-3}	8.1×10^{-4}	6.0×10^{-4}	4.0×10^{-4}	2.0×10^{-4}	$0.0 \times 10^{+0}$
Distance (m)	0.0	10.0	55.0	95.0	120.0	145.0	155.0	190.0	210.0	245.0	260.0	295.0	335.0	555.0

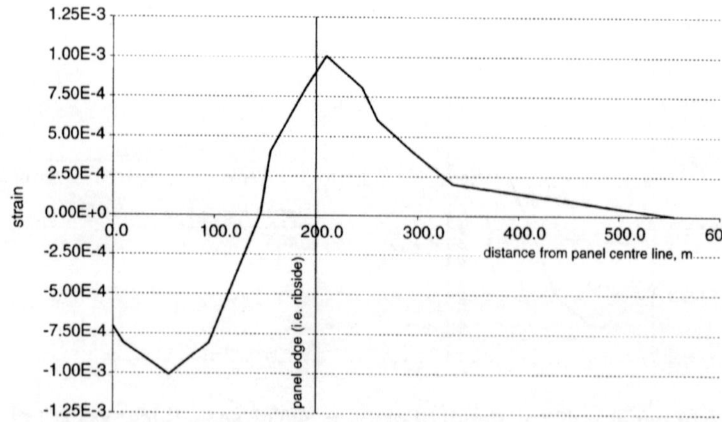

From the results shown above, we can see that — unless there is a constraint on the width of the subsidence trough — extraction of the deeper seam is preferable, because both the maximal subsidence and maximal horizontal strains are significantly less than for the shallow seam. Other factors, such as the magnitudes of differential subsidence (i.e. tilt of the ground), will indicate the same conclusion.

Q20.7 A bauxite room and pillar mine is to be opened at a depth of 100 m in a rock mass whose unit weight is approximately 27 kN/m^3. The design of the mine calls for a factor of safety of 2 against compressive failure of the pillars, and the substantial body of data regarding the design of pillars in this bauxite shows that their compressive strength may be calculated from the expression

$$(\sigma_c)_{\text{PILLAR}} = 10 + 45e^{-0.18A}$$

where *A* is the plan area of the pillar in square metres.

(a) What is the uniaxial compressive strength of the bauxite as measured in the laboratory, and what is the super-REV compressive strength of the pillars?

(b) For operational reasons, the rooms must have a minimum width of 5.0 m. What is the greatest extraction ratio that can be achieved with this size of opening?

(c) If the roof of a room is reinforced with rock bolts, then the maximum achievable opening width is about 8 m. Is it cost effective to do this?

A20.7 (a) When A is very small, the expression $(\sigma_c)_{\text{PILLAR}} = 10{+}45e^{-0.18A}$ tends to a value of 55 MPa — this is the laboratory compressive strength. Similarly, when A becomes very large, the strength tends to 10 MPa, which is the super-REV pillar compressive strength.

(b) Room-and-pillar designs are usually produced using the tributary area theory, and the formula

$$\frac{\sigma_v}{\sigma_p} = \frac{w_p^2}{\left(w_o + w_p\right)^2} \tag{20.9}$$

where σ_v is the vertical stress which, for this case, is given by $\gamma z = 0.027 \times 100 = 2.7$ MPa. Also, the maximal stress that a square pillar can be subjected to is given by

$$\sigma_p = \frac{1}{F}\left(10 + 45e^{-0.18w_p^2}\right) \tag{20.10}$$

where F is the factor of safety.

Eqs. (20.9) and (20.10) can be combined to find a pillar width for a given set of parameters, but the resulting equation has to be solved by iteration. It is more instructive to plot a series of curves for the opening width in terms of the pillar stress and the pillar width. Rearranging Eq. (20.9) gives

$$w_o = w_p\left(-1 \pm \sqrt{\frac{\sigma_p}{\sigma_v}}\right)$$

and hence

$$w_o = w_p\left[-1 \pm \sqrt{\frac{1}{\sigma_v}\frac{10 + 45\exp\left(-0.18w_p^2\right)}{F}}\right].$$

The resulting curves of opening width in terms of pillar width and factor of safety are as shown below.

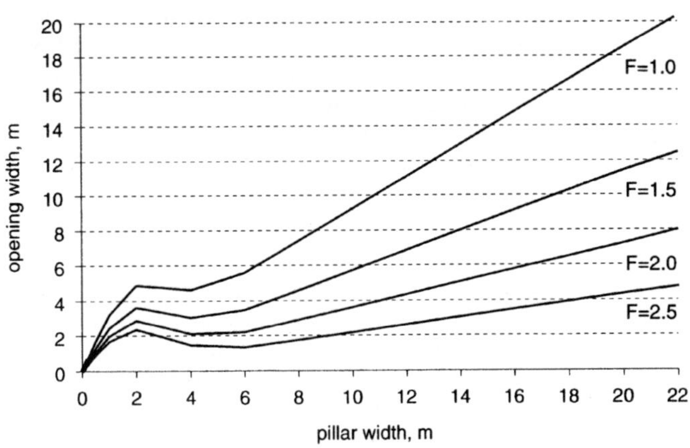

We can see from this that with $F = 2$ the minimal pillar width for an opening of 5 m is about 14 m, and hence the maximal extraction ratio is

$$\frac{(w_o + w_p)^2 - w_p^2}{(w_o + w_p)^2} = \frac{(5+14)^2 - 14^2}{(5+14)^2} = \frac{165}{361} = 0.46. \tag{20.11}$$

(c) If a room can be increased to 8 m wide, then the graph above shows that the pillar width will need to increase to about 22 m. The extraction ratio in this case is

$$r = \frac{(8+22)^2 - 22^2}{(8+22)^2} = \frac{416}{900} = 0.46$$

and so it will not be cost effective: the extraction ratio (and hence the revenue) does not change, yet the reinforcement will have to be paid for.

The way in which extraction ratio changes with pillar width is shown below. In this diagram, the maximal opening width for a given pillar width has been used. It is clear from this figure that no benefit arises from pillars greater than about 6 m wide.

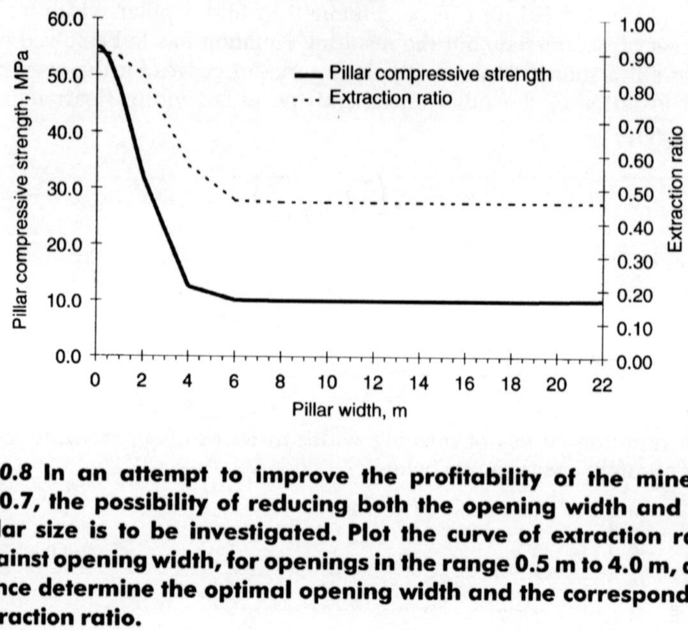

Q20.8 In an attempt to improve the profitability of the mine in Q20.7, the possibility of reducing both the opening width and the pillar size is to be investigated. Plot the curve of extraction ratio against opening width, for openings in the range 0.5 m to 4.0 m, and hence determine the optimal opening width and the corresponding extraction ratio.

If the extraction ratio thus identified is to be kept, what value of the factor of safety is required if the opening width is to be changed to 3.5 m?

A20.8 As the diagram below shows, the extraction ratio will be better than about 0.85 if the opening width can be kept below about 2.5 m. This is a result of the significant reduction in pillar strength with size. An opening width of 2.5 m will require a pillar width of about 2 m but, as the plot immediately above shows, at these sizes the pillar strength is variable and so they are best avoided.

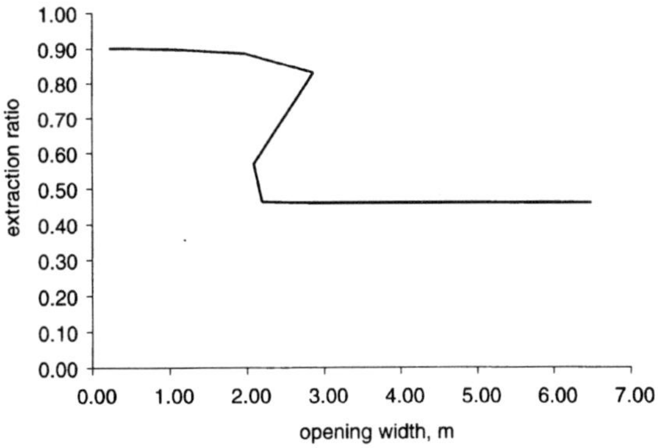

For a given extraction ratio and opening width, rearranging Eq. (20.11) in A20.7 shows that the associated pillar width is given by the solution to

$$r w_{\mathrm{p}}^{2} - 2 w_{\mathrm{o}} (1 - r) w_{\mathrm{p}} - w_{\mathrm{o}}^{2} (1 - r) = 0$$

In this case $r = 0.85$ and $w_{\mathrm{o}} = 3.5$ m, and so the pillar width is found from

$$0.85 w_{\mathrm{p}}^{2} - 1.05 w_{\mathrm{p}} - 1.84 = 0$$

to be 2.2 m. The strength of the pillar is then

$$10 + 45 \exp\left(-0.18 \times 2.2^{2}\right) \times 2.2^{2} = 140 \text{ MN}$$

and the load carried by it is

$$\sigma_{z} \left(w_{\mathrm{o}} + w_{\mathrm{p}}\right)^{2} = 2.7\,(3.5 + 2.2)^{2} = 88 \text{ MN},$$

giving a factor of safety of $140/88 = 1.59$.

Q20.9 A proposal has been made to use an old underground limestone quarry as a storage facility. A site visit to the quarry has revealed that it was mined using the room and pillar method, with a regular rectangular array of pillars. The clear spacing between the pillars is 6 m, the pillars are each 7 m square, and the excavation is at a depth of 80 m.

Examination of the pillars shows that the limestone is horizontally bedded with moderate spacing and gentle undulations. The bedding planes themselves are smooth to touch with slightly weathered surfaces and no visible aperture. Conditions inside the quarry are generally dry. A point load test of the pillar rock conducted at the quarry estimated its uniaxial compressive strength to be 100 MPa; whereas a laboratory triaxial test found that the rock failed when the axial stress in the sample was 110 MPa and the confining pressure was 4 MPa. The unit weight of the limestone is 28 kN/m^3.

Estimate the Rock Mass Rating (*RMR*) for the pillars (see the *RMR* table in Appendix C) and hence determine the Hoek–Brown strength parameters *m* and *s* for the rock mass, by using the equations

$$m = m_{i} \exp\left(\frac{RMR - 100}{28}\right) \quad \text{and} \quad s = \exp\left(\frac{RMR - 100}{9}\right).$$

Use these values together with the Hoek–Brown criterion,

$$\frac{\sigma_1}{\sigma_c} = \frac{\sigma_3}{\sigma_c} + \sqrt{m\frac{\sigma_3}{\sigma_c} + s},$$

to determine the maximal vertical stress the pillars can sustain (1) at their faces and (2) at their centres. Assume that the ratio of horizontal to average vertical stress is 0.075 at the centre of each pillar.

Use the tributary area theory to estimate the average vertical stress in the pillar, and hence determine the factor of safety of the pillars.

A20.9 If we consult a table of *RMR* parameters (see Appendix C), we can assign the following values to the rock mass of the pillars:

Parameter	Comments	Rating
Compressive strength	measured as 100 MPa	4
Groundwater conditions	described as generally dry	12
Spacing	described as 'moderate'	10
RQD	value associated with a spacing of, say, 0.4 m will be close to 100%	20
Discontinuity condition	persistence of bedding planes will be high (0), aperture is zero (6), smooth to touch (1), no infilling (6), slight weathering (5)	18
Total		64

The Hoek–Brown criterion,

$$\frac{\sigma_1}{\sigma_c} = \frac{\sigma_3}{\sigma_c} + \sqrt{m\frac{\sigma_3}{\sigma_c} + s},$$

requires us to know both m and s but, for the case where we are examining the face of a pillar, we can assume $\sigma_3 = 0$ and hence use the reduced form of

$$\sigma_1 = \sigma_c\sqrt{s}.$$

Substituting the relation $s = \exp\left[(RMR - 100)/9\right]$ into this equation we obtain

$$\sigma_1 = \sigma_c\sqrt{\exp\left(\frac{RMR - 100}{9}\right)} = \sigma_c \exp\left(\frac{RMR - 100}{18}\right)$$

from which, with $\sigma_c = 100$ MPa and $RMR = 64$, we find $\sigma_1 = 13.5$ MPa. This is the maximal vertical stress that the rock at the face of the pillar can sustain.

For the case when we are examining the centre of the pillar, where triaxial conditions exist, the laboratory triaxial test data can be used to determine a value for the parameter m. As the laboratory test took place on a specimen of intact rock, we know that $s = 1$. Consequently, rearranging the Hoek–Brown criterion of

$$\frac{\sigma_1}{\sigma_c} = \frac{\sigma_3}{\sigma_c} + \sqrt{m_i\frac{\sigma_3}{\sigma_c} + s}$$

(where m_i is the value of m for intact rock) and substituting $\sigma_1 = 110$ MPa, $\sigma_3 = 4$ MPa and $s = 1$, we find that

$$m_i = \left[\left(\frac{\sigma_1 - \sigma_3}{\sigma_c}\right)^2 - s\right]\frac{\sigma_c}{\sigma_3} = \left[\left(\frac{110-4}{100}\right)^2 - 1\right]\frac{100}{4} = 3.09.$$

Using this value of m_i together with $RMR = 64$, we obtain

$$m = m_i \exp\left(\frac{RMR - 100}{28}\right) = 3.09 \exp\left(\frac{64-100}{28}\right) = 3.09 \times 0.277 = 0.86$$

and

$$s = \exp\left(\frac{RMR - 100}{9}\right) = \exp\left(\frac{64-100}{9}\right) = 0.018.$$

The ratio of horizontal to average vertical stress is 0.075 at the centre of the pillar, and so we can substitute $\sigma_3/\sigma_1 = 0.075$ into the Hoek–Brown criterion to give

$$(1 - 0.075)^2\sigma_1^2 - 0.075m\sigma_c\sigma_1 - s\sigma_c^2 = 0.$$

If we substitute the values of $m = 0.86$, $s = 0.018$ and $\sigma_c = 100$ MPa into this relation and solve the quadratic equation, we find $\sigma_1 = 18.8$ MPa, which is the maximal vertical stress the pillar can sustain at its centre.

The geometry of the tributary area for each pillar is thus:

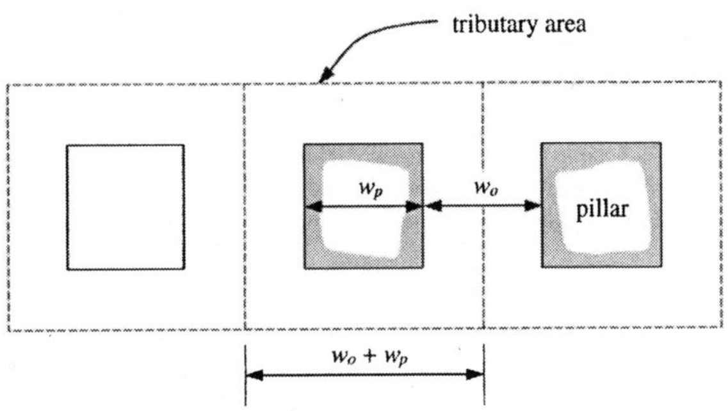

This shows that the force due to the weight of rock supported by each pillar is

$$F_p = \gamma z A = \gamma z (w_o + w_p)^2$$

and so the vertical stress induced in each pillar is

$$\sigma_p = \frac{F_p}{A_p} = \frac{\gamma z (w_o + w_p)^2}{w_p^2}.$$

Substituting the values of $\gamma = 0.028$ MN/m^3, $z = 80$ m, $w_o = 6$ m and $w_p = 7$ m we find that $\sigma_p = 7.7$ MPa.

Finally, we can determine the factor of safety at the pillar faces and the pillar centres. At the face the factor of safety is

$$F_{face} = \frac{\sigma_{face}}{\sigma_p} = \frac{13.5}{7.7} = 1.75$$

and at the centre it is

$$F_{\text{centre}} = \frac{\sigma_{\text{centre}}}{\sigma_{\text{p}}} = \frac{18.8}{7.7} = 2.44$$

This shows that the effect of the confinement offered by the bulk of the pillar is to produce a marked increase in factor of safety at the pillar centres, in comparison to the pillar faces. However, given the uncertainty that will surround the assessment of the geomechanical parameters, both of these factors of safety are very low. Careful consideration would need to be given to the acceptability of these values, and whether reinforcement of the pillars should be considered.

Q20.10 **A gold-bearing quartz vein, 2 m thick and dipping at 90°, is to be exploited by a small cut-and-fill stoping operation. The mining is to take place at a depth of 800 m, and the average unit weight of the granite country rock above this level is 29 kN/m^3. The strike of the vein is parallel to the intermediate principal stress, and the major principal stress is horizontal with a magnitude of 37.0 MPa. The uniaxial compressive strength of the vein material is 218 MPa, and the tensile strength of the country rock is 24 MPa. Poisson's ratio and Young's modulus for the quartz are 0.2 and 48 GPa, respectively. During mining, each blast will extend a stope up-dip by about 2 m.**

Assuming that no stress-induced failure is permissible, what is the maximum height of a stope?

It is considered that the backfill will offer sufficient support to prevent degradation of the side walls of a stope, and that the only stress-induced failure of concern is that in the crown. What is the maximum permissible height of a stope in these circumstances?

A20.10 We will assume that, in the cross-section of the excavation, the stresses induced in the sidewall and the crown of the stope can be approximated using the equations for an elliptical excavation. On the basis that the sidewall stress can be computed using the inscribed ellipse, we have

$$\frac{\sigma_{\text{sidewall}}}{\sigma_{\text{vertical}}} = 1 - k + 2\left(\frac{w}{h}\right)$$

and if the crown is semi-circular we have

$$\frac{\sigma_{\text{crown}}}{\sigma_{\text{vertical}}} = k - 1 + k\sqrt{\frac{2h}{\rho_{\text{crown}}}} = k - 1 + 2k\sqrt{\frac{h}{w}}.$$

Rearranging these gives the height of the excavation as the minimum of

$$h = \frac{2w}{\dfrac{\sigma_{\text{sidewall}}}{\sigma_{\text{vertical}}} + k - 1} \text{ or } h = \frac{w}{4k^2}\left(\frac{\sigma_{\text{crown}}}{\sigma_{\text{vertical}}} + 1 - k\right)^2.$$

The maximal stress that can be sustained by the crown and the sidewall are 218 MPa and −5 MPa, respectively. Note that the sidewall stress is negative because this represents the tensile strength.

The vertical stress is

$$\sigma_{\text{vertical}} = \gamma \cdot z = 0.029 \times 800 = 23.2 \text{ MPa}$$

and hence the ratio of horizontal to vertical stress is

$$k = \frac{37.0}{23.2} = 1.59.$$

The maximal height of a stope such that the compressive strength of the rock in the crown is not exceeded is given by

$$h = \frac{w}{4k^2}\left(\frac{\sigma_{\text{crown}}}{\sigma_{\text{vertical}}} + 1 - k\right)^2 = \frac{2}{4 \times 1.59^2}\left(\frac{218}{23.2} + 1 - 1.59\right)^2 = 15.3 \text{ m}.$$

The maximal height of a stope such that the tensile strength of the rock in the sidewall is not exceeded is given by

$$h = \frac{2w}{\dfrac{\sigma_{\text{sidewall}}}{\sigma_{\text{vertical}}} + k - 1} = \frac{2 \times 2}{\dfrac{-5}{23.2} + 1.59 - 1} = 10.7 \text{ m}.$$

Thus we see that sidewall failure is the limiting condition if no stress-induced failure is allowed. If backfill is used to prevent degradation of the sidewalls, then the height can be increased to about 15 m without failure occurring in the crown.

20.3 Additional points

In the last four chapters we have been able to highlight some of the main rock mechanics techniques used in the design of surface and underground excavations. However, it is not possible to cover all aspects in a book because in a real design case there will be additional factors related to the specific nature of the rock mass, the site and the engineering objective. This means that not only must the engineer be competent in the techniques we have presented but also be able to approach new projects with a **creative eye** and brain.

Much can be achieved in understanding a new site and the requirements for design by thinking through the subjects covered in the book chapters. For example, some of the most **important questions** stimulated by the content of the chapters are

(1) Where can information on previous similar work be obtained?
(2) Is the structural geology of the area understood? Are there any major faults or shear zones present?
(3) Is the regional pattern of *in situ* stress understood for this area?
(4) Are local factors influencing the *in situ* stress distribution at the site?
(5) Is it likely that the rock mass can be modelled as an elastic material?
(6) Is the deformability, strength and failure of the intact rock understood?
(7) Are the geometrical and mechanical properties of the fractures known?
(8) What are the special rock mechanics features of the rock mass at the site?

(9) Is the groundwater understood at the site? Is groundwater likely to be a problem?
(10) Are there any significant anisotropic and/or inhomogeneous rock mass characteristics?
(11) Exactly what type of rock testing is required and why?
(12) Will a rock mass classification approach be useful?
(13) Are there any time-dependent aspects that should be taken into account?
(14) Are there any adverse interactions that could lead to positive feedbacks and hence instabilities?
(15) Are the rock properties used in the design those of the rock mass before or after excavation?
(16) Should rock stabilization be primarily reinforcement or support?
(17) Have all the potential failure mechanisms been identified?

Being observant, being competent in the techniques, systematically going through the rock mechanics factors, and having the courage of your convictions to ensure that all the work is completed properly will significantly reduce the chances of unexpected events occurring when your rock engineering design is implemented.

Part B:

Questions only

The Questions in Part A are reproduced here without the answers for those who wish to attempt the questions without the answers being visible.

Questions 1.1–1.5: introduction

Q1.1 Define the following terms:

- rock mechanics;
- engineering rock mechanics;
- rock engineering;
- structural geology;
- engineering geology;
- soil mechanics;
- geotechnical engineering.

Q1.2 Explain the fundamental purposes of excavation in civil engineering, mining engineering, and petroleum engineering.

Q1.3 The photograph below illustrates construction of the 61 m span, 25 to 50 m deep, underground Gjøvik Olympiske Fjellhall (Olympic

Mountain Hall) in Precambrian gneiss in Norway. This is the largest roofspan public-access civil engineering cavern in the world. Describe the engineering rock mechanics factors that would have to be considered in the design and excavation of such a cavern.

Q1.4 Why do you think that the techniques used in rock mechanics for site characterization, analysis and modelling are not the same as those used in soil mechanics?

Q1.5 How can the subject of 'engineering rock mechanics' be useful to organizations outside the civil and mining engineering professions, e.g. to the petroleum industry, to insurance companies, to environmental engineers?

Questions 2.1–2.10: geological setting

Q2.1 The picture in Fig. Q2.1 shows a limestone slope above a highway in Spain. Comment briefly on the geological factors that could influence rock slope stability at this location.

Q2.2 The picture in Fig. Q2.2 shows the surface of a fault in a hard rock aggregate quarry on which a rock slide has occurred. Explain

(a) why the existence of this fault could indicate that other similar features will be encountered as quarrying continues, and

(b) why encountering an adverse geological feature such as this is likely to be less significant in a quarry than in a road cutting.

Q2.3 The picture in Fig. Q2.3 shows tooth marks from the bucket of a mechanical excavator in the Carboniferous rocks of a near-surface slope

Figure Q2.1

Figure Q2.2

in an opencast coal mine. What evidence is there here of geological disruption to the rock strata?

Q2.4 A site investigation was conducted in a granitic rock mass (Fig. Q2.4). One side of fracture #300 in the core is shown to the right. What does this fracture indicate about the rock mass history and what significance does this have for rock mechanics design of slopes and tunnels in the rock mass?

Q2.5 In the first Glossop Lecture on geology for engineers, Fookes (1997) quotes from Glossop's 1968 Rankine Lecture: "What you look for should be suggested by the natural environment and by the nature of the construction problem to be solved." Explain why this quotation is critically important for rock mechanics and rock engineering.

Figure Q2.3

Figure Q2.4

Q2.6 The quotation at the beginning of the Glossary of Geology is that: "It is not really a mark of distinction for a geologist's writing to be so obscure that a glossary is required for its comprehension." Discuss this in the context of engineering rock mechanics.

Q2.7 When considering the geological setting for a rock engineering project, would an engineer expect information relating to all potential geological hazards to be available?

Q2.8 In an article on geological and geotechnical investigations for tunnelling, Parker (1996) estimates that "even comprehensive exploration programs recover a relatively miniscule drill core volume, less than 0.0005% of the excavated volume of the tunnel". Do you think that sampling only this proportion of the rock mass is enough?

Q2.9 A cavern (165 m long, 22 m wide and 15 m high) is to be excavated in chalk strata beneath the sea. The crown of the cavern will be 35 m below the seabed. What is the main geological information you would like to have before proceeding with the excavation?

Q2.10 The pre-existing stress state in a rock mass is caused by geological processes and is often a critical factor for rock engineering. Why do you think that quantifying the rock stress is important?

Questions 3.1–3.10: stress

Q3.1 Show how the stress state in a solid can be described via the stress components (normal and shear) on an elemental cube of rock. Also, show how these components are listed in the stress matrix. What do the components in a row of the stress matrix have in common? What do the components in a column of the stress matrix have in common?

Q3.2 When is a matrix symmetrical? Why is the stress matrix symmetrical?

Q3.3 Explain the differences between scalar, vector and tensor quantities. Why is stress a tensor quantity?

Q3.4 How are normal and shear stress components plotted on Mohr's circle?

Q3.5 What is a principal stress plane? What is a principal stress?

Q3.6 What are the following stress states: uniaxial stress, biaxial stress, triaxial stress, polyaxial stress, pure shear stress, hydrostatic stress?

Q3.7 Show how to add two tensors and hence how to calculate the mean of n stress states. How would you calculate the mean of n different stress states which were specified by their principal stresses and the associated principal stress directions?

Q3.8 What are the first, second and third stress invariants?

Q3.9 By considering the rates of change of the stress components in the answer to Q3.1, establish force equilibrium in the x, y and z directions and hence write down in differential form the three equilibrium equations for an elemental cube.

Q3.10 Given an elemental cube with a normal stress component and two shear stress components acting on all its faces, it is always possible to find a cube orientation such that the shear stresses disappear on all faces and only normal stresses (the principal normal stresses) remain.

Is it possible to find a complementary orientation such that the normal stresses disappear on all faces and only shear stresses (i.e. principal shear stresses) remain? Explain the reason for your answer.

Questions 4.1–4.10: *in situ* rock stress

Q4.1 There is no internationally agreed terminology for words describing the state of stress in a rock mass. However, describe in one sentence what you think is meant by each of the following terms.

Natural stress	Thermal stress
Induced stress	Palaeostress
Gravitational stress	Near-field stress
Tectonic stress	Far-field stress
Residual stress	Local stress

Q4.2 Add the following 2-D rock stress states, and find the principal stresses and directions of the resultant stress state.

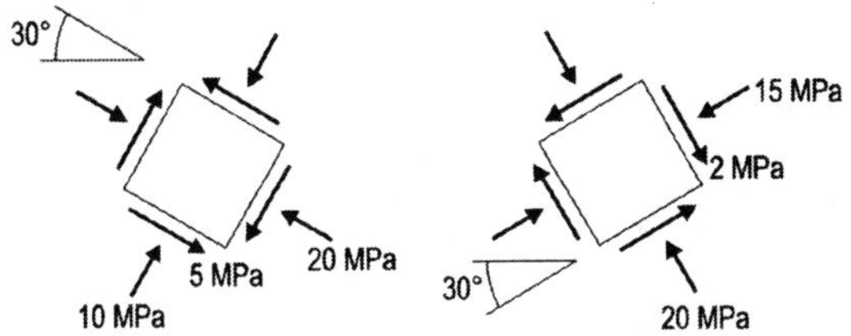

Q4.3 How many experimental set-ups are required to determine the 3-D state of stress in a rock mass, using each of the standard stress measurement methods of flatjack, hydraulic fracturing, USBM gauge and CSIRO gauge? (Assume that the stress field is constant in the vicinity of the test site.)

Q4.4 Three flatjack tests have been made close to each other in the wall of a long, straight tunnel, the axis of which dips at 7°. The measurement position is approximately 250 m below the ground surface and it is assumed that the flatjacks are in the same stress field. The slots for the flat-

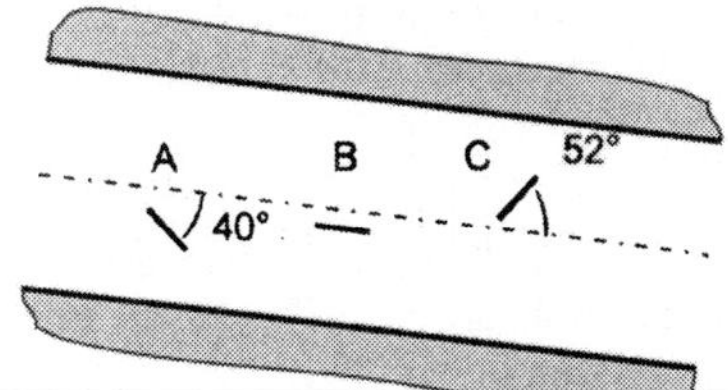

jacks were cut normal to the wall of the tunnel, and were oriented relative to the tunnel axis as shown.

The cancellation pressure for each of the flatjacks A, B and C was 7.56 MPa, 6.72 MPa and 7.50 MPa, respectively. Compute the principal stresses and their directions, and ascertain whether they accord with worldwide trends.

Q4.5 Two further flatjack measurements have been made in the wall of the tunnel considered in Q4.4. These dip at 20° and 90° relative to the tunnel axis, and produced cancellation pressures of 7.38 MPa and 7.86 MPa, respectively. Compute the best estimate of the principal stresses.

Q4.6 The stress in a granitic rock mass has been measured by the hydraulic fracturing technique. Two tests were conducted in a vertical borehole: one test at a depth of 500 m, and the other test at a depth of 1000 m. The results were as follows:

Depth (m)	Breakdown pressure, P_B (MPa)	Shut-in pressure, P_S (MPa)
500	14.0	8.0
1000	24.5	16.0

Given that the tensile strength, σ_t, of the rock is 10 MPa, estimate and list the values of σ_1, σ_2 and σ_3 at the two depths.

State all of the assumptions you have to make in order to produce these estimates. Are any of them doubtful? State whether the two sets of results are consistent with each other, and justify your reasons for the statement. Are the results in agreement with trends exhibited by collated worldwide data?

Q4.7 How are three-dimensional tensors transformed so that the stress components on any plane can be calculated?

Q4.8 Suppose that we have measured the pre-existing stress state in the ground by some means and that the results are as follows:

σ_1, magnitude 15 MPa, plunges 35° towards 085°;
σ_2, magnitude 10 MPa, plunges 43° towards 217°;
σ_3, magnitude 8 MPa, plunges 27° towards 335°.

Find the 3-D stress tensor in the right-handed xyz co-ordinate system with

x, horizontal to the east,
y, horizontal to the north,
z, vertically upwards.

Right-handed co-ordinate system

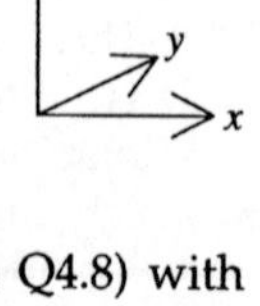

Q4.9 A fault is present in the same rock (continuing from Q4.8) with an orientation of 295°/50°. Determine the stress components in a local co-ordinate system aligned with the fault. Assume for this question that the presence of the fault does not affect the stress field.

Q4.10 The plan below shows a horizontal section through a rock mass. Stress measurements were made from the gallery along the borehole line Ox using the Japanese CCBO technique. The measured principal stresses in the horizontal plane are plotted on the plan. What are the main conclusions that you can draw from the stress variations?

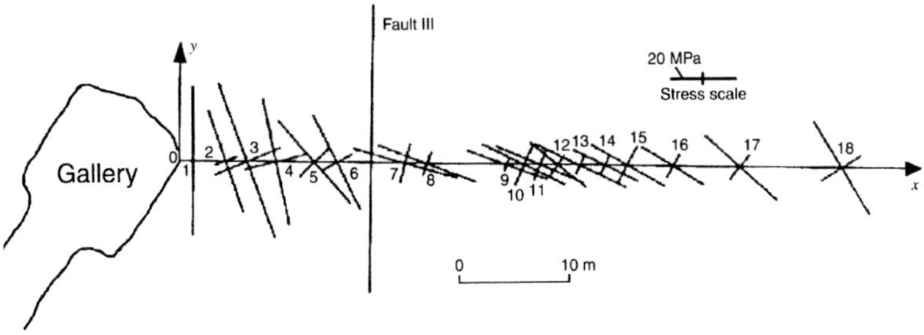

Questions 5.1–5.10: strain and the theory of elasticity

Q5.1 What is the meaning of the first stress invariant and the first strain invariant?

Q5.2 The differential equations of force equilibrium were the subject of Q3.9. The equivalent equations for displacement and strain are the compatibility equations; these equations ensure that the normal and shear strains are compatible, so that no holes, tears or other discontinuities appear during straining. Show that the following compatibility equation is valid:

$$\frac{\partial^2 \varepsilon_{xx}}{\partial y^2} + \frac{\partial^2 \varepsilon_{yy}}{\partial x^2} = \frac{\partial^2 \gamma_{xy}}{\partial x \partial y}.$$

Q5.3 Draw a Mohr circle for strain, indicating what quantities are on the two axes, how to plot a 2-D strain state, and the location of the principal strains, ε_1 and ε_2.

Q5.4 Show why the shear modulus, Young's modulus and Poisson's ratio are related as $G = E/2(1 + \nu)$ for an isotropic material. This equation holds for an isotropic material but not for an anisotropic material — why? Hence explain why five elastic constants are required for a transversely isotropic material rather than six.

Q5.5 (a) How can the strain in a particular direction be found from the strain matrix components and hence how can a strain gauge rosette be used to estimate the state of strain at a point, and hence the state of stress at a point?

(b) Assume that strains measured by a strain gauge rosette are $\varepsilon_P = 43.0 \times 10^{-6}$, $\varepsilon_Q = 7.8 \times 10^{-6}$ and $\varepsilon_R = 17.0 \times 10^{-6}$, and that the gauges make the following angles to the x-direction: $\theta_P = 20^\circ$, $\theta_Q = 80^\circ$ and $\theta_R = 140^\circ$. Determine the principal strains and their orientations and then, using values for the elastic constants of $E = 150$ GPa and $\nu = 0.30$, determine the principal stresses and their orientations.

Q5.6 Explain clearly why an isotropic rock has two independent elastic constants, a transversely isotropic rock has five independent constants and an orthotropic rock has nine independent constants (compared to the general anisotropic case where there are twenty-one independent elastic constants).

Q5.7 Each of the following four rock masses is to be modelled using elasticity theory. State whether you think that an isotropic rock assumption is justified, or whether one of the anisotropic assumptions would be more appropriate. Assume that the fractures have a significant effect on the rock deformability.

(a) A limestone with effectively three fracture sets, i.e. the bedding with mean strata thicknesses of 1 m, plus fracture set 1 (perpendicular to the bedding) with two fractures/m, and fracture set 2 (perpendicular to both the bedding and fracture set 1) with five fractures/m.

(b) A welded volcanic tuff with five fracture sets.
Fracture set 1: dip direction 089°; dip 50°; frequency 2.9/m.
Fracture set 2: dip direction 278°; dip 88°; frequency 1.3/m.
Fracture set 3: dip direction 224°; dip 08°; frequency 0.9/m.
Fracture set 4: dip direction 169°; dip 23°; frequency 2.1/m.
Fracture set 5: dip direction 113°; dip 70°; frequency 0.7/m.

(c) A strong sandstone with almost no fractures present.

(d) A granodiorite with three fracture sets.
Fracture set 1: dip direction 314°; dip 35°; frequency 1.2/m.
Fracture set 2: dip direction 048°; dip 43°; frequency 1.3/m.
Fracture set 3: dip direction 089°; dip 79°; frequency 0.9/m.

Q5.8 (a) At the time of writing this book, most elastic analyses that have been conducted for rock engineering design purposes have assumed that the rock is perfectly isotropic with two elastic constants. Why do you suppose that is, given that most rock masses are clearly not isotropic?

(b) Conversely, no elastic analysis for rock mechanics has been conducted assuming that the rock mass is fully anisotropic with 21 elastic constants? Why is that?

(c) In this context, what do you think will happen in future analyses?

Q5.9 (a) How do you know if a material is elastic?

(b) How is time taken into account in the theory of elasticity?

Q5.10 When a rock specimen fails in a compressive test in the laboratory or when an *in situ* rock mass fails due to natural compression (see Fig. Q5.1), is 'stress' or 'strain' the cause of failure?

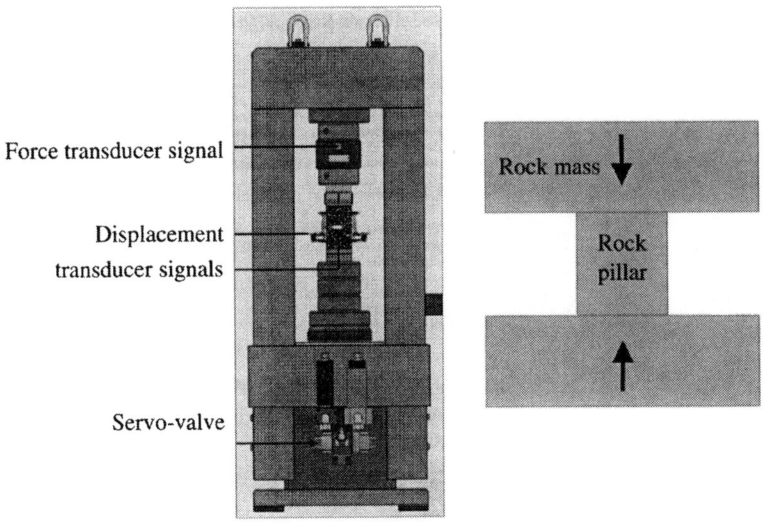

Figure Q5.1 *Left:* Laboratory rock testing controlled using load or displacement feedback (courtesy of MTS Systems Corp.). *Right:* An underground rock pillar being stressed or strained?

Questions 6.1–6.10: intact rock

Q6.1 Predict how the volume of a specimen will change during a uniaxial compression test, and sketch the variation of volumetric strain versus axial strain. What physical processes are occurring in the sample as the curve manifests significant gradient changes?

Q6.2 What causes a rock to break when it is compressed uniaxially to failure? Is it when the stress reaches a certain value or when the strain reaches a certain value? Or is it when some other parameter reaches a critical value, such as the energy input per unit volume, or the microcrack density per unit volume?

Q6.3 The geometry of the linear Mohr–Coulomb envelope is such that a number of useful relations between strength parameters can be drawn from it. Derive an expression for the uniaxial compressive strength of rock in terms of the cohesion and angle of internal friction.

Q6.4 The linear Mohr–Coulomb envelope with a tensile cut-off sets a definite limit on the maximal uniaxial tensile strength of a material. By considering the largest uniaxial tensile Mohr circle that can be drawn, determine this tensile strength limit in terms of σ_c and ϕ.

Q6.5 A firm whose judgement has been questioned on a previous occasion has been entrusted with the strength testing of rock in a site investigation project. During their first uniaxial compression test, the equipment failed to measure the peak axial load, but the technician did record that the specimen failed by the formation of a single fracture inclined at 20° to the loading axis. In a subsequent triaxial test, as the confining pressure was being increased before application of the axial stress, the specimen failed prematurely when the confining pressure in the Hoek cell was 85 MPa.

On the basis of these results, propose a failure criterion of the form $\sigma_1 = a\sigma_3 + b$ for the rock.

Q6.6 A servo-controlled compression test has been conducted on a weak soapstone such that the specimen length remained unchanged throughout: as the axial stress, σ_a, was increased, so the confining pressure, p, was increased so that no net axial strain resulted. A plot of axial stress (vertical axis) against confining pressure (horizontal axis) gave an initial straight line passing through the origin. At a critical confining pressure of $p = 85$ MPa (when $\sigma_a = 39.1$ MPa), the slope of the σ_a–p plot suddenly changed to 29° and remained constant for the remainder of the test. This change in slope may be taken to represent the onset of yield.

(i) Determine an elastic constant from the slope of the initial portion of the σ_a–p curve.

(ii) Assuming that the Mohr–Coulomb criterion is applicable, determine σ_c, c and ϕ for the rock.

Q6.7 Laboratory tests on specimens of a limestone have produced unconfined compressive and tensile strengths of 80 MPa and 10 MPa, respectively. Using the Hoek–Brown and plane Griffith criteria, estimate the maximum principal stress at failure for two biaxial tests in which $\sigma_2 = 20$ MPa and $\sigma_2 = 40$ MPa. Which of these two criteria would best predict peak strength under these conditions?

Q6.8 Comment on the applicability of each of the Griffith, Mohr–Coulomb, and Hoek–Brown criteria for the following triaxial test results on quartzite:

$(\sigma_1 + \sigma_3)/2$	−6.65	100	135	160	200	298	435	MPa
$(\sigma_1 - \sigma_3)/2$	6.65	100	130	150	180	248	335	MPa

Q6.9 In 1926, in a paper on tensile tests for cotton yarns (Pierce, 1926), Pierce stated, "it is a truism, of which the mathematical implications are of no little interest, that the strength of a chain is that of its weakest link". What is the relevance of this statement to the tensile strength of intact rock?

Q6.10 The two marble panels illustrated opposite are from the Greek Parthenon frieze which depicts the four-yearly procession of the 'Great Panathenaia'. They were sculpted in light relief under the direction of Pheidias and placed in position around 450 BC. One panel is from the north side of the Parthenon and one panel is from the south side. Which is which?

Panel A

Panel B

Questions 7.1–7.10: fractures and hemispherical projection

Q7.1 The overall *RQD* for 134 m of borehole core was found to be 58%.

(a) Compute estimates for the total number of pieces of core, and the total length of those pieces of core that could be expected to be greater than 0.1 m long.

(b) How many pieces of core could be expected to have a length greater than 0.2 m, and what is their mean length?

Q7.2 Based on a sample of 128 fracture spacing values which gave a mean spacing of 0.215 m, estimate the range of the population mean fracture spacing and frequency at the 80% confidence level. How many fractures should be in the sample for an error of ±10% at the 90% and 95% confidence levels?

Q7.3 The mean fracture frequency in a vertical direction in a sandstone rock mass is 1.22 m^{-1}, and a total of 500 vertical 3-m-long rockbolts are to be installed to stabilize the roof of an underground excavation in this rock mass. How many rockbolts would you expect to:

- intersect no fractures;
- intersect less than 3 fractures; and,
- intersect more than 4 fractures?

What length should the rockbolts be if 95% of them are required to intersect at least 3 fractures, i.e. extend to the fourth rock block back into the rock mass?

Q7.4 A fault plane with orientation 234/56 has been discovered during a site investigation. Closer inspection shows that it has surface lineations which have a pitch of 78° measured from the northwest strike line. What is the trend and plunge of these lineations?

Q7.5 The line of intersection between two planes trends approximately northwest and plunges at 38°. The orientation of one of the planes is

256/50, and the strike of the other is 132°. What is the trend of the line of intersection and what is the dip of the second plane?

Q7.6 Surveys have revealed that a rock mass contains 3 fracture sets, the dip directions/dip angles and fracture frequencies of which are 161/23 and 7.72 m^{-1}, 218/58 and 3.07 m^{-1}, and 100/75 and 5.34 m^{-1}.

(a) What will be the fracture frequency and mean length of the recovered pieces of core in:

- a vertical borehole;
- a horizontal tunnel heading due north; and
- an inclined borehole with a trend/plunge of 280/35?

Consider the engineering implications of this variation in terms of site investigation procedures and subsequent engineering design.

(b) Using the frequency values determined above, compute the theoretical *RQD* values that would be encountered in this rock mass using the customary threshold value of 0.1 m. Adopting a threshold value given by

$$t^* = \frac{2}{\lambda_{max} - \lambda_{min}} \ln\left(\frac{\lambda_{max}}{\lambda_{min}}\right)$$

will maximize the range of *RQD* values. How do the *RQD* values computed with this threshold compare to the earlier values?

Q7.7 A rock mass is known to contain two sets of fractures, the orientations of which are 265/42 and 003/69. What borehole orientations will allow one to drill through the rock mass and intersect all of the fractures at an angle of 45° (measured between the borehole axis and the normals to the fractures)? What is the minimum angle that all fractures can be intersected at?

Q7.8 A petroleum reservoir is known to contain numerous fractures which are highly conductive. These fractures dip almost vertically in a northeasterly direction. To maximize production from the reservoir, the production wells are to be deviated to run sub-horizontally, so that they intersect the fractures as close to perpendicular as possible.

In order to determine the optimal orientation of the production holes, two test holes have been drilled to orientate the fractures. The orientations of these test holes within the reservoir are 011/09 and 087/03, and impression packers which were run down these holes show that in both holes the angle between the normal to the fractures and the hole axis is 40°.

Determine the orientation of the fractures, and the required trend of the production holes.

Q7.9 A length of core, from a borehole whose orientation is 143/68, contains a fracture plane of 204/47. The core has rotated through a clockwise angle (looking down the borehole) of 140° during retrieval. What will be the apparent orientation of the fracture as it emerges from the borehole?

Q7.10 A borehole of orientation 136/55 intersected six fractures belonging to the same sub-parallel set, the orientations of which were 201/39, 213/50, 215/63, 230/52, 247/42 and 253/28. Compute the mean orientation of the set,
(a) without accounting for sampling bias, and
(b) accounting for sampling bias.

Questions 8.1–8.10: rock masses

Q8.1 For a simple sedimentary rock mass in which the only effective fractures are the bedding planes, the elastic modulus of the rock mass can be found from the addition of the displacements due to both the intact rock and the fractures, noting that the rock mass can comprise more than one stratum, each containing bedding plane fractures with different frequencies.

Unfractured strata. Consider firstly the case of n strata of intact rock, each with a thickness t_i and modulus of elasticity E_i. Derive an expression for the composite elastic modulus, E_m, of the rock mass in a direction normal to the strata by considering the total displacement (and hence strain) of the total thickness of the rock mass due to the applied stress. Write the expression in terms of E_i and t_i, and assume that the interfaces between adjacent units have no mechanical effect.

Strata with bedding plane fractures. Now consider the case where each stratum of rock contains a set of bedding plane fractures parallel to the stratum boundaries. The fracture frequency of the set within each stratum is unique — stratum i possesses a frequency λ_i; similarly, the modulus of deformation (i.e. applied stress/displacement) within each stratum is unique and for unit i is E_{di}. Extend the expression for E_m to include t_i, E_i, λ_i and E_{di}.

Q8.2 For the unfractured and fractured stratified rock mass geometries described in Q8.1, develop expressions for the composite shear modulus of a rock mass, G_m, using a shear stress τ and the parameters t_i, G_i, λ_i, and G_{di}.

Q8.3 When the application of stress is not perpendicular to the fractures, as in Q8.1 and Q8.2, it is necessary to transform the stress components in order to establish rock mass deformation moduli using the fracture stiffnesses or compliances. This results in equations for the rock mass modulus, E_m, of the type (Wei and Hudson, 1986)

$$\frac{1}{E_m} = \frac{1}{E} + \lambda_1 s_{11}^1 \cos^4\alpha + \lambda_1 s_{22}^1 \cos^2\alpha \sin^2\alpha + \lambda_2 s_{22}^2 \cos^2\alpha \sin^2\alpha + \lambda_2 s_{11}^2 \sin^4\alpha$$

for a 2-D analysis with uniaxial loading.

The equation is for two orthogonal fracture sets, where E is the modulus of the intact rock, λ_1 and λ_2 are the fracture frequencies of the two sets, s_{11}^1, s_{22}^1, s_{11}^2, and s_{22}^2 are the normal and shear compliances for the fractures in each set, and α is the angle between the applied normal stress and the normal to the first set.

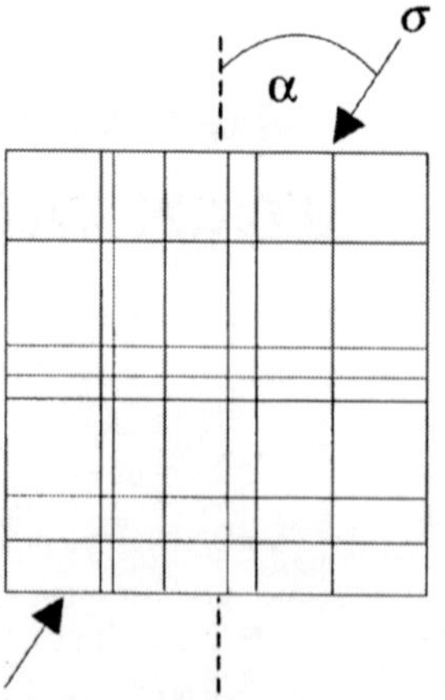

The equation reduces for one set (i.e. $\lambda_2 = 0$) to

$$\frac{1}{E_m} = \frac{1}{E} + \lambda s_{11}^1 \cos^4 \alpha + \lambda s_{22}^1 \cos^2 \alpha \sin^2 \alpha$$

and if we put $s_{11}^1 = s$ and $s_{22}^1 = ks$, and then rearrange, we obtain

$$\frac{1}{E_m} = \frac{1}{E} + \lambda s \cos^2 \alpha \left(\cos^2 \alpha + k \sin^2 \alpha\right)$$

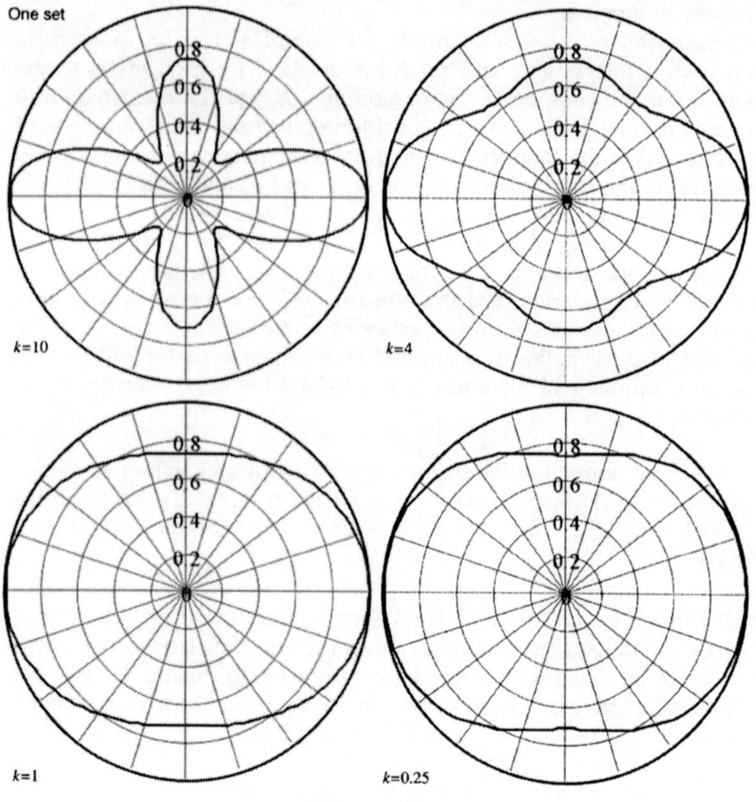

Diagram Q8.1

from which we find

$$\frac{E_m}{E} = 1 - E_m\lambda s \cos^2\alpha\left(\cos^2\alpha + k\sin^2\alpha\right).$$

Using $E = 2.75$ GPa, $\lambda = 2$ m^{-1} and $s = 0.05$ m/GPa, the plots as shown in Diagram Q8.1 result.

For two sets with $\lambda_1 = \lambda_2 = \lambda$, $s_{11}^1 = s_{11}^2 = s$ and $s_{22}^1 = s_{22}^2 = ks$, the basic equation reduces to

$$\frac{E_m}{E} = 1 - E_m\lambda s\left(\cos^4\alpha + 2k\cos^2\alpha\sin^2\alpha + \sin^4\alpha\right).$$

Using $E = 2.75$ GPa, $\lambda = 2$ m^{-1} and $s = 0.05$ m/GPa, the plots as shown in Diagram Q8.2 result.

Explain why terms such as $\cos^4\alpha$ appear in the formula and comment on any general principles that are apparent from these illustrative plots.

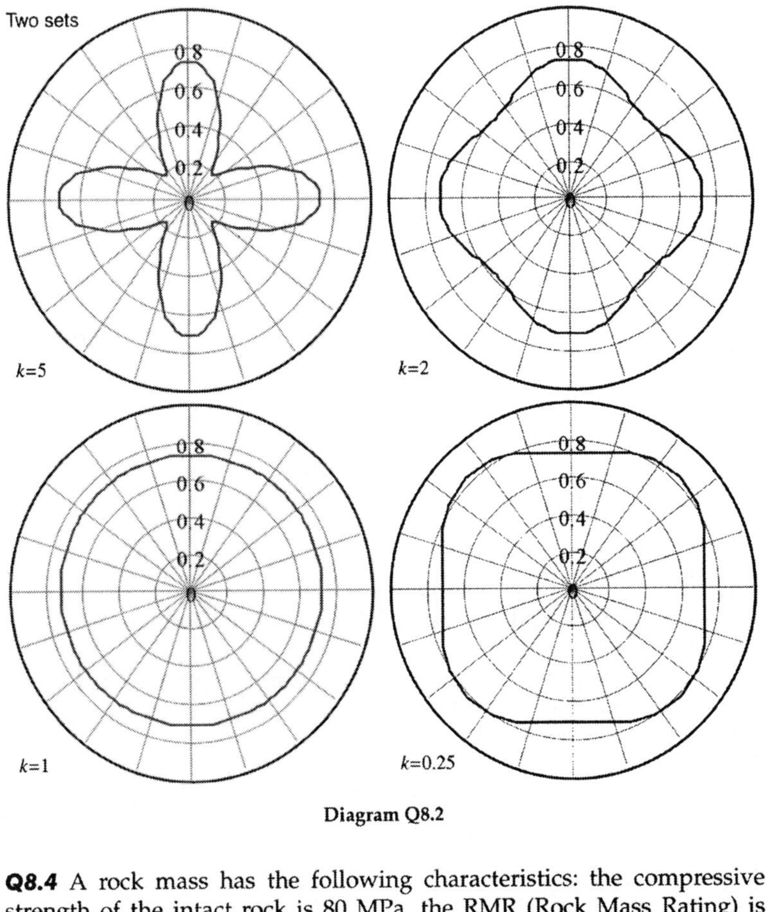

Diagram Q8.2

Q8.4 A rock mass has the following characteristics: the compressive strength of the intact rock is 80 MPa, the RMR (Rock Mass Rating) is 62%, and the GSI (Geological Strength Index) is 50. Estimate the *in situ* deformation modulus, E_m.

Q8.5 To study the effect of a fracture on the rock strength, plot a graph of $(\sigma_1 - \sigma_3)$ vs. β_w using the single plane of weakness formula included in Section 8.1. Assume $\sigma_3 = 10$ MPa, $c = 0$ and $\phi_w = 35^\circ$, and that the intact rock strength is given by $\sigma_1 = 75 + 5.29\sigma_3$. Also, explain the form of the resulting graph.

Q8.6 If a rock mass contains more than one fracture set, we can apply the single plane of weakness theory to each set, and superimpose the results to find a lowest-bound envelope of strength.

(a) Plot the 2-D variation in strength for a rock mass containing two orthogonal sets of fractures, A and B, the strengths of which are $c_A = 100$ kPa, $\phi_A = 20^\circ$ and $c_B = 0^\circ$, $\phi_B = 35^\circ$, when the minor principal stress has the value 10 MPa. The intact rock strength is again given by $\sigma_1 = 75 + 5.29\sigma_3$.

(b) How would this strength variation change if the minor principal stress were reduced to zero?

Q8.7 When a rock mass contains a large number of fracture sets and each set has similar strength properties, the rock mass strength can tend to become isotropic, with the multiple plane of weakness theory generating an approximately isotropic strength criterion. Develop such a criterion for the 2-D case of a rock mass that contains four sets of fractures mutually inclined at 45°, the shear strengths of which are given by a linear Mohr–Coulomb criterion with $c = 100$ kPa and $\phi_w = 30^\circ$. The intact rock strength is given by $\sigma_1 = 75 + 5.29\sigma_3$. Assume that the minor principal stress, σ_3, is 10 MPa.

Q8.8 How does the significant effect of fractures on the rock mass strength indicate some of the differences between rock mechanics and soil mechanics?

Q8.9 The peak strength of a poor-quality, closely jointed and weathered granite may be represented by the Hoek–Brown criterion,

$$\sigma_1 = \sigma_3 + \sqrt{m\sigma_c\sigma_3 + s\sigma_c^2}, \text{ with } m = 1.3 \text{ and } s = 0.00001.$$

The uniaxial compressive strength of the intact rock material is estimated as $\sigma_c = 40$ MPa, and the residual strength of the fractures is given by $c_r = 0$, $\phi_r = 15^\circ$.

(a) Plot, in τ–σ space, the expected peak shear strength envelopes for the intact rock material, a fracture at residual strength, and the jointed rock mass for normal stresses up to 10 MPa.

(b) The Hoek–Brown criterion is to be used for the analysis of potential circular slopes cut in the rock mass. Determine values of instantaneous cohesion and friction angle for normal stresses up to 10 MPa.

Q8.10 To study the influence of fracture persistence, consider a rock mass containing two fracture sets, A and B, mutually inclined at 45°, as shown to the right. Fracture set A is continuous and contains clay infilling, while fracture set B is rough, clean and intermittent with a 'two-dimensional proportional extent of fracturing' of 0.5.

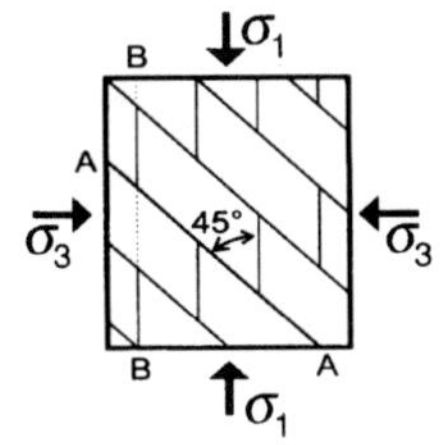

Plot the expected peak strength of this rock mass as a function of the orientation of the principal stresses to the fracture directions, for a minor principal stress of 5 MPa and given the following peak strength characteristics:

Intact rock: $\sigma_1 = 75 + 5.29\sigma_3$

Fracture set A: $c = 100$ kPa, $\phi = 20°$,

Fracture set B: $c = 0$, $\phi = 35°$

Questions 9.1–9.10: permeability

Q9.1 How long does it takes for water, subjected to a 10-m head difference, to pass horizontally through

(a) a 5 m length of intact granite which has an isotropic hydraulic conductivity, K, of 1×10^{-12} m/s, and

(b) through a 5 m length of fractured limestone with an isotropic hydraulic conductivity of 1×10^{-4} m/s?

Q9.2 The hydraulic conductivity of an array of parallel fractures (Hoek and Bray, 1977) in the direction parallel to the plane of the fractures is given by the equation, $K = \lambda g e^3/12\nu$ where λ is the fracture frequency, g is gravitational acceleration, e is the fracture aperture, and ν is the kinematic viscosity of the fluid.

For a rock mass with a fracture frequency of one fracture per metre and with fracture apertures of 0.01 mm, the hydraulic conductivity is 8.3×10^{-10} m/s.

(a) What is the hydraulic conductivity of a second rock mass which has ten fractures per metre and fracture apertures of 1 mm?

(b) What is the main factor contributing to the difference in the hydraulic conductivity of the two rock masses?

Q9.3 In question Q9.2, the hydraulic aperture was used. For a fracture with planar and parallel sides, this aperture is the perpendicular distance between the two sides (see Fig. Q9.1).

(a) Explain the meaning of fracture aperture when the fracture surfaces are rough.

(b) Do you think that the mechanical aperture and the hydraulic aperture of a fracture have the same value? If not, which is greater?

Q9.4 What is meant by the term 'transmissivity'?

Q9.5 The horizontal section below shows two sets of rock fractures in a 10 m square rock block of unit thickness. On the left-hand side, there is a hydraulic head of 3 m; on the right-hand side, the head is 1 m; and

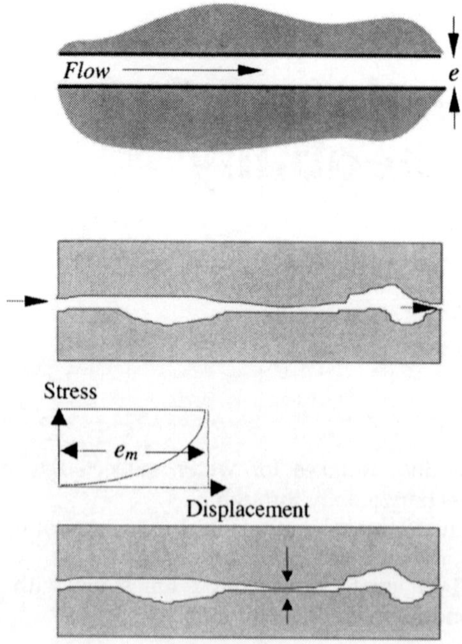

Figure Q9.1

along the top and bottom edges the head linearly decreases from the left to the right.

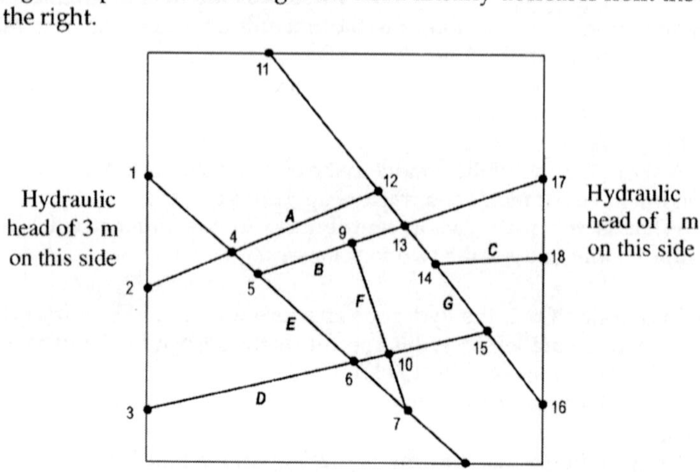

The (x, y) co-ordinates (in m) of the fracture intersections, i.e. the numbered nodes, relative to an origin at the bottom left of the diagram, are given in the table below.

Assuming that all the fractures have the same aperture, 0.12 mm, and that there is no variation in flow throughout the thickness of the rock block, determine the nodal heads and hence the direction and magnitude of flow in each fracture segment.

1	2	3	4	5	6	7	8	9	10	11	12	13	14	15	16	17	18
0.00	0.00	0.00	2.10	2.74	5.16	6.61	8.06	5.16	6.13	3.06	5.81	6.61	7.42	8.71	10.00	10.00	10.00
6.94	4.19	1.29	5.16	4.52	2.42	1.13	0.00	5.32	2.58	10.00	6.61	5.65	4.84	3.23	1.45	6.94	5.00

Q9.6 Determine the nodal heads and hence the direction and magnitude of flow in each rock mass fracture segment for the same case as Q9.5, but with different fracture apertures, as given in the matrix below.

	1	2	3	4	5	6	7	8	9	10	11	12	13	14	15	16	17	18
1	■			0.07														
2		■		0.16														
3			■			0.24												
4				■	0.07							0.16						
5					■	0.07			0.18									
6						■	0.07			0.24								
7							■	0.07		0.09								
8								■										
9									■	0.09			0.18					
10										■					0.24			
11											■	0.06						
12												■	0.06					
13													■	0.06			0.18	
14														■	0.06			0.13
15															■	0.06		
16																■		
17																	■	
18																		■

Note that this matrix is symmetric, and we have suppressed the values in the lower left of the matrix. The units are millimetres.

Q9.7 The following rock mass hydraulic conductivity data were obtained from four sets of tests on fractured rock samples with volumes 1 m^3, 5 m^3, 10 m^3, and 15 m^3. In each case it was assumed that the hydraulic conductivity was a scalar quantity. The data values are in units of 1×10^{-6} m/s. As the test volume increased, so did the difficulty and the cost of the tests, so there are fewer data for the tests at larger volumes.

Data for test size of 1 m^3
2.122 1.156 3.696 1.165 1.649 2.886 1.652 2.876 1.197 2.593
2.114 2.771 16.214 2.529 1.700 7.658 10.928 0.627 9.854 0.847
0.670 1.623 2.274 5.163 2.209

Data for test size of 5 m^3
1.630 1.981 2.436 3.700 1.215 1.767 0.909 0.450 3.512 1.314

Data for test size of 10 m^3
1.369 1.188 2.037 1.688

Data for test size of 15 m^3
1.487 1.343 1.473 1.738

Using the data above, estimate the REV value for the hydraulic conductivity of the fractured rock mass tested.

Q9.8 The values and directions for the principal hydraulic conductivities in a dolomitic rock, shown below, were obtained from back analysis of induced drainage discharges at the Morro da Usina Mine, Vazante District, State of Minas Gerais, Brazil[1]. These are the mean principal hydraulic conductivities for a rock mass volume greater than 3 km^3. What geological circumstances could cause these values and orientations?

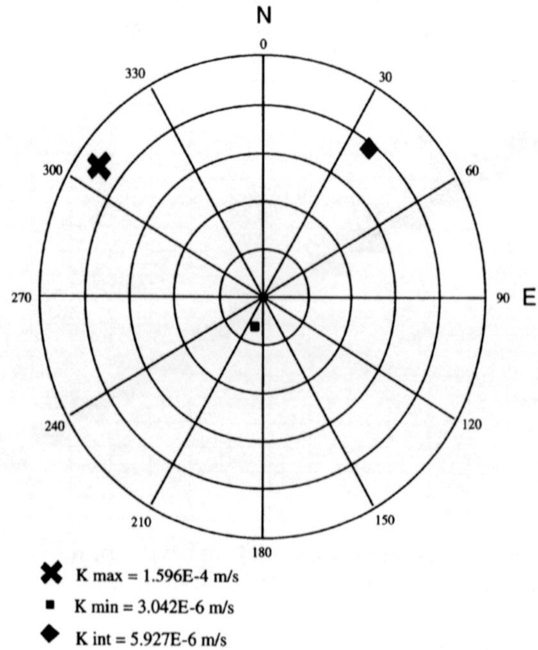

Q9.9 (a) Explain the term 'effective stress'.

(b) If a fracture contains water under pressure and the stresses at the fracture surfaces are being considered, what would be the effect on the normal stress and on the shear stress in the rock of steadily increasing the water pressure?

Q9.10 In the photograph below, the water pressure in a borehole in marble has been steadily increased. In addition to the normal pressure created at the rock surface, the water pressure in this configuration also creates a circumferential tensile stress of the same magnitude in the rock, as in Q4.6. A crack developed and was propagated under servo-controlled conditions from A to B. The water not only penetrated the crack but also the intact marble, forming the dark inverted V visible

[1] The authors are grateful to Prof. E. Quadros and Dr. F. O. Franciss for permission to use this example.

in the photograph. How do we deal with the effective stress under these circumstances?

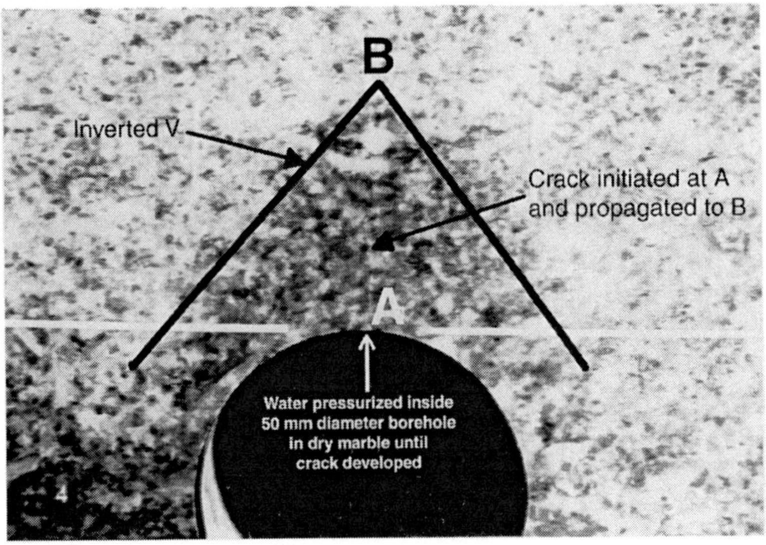

Questions 10.1–10.10: anisotropy and inhomogeneity

Q10.1 On the rock mass photograph (see Fig. Q10.1), identify example features that indicate that the rock mass is more DIANE than CHILE.

Q10.2 With reference to the photograph in Fig. Q10.1, explain why inhomogeneity and anisotropy can be related.

Q10.3 The following data are pairs of point load strength (*PL*) and uniaxial compressive strength (*UCS*) values for a particular rock type.

UCS (MPa)	52.4	60.7	44.6	66.6	47.6	56.3	61.5	48.2	49.6	47.2	56.7	48.4	61.5	52.9
PL (MPa)	2.51	2.87	2.14	3.04	2.31	2.64	2.90	2.21	2.25	2.18	2.55	2.23	2.75	2.55

We wish to correlate these strength values, and can do so either in the form $PL = a\sigma_c + b$ or in the form $\sigma_c = cPL + d$. On the basis of the best independent variable, which of these forms is appropriate? Determine values for the appropriate constants (i.e. either a and b, or c and d).

Q10.4 Imagine that a new index test for determining the tensile strength of specimens of intact rock is under development in the Rock Mechanics Laboratory at Imperial College. This test involves bonding a steel rod to the surface of a specimen with high strength adhesive, and then measuring the tensile load required to pull the rod together with a small piece of rock away from the main block of rock. Four test configurations are under consideration, and for each of these a theoretical relation between rock strength and pull-off force has been developed. Test results for the four configurations, together with the appropriate theoretical relation, are given below. The first row in each table is the load measured in the new test. The second row in each table is the tensile strength of the rock as measured by a standard method.

Classify each configuration in terms of accuracy and precision, and hence recommend which configuration(s) should be retained for further development.

Figure Q10.1

Configuration 1: strength = 0.049 × load

Load (*N*)	67.3	76.8	83.9	104.8	153.7	168.9	191.2	194.7	237.5	258.3
Strength (MPa)	4.2	4.8	5.2	6.5	9.6	10.6	11.7	12.2	14.6	16.1

Configuration 2: strength = 0.066 × load

Load (*N*)	68.9	105.3	106.2	120.1	148.5	164.8	197.4	220.5	232.8	236.9
Strength (MPa)	4.7	2.4	3.3	7.4	6.7	10.9	6.3	7.8	8.2	7.6

Configuration 3: strength = 0.074 × load

Load (*N*)	83.5	95.0	111.7	151.6	170.0	189.5	190.2	193.9	201.1	205.3
Strength (MPa)	6.1	7.1	8.3	11.4	12.7	13.8	14.3	14.3	14.9	15.3

Configuration 4: strength = 0.094 × load

Load (*N*)	68.9	105.3	106.2	120.1	148.5	164.8	197.4	220.5	232.8	236.9
Strength (MPa)	5.5	10.1	10.5	11.6	14.4	12.3	20.1	22.9	20.9	21.8

Q10.5 A vertical site investigation borehole intersects a stratum of sandstone which is dipping at 17°. A length of intact core from this borehole was taken to the laboratory for hydraulic conductivity testing, and two small-diameter plugs drilled from it: one axially and one radially. These plugs were then tested in a permeameter, and the values of the hydraulic conductivity were found to be 1.728×10^{-7} m/s and 1.557×10^{-6} m/s in the axial and the radial directions, respectively. Stating any assumptions that you make, compute estimates of the hydraulic conductivity normal and parallel to the bedding.

Q10.6 The diagram shows a polar plot of the variation in hydraulic conductivity, K, of a sandstone, with the maximal and minimal values occurring parallel and perpendicular to the bedding.

Why is there a cusp in the (K, θ) locus on the line representing the direction normal to the bedding?

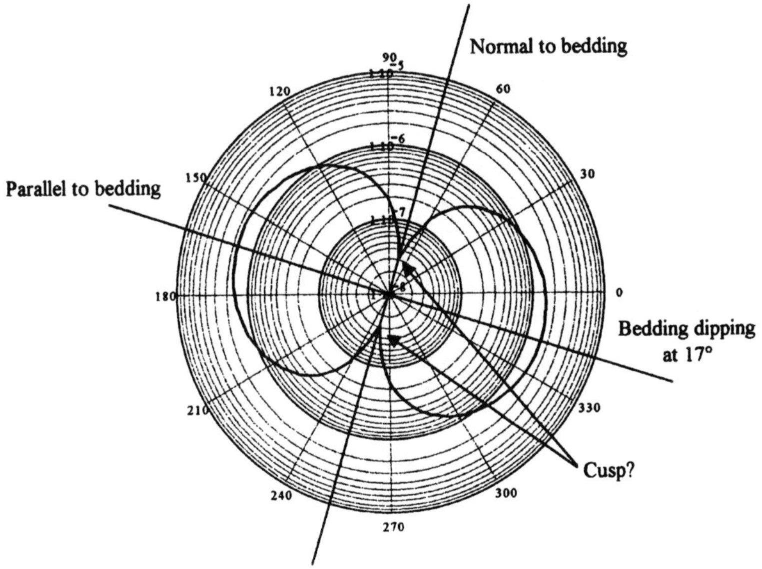

Q10.7 What does the term 'structural domain' mean in the context of engineering rock mechanics?

Q10.8 During a site investigation for quarry development, geometrical properties of the fractures on a large surface rock exposure were measured. Eight sampling squares, each 100 m × 100 m, were established on the rock exposure. One of the set of statistics produced during the site investigation was the number of fractures with a mean aperture exceeding 10 mm in each sampling square. These results were as follows:

Square:	1	2	3	4	5	6	7	8
	461	397	453	362	389	421	382	423

On the basis of these data alone, is there sufficient evidence to conclude that the sampling squares should be regarded as different structural domains?

Q10.9 The following list of fracture locations, quoted in metres, is taken from the fracture log of a borehole core which transects a well known stratigraphic boundary between two units of limestone. Evidently, this boundary is clear to sedimentologists, but not to geotechnical engineers.

5.780, 6.391, 6.761, 7.105, 7.180, 7.401, 7.478, 8.142, 8.455, 9.139, 10.072, 10.264, 10.470, 10.539, 10.678, 11.421, 11.541, 12.178, 12.596, 12.620, 12.736, 12.936, 13.134, 13.325, 13.430

Use the concept of moving averages to help locate this boundary.

Q10.10 The subject of geostatistics deals with the variation of property values in space, and so anisotropy and inhomogeneity can be quantitatively studied. The basic device of geostatistics is the semi-variogram, defined as

$$\gamma(h) = \left(\frac{1}{2n}\right)\sum_{i=1}^{n}\left[p(x) - p(x+h)\right]^2$$

where $\gamma(h)$ is the semi-variogram statistic for samples distance h apart, n is the number of sample pairs, $p(x)$ is the rock property value at location x, and $p(x+h)$ is the value at location $x+h$.

Using this statistic, a graph can be constructed of $\gamma(h)$ *versus* h to indicate the variation in rock property values as a function of distance between the observations.

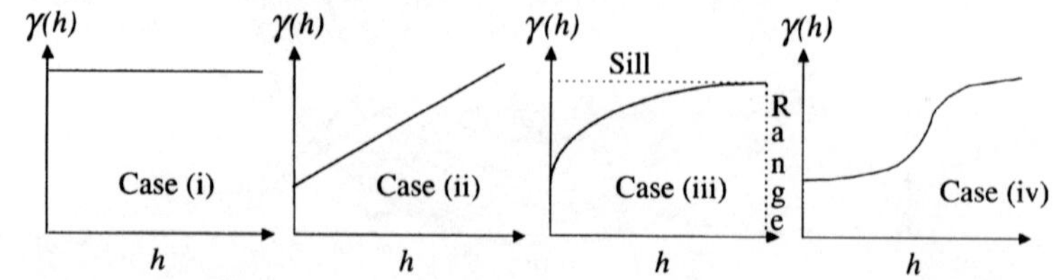

In the diagram above, there are four examples of such semi-variogram curves for a rock mass. In each case, the length of the h axis represents 50 m.

(a) Explain what type of variation in rock property values, or inhomogeneity, each type of semi-variogram represents.

(b) For each type of semi-variogram behaviour, how far away from a borehole would you feel confident in extrapolating results obtained from a borehole core?

Questions 11.1–11.10: testing techniques

Q11.1 The section of site investigation borehole core shown in the photograph is from a vertical borehole and contains three stress-induced fractures. The top of the core is a stress fracture of the same kind. The bottom end is a drilling break. Assuming that the strip of translucent tape (adjacent to the numbers written on the core in the photograph) is on the northern side of the core, in which horizontal directions do you think the major and minor principal stresses act?

Q11.2 With reference to fracture property measurements made during a site investigation

- on borehole rock core,
- on the borehole walls, and
- on rock exposures,

complete the table below indicating your opinion of how well you think that the listed fracture properties can be measured or estimated. The first column of the table represents the ten fracture measurements recommended by the ISRM, as in Fig. Q7.7. Use the letters G for Good, M for Medium, and P for Poor.

Characteristic	Measurement method	Core	B/H wall	Exposure
Orientation	Compass-clinometer			
Spacing	Measuring tape			
Persistence	Measuring tape			
Roughness	Against reference chart			
Wall strength	Schmidt hammer			
Aperture	Feeler gauge			
Filling	Visual			
Seepage	Timed observations			
Number of sets	Stereographic projection			
Block size	3-D fracture frequency			

Q11.3 The results of a series of scanline surveys at a particular site are as follows:

Scanline trend (°)	000	355	085	153	216	271
Scanline plunge (°)	90	35	28	51	05	12
Fracture frequency (m^{-1})	5.54	7.93	6.02	7.00	6.99	7.65

Analysis of the fractures intersected by the scanlines has shown that the rock mass contains four fracture sets, with orientations 145/08, 148/88, 021/76 and 087/69 (given as dip direction/dip angle). What is the best estimate of the frequency of each fracture set?

Q11.4 As part of a site investigation study, a rock mass was found to contain four fracture sets with dip/dip direction and frequencies as follows:

Set 1: 08/145, 3.48/m
Set 2: 88/148, 3.91/m
Set 3: 76/021, 3.58/m
Set 4: 69/087, 3.26/m

In order to establish in which directions through the rock mass an excavation will encounter the minimal and maximal numbers of fractures, the fracture frequency in different directions through the rock mass, λ_s, was calculated using the formula $\lambda_s = \Sigma_{i=1}^{n} |\lambda_i \cos\theta_i|$ (see Q7.6 and A7.6). The results are presented below on a hemispherical projection, with the contouring corresponding to the fracture frequency values in the different directions.

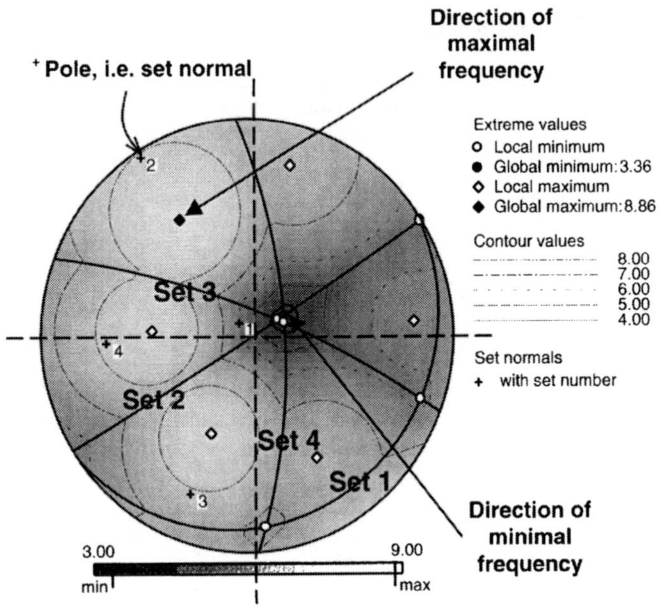

Explain from first principles why the directions of the minimal and maximal frequencies occur where they do.

Q11.5 When cyclic deformability tests are conducted on rock masses, the typical force–displacement curve is as shown below (Schneider, 1967; Goodman, 1989).

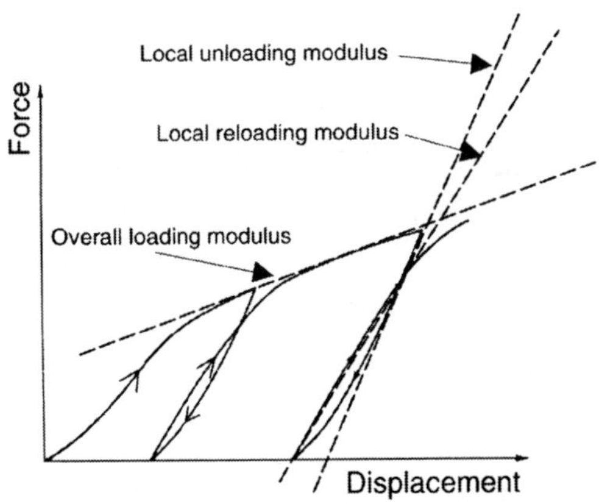

Explain why, with cycles of repeated unloading and reloading, the curve manifests different unloading and reloading moduli, permanent deformations and hysteresis.

Q11.6 The tensile strength of an architectural granite was measured to ensure that the granite would be strong enough to form the structural elements of a pedestrian bridge in a shopping mall. Ten specimens were tested in each of four test configurations, and the values obtained were as follows.

Type of test	Mean value (MPa)	Standard deviation (MPa)
Direct tension test	8.4	3.2
Point load test	9.6	3.8
Beam test	10.4	4.5
Ring test	12.9	6.7

Are these results consistent with what you know about tensile strength variation and which value would you use for the structural calculations?

Q11.7 The following table shows data obtained from a single-stage triaxial compression test on a cylindrical rock sample, conducted with closed-loop servo-control, at a confining stress of 10.0 MPa, and at zero pore pressure.

Total axial load (kN)	Sample height (mm)	Sample diameter (mm)
0.00	100.84	50.20
19.89	100.80	50.20
39.60	100.77	50.20
63.40	100.74	50.20
88.67	100.71	50.21
116.18	100.68	50.21
144.68	100.65	50.22
162.38	100.63	50.22
185.23	100.58	50.24
190.62	100.56	50.25
191.99	100.54	50.25
180.22	100.52	50.26
137.56	100.49	50.26
115.79	100.46	50.27
101.93	100.43	50.28
97.97	100.40	50.28
96.98	100.37	50.28

Estimate values for the following:
(i) yield strength σ_y;
(ii) peak strength σ_{max};
(iii) residual strength σ_r;
(iv) tangent Young's modulus E_{tan} at 50% peak axial stress; and
(v) tangent Poisson's ratio ν_{tan} at 50% peak axial stress.

Q11.8 Explain why the introduction of servo-controlled testing machines in the early 1970s revolutionized rock mechanics laboratory testing, and why we are now able to test rocks under virtually any loading conditions.

Q11.9 (a) The results in the table below represent shear displacement and shear stress recorded during a direct shear test on a fracture in slate. The shear displacement range was from 0 to 15 mm as shown in the table below. The normal stress during the test was 0.34 MPa.

Shear displacement (mm)	0.0	0.5	1.0	1.5	2.0	2.5	3.0	3.5
Shear stress (kPa)	0	281	344	344	328	281	281	297
Shear displacement (mm)	4.0	4.5	5.0	5.5	6.0	6.5	7.0	7.5
Shear stress (kPa)	281	281	266	266	266	281	281	281
Shear displacement (mm)	8.0	8.5	9.0	9.5	10.0	10.5	11.0	11.5
Shear stress (kPa)	297	297	297	313	313	313	313	313
Shear displacement (mm)	12.0	12.5	13.0	13.5	14.0	14.5	15.0	
Shear stress (kPa)	313	313	313	313	313	313	313	

Use these results to determine the residual shear strength of the fracture.

(b) A series of direct shear tests was undertaken at different normal stress values on samples from the fracture, and the peak shear stress encountered during each test was recorded, as shown in the table below.

Normal stress (kPa)	336	648	961	1273	1586
Peak shear stress (kPa)	344	516	719	953	1156

Use these results to determine the basic friction angle, ϕ, and the asperity angle, i, for the fracture. Also comment on the validity of the bi-linear approximation for the failure locus.

Q11.10 The diagram below shows example results from a numerical modelling code for predicting the elastic displacements (indicated by the

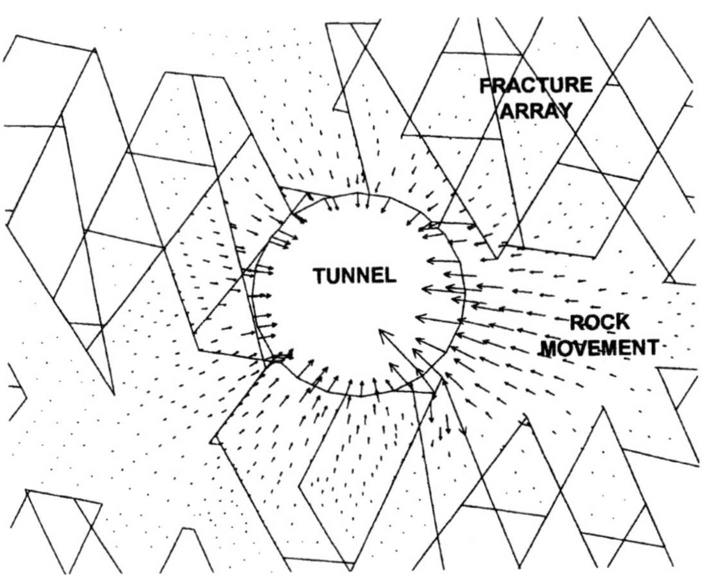

arrows) of a 2-D assemblage of distinct rock blocks through which a tunnel has been excavated. The plot shows the displacement vectors.

(a) Write down a list of rock properties that you think would be required as input to such a modelling exercise.

(b) Indicate which of these are likely to be practicably measureable.

Questions 12.1–12.10: rock mass classification

Q12.1 A mudstone rock mass at a depth of 200 m contains three fracture sets. One set comprises bedding planes; these are highly weathered, slightly rough surfaces, and are continuous with an orientation of 180/10. Another set is jointing; these joints are slightly weathered, slightly rough, and have an orientation of 185/75. The third set is also jointing; again, the joints are slightly weathered and slightly rough, and have an orientation of 090/80. The strength of the intact rock has been assessed as 55 MPa, and values for the *RQD* and mean fracture spacing are reported as 60% and 0.4 m, respectively.

Use the *RMR* system to classify this rock mass, and assess the stability of a 10 m wide excavation being driven from east to west.

Q12.2 A 7-m-diameter tunnel is to be driven through a sequence of shale and basalt rock at a maximum depth of 61 m. The shales dip towards the east, and the basalts form sub-vertical dykes. The bedding dips between 15° and 20°, the joints dip between 70° and 90°. The joints in the shale are rough, and most of them are thin and healed with calcite, but overall the rock is described as 'blocky and seamy'. The groundwater level is about 50 m above the invert of the tunnel. The average uniaxial compressive strength of the shale is 53 MPa, of the basalt it is 71 MPa. The vertical stress is about 1.0 MPa, and the horizontal stress is about 3.4 MPa. The snaking nature of the tunnel's route means that at some place along its length it will head in all directions between 090° and 180°.

Use the *RMR* system to predict how the rock will behave in the excavation.

Q12.3 Use the *Q* system to assess the stability of the rock mass as described in Q12.1.

Q12.4 Use the *Q* system to assess the stability of the rock mass described in Q12.2.

Q12.5 Using your assessments of *RMR* and *Q* for questions Q12.1, Q12.2, Q12.3 and Q12.4, investigate the relation between the *Q* and *RMR* values. Do your results correspond with a generally accepted relation, $RMR = 9 \ln Q + 44$?

Q12.6 The diagram below (Singh and Goel, 1999) shows *RMR*–*Q* correlations for case studies in India, Scandinavia, UK and USA.

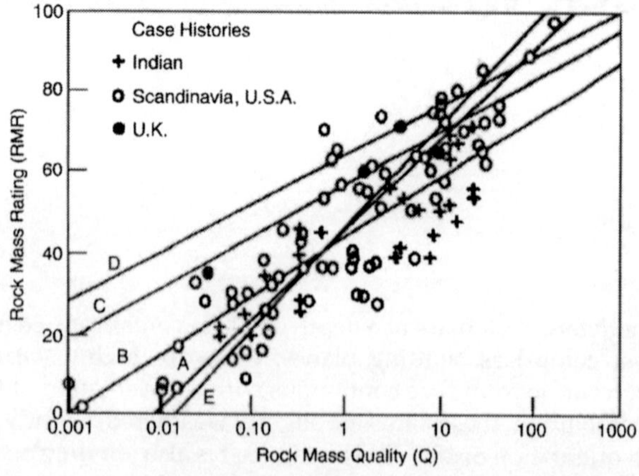

The suggested *RMR*–*Q* correlation lines shown on the diagram are

A	$RMR = 9 \ln Q + 44$
B	$RMR = 5.9 \ln Q + 43$
C	$RMR = 5.4 \ln Q + 55.2$
D	$RMR = 5 \ln Q + 60.8$
E	$RMR = 10.5 \ln Q + 41.8$

For a rock engineering design project where a correlation between *RMR* and *Q* is required to support the design, which of the correlations would you choose?

Q12.7 Imagine that a rock mass classification system is required to assess the instability of natural slopes in the Italian Alps. List 15–25 parameters that you think would be most useful for such a classification scheme.

Q12.8 A rock mass classification system is required for assessing the suitability of different rock formations for storing compressed domestic gas in unlined rock caverns along the route of a main gas transmission line. List the rock parameters that you would use in a rock mass–rock engineering classification scheme for this objective.

Q12.9 The following parameter values have been determined for three rock mass types found along the route of a major new highway tunnel that passes at a high level through the flank of a mountain range:

	Strength (MPa)	*RQD* (%)	Mean fracture spacing (m)
Sandstone	80	45	0.4
Mudstone	20	75	0.3
Syenite intrusions	250	10	0.2

The fractures within each rock mass type have the properties shown in the following table:

	Persistence (m)	Aperture (mm)	Roughness	Infilling	Weathering
Sandstone	5–8	~1.5	rough	none	none
Mudstone	1.5–2.5	~0.5	slight	none	slight
Syenite	2	~6	very	none	none

Write down a description for each of these three rock mass types, and describe their likely engineering behaviour.

Then apply the *RMR* system to these rock mass types, and compare the assessment of their engineering behaviour made in this way with the description you wrote down earlier.

What do you conclude from this exercise about the ability of *RMR* to discriminate between the engineering behaviour of these particular rock mass types?

Q12.10 The following measurements of mean fracture spacing (in metres) have been made on core from 12 boreholes as part of a site investigation project:

0.259	0.304	0.875	0.292	0.467	0.412	0.350	0.368	0.438	0.389	0.280	0.318

As the rock mass is to be characterized using the Q system, the following parameters have also been determined: $J_n = 9$; $J_r = 1.5$; $J_a = 2$; $SRF = 2.5$; and $J_w = 1$.

(a) Using the frequency measurements to determine *RQD* values and thence Q values with the additional parameters given, comment on the inhomogeneity of the rock mass in terms of:

(i) fracture frequency, and

(ii) Q.

(b) A technique for increasing the range of *RQD* values in a given rock mass is to adopt a different *RQD* threshold value (from the usual value of 0.1 m) computed using $t^* = 2\ln(\lambda_{max}/\lambda_{min})/(\lambda_{max} - \lambda_{min})$, where λ_{max} and λ_{min} are the extreme values of the fracture frequency occurring in the rock mass. Use this technique to compute new values of Q, and compare the results with those found in Part (a).

Questions 13.1–13.10: rock dynamics and time dependency

Q13.1 There is no time component in the theory of elasticity. Why then does Young's modulus, expressed in units of stress, have time in its dimensions: $L^{-1}MT^{-2}$?

Q13.2 A 10-mm-diameter core of intact marble is carefully drilled out to a length of 1 m. The core is suspended horizontally by steel wires and then struck gently at one end to produce a longitudinal stress wave through the bar, as shown below. This is known as the Hopkinson bar experiment, used to study the transmission of stress waves.

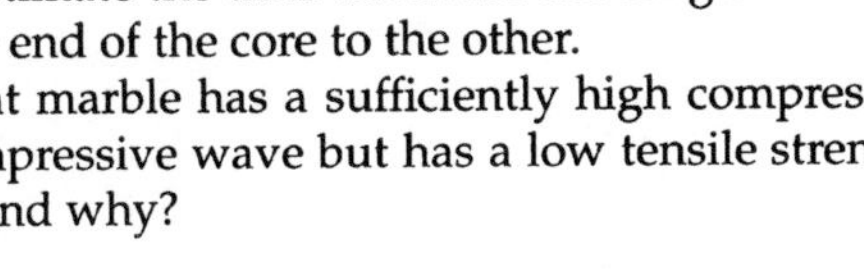

(a) If Young's modulus of the marble is 50 GPa and the unit weight is 27 kN/m^3, estimate the time taken for the longitudinal stress wave to travel from one end of the core to the other.

(b) Given that marble has a sufficiently high compressive strength to sustain the compressive wave but has a low tensile strength, where will the bar break, and why?

Q13.3 What is the ratio V_p^2/V_s^2 in terms of the elastic rock constants and what is the specific value of the expression for a rock with $\nu = 0.27$?

Q13.4 A 100-mm-long rock specimen is to be tested in uniaxial compression using strain control in a servo-controlled testing machine.

The Young's modulus of the rock is 60 GPa and the compressive strength is 200 MPa. We should like to reach the compressive strength in the test in about 10 minutes. What displacement rate should be used for the testing machine program, and what is the corresponding rock strain rate?

Q13.5 The results in the table below show the axial displacement and radial strain induced in a cylindrical specimen of weak chalk during a uniaxial creep test. In this test, the specimen was initially 250 mm high and was subjected to an axial stress of 55 MPa. After 3 hours the test was stopped, at which stage creep had ceased and the displacement had become constant at 0.4545 mm.

Time (min):	0	1	2	3	4	5	6	7
Axial displacement (mm):	0.409	0.414	0.419	0.423	0.427	0.430	0.433	0.435
Radial strain ($\times10^{-6}$):	−451	−461	−471	−479	−487	−493	−499	−504
Time (min):	8	9	10	11	12	13	14	15
Axial displacement (mm):	0.438	0.440	0.441	0.443	0.444	0.445	0.447	0.447
Radial strain ($\times10^{-6}$):	−509	−513	−516	−519	−522	−524	−526	−528

On the basis of these results, select a simple viscoelastic model for the rock, and determine values for the various viscoelastic constants.

Q13.6 On a sketch of the complete stress–strain curve for rock in uniaxial compression, draw lines illustrating creep, relaxation and intermediate time-dependent straining along a line of slope $-K$

(a) on the ascending and descending sides of the curve, and

(b) then comment on the significance of your diagram for rock mass stability during time-dependent deformations for a single rock pillar of intact rock and for an abandoned chalk mine excavation (Smith and Rosenbaum, 1993) as shown below.

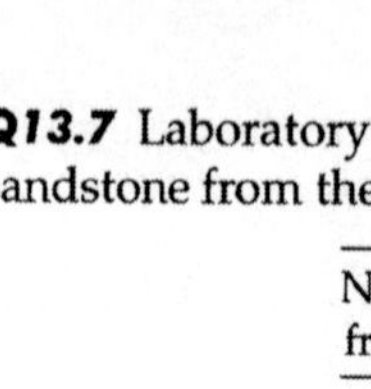

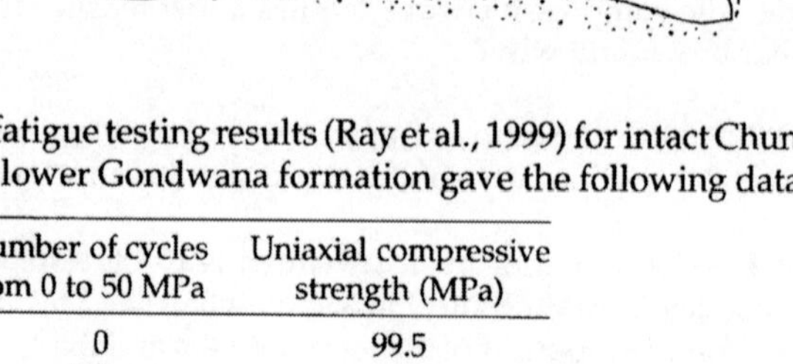

Q13.7 Laboratory fatigue testing results (Ray et al., 1999) for intact Chunar sandstone from the lower Gondwana formation gave the following data.

Number of cycles from 0 to 50 MPa	Uniaxial compressive strength (MPa)
0	99.5
100	90.1
200	77.1
300	71.6
400	67.6
500	65.9
600	65.1

The rock samples were cycled for the number of times given in the left-hand column, and then tested for their uniaxial compressive strength, giving the values in the right-hand column.

Plot these results (number of cycles on x-axis, UCS on y-axis) and comment on the trend.

Q13.8 How do fractures respond to stress waves? What do you think are the influences of a dynamic load on the fracture shear behaviour? What happens if the fractures pictured below are subjected to repeated shear movements?

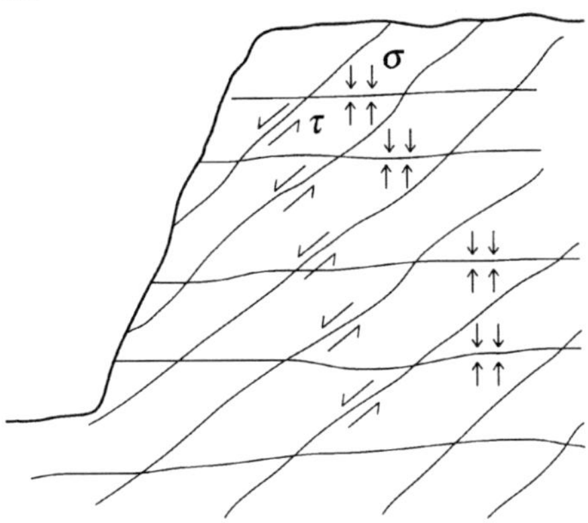

Q13.9 During the progression of a longwall mining face, bedded rock strata adjacent to the coal mine are being subjected to an extensile strain rate of 1×10^{-5} s^{-1} normal to the bedding planes. Assume that all the strain accumulated in the rock mass is concentrated in opening the bedding planes. How long will it be before the flow of water along the bedding planes is doubled?

Q13.10 The diagram below (Tan, 1993) illustrates deformations in two mine rail tunnels (originally horseshoe-shaped) in China, which are located at 430 m depth in strongly fractured granite.

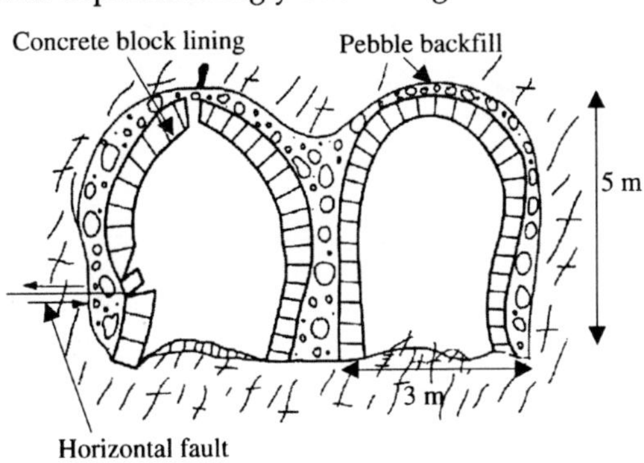

The tunnel axes are oriented perpendicular to the high intraplate horizontal tectonic stress, σ_H. The tunnel deformations have been described as follows:

> "Directly after excavation, the sidewalls started to bulge, horizontal cracks occurred which were growing in width and length; bottom heave was obvious. These processes increased in intensity with time. Void formation in the dilatancy process increasing with time leads to serious overall weakening of the rock structure; in addition it is aggravated by the suction and seepage of underground water. The horizontal displacements... were in the order of 50 cm within 90–150 days with an average rate of 5.4 mm/day. Usually the inwards motion of the corner areas near the bottom are largest, whereas the roof-top moved upwards so gradually the tunnel cross-section is squeezed into the 'peach' form... Bottom upheavals at some locations were about 40 cm within 131 days. Serious lateral motions of one sidewall have been observed."

Explain these observations in terms of the basic mechanisms involved.

Questions 14.1–14.10: rock mechanics interactions and rock engineering systems

Q14.1 Explain with a one-sentence answer how, in a rock mass,
(a) the rock stress can affect water flow in fractures,
(b) the water pressure in fractures can affect the rock stress,
(c) the fractures can affect water flow,
(d) water flow can affect the fractures,
(e) the fractures can affect rock stress, and
(f) the rock stress can affect the fractures.

Q14.2 Draw an interaction matrix having Fractures, Rock Stress, and Water Flow as the leading diagonal terms. Insert the letters representing the six mechanisms in Q14.1 in the appropriate off-diagonal boxes of the interaction matrix.

Q14.3 Under what conditions is a matrix symmetrical? Why will almost all rock mechanics interaction matrices be asymmetrical?

Q14.4 In order to consider also the interaction of rock engineering activities with the rock mechanics factors, extend the 3 × 3 interaction matrix of Q14.2 to a 4 × 4 interaction matrix by adding the extra leading diagonal term CONSTRUCTION. Write a one-phrase explanation of the content of each of the six extra off-diagonal interaction boxes introduced by adding the leading diagonal term, CONSTRUCTION.

Q14.5 The use of the interaction matrix is helpful for thinking about any problem involving manifold factors. The principal factors, located on the leading diagonal, are established first. Then, the interactions, i.e. the mechanisms linking the principal factors, are considered in the off-diagonal positions.

For example, when studying natural fractures in the rock mass, assume that we are interested in the four parameters of fracture orientation, spacing, extent (or persistence) and roughness. Using these four parameters as the leading diagonal terms of a 4 × 4 interaction matrix, identify the content of the 12 off-diagonal boxes and hence show that the parameters are likely to be related.

Q14.6 Each off-diagonal term in the interaction matrix represents a mechanism linking two principal factors. If a mechanism involves three or more principal factors, their combined interaction is represented by pathways through the interaction matrix. With reference to the interaction matrix required by Q14.4, show the example pathway representing how Construction blasting affects the Rock Stress around an excavation, which allows Fractures to open, leading to increased Water Flow which then affects Construction.

Q14.7 Assume that you have been asked to undertake a technical audit of a rock engineering design for a DIANE rock mass, where a CHILE numerical code has been used to determine the elastic rock stresses and displacements around an underground excavation. List briefly the likely sources of error in the determination of the rock stresses and displacements.

Q14.8 Assume that the interactions in the matrix required by Q14.4 (which has the principal factors of Fractures, Rock Stress, Water Flow, and Construction) have been considered for a specific rock mass and engineering objective. Each interaction has been assigned a number according to the following scheme:

0 – no interaction;
1 – weak interaction;
2 – medium interaction;
3 – strong interaction;
4 – critical interaction.

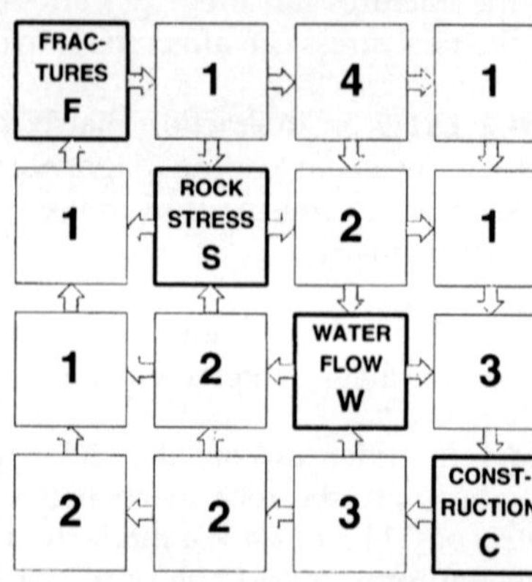

These numbers are shown in the matrix to the right.

For each principal factor, develop its 'Cause–Effect' (C, E) co-ordinates. These are the sums of the values in the row and column through each principal factor.

For example, the (C, E) co-ordinates for principal factor F are $C = 1 + 4 + 1 = 6$ and $E = 1 + 1 + 2 = 4$, i.e. (6,4). Hence establish the interaction intensity, $C + E$, and dominance, $C - E$, of each principal factor in the interactive system. Then plot the four principal factors using Cause and Effect axes.

Q14.9 In developing a rock mass classification system for estimating the relative instability of natural rock slopes in the Italian Alps (Mazzoccola and Hudson, 1996), the following parameters were chosen.

P1. Geology	P2. Folds	P3. Faults	P4. Rainfall
P5. Freeze and thaw	P6. Previous instability	P7. Intact rock strength	P8. Weathering
P9. Number of sets	P10. Orientation	P11. Aperture	P12. Persistence
P13. Spacing	P14. Mechanical properties	P15. Rock mass strength	P16. Hydraulic conditions
P17. Slope orientation	P18. Slope dimensions	P19. *In situ* stress	P20. Potential instability

The 20th parameter, 'Potential instability', was added in a similar way to the CONSTRUCTION term in Q14.4 in order to consider the effects of the parameters on potential instability.

An interaction matrix was generated using these 20 parameters as the leading diagonal terms. The interactions in the off-diagonal terms were then assigned a value according to the same scheme as in Q14.8:

0 – no interaction; 1 – weak interaction; 2 – medium interaction;
3 – strong interaction; 4 – critical interaction.

This gave (C, E) co-ordinates and values of $C + E$ and $C - E$ as follows.

	P1	P2	P3	P4	P5	P6	P7	P8	P9	P10
C	24	34	52	17	21	22	15	9	24	28
E	5	4	3	4	17	42	8	27	13	12
C + E	29	38	55	21	38	64	23	36	37	40
C − E	19	30	49	13	4	−20	7	−18	11	16

	P11	P12	P13	P14	P15	P16	P17	P18	P19	P20
C	27	26	18	9	9	21	21	8	18	0
E	22	19	20	26	40	29	17	19	16	60
C + E	49	45	38	35	49	50	38	27	44	60
C − E	5	7	−2	−17	−31	−8	4	−11	2	−60

Plot the 20 parameters using C–E axes, and hence illustrate for the natural slope system which three parameters are the most interactive, which three are the most dominant, and which three are the most subordinate.

Q14.10 One of the most important aspects of the systems approach is establishing when positive feedback in the system can occur, because positive feedback loops can cause instabilities.

(a) Explain the meaning of the terms 'negative feedback' and 'positive feedback', and give examples of each in engineering rock mechanics.

(b) Also explain how engineering actions can mitigate or eliminate positive feedback (equivalent to altering the mechanisms in the off-diagonal boxes of the interaction matrix).

Questions 15.1–15.10: excavation principles

Q15.1 The complete stress–strain curve has been obtained for a cylindrical specimen of intact granite tested in uniaxial compression. The specimen is 100 mm long and 50 mm in diameter. Assume that, for the purposes of calculation, the curve can be approximated to the bilinear form below. The uniaxial compressive strength is reached at 0.1% strain and 200 MPa stress. When the curve reaches 0.2% strain, the rock microstructure has been destroyed and all that remains are small flakes of crushed mineral grains.

(a) Calculate the energy under the equivalent complete force–displacement curve. This is the energy required to completely destroy the rock specimen. Express your answer in joules.

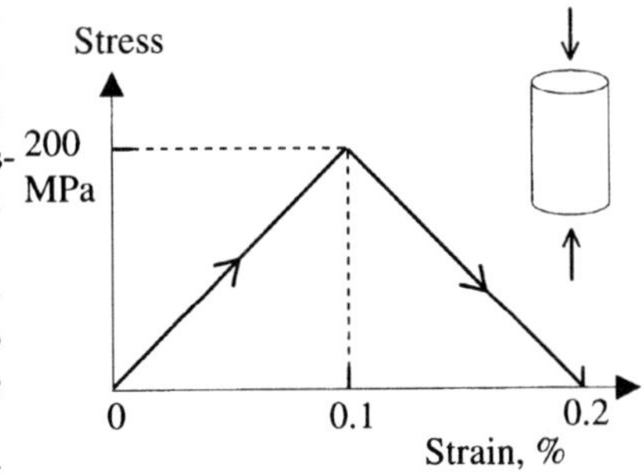

(b) For how long would a domestic 100-W light bulb have to be illuminated to use up the same amount of energy?

(c) What is the specific energy of the rock, i.e. the energy required/unit volume to pulverize the rock?

Q15.2 During bench blasting in a quarry, it was found that 48.5 kg of explosive is required to break 125 m^3 of marble. Given that the explosive used was ammonium nitrate–fuel oil (ANFO) with an explosive energy of 3.92 MJ/kg, calculate the specific energy for these circumstances, state whether this is greater or less than the answer to Q15.1, and explain why there is a difference.

Q15.3 A 5-m-diameter tunnel is being excavated in limestone by a full-face tunnel boring machine (TBM), which operates by exerting a torque and a thrust.

(a) Calculate the specific energy required to break the rock in the circumstances given by the data in (i) and (ii) below.

(i) For the case where the TBM has four 500-V electrical motors providing the torque to drive the cutting head. Each motor uses 10 A when rotating freely, and 110 A when the head is cutting. Neglect the thrust for this case and take the cutting rate as 3 m/h.
(ii) For the same cutting information in part (i) but with an additional continuous thrust of 2.7 MN and an increased cutting rate of 3.6 m/h.

(b) Does the work associated with the thrust contribute greatly to the specific energy required?

Q15.4 Comment on the magnitudes of the specific energy values obtained in Q15.1, Q15.2 and Q15.3 for the different circumstances of a laboratory compression test, blasting, and using a tunnel boring machine?

Q15.5 The objective of 'pre-splitting' when blasting is to create a continuous new fracture plane through the rock mass (which will become the final rock surface) before bulk blasting removes the rock up to the pre-split plane.

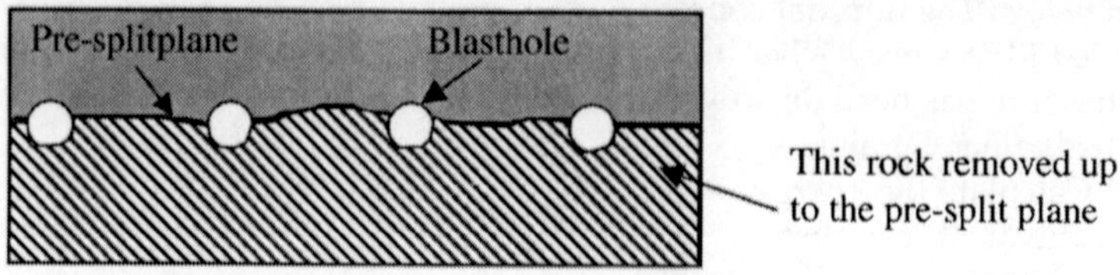

(a) List the most important factors that need to be controlled to ensure that the pre-split blasting techniques will indeed create a continuous pre-split fracture.

(b) The three photographs opposite show different rock slopes that have been pre-split. The pre-split plane has been created first; then bulk blasting has removed the rock up to the pre-split plane, as indicated in the sketch above. Thus, the photographs show one side of the pre-split 'plane' which is the final slope.

Each of the photographs illustrates one of the following:

- an acceptable pre-split rock face;
- a pre-split face where the fractures have adversely affected the pre-splitting; and
- a pre-split rock face where one or more of the necessary factors intimated in (a) above has not been properly controlled.

Which photograph is which case?

Q15.6 In order to be able to estimate the blasting specific energy for different rock masses (and hence the different amounts of explosive required for blasting) a Blastability Index (*BI*) has been proposed (Jimeno et al., 1995) using five rock properties. The *BI* is shown in the table below, with the various ratings related to the specific rock mass properties.

Photo 1

Photo 2

Photo 3

Rock Mass Description (*RMD*)	
Powdery or Friable	10
Blocky	20
Totally massive	50
Joint Plane Spacing (*JPS*)	
Close (<0.1 m)	10
Intermediate (0.1 to 1 m)	20
Wide (>0.1 m)	50
Joint Plane Orientation (*JPO*)	
Horizontal	10
Dip out of face	20
Strike normal to face	30
Dip into face	40
Specific Gravity Influence (*SGI*)	
$SGI = (25SG - 50)$ where SG is in tonnes/m^3	
Hardness (*H*)	
H = hardness on scale 1–10	

The value of the Blastability Index is given as

$$BI = 0.5\,(RMD + JPS + JPO + SGI + H)$$

and the amount of explosive required is 0.004 BI in kg ANFO/t, which is therefore a measure of specific energy.

(a) Calculate the specific energy that this index predicts for the following conditions: a highly laminated, soft ferruginous shale with horizontal to sub-horizontal bedding and BI rating values of $RMD = 15$, $JPS = 10$, $JPO = 10$, $SGI = 10$, $H = 1$. With reference to the values discussed in Q15.4, comment on the value obtained here.

(b) Comment on the advantages and disadvantages of this index.

Q15.7 The rock engineering systems (RES) approach has been used to develop a comprehensive Blastability Designation (BD) index for rock masses (Latham and Lu Ping, 1999), the purpose of which is to provide a quantitative assessment of the resistance of a rock mass to blasting. The index value is given as $BD = \sum_{j=1}^{n} W_j R_j$ where W_j is a parameter weighting value, R_j is the parameter rating, with the summation over the j parameters. Both the W_j and R_j values are evaluated for the rock mass under consideration. The greater the value of BD, the more difficult the rock is to blast. The twelve parameters used in the BD index are:

(1) strength (uniaxial compressive strength, UCS, MPa, also via Point Load Index);
(2) resistance to fracturing (uniaxial tensile strength, σ_t, MPa);
(3) sturdiness of the rock (density, ρ, t/m^3);
(4) elasticity of rock (Young's modulus, E, GPa);
(5) resistance of rock to dynamic loading (P-wave velocity, V_p, km/s);
(6) hardness of rock (Schmidt rebound hardness value, SHV, rebound height scale);
(7) deformability (Poisson's ratio, ν, dimensionless);

(8) resistance of rock to breaking (fracture toughness, K_{Ic}, MPa-m$^{1/2}$);
(9) *in situ* block sizes (mean of block size distribution, mean);
(10) fragility of rock mass (fractal dimension of rock block sizes, D);
(11) integrity of rock mass (ratio of field: lab P-wave velocities, R_v, dimensionless);
(12) fracture plane's strength (cohesion, c, MPa and friction angle, ϕ, degrees).

Compile a table of these twelve parameters in three columns with headings Intact Rock Properties, Fracture Properties and Rock Mass Properties, and then comment on how well the rock mass has been characterized.

Q15.8 In relation to rock tunnel boring machines (TBMs), two progress rates can be defined:

Cutting rate, C: the rate at which a tunnel is constructed *when the tunnelling machine is operating*; and
Advance rate, A: the rate at which a tunnel is constructed.

These two rates are not the same because the TBM is not continuously cutting; there is a *Utilization factor, U*, which is the amount of time (expressed as a proportion) that the machine is cutting, i.e. $A = U \times C$.

Over a short period, say 20 min, A can be equal to C. Over longer periods, of days, weeks or months, there can be a greater and greater discrepancy between C and A, as indicated in the diagram below (Barton, 1999).

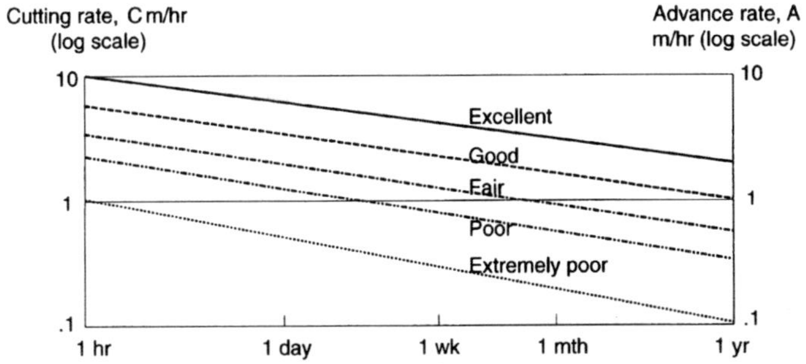

The terms 'Excellent' through to 'Extremely poor' refer to the rates, and Barton noted that the majority of TBM tunnels lie between the 'Good' and 'Poor' lines.

Give at least five reasons why such lines occur, i.e. why the value of U decreases with time.

Q15.9 When an underground excavation is made in a rock mass, there are three primary effects, all three of which result from Newton's 3rd law, i.e. for every action, there is an equal and opposite reaction. Thus, when new underground space is created, the resistance of the rock previously occupying the space becomes zero, forces cannot be transmitted from the rock into the space, and the water pressure is reduced to atmospheric pressure.

Draw a sketch of these three effects and explain why they cause a so-called Excavation Disturbed Zone (EDZ).

Q15.10 (a) Can drill and blast excavation rates be significantly increased from those currently achieved?

(b) Can tunnel boring machine penetration rates be significantly increased?

(c) Are new methods of excavation likely to be introduced?

Questions 16.1–16.10: rock reinforcement and rock support

Q16.1 When rockbolts are used for surface or underground excavations to reinforce a fractured rock mass, the rockbolts will be subjected to tension, shear and compressive forces. The diagram below (from Windsor, 1997) indicates the types of forces on rockbolts a to f.

Recognizing that rockbolts are designed to operate in tension with little shear force applied, rate these rockbolt configurations in terms of their suitability on this criterion alone.

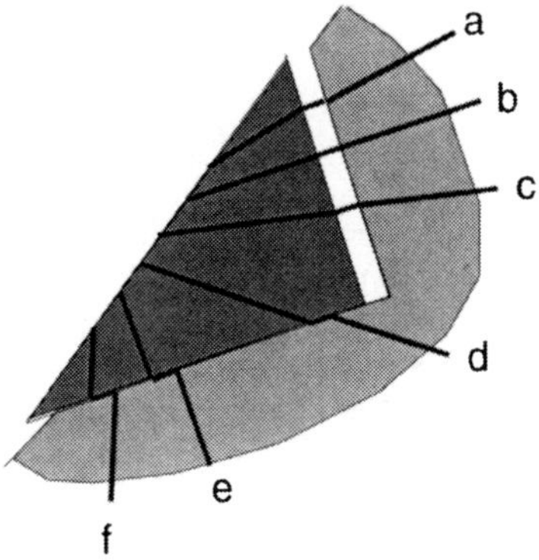

Q16.2 A rock block of weight W is bolted to a plane inclined at angle α to the horizontal. Develop an expression for the bolt tension, T, in terms of the angle, β, at which the bolt is installed relative to the plane, the angle of inclination of the plane and the Mohr–Coulomb friction properties, c and ϕ, of the block/plane interface. Hence, by differentiation, find the rockbolt angle at which the bolt tension is minimized.

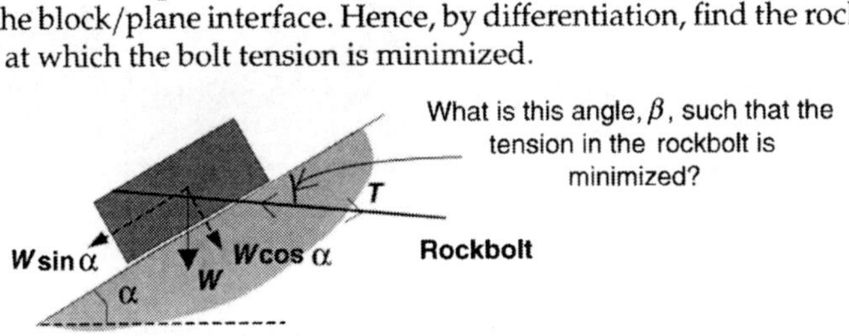

Q16.3 The sketch below shows a rock block resting on a slope and subject to a stabilizing force from a rockbolt. The strength of the interface between the block and the slope is purely frictional, with an angle of friction ϕ.

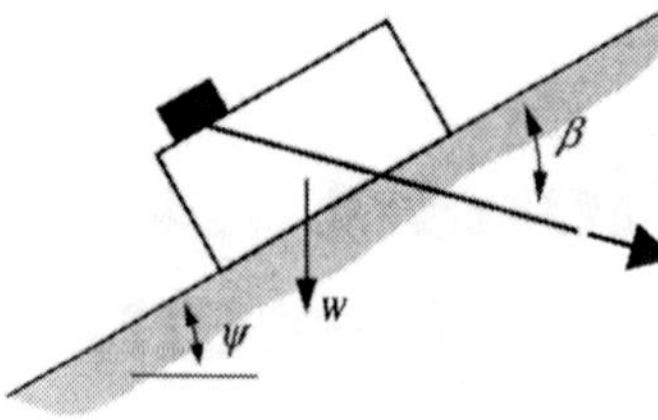

(a) The component of the bolt force acting parallel to the slope can be considered as either a positive restraining component or as a negative actuating component. Derive an expression for the factor of safety in both these cases.

(b) For the situation when $W = 1000$ kN, $\beta = 15°$, $\psi = 42°$ and $\phi = 36°$, what bolt tension is required to give a factor of safety of unity for each of the cases?

(c) Examine the behaviour of the two expressions for factor of safety as the bolt tension varies from 25 kN less than, to 25 kN more than, the force computed for each case above. Which of these two expressions should we use for design purposes, when we require a factor of safety in excess of 1?

(d) Rock bolts function by developing a tensile force within them, and are sometimes tensioned at the time of installation in order to generate this force. In the case of untensioned bolts, how is this force generated? Should we regard untensioned bolts as providing a positive restraining component or a negative actuating component?

(e) Imagine that we require a factor of safety of three for the rock block shown above. Should we use tensioned or untensioned bolts?

Q16.4 A circular tunnel is being excavated in a blocky rock mass by drilling and blasting. There is an Excavation Disturbed Zone (EDZ) around the excavated tunnel (defined on the basis of a blast-disturbed zone where there are loosened blocks which can fall into the tunnel under the action of gravity) which extends 0.75 m into the rock from the excavation surface.

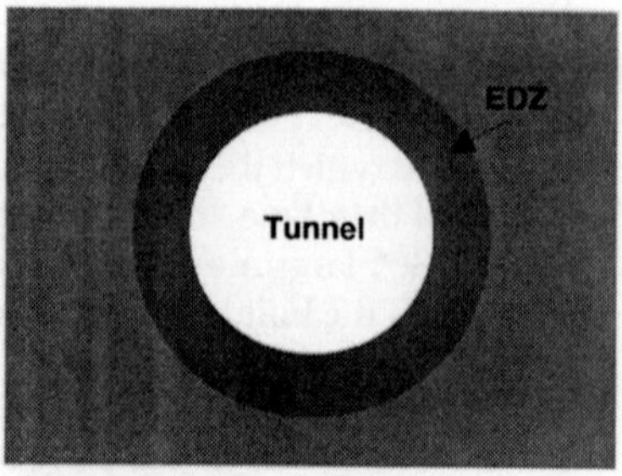

What support pressure is required at the crown to stabilize the loose blocks of the EDZ given that the unit weight of the rock, γ, is 25 kN/m^3?

Q16.5 (a) If the EDZ in Q16.4 is to be stabilized by the use of rockbolts inserted into the roof as a supporting method, and the working capacity of each bolt, T, is 150 kN, what area of the roof will each bolt support?

(b) Would you anticipate any ancillary problems with this bolting pattern?

(c) In the circular tunnel, in which directions and at which locations would you install the bolts?

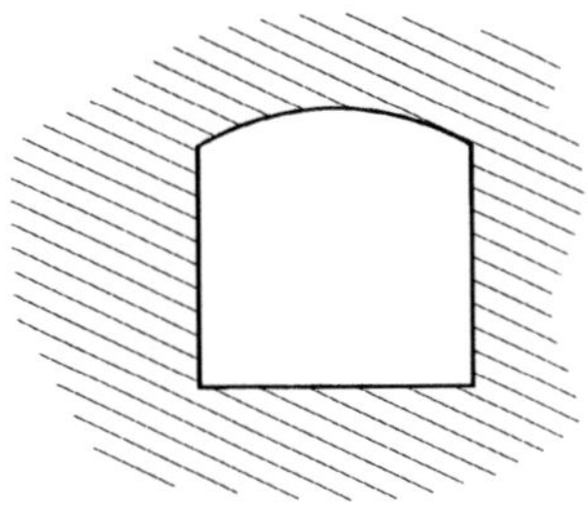

Q16.6 A long underground chamber is to be constructed in a rock mass with fractures dipping at 25° as shown. This problem is included in Goodman (1989).

(a) Assuming the friction angle of the fractures, ϕ_j, is 50°, use the ϕ_j geometrical construction in two dimensions to locate zones of potential distress around the opening.

(b) Repeat this procedure with $\phi_j = 20°$ to investigate the effect of rock deterioration on interlayer slip around the excavation.

Q16.7 In a paper discussing the potential slip on fractures in the rock around a tunnel or shaft, Daemen (1983) uses the solution for the stresses around a circular opening in a CHILE rock to establish the normal and shear stresses on the fractures. Applying the Mohr–Coulomb criterion to these stresses enables the potential zones of distress around the excavation to be established.

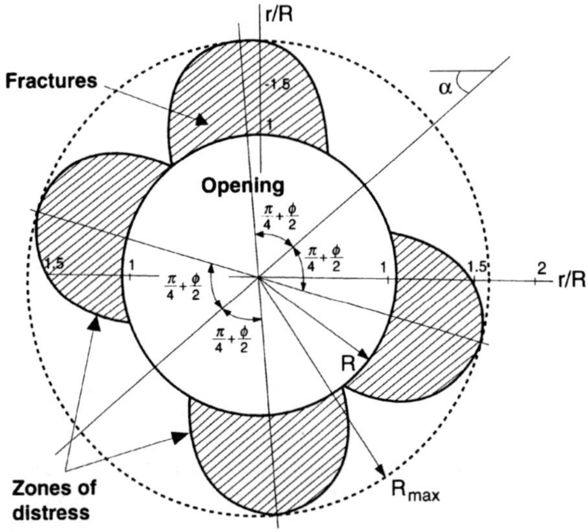

The zones of distress shown in Daemen's diagram above are for an isotropic stress field and an angle of friction of $\phi = 20°$.

Show that the same result for the zone of distress at the excavation *boundary* can also be obtained using the ϕ_j theory.

Q16.8 Two types of 'ground response curve', A and B, are illustrated in the diagram below. The curves indicate the amount of support pressure required to maintain an excavated rock surface at a given value of displacement.

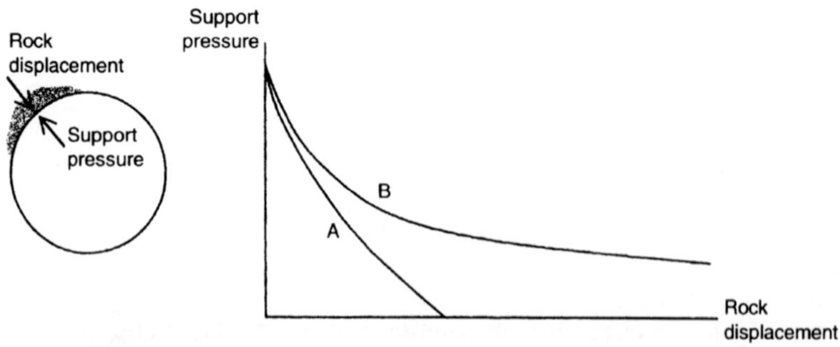

With reference to these curves, explain:

(a) why the *in situ* stress condition before excavation cannot be re-established by the support pressure applied after excavation;

(b) why it is inappropriate for an engineer to insist that the rock should be supported in such a way that no displacement whatsoever occurs.

Q16.9 A tunnel of radius 1.85 m is excavated in rock subjected to an initial hydrostatic stress field of 20 MPa and provided with a concrete lining of internal radius 1.70 m. Assuming elastic behaviour of the rock and lining, calculate the radial pressure and the radial displacement at the rock-lining interface if:

(a) the lining is installed at the face before any displacement of the rock has occurred; and

(b) the lining is installed following a radial convergence of 1 mm.

Q16.10 The sketch shows a pillar of rectangular cross section, which is very long in the y-direction (i.e. out of the plane of the paper) and is made of a linearly elastic isotropic rock. A support pressure is applied to the horizontal sides of the pillar through the action of springs that pass horizontally through the pillar.

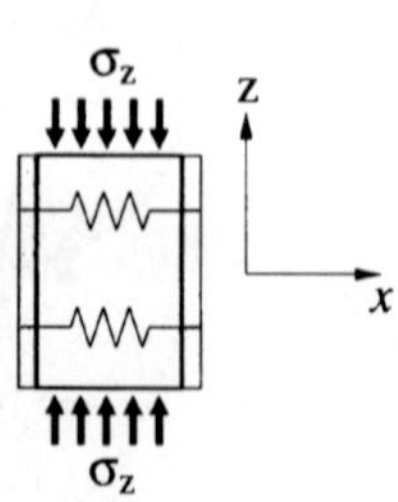

(a) Develop an expression linking the vertical stress, σ_z, to the support pressure, p, in terms of the elastic constants of the pillar material and the spring stiffness. Use this expression to show that, for zero horizontal strain of the pillar, the ratio of vertical stress to support pressure is $(1 - \nu)/\nu$.

(b) If the strength of the pillar can be described by the Hoek–Brown criterion with $\sigma_c = 37$ MPa, $s = 1$ and $m = 15$, and the elastic modulus and Poisson's ratio of the rock are 75 GPa and 0.27, respectively, what spring stiffness is required if the pillar is to withstand a vertical stress of 55 MPa?

Questions 17.1–17.10: foundation and slope instability mechanisms

Q17.1 The diagram below (Serrano and Olalla, 1998) shows six basic potential mechanisms by which failure can occur beneath a foundation on a fractured rock mass containing one fracture set, indicated by the fine shading. Failure can occur through the intact rock, indicated by the letter R, and/or along a fracture plane, indicated by the letter P.

Describe the six basic mechanisms illustrated.

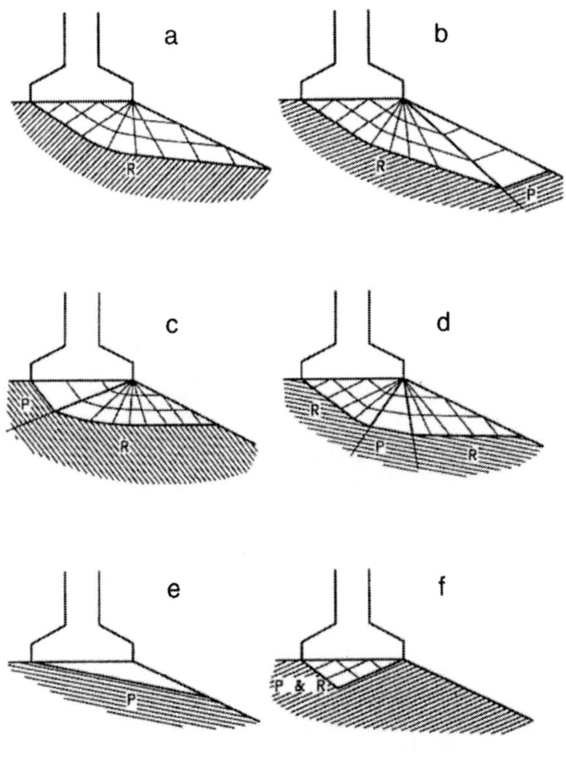

Q17.2 Consider extending the method of equilibrium analysis for plane failures (see Q16.3) to the case of retaining walls. The failing block, ABC in the sketch below, is restrained in three cases by the application of load to the vertical face, AB:

(a) by a horizontal force of magnitude T acting through the centroid of the block;
(b) by a horizontally acting uniform pressure distribution p from the top of the face to the point where the failure plane daylights; and
(c) by a pressure distribution varying linearly from zero at the top of the face to q at the point where the failure plane daylights.

In each case, derive an expression for the factor of safety and, for the special case of $F = 1.0$, give an expression for T, p or q as appropriate.

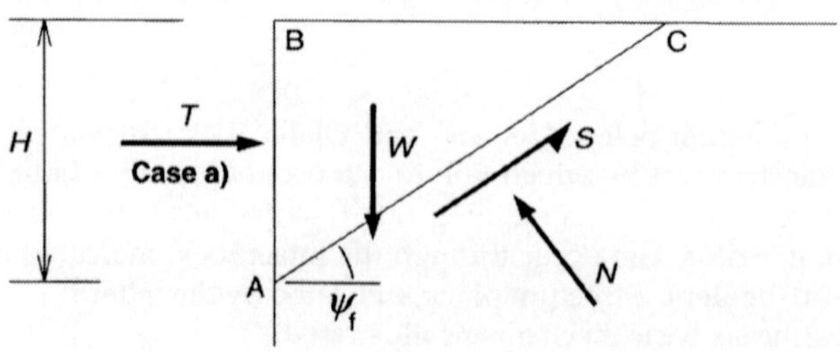

Q17.3 For the case of wedge instability in rock slopes, the factor of safety can be related to that of an equivalent plane instability (i.e. plane sliding in the same direction as that of the wedge) by

$$F_W = k_W \times F_P$$

where the wedge factor, k_W, is computed from $k_W = \sin\beta / \sin\frac{1}{2}\xi$, and the angles β and ξ are defined as shown below.

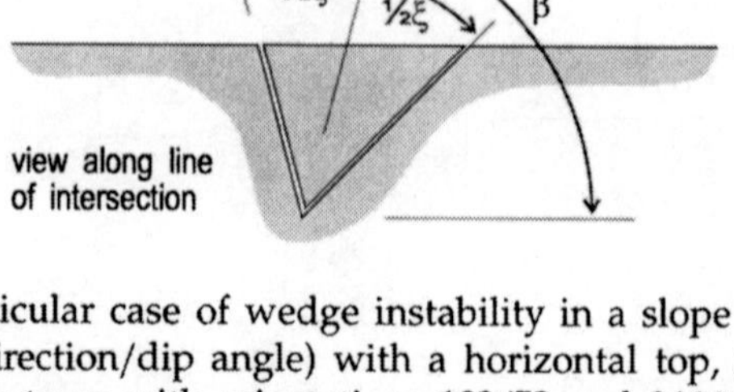

For the particular case of wedge instability in a slope of orientation 124/63 (dip direction/dip angle) with a horizontal top, intersected by two sets of fractures with orientations 182/52 and 046/69 and friction angle 29°, determine F_W.

Q17.4 Determine an upper bound for the collapse pressure, p, for the foundation shown below consisting of three rock wedges formed by the fracture sets in the rock mass.

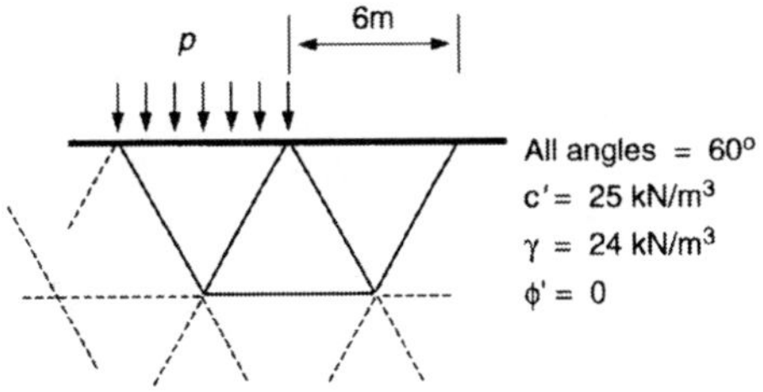

Q17.5 A cross-section through a rock slope is shown below. The rock is impermeable with a unit weight of 22 kN/m^3 and the fractures AC and DE have an aperture of 1 mm. A drainage gallery at E reduces the water pressure at that point to the atmospheric pressure value. The drainage gallery, the fractures and the slope face all have the same strike.

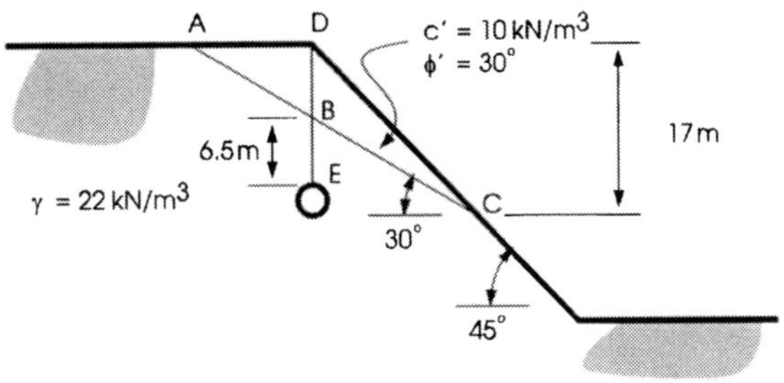

Assuming conditions of heavy rainfall, determine the water pressure at B, and hence the factor of safety against sliding of the block DBC. Take the kinematic viscosity of water to be 1.3×10^{-6} m^2/s and assume block ABD is stable.

Q17.6 Assume the same circumstances in Q17.5 and consider the case when the block DBC has moved downhill a limited amount, resulting in a considerable increase in the aperture of DB. Calculate the factor of safety against sliding of the block DBC.

Q17.7 A rectangular foundation base, 6 m × 10 m in plan, carries a vertical load of 180 MN which it transmits to the rock as a uniform bearing pressure. The rock can be taken to be CHILE, with a modulus of elasticity $E = 800$ MPa and Poisson's ratio $\nu = 0.2$.

Ignoring the weight of the rock, evaluate the vertical settlement at the centre, C, of the base and the vertical stress at a point 5 m below C.

Q17.8 The expression for the radial stress induced beneath a line load of magnitude P acting normal to the surface of a semi-infinite half space is

$$\sigma_r = \frac{2P \sin\theta}{\pi r}.$$

(a) Show on a sketch the geometry of this problem, indicating clearly the angle θ and the orientation of the stress components σ_θ and σ_r. What are the magnitudes of σ_θ and $\tau_{r\theta}$ in this problem?

(b) Sketch the locus of constant σ_r for varying θ, and show how the expression above leads to the concept of a 'bulb of pressure'. For the case when the applied line load is not normal to the surface, show on a sketch how the bulb of pressure changes its nature, indicating clearly any zones where tensile radial stresses may be induced.

(c) This analysis is for CHILE materials. By making use of the ϕ_j theory, show qualitatively how the bulb of pressure may be expected to change when the ground beneath the load consists of a layered discontinuous rock mass dipping at some arbitrary angle.

(d) On the basis of these results, what are the ramifications this analysis has for site investigation in layered rocks?

Q17.9 The system of rock blocks shown in the sketch below is to be used in the verification procedure of a computer code for analysing progressive failure of rock slopes, and for this a manually derived solution is required.

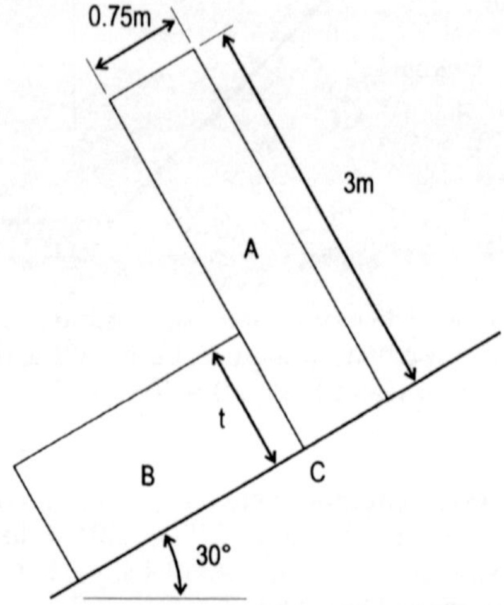

The system is in limiting equilibrium with block A tending to topple about the corner C, while block B is on the point of sliding downhill. The shear resistance on all surfaces is purely frictional with $\phi = 35°$.

Given that B is twice as heavy as A, determine the thickness t of block B. Also show that there is no tendency for block A to slip at the corner C.

Q17.10 Solve Q17.5 numerically using the sector method, and comment on any discrepancies between the numerical solution and the analytical solution.

Questions 18.1–18.10: design of surface excavations

Q18.1 A foundation is proposed on a rock mass consisting of a series of horizontal strata. The results of a site investigation and associated laboratory testing programme are shown in the table below.

Depth (m)	Description	*RQD* (%)	Intact rock elastic modulus (GPa)	Fracture deformation modulus (GPa/m)
0	Moderately weathered sandstone	61	4.0	18.0
3	Slightly weathered sandstone	98	5.0	20.0
8	Slightly weathered limestone	80	39	115
10	Fresh, massive limestone	100	40	120

End of hole at 12 m

(a) Calculate the effective elastic modulus for the uppermost 10 m of the rock mass.

(b) A plate-loading test using a 0.5 m diameter rigid plate at the ground surface has indicated an elastic modulus of 1.0 GPa. By considering the moduli of the various strata, explain why this might be. What value of modulus should be used in the design?

(*Aide memoire*: For a rock mass consisting of i strata, each of which contains a single set of planar, parallel fractures, oriented parallel to the strata boundaries, the effective elastic modulus normal to the fractures, E_m, may be calculated using the expression

$$E_{\mathrm{m}} = \frac{\sum t_i}{\sum \left(\frac{1}{E_i} + \frac{\lambda_i}{E_{\mathrm{d}i}} \right) t_i}$$

where, for each of the i strata, t_i = thickness, E_i = Young's modulus of the intact rock, λ_i = fracture frequency, $E_{\mathrm{d}i}$ = modulus of deformation of the fractures.)

Q18.2 In a rock mass containing a single set of vertically dipping fractures, consider how the fractures might adversely affect the profile of a vertical slope excavated using pre-split blasting.

Q18.3 A quarry is to be opened in a rock mass which contains four fracture sets with dip directions and dip angles as follows:

Set	1	2	3	4
Dip direction/Dip angle	292/64	151/37	052/76	020/16

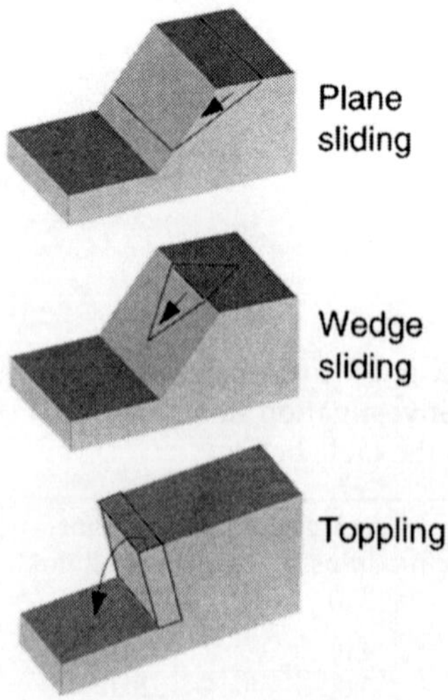

The rock mass can be considered dry, and the angle of friction for all fractures is 30°.

Consider the primary potential modes of instability (plane, wedge and direct toppling) at 15° intervals of dip direction (i.e. 0°, 15°, 30°, ..., 345°, 360°) and use kinematic feasibility techniques to prepare a table showing the steepest safe slope and the respective critical failure mode at each azimuth.

Q18.4 Prepare a similar table as in Q18.3 but for the case when the friction angle is not known.

Q18.5 Use the results from Q18.3 to draw a plan of the slope crest around the quarry excavation assuming that the floor of the quarry is circular.

Determine the best orientation for a *radial* access road to the quarry floor (assume that the road can be constructed in any direction and that the optimal orientation is for a road with the steepest possible side slopes — so that excavation associated with the road is minimized).

Repeat part (b) for an unknown friction angle, as was the case in Q18.4.

Q18.6 Add toppling due to interlayer slip to the analysis of Q18.3.

Q18.7 A trial design for a proposed 100 m deep roadstone quarry can be approximated in plan by a circle. The quarry has been designed with a number of 10 m high benches, such that the overall slope angle is 45° and the individual bench angles are 60°.

Preliminary mapping of the site has established that the rock mass contains four dominant sets of fractures, having the following mean orientations:

Feature	Cleavage	Joint set A	Joint set B	Bedding
Dip	36°	75°	45°	28°
Dip direction	015°	330°	080°	260°

(a) Using kinematic feasibility techniques, determine whether or not conditions exist that could lead to instability in the proposed slopes.

(b) Indicate, in general terms, the design changes required to overcome any predicted instability problems.

Q18.8 A rock slope with a face angle of ψ_f is cut into a rock mass containing a single set of fractures dipping into the slope at an angle β. The strength of the fractures is purely frictional. Assume that $\tan\phi$, ψ_f and β can all be considered as normally distributed variables with the following parameters:

Variable	$\tan\phi$	ψ_f	β
Mean	0.55	50°	60°
Standard deviation	0.15	5°	10°

Investigate the variability of the factor of safety due to interlayer slip for this slope, using either standard normal random values obtained from statistical tables, or values produced by your computer/calculator. Perform as many trials as you have the patience for (but at least 35).

Q18.9 A proposed dam will exert a vertical force of 9400 kN/m and a horizontal force of 4400 kN/m on its horizontal foundation, as shown in the figure below. At a depth of 6 m below the founding surface there is a horizontal bedding plane, the strength of which is purely frictional with $\phi = 32°$. The unit weight of the rock above this plane is 22 kN/m^3.

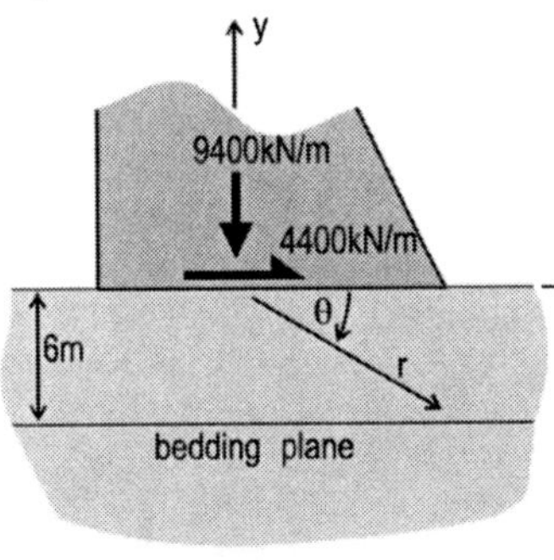

For a polar co-ordinate system, the origin of which coincides with the rectangular co-ordinate system as shown in the diagram, the components of radial stress due to a vertical line load P and a horizontal line load Q may be calculated from $\sigma_{r(P)} = (2P\sin\theta)/\pi r$ and $\sigma_{r(Q)} = (2Q\cos\theta)/\pi r$, respectively.

At what value of x-ordinate is the radial stress on the bedding plane zero, and will the stress state on the bedding plane to the left of this point be sustainable? Over what range of x-ordinate values to the right of this point will slip occur on the bedding plane? What conclusions can be drawn about the overall stability of the proposed dam?

Q18.10 Using the principle of virtual work, determine the value of P for limiting equilibrium for the foundation shown below.

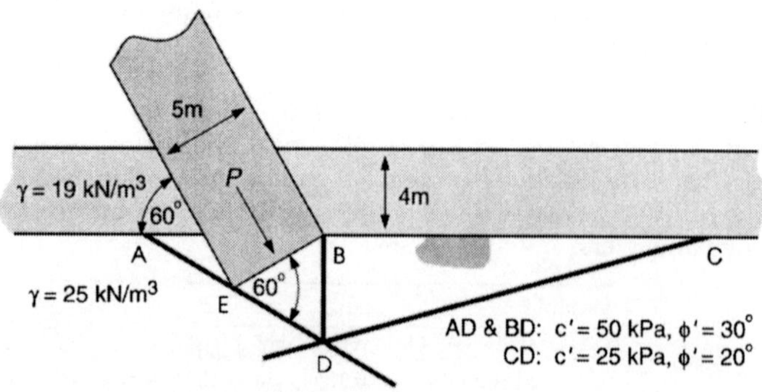

Questions 19.1–19.10: underground excavation instability mechanisms

Q19.1 The diagram to the right shows the cross-section of an underground machine hall. The hall is to be excavated in a sequence of metamorphosed argillaceous rocks, of unit weight 22 kN/m^3, which contains five fracture sets with the following orientations:

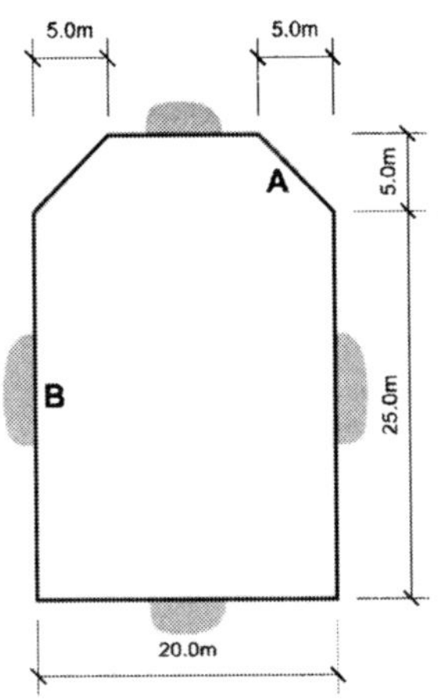

Set	Dip direction	Dip angle
1	058°	54°
2	195°	70°
3	127°	81°
4	160°	32°
5	335°	64°

Each set has a friction angle of approximately 30° and zero cohesion. It is proposed that the axis of the machine hall will be oriented on a heading 030° and will be horizontal.

Determine the kinematic feasibility of all the tetrahedral blocks formed by the intersection of the fracture sets and the *roof*.

Q19.2 For the unstable roof blocks established in the answer to Q19.1, determine the direction of movement, and plane(s) of sliding where relevant.

Q19.3 For the unstable roof blocks established in the answer to Q19.1, determine the volume of the largest block that could fall from the roof and hence calculate the necessary support pressure to ensure stability.

Q19.4 Repeat Qs 19.1 and 19.2 for the haunch A (as illustrated in Q19.1).

Q19.5 Repeat Qs 19.1 and 19.2 for side wall B (as illustrated in Q19.1).

Q19.6 For both the haunch A and the side wall B studied in Qs 19.4 and 19.5, and with the information given in Q19.1, calculate the factor of safety of the block formed by fracture sets 1, 2 and 5.

Q19.7 At a depth of 450 m, a 3 m diameter circular tunnel is driven in rock having a unit weight of 26 kN/m^3 and uniaxial compressive and tensile strengths of 60.0 MPa and 3.0 MPa, respectively. Will the strength of the rock on the tunnel boundary be reached if
(a) $k = 0.3$, and
(b) $k = 2.5$?

A second tunnel, of 6 m diameter, is subsequently driven parallel to, and at the same centre line level as, the first such that the centre line spacing of the two tunnels is 10 m.

Comment on the stability of the tunnels for the field stresses given by (a) and (b) above.

Q19.8 The diagram shows the relative positions of two parallel horizontal tunnels, each 3 m in diameter. Prior to excavation, the principal stresses in the area were $p_x = p_y = p_z = 11$ MPa.

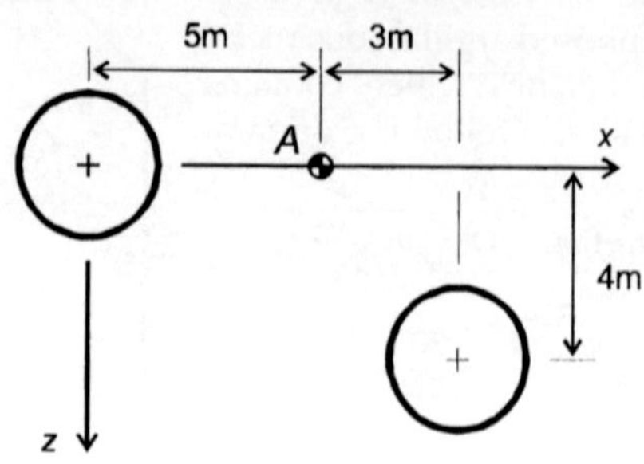

(a) Determine the principal stresses and their directions at point A after excavation has been completed.

(b) A horizontal fault coincident with the x-axis passes through A. If the shear strength of the fault is purely frictional with $\phi = 20°$, will slip on the fault occur at A?

Q19.9 An ovaloid excavation at a depth of 750 m has in vertical section its major axis horizontal, and the ratio of its width to height is 4. The radius of curvature of its ends is equal to half its height. Assume that the *in situ* stress state can be calculated on the basis of complete lateral restraint in a CHILE medium (ERM 1, Section 4.6.2) with $\gamma = 28.0$ kN/m^3 and $\nu = 0.3$.

An elastic boundary element analysis for $k = 0$ shows that the stress in the centre of the roof is −20.5 MPa, and in the side wall is 104 MPa. An analysis with $k = 1$ gives corresponding stresses of 4.59 MPa and 84.2 MPa. What stresses would the boundary element analysis give for the *in situ* stress state?

Using the equations for stresses in terms of radius of curvature:

(a) compare the stress in the centre of the roof of the excavation with that for an elliptical excavation with the same width/height ratio; and
(b) compare the stress in the side wall with that for a boundary of equal radius of curvature.

Q19.10 The axes of a long horizontal elliptical excavation are inclined at $\pm 45°$ to the horizontal, and their lengths are in the ratio 2.5 : 1. The vertical and horizontal principal field stresses are 8.5 MPa and 25.5 MPa, respectively.

(a) Calculate the maximum and minimum elastic stress values induced on the boundary of the excavation.

(b) Investigate the possibility of failure of the rock on the excavation boundary, on the basis that the strength of the rock in compression is given by $c = 30$ MPa and $\phi = 40°$, and in tension is zero.

Questions 20.1–20.10: design of underground excavations

Q20.1 A vertical shaft is to be sunk through a rock mass in which the magnitudes of the two horizontal principal stresses have a ratio of 3:1. A clear space of 5 m × 5 m must be maintained within the shaft for equipment installation. The three excavation cross-sections shown in the sketch are under consideration for the shaft design.

Discuss the relative design merits of each cross-section in terms of rock mechanics principles, and hence provide recommendations for the optimal shape (and in the case of profile (b) and (c), optimal orientation) for the excavation.

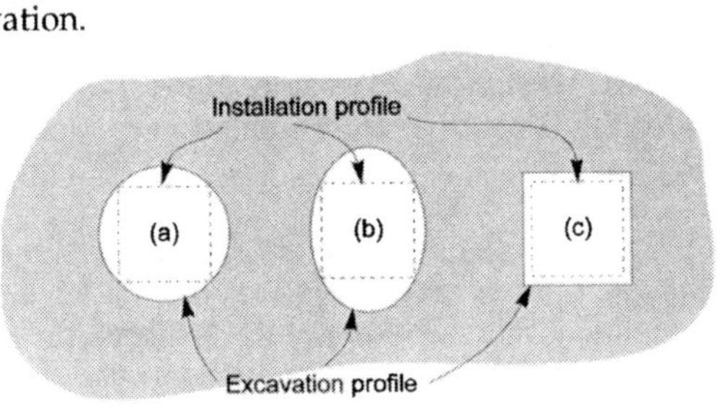

Q20.2 You have been asked for an initial decision on the orientation of an underground structure. The structure comprises ten parallel tunnels, each 5 km long, to be built in a rock mass containing definite fracture sets. You have also been asked about the best shape for the cross-section of the tunnels.

The dip direction and dip of the fracture sets are 270/45 and 90/45; the direction of the maximum principal stress is horizontal, west–east, and its magnitude is twice the value of both the vertical and the other horizontal stress component. The ratio of the maximum principal stress component to the intact rock uniaxial compressive strength is 0.5.

What would be your initial design based on this information alone?

Q20.3 A site is under investigation for use in a geothermal energy project. It is planned to drill a number of boreholes to a depth of about 2000 m and use these for injecting water at high pressure into the naturally fractured rock mass. The water will take heat from the surrounding rock, and will then be extracted from production boreholes and used for heating. A key question is the stability of the fractured rock mass during this process.

Extensive site investigations have revealed that the rock mass contains three pervasive sets of fractures, oriented at 000/00, 165/90 and 245/90, all of which have a friction angle of 44°. Also, a stress determination programme has established that the three principal stresses have orientations 000/90, 040/00 and 130/00. Their magnitudes (in MPa) are given by $26z$, $6 + 12z$ and $15 + 28z$ respectively, where z is the depth below ground surface. Although the intact rock is essentially impermeable, the rock mass is so highly fractured that it is naturally saturated with groundwater and hence effective stress conditions exist within the fractures.

The water injection holes have to be located relative to the production hole such that the pressurized water will flow along the natural fractures between them. For the target depth, examine the effective normal stress induced on a fracture from each of the three sets, and identify which set will be jacked open by the injected water. Hence, propose a layout for the injection boreholes. Bear in mind that the boreholes can be deviated during drilling such that they run horizontally, so that a vertical flow regime can be established if necessary.

Compute the water pressure required to open the critical fracture set (1) 250 m above, (2) 250 m below and (3) at the target horizon.

Using these results, predict the flow regime and overall stability of the rock mass when water is injected at the target horizon.

Does this affect the layout of the boreholes and operation of the system, and if so, how?

Q20.4 A circular tunnel of radius 4 m is to be driven in a weak rock mass subjected to an *in situ* hydrostatic stress field of 9 MPa. The triaxial compressive strengths of the rock mass in its initial and fractured states are given by $\sigma_1 = 8 + 4\sigma_3$ (MPa) and $\sigma_1 = 3\sigma_3$, respectively.

A preliminary analysis for design is to determine the relative instabilities of the roof, sidewalls and floor.

Plot the ground response curves for the roof, sidewalls and floor, taking $G = 2.1$ GPa, $f = 1.4$ and $\gamma = 25$ kN/m^3, and comment on the appropriate support pressure.

Q20.5 For the design of part of a large underground civil defence facility in a rock mass, there are two competing excavation geometries, as shown in the sketches of the vertical cross-sections given below. Both geometries consist of excavated rooms separated by rock pillars.

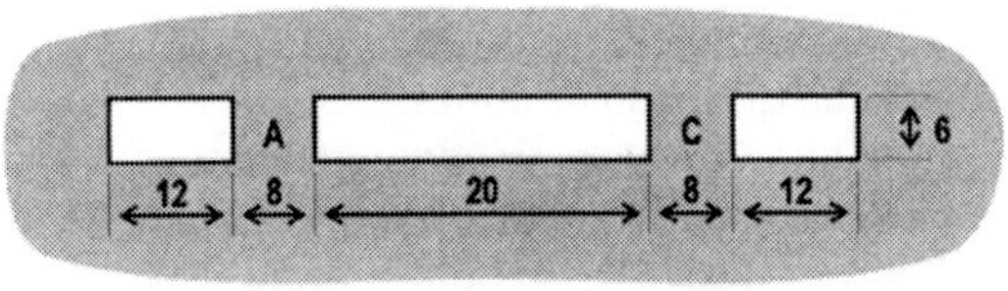

$$\delta_A = (56.20 - 3.304p_A - 0.533p_C) \times 10^{-3}$$
$$\delta_C = (56.20 - 0.533p_A - 3.304p_C) \times 10^{-3}$$

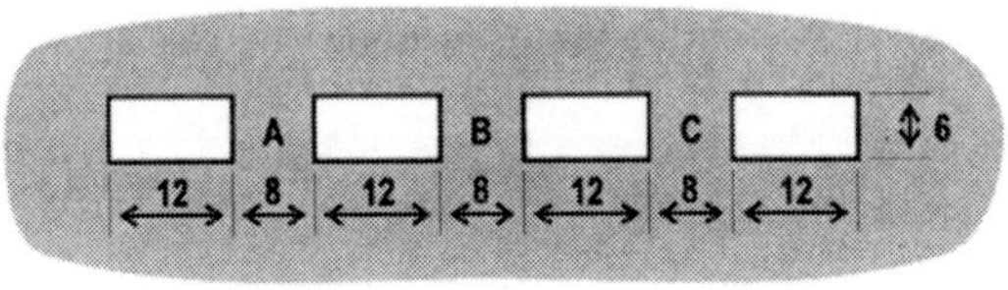

$$\delta_A = (56.20 - 3.304p_A - 1.121p_B - 0.533p_C) \times 10^{-3}$$
$$\delta_B = (66.52 - 1.121p_A - 3.602p_B - 1.121p_C) \times 10^{-3}$$
$$\delta_C = (56.20 - 0.533p_A - 1.121p_B - 3.304p_C) \times 10^{-3}.$$

Analysis of these geometries has been undertaken using a CHILE boundary element program, in order to determine the relation between displacement and support pressure at the various pillar locations. These relations are given below the sketches with δ representing displacement and p representing support pressure.

The support pressure is to be supplied by the rock pillars. Perform a rock-support interaction analysis for each of the two geometries in order to determine which is the preferred design. The stress–strain characteristics of the two pillars are as given below:

σ (MPa)	5.0	8.0	10.0	11.1	11.4	10.9	10.0	8.7	3.0	0.5
$\varepsilon \times 10^{-3}$	0.5	1.0	1.5	2.0	2.5	3.0	3.5	4.0	7.0	12.0

Q20.6 A new mining operation is planned which can extract from one of two coal seams. The first seam is at a depth of 250 m and has a thickness of 3 m, and the second seam is at a depth of 500 m with a thickness of 1 m. A preliminary design for the longwall mining panels calls for the upper seam to be mined with a panel width of 200 m, and the lower seam to have a panel width of 400 m (see the sketch below for typical geometry, in which the mining direction is into the page).

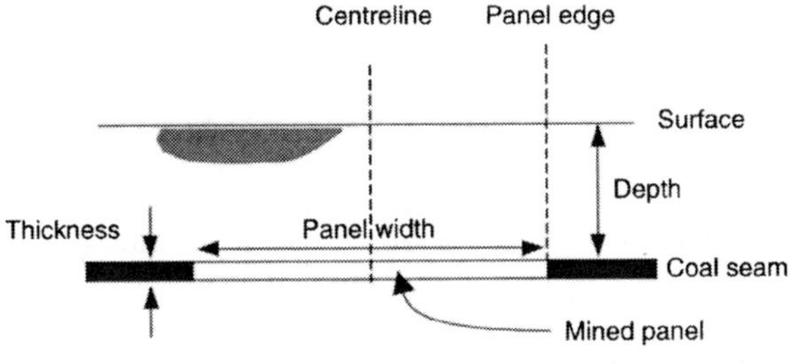

It has been suggested that the *Subsidence Engineers' Handbook* (SEH) (National Coal Board, Mining Department, 1975) can be used to provide estimates of the surface subsidence profile and horizontal strain distribution transverse to the direction of mining.

Assess the suitability of the extraction geometries for each seam (based on estimates of surface subsidence and horizontal strain), using the following data taken from the SEH.

The SEH gives values of the ratio of maximal subsidence to extraction thickness of about 0.75 and 0.72 for the upper and lower seams, respectively. The proportion of subsidence that occurs at a given position along the transverse profile for both seams is given by the table below, taken from the SEH.

Subsidence proportion	0.95	0.90	0.80	0.7	0.6	0.5	0.4	0.3	0.2	0.1	0.0
Ratio of distance from panel centre to depth	0.08	0.1	0.17	0.2	0.25	0.29	0.32	0.38	0.42	0.62	1.11

Similarly, the strain proportion that occurs at a given position along the transverse profile for both seams is given by the table below, also taken from the SEH.

Strain proportion	−0.70	−0.80	−1.00	−0.80	−0.40	0.00	0.40	0.80	1.00	0.80	0.60	0.40	0.20	0.00
Ratio of distance from panel centre to depth	0.00	0.02	0.11	0.19	0.24	0.29	0.31	0.38	0.42	0.49	0.52	0.59	0.67	1.11

In addition, the multipliers for the maximum contractile and extensile strains for these panel width to panel depth ratios are both equal to about 0.7.

Q20.7 A bauxite room and pillar mine is to be opened at a depth of 100 m in a rock mass whose unit weight is approximately 27 kN/m^3. The design of the mine calls for a factor of safety of 2 against compressive failure of the pillars, and the substantial body of data regarding the design of pillars in this bauxite shows that their compressive strength may be calculated from the expression

$$(\sigma_c)_{\text{PILLAR}} = 10 + 45e^{-0.18A}$$

where A is the plan area of the pillar in square metres.

(a) What is the uniaxial compressive strength of the bauxite as measured in the laboratory, and what is the super-REV compressive strength of the pillars?

(b) For operational reasons, the rooms must have a minimum width of 5.0 m. What is the greatest extraction ratio that can be achieved with this size of opening?

(c) If the roof of a room is reinforced with rock bolts, then the maximum achievable opening width is about 8 m. Is it cost effective to do this?

Q20.8 In an attempt to improve the profitability of the mine in Q20.7, the possibility of reducing both the opening width and the pillar size is to be investigated. Plot the curve of extraction ratio against opening width, for openings in the range 0.5 m to 4.0 m, and hence determine the optimal opening width and the corresponding extraction ratio.

If the extraction ratio thus identified is to be kept, what value of the factor of safety is required if the opening width is to be changed to 3.5 m?

Q20.9 A proposal has been made to use an old underground limestone quarry as a storage facility. A site visit to the quarry has revealed that it was mined using the room and pillar method, with a regular rectangular array of pillars. The clear spacing between the pillars is 6 m, the pillars are each 7 m square, and the excavation is at a depth of 80 m.

Examination of the pillars shows that the limestone is horizontally bedded with moderate spacing and gentle undulations. The bedding planes themselves are smooth to touch with slightly weathered surfaces and no visible aperture. Conditions inside the quarry are generally dry. A point load test of the pillar rock conducted at the quarry estimated its uniaxial compressive strength to be 100 MPa; whereas a laboratory triaxial test found that the rock failed when the axial stress in the sample was 110 MPa and the confining pressure was 4 MPa. The unit weight of the limestone is 28 kN/m^3.

Estimate the Rock Mass Rating (*RMR*) for the pillars (see the *RMR* table in Appendix C) and hence determine the Hoek–Brown strength parameters m and s for the rock mass, by using the equations

$$m = m_i \exp\left(\frac{RMR - 100}{28}\right) \text{ and } s = \exp\left(\frac{RMR - 100}{9}\right).$$

Use these values together with the Hoek–Brown criterion,

$$\frac{\sigma_1}{\sigma_c} = \frac{\sigma_3}{\sigma_c} + \sqrt{m\frac{\sigma_3}{\sigma_c} + s},$$

to determine the maximal vertical stress the pillars can sustain (1) at their faces and (2) at their centres. Assume that the ratio of horizontal to average vertical stress is 0.075 at the centre of each pillar.

Use the tributary area theory to estimate the average vertical stress in the pillar, and hence determine the factor of safety of the pillars.

Q20.10 A gold-bearing quartz vein, 2 m thick and dipping at 90°, is to be exploited by a small cut-and-fill stoping operation. The mining is to take place at a depth of 800 m, and the average unit weight of the granite country rock above this level is 29 kN/m^3. The strike of the vein is parallel to the intermediate principal stress, and the major principal stress is horizontal with a magnitude of 37.0 MPa. The uniaxial compressive strength of the vein material is 218 MPa, and the tensile strength of the country rock is 24 MPa. Poisson's ratio and Young's modulus for the quartz are 0.2 and 48 GPa, respectively. During mining, each blast will extend a stope up-dip by about 2 m.

Assuming that no stress-induced failure is permissible, what is the maximum height of a stope?

It is considered that the backfill will offer sufficient support to prevent degradation of the side walls of a stope, and that the only stress-induced failure of concern is that in the crown. What is the maximum permissible height of a stope in these circumstances?

References

Amadei B. (1988) Strength of a regularly jointed rock mass under biaxial and axisymmetric loading. *Int. J. Rock Mech. Min. Sci.*, **25**, 1, 3–13.

Bankwitz P. and Bankwitz E. (1995) Fractographic Features on Joints of KTB Drill Cores (Bavaria, Germany), in *Fractography, Fracture Topography as a Tool in Fracture Mechanics and Stress Analysis* (M. S. Ameen, ed.). Geological Society Special Publication No. 92, pp. 39–58.

Barton, N., Lien R. and Lunde J. (1974) Engineering classification of rock masses for the design of tunnel support. *Rock Mech.*, **6**, 183–236.

Barton N. (1978) Suggested methods for the quantitative description of discontinuities in rock masses. *Int. J. Rock Mech. Min. Sci. Geomech. Abstr.*, **15**, 6, 319–368.

Barton N. (1999) TBM performance estimation in rock using Q_{TBM}. *Tunnels and Tunnelling*, September 1999, 30–34.

Bates R. L. and Jackson J. A. (eds) (1980) *Glossary of Geology*. American Geological Institute, USA, 749pp.

Bear J. (1979) *Hydraulics of Groundwater*. McGraw-Hill, New York.

Beer G. and Watson J. O. (1992) *Introduction to Finite Element and Boundary Element Methods for Engineers*. John Wiley and Sons, Chichester.

Bhasin R. and Løset F. (1992) Norway's Olympic Cavern. *Civ. Eng.*, December, 60–61.

Bieniawski Z. T. (1989) *Engineering Rock Mass Classifications*. Wiley, Chichester, 251pp.

Borowski E. J. and Borwein J. M. (1989) *Dictionary of Mathematics*. Harper Collins, London, 659pp.

Bromhead E. N. (1992) *The Stability of Slopes*. Blackie Academic, London, 2nd edn.

Cheng A. H.-D. (1998) On generalized plane strain poroelasticity. *Int. J. Rock Mech. Min. Sci.*, **35**, 2, 183–193.

Committee on Fracture Characterization and Fluid Flow, US National Committee for Rock Mechanics, (1996) *Rock Fractures and Fluid Flow*. National Academy Press, Washington, DC.

Cristescu N. D. and Hunsche U. (1998) *Time Effects in Rock Mechanics*. Wiley, New York, 342pp.

Daemen J. J. K. (1983) Slip zones for discontinuities parallel to circular tunnels of shafts. *Int. J. Rock Mech. Min. Sci. Geomech. Abs.* **20**, 135–148.

Davis J. C. (1973) *Statistics and Data Analysis in Geology*. Wiley, New York, 550pp.

Fairhurst C. E. and Hudson J. A. (1999) Draft ISRM suggested method for the complete stress–strain curve for intact rock in uniaxial compression. *Int. J. Rock Mech. Min. Sci.*, **36**, 3, 279–289.

Fjaer E., Holt R. M., Horsrud P., Raaen A. M. and Risnes R. (1992) *Petroleum Related Rock Mechanics*. Elsevier, Oxford, 338pp.

Fookes P. G. (1997) Geology for Engineers: the geological model, prediction and performance. *Q. J. Eng. Geol.*, **30**, 293–424.

Giani G. P. (1992) *Rock Slope Stability Analysis*. Balkema, Rotterdam, 361pp.

Goodman R. E. (1989) *Introduction to Rock Mechanics*, 2nd edn., Wiley, New York, 562pp.

Goodman R. E. (1998) *The Engineer as Artist*. ASCE Press, Reston, VA, 340pp.

Haimson B. C. (1987) Measurement of *in situ* stress, in *Geophysics, Methods of Experimental Physics, Vol. 24B* (C. G. Sammis and T. L. Henyey (eds), Academic Press, New York, pp. 377–408.

Haimson B. C. (1997) Borehole Breakouts and Core Disking as Tools for Estimating *In Situ* Stress in Deep Holes, in *Proc. of the Rock Stress Conference* (K. Sugawara and Obara Y., eds). Balkema, Rotterdam, pp. 35–42.

Harrison J. P. (1999) Selection of the RQD threshold value in RQD assessments. *Int. J. Rock Mech. Min. Sci.*, **36**, 5, 673–685.

Hoek E. (1990) Estimating Mohr–Coulomb friction and cohesion values from the Hoek–Brown failure criterion. *Int. J. Rock Mech. Min. Sci.*, **27**, 3, 227–229.

Hoek E. and Bray J. W. (1977) *Rock Slope Engineering*. Institution of Mining and Metallurgy, London, 402pp.

Hoek E. and Brown E. T. (1980) *Underground Excavations in Rock*. Institution of Mining and Metallurgy, London, 527pp.

Hoek E. and Brown E. T. (1988) Hoek–Brown failure criterion — a 1988 update. *Proc. 15th Canadian Rock Mech. Symp*. University of Toronto, ON, pp. 31–38.

Hoek E. and Brown E. T. (1997) Practical estimates of rock mass strength. *Int. J. Rock Mech. Min. Sci.*, **34**, 8, 1165–1186.

Hoek E., Bawden W. F. and Kaiser P. K. (1995) *Support of Underground Excavations in Hard Rock*. Balkema, Rotterdam, 215pp.

Hudson J. A. (1989) *Rock Mechanics Principles in Engineering Practice*. CIRIA Report, Butterworths, London, 72pp.

Hudson J. A. (1992) *Rock Engineering Systems: Theory and Practice*. Ellis-Horwood, Chichester, 185pp.

Hudson J. A. (1999) Keynote lecture: technical auditing of rock mechanics modelling and rock engineering design, in *Proc. 37th US Rock Mechanics Symposium* held at Vail, CO, 1999 (B. Amadei, R. L. Kranz, G. A. Scott and P. H. Smeallie, eds). Balkema, Rotterdam, pp. 3–12.

Hudson J. A. and Harrison J. P. (1997) *Engineering Rock Mechanics: An Introduction to the Principles*. Elsevier, Oxford, 444p. This book is referred to in the text as ERM 1.

Hudson J. A. and Jiao Y. (2001) *Analysis of Rock Engineering Projects*. Imperial College Press, London (in preparation).

Jaeger J. C. and Cook N. G. W. (1979) *Fundamentals of Rock Mechanics*. Chapman and Hall, London, 3rd edn., 593pp.

Jiao Y. and Hudson J. A. (1998) Identifying the critical mechanisms for rock engineering design. *Géotechnique*, **48**, 3, 319–335.

Jimeno C. L., Jimeno E. L. and Carcedo F. J. A. (1995) *Drilling and Blasting of Rocks*, Balkema, Rotterdam, 391pp.

Judd W. R. (1965) Some rock mechanics problems in correlating laboratory results with prototype reactions. *Int. J. Rock Mech. Min. Sci.*, **2**, 197–218.

Kim K. and Franklin J. A. (1987) Suggested methods for rock stress determination. *Int. J Rock Mech. Min. Sci. Geomech. Abstr.*, **24**, 1, 53–73.

Langefors U. and Kihlström B. (1963) *The Modern Technique of Rock Blasting*. Wiley, New York, 405pp.

Latham J.-P. and Lu Ping (1999) Development of an assessment system for the blastability of rock masses. *Int. J. Rock Mech. Min. Sci.*, **36**, 41–55.

Long J. C. S. (1983) *Investigation of equivalent porous medium permeability in networks of discontinuous fractures*. PhD Thesis, University of California, Berkeley.

Mazzoccola D. F. and Hudson J. A. (1996). A comprehensive method of rock mass characterization for indicating natural slope instability. *Q. J. Eng. Geol.*, **29**, 37–56.

Miles R. E. (1964) Random polygons determined by random lines in a plane. *Proc. Natl. Acad. Sci. USA*, **52**, 901–907.

National Coal Board, Mining Department (1975) *Subsidence Engineer's Handbook*. National Coal Board, Mining Department, London, 2nd edn. (revised).

Pan X. D. (1988) *Numerical Modelling of Rock Movements Around Mine Openings*. PhD thesis. University of London, 375pp.

Panda B. B. and Kulatilake P. H. S. W. (1999) Effect of joint geometry and transmissivity on jointed rock hydraulics. *J. Eng. Mech.*, **125**, 1, 41–50.

Panet M. (1993) Understanding Deformations in Tunnels, in *Comprehensive Rock Engineering* (J. A. Hudson, ed.), Vol. 1. Pergamon Press, Elsevier, Ch. 27, pp. 663–690.

Parker H. W. (1996) Geotechnical Investigations, in *Tunnel Engineering Handbook* (J. O. Bickel, T. R. Kuesel and E. H. King, eds). Chapman and Hall, New York, 544pp.

Pierce F. T. (1926) Tensile tests for cotton yarns, V. The weakest link, theorems on the strength of long and composite specimens. *J. Tex. Inst.*, **17**, 355–368.

Press W. H., Teukolosky S. A., Vetterling W. T. and Flannery B. P. (1992) *Numerical Recipes in FORTRAN: The Art of Scientific Computing*. Cambridge University Press, Cambridge, 2nd edn., 963pp.

Price N. J. and Cosgrove J. W. (1990) *Analysis of Geological Structures*. Cambridge University Press, Cambridge, 502pp.

Priest S. D. (1993) *Discontinuity Analysis for Rock Engineering*. Chapman and Hall, London, 473pp.

Priest S. D. and Brown E. T. (1983) Probabilistic stability analysis of variable rock slopes. *Trans. Inst. Min. Metall.* **92**, A1–12.

Ramsay J. G. and Huber M. I. (1983) *The Techniques of Modern Structural Geology. Vol. 1: Strain Analysis*. Academic Press, London, 307pp.

Ray S. K., Sarkar M. and Singh T. N. (1999) Effect of cyclic loading and strain rate on the mechanical behaviour of sandstone. *Int. J. Rock Mech. Min. Sci.*, **36**, 543–549.

Roberts A. (1977) *Geotechnology*. Pergamon Press, Oxford, 347pp.

Romana M. R. (1993) A geomechanical classification for slopes: slope mass rating, in *Comprehensive Rock Engineering*, Vol. 3, Ch. 23 (J. A. Hudson, ed.), Pergamon Press, Oxford, pp. 575–599.

Schneider B. (1967) Moyens Nouveaux de Reconnaissance des Massifs Rocheux. *Supp. to Annales de l'Inst. Tech. de Batiment et des Travaux Publics*, **20**, 235–236, 1055–1093 (as illustrated in Goodman R. E. (1989) Introduction to Rock Mechanics, 2nd edn., John Wiley and Sons, New York, 562pp.

Serrano A. and Olalla C. (1998) Ultimate bearing capacity of an anisotropic discontinuous rock mass, Part I. Basic modes of failure. *Int. J. Rock Mech. Min. Sci.*, **35**, 3, 301–324.

Singh B. and Goel R. K. (1999) *Rock Mass Classification*. Elsevier, Oxford, 267pp.

Smith G. J. and Rosenbaum M. S. (1993) Recent underground investigations of abandoned chalk mine workings beneath Norwich City, Norfolk. *Eng. Geol.*, **36**, 37–78.

Sonmez H. and Ulusay R. (1999) Modifications to the geological strength index (GSI) and their applicability to stability of slopes. *Int. J. Rock Mech. Min. Sci.*, **36**, 743–760.

Stephansson O. Hudson J. A. Tsang C.-F. Jing L. and Andersson J. (1999) *DECOVALEX II Project. Coupled THM issues related to repository design and performance task 4.* SKI Report 99:7, Stockholm.

Su D. W. H. and Hasenfus G. J. (1995) Regional Horizontal Stress and Its Effect on Longwall Mining in the Northern Appalachian Coal Field. *Proceedings of the 14th International Conference on Ground Control in Mining* (S. S. Peng, ed.) West Virginia University, Morgantown, pp. 39–43.

Sugawara K. and Obara Y. (1999) *In situ* stress measurement using the compact conical-ended borehole overcoring (CCBO) technique. *Int. J. Rock Mech. Min. Sci.*, **36**, 3, 307–322.

Tan T. J. (1993) The importance of creep and time-dependent dilatancy as revealed from case records in China, in *Comprehensive Rock Engineering* (J. A. Hudson, ed.). Vol. 3. Pergamon Press, Elsevier, Ch. 31, 709–744.

Vervoort A. and de Wit K. (1997) Use of rock mass classifications for dredging. *Int. J. Rock Mech. Min. Sci.*, **34**, 5, 859–864.

Villaescusa E., Windsor C. R. and Thompson A. G. (eds) (1999) *Rock Support and Reinforcement Practice in Mining.* Proceedings of the International Symposium on Ground Support, Kalgoorlie, Western Australia. Balkema, Rotterdam, 437pp.

Wang Tongliang and Fan Qiuyan (2000) Optimization of soft rock engineering with particular reference to coal mining. *Int. J. Rock Mech. Min. Sci.* **37**, 3, 535–542.

Warren C. and Varley P. (1995) Geology, in *Engineering the Channel Tunnel* (C. J. Kirkland and F. N. Spon, eds). Chapman and Hall, London, p. 334.

Wei Z. Q. and Hudson J. A. (1986) The influence of joints on rock modulus. *Proc. Int. Symp. Engineering in Complex Rock Formations* (T. K. Tan, ed.), Pergamon Press, Beijing, pp. 54–62.

Whittaker B. N. and Russell C. F. (1990) *Tunnelling: Design, Stability and Construction.* Institution of Mining and Metallurgy, London, 460pp.

Wilson D. (1991) *Breakthrough.* Random Century Group, London, 144pp.

Windsor C. R. (1997) Rock Reinforcement Systems. *Int. J. Rock Mech. Min. Sci.* **34**, 6, 919–951.

Wittke W. (1999) Stability analysis of tunnels of the new high speed Cologne–Rhine/Main Railway Line. *ISRM News J.*, **5**, 3, 26–33.

Yoshinaka R. and Kikuchi K. (eds) (1995) *Rock Foundation.* Proceedings of the International Workshop on Rock Foundation, Tokyo, Balkema, Rotterdam, 457pp.

Appendix A: 3-D stress cube model

This cut-out can be copied and made into a cube. It is provided as an aide-memoire for the 3-D stress state, and especially to indicate the sign of the shear stresses using the right-hand rule (see Footnote 6 in Chapter 4). For ease of use, we recommend that an enlarged photocopy of this is made.

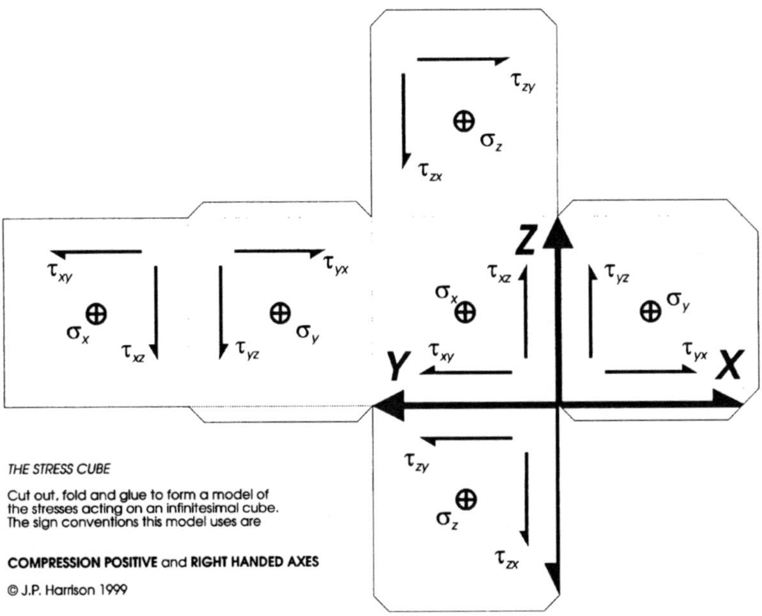

Appendix B: Hemispherical projection sheet

The equal-angle hemispherical projection below can be copied and enlarged for use with the problems in this book. The same projection can be obtained from the Elsevier website:
http://www.elsevier.com/locate/isbn/0080430104

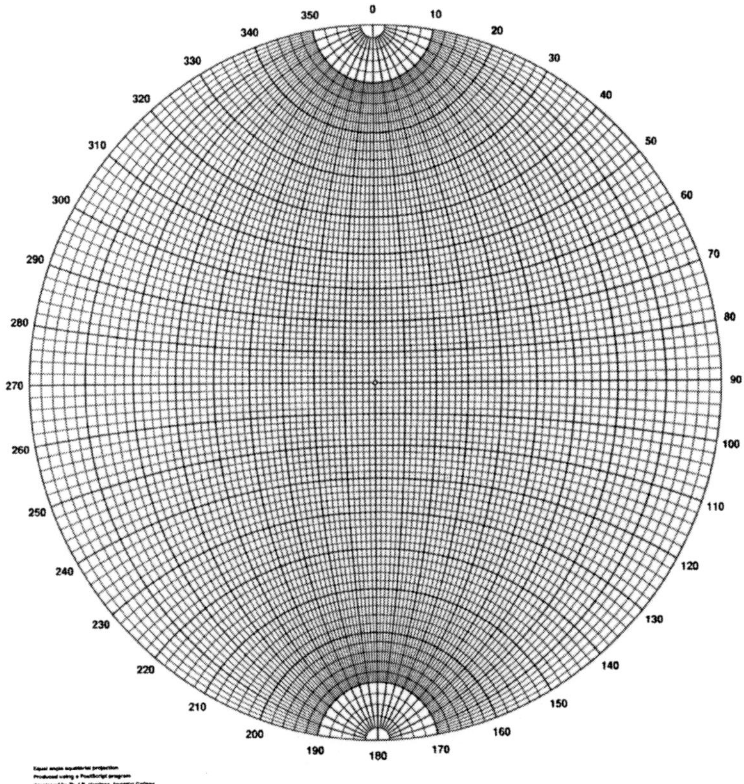

Appendix C: Rock mass classification tables — *RMR* and Q

The following tables are provided to support the questions in Chapter 12. Note that although we have used the term 'fracture' in this book, the originators of the *RMR* and *Q* classifications used the words 'discontinuity' and 'joint', respectively. To avoid confusion with the original publications[1], we have retained these words in the tables below.

Rock mass rating system[1]

$$RMR = \sum(\text{classification parameters}) + \text{discontinuity orientation adjustment}$$

(A) Classification parameters and ratings

Strength of intact rock material							
Point-load strength index (MPa)	*	*	*	1–2	2–4	4–10	>10
Uniaxial compressive strength (MPa)	<1	1–5	5–25	25–50	50–100	100–250	>250
Rating:	0	1	2	4	7	12	15

* For this low range, uniaxial compressive test is preferred.

Groundwater					
Inflow per 10-m tunnel length (l/min)	>125	25–125	10–25	<10	none
Ratio $\frac{\text{joint water pressure}}{\text{major principal stress}}$	>0.5	0.2–0.5	0.1–0.2	<0.1	0
General conditions	flowing	dripping	wet	damp	completely dry
Rating:	0	4	7	10	15

[1] Bieniawski Z. T. (1989) *Engineering Rock Mass Classifications*. Wiley, Chichester, 251pp.

Drill core quality *RQD* (%)

	<25	25–50	50–75	75–90	90–100
Rating:	3	8	13	17	20

Spacing of discontinuities

	<60 mm	60–200 mm	200–600 mm	0.6–2 m	>2 m
Rating:	5	8	10	15	20

Condition of discontinuities

Use Table (B), or the following guidelines

	Soft gouge > 5 mm thick *or* separation > 5 mm, continuous	Slickensided surfaces *or* gouge < 5 mm thick *or* separation 1–5 mm, continuous	Slightly rough surfaces, separation < 1 mm, highly weathered wall rock	Slightly rough surfaces, separation < 1 mm, slightly weathered wall rock	Very rough surfaces, no separation, unweathered wall rock, not continuous
Rating:	0	10	20	25	30

(B) Guidelines for classification of discontinuities

Discontinuity length (persistence)

	<1 m	1–3 m	3–10 m	10–20 m	>20 m
Rating:	6	4	2	1	0

Separation (aperture)

	none	<0.1 mm	0.1–1.0 mm	1–5 mm	>5 mm
Rating:	6	5	4	1	0

Roughness

	very rough	rough	slightly rough	smooth	slickensides
Rating	6	5	3	1	0

Infilling (gouge)

	hard filling			soft filling	
	none	<5 mm	>5 mm	<5 mm	>5 mm
Rating:	6	4	2	2	0

Weathering

	unweathered	slightly weathered	moderately weathered	highly weathered	decomposed
Rating:	6	5	3	1	0

(C) Effect of discontinuity orientations in tunnelling

Strike perpendicular to tunnel axis

drive with dip		drive against dip	
dip 45–90 very favourable	dip 20–45 favourable	dip 45–90 fair	dip 20–45 unfavourable

Strike parallel to tunnel axis		**Irrespective of strike**
dip 20–45 fair	dip 45–90 very unfavourable	dip 0–20 fair

(D) Rating adjustment for discontinuity orientations

Effect of discontinuity orientation (from Table C)	Very favourable	Favourable	Fair	Unfavourable	Very unfavourable
Ratings:					
Tunnels and mines	0	−2	−5	−10	−12
Foundations	0	−2	−7	−15	−25
Slopes	0	−5	−25	−50	−60

(E) Rock mass classes determined from total ratings

Rating	100–81	80–61	60–41	40–21	<20
Class no.	I	II	III	IV	V
Description	very good rock	good rock	fair rock	poor rock	very poor rock

(F) Interpretation of rock mass classes

Class no.:	I	II	III	IV	V
Average stand-up time	20 yr for 15-m span	1 yr for 10-m span	1 wk for 5-m span	10 h for 2.5-m span	30 min for 1-m span
Cohesion of rock mass (kPa)	>400	300–400	200–300	100–200	<100
Friction angle of rock mass (deg)	>45	35–45	25–35	15–25	<15

Q-system of Rock Mass Classification [2]

$$Q = \frac{RQD}{J_n} \times \frac{J_r}{J_a} \times \frac{J_w}{SRF}$$

Rock Quality Designation (*RQD*)

0–25	(a) Very poor
25–50	(b) Poor
50–75	(c) Fair
75–90	(d) Good
90–100	(e) Excellent

Note: Where *RQD* is reported or measured as 10 (including 0), a nominal value of 10 is used to evaluate *Q*. *RQD* intervals of 5, i.e. 100, 95, 90, etc., are sufficiently accurate.

Joint Set Number (J_n)

0.5–1.0	(a) Massive, none or few joints
2	(b) One joint set
3	(c) One joint set plus random
4	(d) Two joint sets
6	(e) Two joint sets plus random
9	(f) Three joint sets
12	(g) Three joint sets plus random
15	(h) Four or more joint sets, random, heavily jointed 'sugar cube', etc.
20	(j) Crushed rock, earth-like

Note: For intersections, use ($3.0 \times J_n$); for portals, use ($2.0 \times J_n$).

Joint Roughness Number (J_r)

	(i) Rock wall contact and	
	(ii) Rock wall contact before 10 cm shear	
4	(a) Discontinuous joint	
3	(b) Rough or irregular, undulating	small-scale
2.0	(c) Smooth, undulating	
1.5	(d) Slickensided, undulating	scale
1.5	(e) Rough or irregular, planar	increasing ↓
1.0	(f) Smooth, planar	
0.5	(g) Slickensided, planar	intermediate-scale
	(iii) No rock wall contact when sheared	
1.0	(h) Zone containing clay minerals thick enough to prevent rock wall contact	
1.0	(j) Sandy, gravelly, or crushed zone thick enough to prevent rock wall contact	

Notes: Add 1.0 if the mean spacing of the relevant joint set is greater than 3 m.
$J_r = 0.5$ can be used for planar slickensided joints having lineation, provided the lineations are favourably orientated.
Descriptions (b) to (g) refer to small-scale features and intermediate-scale features, in that order.

[2] Barton, N., Lien R. and Lunde J. (1974) Engineering classification of rock masses for the design of tunnel support. *Rock Mech.*, **6**, 183–236.

Joint Alteration Number (J_a)

		ϕ_r (approx.)*
	(i) Rock wall contact	
0.75	(a) Tightly healed, hard, non-softening, impermeable filling, i.e. quartz or epidote	
1.0	(b) Unaltered joint walls, surface staining only	25–35°
2.0	(c) Slightly altered joint walls. Non-softening mineral coatings, sandy particles, clay-free disintegrated rock, etc.	25–30°
3.0	(d) Silty or sandy clay coatings, small clay fraction (non-softening)	20–25°
4.0	(e) Softening or low-friction clay mineral coatings, i.e. kaolinite, mica. Also chlorite, talc, gypsum, and graphite, etc., and small quantities of swelling clays (discontinuous coatings, 1–2 mm or less in thickness)	8–16°
	(ii) Rock wall contact before 10 cm shear	
4.0	(f) Sandy particles, clay-free disintegrated rock, etc.	25–30°
6.0	(g) Strongly over-consolidated, non-softening clay mineral fillings (continuous, <5 mm in thickness)	16–24°
8.0	(h) Medium or low over-consolidation, softening, clay mineral fillings (continuous, <5 mm in thickness)	12–16°
8.0–12.0	(j) Swelling clay fillings, i.e. montmorillonite (continuous, <5 mm in thickness). Value of J_a depends on percentage of swelling clay-sized particles, and access to water, etc.	6–12°
	(iii) No rock wall contact when sheared	
6.0, 8.0 or 8.0–12.0	(k) Zones or bands of disintegrated or rock and clay — see (g), (h), (j) for description of clay condition	6–24°
5.0	(l) Zones or bands of silty clay, small clay fraction (non-softening)	
10.0, 13.0 or 13.0–20.0	(m) Thick, continuous zones or bands of clay — see (g), (h), (j) for description of clay condition	

* Values of ϕ_r are intended as an approximate guide to the mineralogical properties of the alteration products, if present.

Stress Reduction Factor (*SRF*)

	(i) Weakness zones intersecting excavation, which may cause loosening of rock mass when tunnel is excavated		
10.0	(a) Multiple occurrences of weakness zones containing clay or chemically disintegrated rock, very loose surrounding rock (any depth)		
5.0	(b) Single weakness zones containing clay or chemically disintegrated rock (excavation depth <50 m)		
2.5	(c) Single weakness zones containing clay or chemically disintegrated rock (excavation depth >50 m)		
7.5	(d) Multiple shear zones in competent rock (clay-free), loose surrounding rock (any depth) [a]		
5.0	(e) Single shear zones in competent rock (clay-free) (depth of excavation <50 m) [a]		
2.5	(f) Single shear zones in competent rock (clay-free) (depth of excavation >50 m) [a]		
5.0	(g) Loose open joints, heavily jointed or 'sugar cube', etc. (any depth) [a]		
	(ii) Competent rock, rock stress problems [b]	σ_c/σ_1	σ_t/σ_1
2.5	(h) Low stress, near surface	>200	>13
1.0	(j) Medium stress	200–10	13–0.66
0.5–2.0	(k) High-stress, very tight structure (usually favourable to stability, may be unfavourable for wall stability)	10–5	0.66–0.33
5–10	(l) Mild rock burst (massive rock)	5–2.5	0.33–0.16
10–20	(m) Heavy rock burst (massive rock)	<2.5	<0.16
	(iii) Squeezing rock; plastic flow of incompetent rock under the influence of high rock pressures		
5–10	(n) Mild squeezing rock pressure		
10–20	(p) Heavy squeezing rock pressure		
	(iv) Swelling rock; chemical swelling activity depending on presence of water		
5–10	(q) Mild swelling rock pressure		
10–15	(r) Heavy swelling rock pressure		

[a] Reduce these *SRF* values by 25–50% if the relevant shear zones only influence, but do not intersect, the excavation.

[b] Few case records available where depth of crown below surface is less than span width; suggest *SRF* increase from 2.5 to 5 for such cases. For strongly anisotropic stress field (if measured): $5 \leq \sigma_1/\sigma_3 \leq 10$: reduce σ_c and σ_t to $0.8\sigma_c$ and $0.8\sigma_t$; $\sigma_1/\sigma_3 > 10$: reduce σ_c and σ_c to $0.6\sigma_c$ and $0.6\sigma_t$ (σ_c = unconfined compressive strength, σ_t = tensile strength (point load), σ_1 and σ_3 are major and minor principal stresses).

Joint Water Reduction Factor (J_w)

		Approx. water pressure (kg/cm^2)
1.0	(a) Dry excavations or minor inflow, e.g. 5 l/min locally	<1
0.66	(b) Medium inflow or pressure, occasional outwash of joint fillings	1.0–2.5
0.5	(c) Large inflow or high pressure in competent rock with unfilled joints	2.5–10.0
0.33	(d) Large inflow or high pressure, considerable outwash of joint fillings	2.5–10.0
0.2–0.1	(e) Exceptionally high inflow or water pressure at blasting, decaying with time	>10.0
0.1–0.05	(f) Exceptionally high inflow or water pressure continuing without noticeable decay	>10.0

Note: Factors (c) to (f) are crude estimates. Increase J_w if drainage measures are installed. Special problems due to ice formation are not considered.

When making estimates of Q, the Rock Mass Quality, the following guidelines should be followed, in addition to the notes in the tables.

(1) When borehole core is unavailable, for the case of clay-free rock masses *RQD* can be estimated from $RQD = 115 - 3.3J_v$ (approx.) where J_v = total number of joints per m^3 ($RQD = 100$ for $J_v < 4.5$). J_v is evaluated as the sum of the number of joints per metre for each joint set.

(2) The parameter J_n, representing the number of joint sets, will often be affected by foliation, schistosity, slaty cleavage or bedding, etc. If strongly developed, these features should be counted as a complete joint set: if they are poorly developed or rarely visible, then it will be more appropriate to count them as 'random joints' when evaluating J_n.

(3) The parameters J_r and J_a (representing shear strength) should normally be relevant to the weakest significant joint set or clay-filled discontinuity in a given zone, but the value of J_r/J_a should relate to the surface most likely to allow failure to initiate. Thus, if the joint set or discontinuity with the minimum value of J_r/J_a is favourably orientated for stability, then a second, less favourably orientated joint set or discontinuity may sometimes be more significant, and its higher value of J_r/J_a should be used when evaluating Q.

(4) When a rock mass contains clay, the factor *SRF* appropriate to 'loosening loads' should be evaluated. In such cases the strength of the intact rock is of little interest. However, when jointing is minimal and clay is completely absent, the strength of the intact rock may become the weakest link, and the stability will then depend on the ratio rock stress/rock strength. A strongly anisotropic stress field is unfavourable for stability and is roughly accounted for as in the note in the table for *SRF* evaluation.

(5) The compressive and tensile strengths (σ_c and of σ_t) the intact rock

should be evaluated in the saturated condition if this is appropriate to present or future *in situ* conditions. A conservative estimate of strength should be made for those rocks that deteriorate when exposed to moist or saturated conditions.

Index

Lightning Source UK Ltd.
Milton Keynes UK
21 November 2009

146566UK00001B/14/P

9 780080 430102